D0161553

COASTAL DYNAMICS AND LANDFORMS

Coastal Dynamics and Landforms

A. S. TRENHAILE

CLARENDON PRESS · OXFORD
1997

Oxford University Press, Great Clarendon Street, Oxford OX2 6DP

Oxford New York
Athens Auckland Bangkok Bombay
Calcutta Cape Town Dar es Salaam Delhi
Florence Hong Kong Istanbul Karachi
Kuala Lumpur Madras Madrid Melbourne
Mexico City Nairobi Paris Singapore
Taipei Tokyo Toronto

and associated companies in
Berlin Ibadan

Oxford is a trade mark of Oxford University Press

Published in the United States
by Oxford University Press Inc., New York

© Alan Trenhaile 1997

All rights reserved. No part of this publication may be reproduced,
stored in a retrieval system, or transmitted, in any form or by any means,
without the prior permission in writing of Oxford University Press.
Within the UK, exceptions are allowed in respect of any fair dealing for the
purpose of research or private study, or criticism or review, as permitted
under the Copyright, Designs and Patents Act, 1988, or in the case of
reprographic reproduction in accordance with the terms of the licences
issued by the Copyright Licensing Agency. Enquiries concerning
reproduction outside these terms and in other countries should be
sent to the Rights Department, Oxford University Press,
at the address above

British Library Cataloguing in Publication Data
Data available

Library of Congress Cataloging in Publication Data
Trenhaile, Alan S.
Coastal dynamics and landforms / A. S. Trenhaile.
Includes bibliographical references and index.
1. Coast changes   I. Title.
GB451.2.T735 1997   551.4'57—dc20   96–38341
ISBN 0–19–823353–1

1 3 5 7 9 10 8 6 4 2

Typeset by Alliance Phototypesetters
Printed in Great Britain
on acid-free paper by
Bookcraft (Bath) Ltd.,
Midsomer Norton, Somerset

LEDL
CIRC
GB
451.2
.T735
1997

846226

Preface

This book developed from a fourth-year coastal geomorphology class that I teach at the University of Windsor. It is designed as a reference text for senior undergraduates, graduate students, faculty, and working professionals in the coastal field. It is generally assumed that the reader has already acquired some basic coastal knowledge from introductory geomorphology courses and texts. The book is not issue-oriented, nor is it a coastal management book, although anthropological influences and managerial implications are discussed in several of the chapters where they are considered to be particularly relevant. Although variations in our present state of knowledge create some differences in the level of analysis of each topic, the alternative, which would have been to reduce the discussion of some topics to a common level, was considered to be less desirable.

I intended, from the start, to make a detailed survey of a broad spectrum of coastal types. There is an enormous coastal literature, however, and the manuscript grew, inexorably, to unmanageable proportions. Although some material was removed from the text, most of the necessary reduction was accomplished by eliminating about 40 percent of the original references. Although the list is still long, it is now strongly biased towards very recent literature—the interested reader will be able to find references to older material in these articles. Nevertheless, I must thank the numerous coastal workers who have made this work possible, and apologize for having had, in many cases, to omit direct reference to their work.

I thank two anonymous reviewers for their helpful comments, and Andrew Lockett, Michael Belson, and other staff at Oxford University Press for their editorial assistance. My colleagues, Dr Kirsty Duncan and Dr Chris Lakhan were a constant source of encouragement and enthusiasm. Mr Ron Welch drew the diagrams. I am especially grateful to my family, Sue, Rhys, and Lynwen, who have accompanied me to many coastal regions.

The author also thanks the following for permission to reproduce figures: John Wiley and Sons for Figs. 2.18, 2.22, 3.2, 3.4, 3.5, 3.12, 3.13, 6.1, 6.2, 6.12, 7.3, 7.4, 8.3, and 10.12; Elsevier Science for Figs. 2.11, 3.6, 4.3, 4.6, 4.7, 5.1, 5.3, 5.7, 5.8, 5.11, 5.16, 6.10, 7.6, 8.2, 8.9, 9.5, 11.10, and 12.6; Society Econ. Paleont. Mineral. for Figs. 4.12, 5.10, 6.7, 7.2, 7.7, 10.8, 12.4; Centre Biologie Ecologie Tropicale for Figs. 10.5 and 10.11; Methuen for Fig. 6.3; A. A. Balkema for Fig. 6.11; Les Presses de l'Universite de Montreal for 12.10; Springer-Verlag for Figs. 8.1 and 9.8; Institute of Physics for a part of Fig. 2.2; Royal Irish Academy for Fig. 6.14; CRC Press for Fig. 1.4; New Zealand Geographical Society for Fig. 5.14; Instituto di Geologia Applicata e Geotecnica for Figs. 11.11, 11.12, and 11.13; Houston Geological Society for Figs. 9.4 and 9.6; American Geophysical Union for part of Fig. 2.21, and for Figs. 2.3a, 2.3b, 2.4, 2.8, and 3.8; Academic Press for Figs. 2.13, 5.2, 5.9, 5.12, 7.8, 10.6, and 12.2; Nature for Figs. 1.3 and 10.2; Masson S.A. for Fig. 12.13; Journal of Coastal Research for Figs. 4.9 and 12.3; Blackwell Science for Figs. 3.10, 3.11, 5.4, 5.15, 6.6, 6.15, and 9.10; University of Chicago Press for Figs. 1.2 and 3.3; Oxford University Press for Figs. 2.5, 2.6, 11.6, 11.7, and 11.8; Natural Resources Canada (Ministry of Supply and Services) for Figs. 12.7 and 12.9; Edward Arnold for Fig. 8.7; Cambridge University Press for Figs. 2.12, 11.4, and 11.5; Universitetsforlaget for Fig. 12.12; Quaternary Research for Figs. 2.23, 2.24, and part of Fig. 2.20; the Geological Society of America for Figs. 5.5, 8.8, 9.2, 9.9, 11.3, and part of Fig. 2.20; Massachusetts Institute of Technology Press for Fig. 5.13; American Association of Petroleum Geologists for Figs. 5.18, 9.3, and 10.3; Hodder Headline for Fig. 2.16; The Department of the Navy (U.S.) For Fig. 2.15; Dr M. E. Brookfield for Fig. 6.8; Drs J. D. Hansom and R. M. Kirk for Figs. 12.1, 12.5, and 12.8; Dr N. J. Shackleton for part of Fig. 2.19; Dr M. O. Hayes for Figs. 2.11 and 7.8; Dr

S. P. Leatherman for Fig. 5.2; Dr D. Hopley for Fig. 10.12; and Mrs B. Kinsman Brown for Fig. 2.1.

Fees have been paid to the Royal Geographical Society for permission to use Figs. 1.5, 11.2, and 12.11; Prentice-Hall for Fig. 2.15; Blackwell Publishers for Fig. 6.4; American Society of Civil Engineers for Figs. 2.7, 2.10, 3.1, 3.7, 4.8, and 11.14, and American Association for the Advancement of Science for Fig. 2.19.

In a few cases, it was not possible to trace the copyright holders of figures used in this text. Apologies must therefore be offered for any copyright infringements that may have occurred.

A.S.T.

Contents

List of Figures

List of Tables

List of Symbols

Some symbols that are used only once in the text have been omitted. These symbols are defined where they appear.

A_b	Excursion amplitude of the horizontal water motion at the bottom
B_r	Relative bed roughness
C	Phase celerity or speed of propagation of a wave
C_D	Bottom drag coefficient
C_{Dz}	Drag coefficient at a height z above the bed
Cn	Wave group celerity
C_o	Fractional volumetric concentration—volume of grains per unit volume of grain-water mixture.
CSF	Corey shape factor
c	Constant in Stokes's Law
D	Grain diameter
D_ϕ	Grain diameter in phi units
D_{mm}	Grain diameter in millimetres
D_n	Grain diameter coarser than n per cent of the sample
D_s	Diameter of the short axis of a grain
D_i	Diameter of the intermediate axis of a grain
D_1	Diameter of the long axis of a grain
do	Length of the horizontal water motion at the bottom ($2A_b$)—the orbital diameter
E	Wave energy per unit of crest
ESF	E-shape factor
e_b	Bedload efficiency factor in Bagnold's transport equation
e_s	Suspended load efficiency factor in Bagnold's transport equation
ex	Subscript referring to seawords flow
f_w	Friction factor
G	The universal gravitational constant ($6.67.10^{-11}$ N m^2 kg^{-2})
g	Acceleration due to gravity
H	Wave height
H_b	Height of the breaking waves
$H_{1/3}$	Significant wave height
H_r	Ripple height
h	Depth of water
h_b	Depth at the breakers
h_o	Depth of water to the centre of a wave orbit
I_s	Suspended load immersed weight longshore transport rate
I_t	Total immersed weight longshore transport rate
i	Immersed weight cross-shore transport rate
in	Subscript referring to shorewards flow
i_{net}	Net sediment transport
k	Wave number ($2\#/L$)
k_s	Nikuradse roughness length or equivalent roughness of the bed
i_t	Total immersed weight sediment transport rate across unit width of bed
L	Wavelength of a wave
L_r	Ripple wavelength
$\mathbf{M}$	The Mobility number
m_e	The mass of the Earth
m_m	The mass of the Moon
O_d	Diameter of wave orbit beneath the surface
P	Wave power or energy flux
Q_s	Volume longshore transport rate
q	Volume cross-shore transport rate
$\tilde{q}$	Average cross-shore transport rate in a half-wave cycle
q_{net}	Net cross-shore transport rate
$q(t)$	Instantaneous cross-shore transport rate
R	Distance between the centres of the Earth and Moon
r	Radius of the Earth
Re_g	Grain or boundary Reynolds number
Re_s	Settling Reynolds number
Re_w	Wave or amplitude Reynolds number

T	Wave period	β	Angle of beach or bed slope
$T_{1/3}$	Significant wave period	γ	The ratio of breaker height to breaker depth
U_r	The Ursell parameter (HL^2/h^3)	δ	Thickness of the boundary layer
u	Horizontal current velocity	δ_v	Thickness of the viscous sublayer
u_m	Maximum orbital velocity near the bed	ε	Surf scaling parameter
u_z	Fluid velocity at height z above the bed	η	Eddy viscosity
u_{zcr}	Critical (threshold) fluid velocity at height z above the bed	Θ	Shields parameter
		Θ_{cr}	Threshold (critical) Shields parameter
u_*	Shear or drag velocity	κ_n	von-Karman constant
u_{*cr}	Threshold (critical) shear velocity	μ	Molecular viscosity of the fluid
u_{crb}	Threshold (critical) shear velocity on a sloping bed	v	Kinetic viscosity of the fluid
V_1	Average longshore current velocity in the surf zone	ξ	Surf similarity parameter
		ξ_b	Iribarren number
w_s	Settling or fall velocity of a spherical grain	ρ	Density of the fluid
w_n	Settling or fall velocity of a natural grain	ρ_a	Density of air
w_o	Settling or fall velocity in clear water	ρ_s	Density of the grain
w_c	Settling or fall velocity in a uniform sediment concentration	τ_l	Viscous shear stress for laminar flow
X_b	Width of the surf zone	τ_t	Viscous shear stress for turbulent flow
z_o	Bed roughness length	τ_{cr}	Threshold (critical) shear stress
du/dz	Vertical velocity gradient or viscous shear strain	τ_m	Maximum shear stress
		τ_o	Shear stress on the bed
α_b	Breaker angle	ϕ	Angle of repose
α_{df}	Dynamic friction angle	ω	Wave radian frequency ($2\pi/T$)

1 Introduction

From about half to two-thirds of the world's population lives near the coast, and their industrial, recreational, agricultural, and transportational activities are placing enormous pressure on coastal resources. To manage these activities in the least detrimental ways, we need to understand how terrestrial and marine processes operate and interact in the coastal zone, and to recognize and accommodate the dynamic nature of its landforms.

The small-scale elements of depositional coasts, which can experience rapid changes in morphology, are usually self-regulating. They attain a rough state of balance with their environmental controls via negative feedback mechanisms, which cause the system to counteract the effects of changes in external conditions. Other coastal elements, particularly on hard rock coasts, may require long relaxation or recovery times, while they adjust to changing conditions. The development of these features may be cyclical or continuous. Even if all other environmental controls remain constant, individual coastal landforms have to adjust continuously to slow changes in the morphology of the coast itself. The profiles of sandy beaches, for example, respond fairly quickly to changing wave conditions, but they may also have to adjust slowly to long-term changes in coastal configuration, sediment budgets, offshore gradients, climate, and sea level, and increasingly to the effects of human interference.

Coastal regions represent a mosaic of diverse elements. Some are contemporary, whereas others are vestiges of periods when climate and sea level were often quite different from today. Resistant rock coasts have probably evolved very slowly, during successive periods of high interglacial sea level. The evidence of past sea levels and climates is more easily obliterated in unconsolidated coastal deposits, although many sandy coasts retain sedimentary and morphological elements of former environmental conditions. Because interglacial sea levels were similar to today, barriers, dunes, and other contemporary coastal features often formed close to, or were superimposed on top of, their ancient counterparts. Coastal deposits from the last interglacial stage are also being cannibalized to provide sediment for the construction and maintenance of modern coastal features. Nevertheless, because of tectonic uplift and slightly higher sea levels, wave-built and wave-cut terraces, coral reefs, beaches, and other coastal landforms can extend from less than a metre up to hundreds of metres above the present level of the sea.

COASTAL CLASSIFICATIONS

Although there have been many attempts to classify coasts, none is entirely satisfactory (Cotton 1954). Tanner (1960) examined fourteen classifications and found that they use seventeen basic criteria. Nevertheless, almost all classifications employ at least two of three basic variables: the shape of the coast; changes in relative sea level; and the effect of marine processes (King 1972). Some classifications are genetic, others are descriptive, and others combine the two approaches. Genetic classifications are hindered by a lack of relevant data, however, and descriptive classifications, which have to accommodate an enormous variety of coastal types, tend to be cumbersome. Two classifications, which consider the nature of coastal environments and the effect of plate tectonics on coastal development, are particularly useful.

The Morphogenic Classification

Davies (1972) proposed that coastal processes are strongly influenced by morphogenic factors that vary in a fairly systematic way around the world. Davies's classification was based upon four major wave climates, although differences in coastal characteristics also reflect variations in tidal range, climate, and many other factors (Fig. 1.1).

With the exception of areas experiencing tropical cyclones, the highest waves are generated in the storm belts of temperate latitudes, where the winds are strong and variable. Coasts experience a high proportion of short, high-energy waves of varying direction, generated in local waters. As westerlies are dominant in these latitudes, the western sides of continents are subjected to more frequent, and probably higher, storm waves than the eastern sides. In the northern hemisphere, storm waves occur most frequently in winter, but they are less seasonal in the southern hemisphere, and swell waves form a consistent background. Storm-wave environments are much more extensive in the northern than in the southern hemisphere, where,

because of the lack of land masses in the storm belt, they are restricted to the most southerly part of South America—although Tasmania and South Island New Zealand are marginal environments.

Beaches in storm-wave environments tend to have dissipative or gently sloping and barred profiles, and the major constructional features are often composed of coarse clastic material (Chapter 4). Constructional features are oriented more by local fetch than by the variable direction of the deep-water waves, and mechanical wave erosion is important in the formation of cliffs and shore platforms.

Swell-wave environments lie between the northern and southern storm belts. Swell-wave environments are dominated by long, low waves that fan out from the storm belts and travel considerable distances along great circle routes. As the westerly component in temperate gales is much

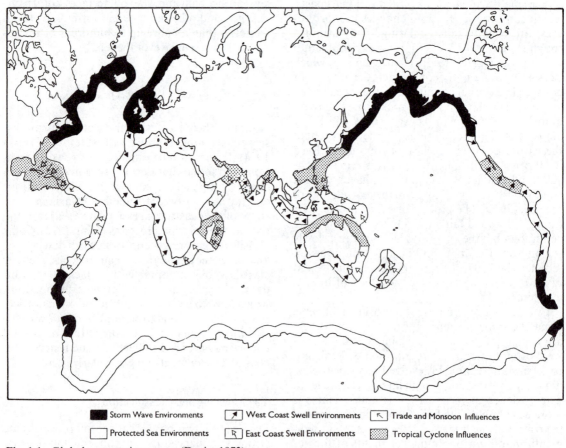

■ Storm Wave Environments	↗ West Coast Swell Environments	↖ Trade and Monsoon Influences
☐ Protected Sea Environments	↘ East Coast Swell Environments	⣿ Tropical Cyclone Influences

Fig. 1.1. Global wave environments (Davies 1972).

more important than the easterly, south-westerly swell dominates in the southern hemisphere and north-westerly swell in the northern hemisphere. West-facing coasts between the temperate storm belts therefore experience strong and persistent swell, whereas more important waves may be generated on east-facing coasts by north-easterly and south-easterly Trade Winds, blowing in the same direction as the swell. Although waves generated by Trade Winds are of great importance on some islands, they are only consistently important throughout the year on extensive stretches of the continental coasts of tropical Brazil and eastern Africa. Monsoon-generated winds are also important along the coasts of India and south-eastern Asia, where there is sufficient fetch, and cyclones can subject tropical coasts to extremely high energy inputs for brief periods, at infrequent and irregular intervals.

Flat, constructional waves dominate swell environments. The beaches have berms, and they tend towards the steeper, reflective, non-barred end of the spectrum (Chapter 4). Large, sandy constructional features are oriented towards the approaching swell, which is fairly consistent in frequency of occurrence and direction, especially in the southern hemisphere. The direction of longshore currents is therefore more constant than in storm wave environments. Mechanical wave erosion of cliffs and platforms is probably slower than in storm-wave environments, and this, combined with warmer climates, makes chemical and biological weathering more important in swell-wave environments.

Low-energy environments occur in sheltered, enclosed seas and ice-infested waters. In addition to the damping effects of ice, wave energy is low in Antarctica because the swell generated by gales in the southern hemisphere tends to take northward tracks, towards the equator. Violent storms can occasionally occur in low-energy environments, but they are infrequent and the fetch distances are generally short. Waves are therefore flat and constructional, and beaches have prominent berms. The orientation of sandy constructional features, which are common in partially enclosed seas, is largely determined by local fetch.

The Geotectonic Classification

Inman and Nordstrom (1971) proposed that the morphology of the largest, or first order, coastal elements can be attributed to their position on moving tectonic plates. Three main classes and several subclasses were identified (Fig. 1.2):

a) Collision coasts are formed by plate convergence. They include:
   i) Collision coasts along continental margins, and
   ii) Collision coasts along island arcs.
b) Plate-imbedded or trailing edge coasts face spreading centres. There are three main types:
   i) Neo-trailing edge coasts formed near new separation centres and rifts.
   ii) Amero-trailing edge coasts where the opposite coast of the continent is a collision coast.
   iii) Afro-trailing edge coasts where the opposite coast of the continent is also trailing.
c) Marginal sea coasts form where island arcs separate and protect continental coasts from the open ocean. They are the most diverse sea coasts, ranging from low-lying to hilly, with wide to narrow shelves, and they are often modified by large rivers and deltas.

Continental and island arc collision coasts occur along the edges of converging plates, especially around the margins of the oceanic Pacific plates. Their structural grain is parallel to the shore and they are therefore fairly straight and regular, although indentations may develop where the sea or rivers break through resistant strata and expose the weaker materials behind. Tectonically mobile collision coasts have narrow continental shelves and high, steep hinterlands, often with flights of raised terraces. The high relief provides an abundant supply of sediment to the coast.

Plate-imbedded, trailing edge coasts usually have hilly, plateau, or low hinterlands, and wide continental shelves. The structural grain may be at high angles to the coast, which can therefore be very indented. Young neo-trailing edge coasts have little or no continental shelf and few continuous beaches. Although they are close to spreading centres and experience some volcanic and seismic activity, it is usually at a lower level than on collision coasts. Amero-trailing coasts have collision coasts and high mountains on the opposite side of the continent, and the largest drainage basins on the trailing edge side. Much more sediment is therefore delivered to Amero- than Afro-trailing edge coasts. Morphologically mature Amero-trailing edge

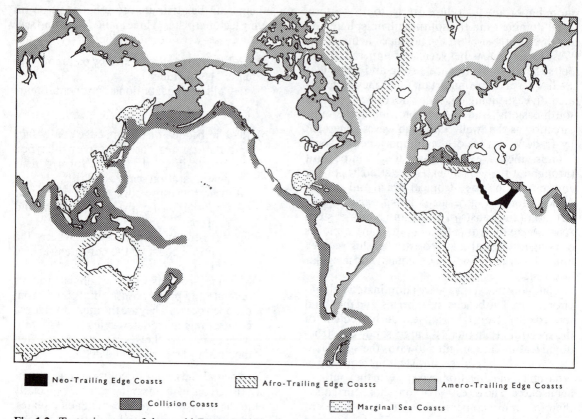

Neo-Trailing Edge Coasts   Afro-Trailing Edge Coasts   Amero-Trailing Edge Coasts

Collision Coasts   Marginal Sea Coasts

Fig. 1.2. Tectonic coasts of the world (Inman and Nordstrom 1971; Davies 1972)

coasts also have the lowest lying coastal landforms and the widest continental shelves. Afro-trailing edge coasts have more mature drainage and erosional features than neo-trailing coasts, and narrower shelves and higher coastal landforms than Amero-trailing coasts.

Plate tectonics provide a partial explanation for the distribution of a variety of coastal elements, although the degree of explanation decreases with the decreasing size of the feature. In general, formation of erosional features, including stacks and cliffs, may be enhanced by narrow shelves and a lack of sediment, whereas barriers, dune ridges, and other depositional features develop where there are wide shelves and an abundance of sandy sediments (Chapters 5 and 6).

GLOBAL WARMING

It is generally accepted that global warming will result from human activities that are increasing the concentration of carbon dioxide, water vapour, methane, nitrous oxide, chlorofluorocarbons, tropospheric ozone, and other greenhouse gases in the atmosphere. There is increasing concern that strong feedback relationships could cause unidirectional changes in climate and sea level to accelerate over the next century, to the point where they would dominate natural changes and exceed the capacity of natural systems to adapt (Warrick 1993). There have been many attempts to model the effects of global warming, but the conclusions tend to be very broad and general because of a lack of data, the assumptions that are necessary, and the complexity and difficulty in predicting system interactions. Therefore, although higher temperatures will probably cause higher sea levels, there is continuing debate over the rate and magnitude of the changes that are to be expected (Barnett 1990; Warrick and Farmer 1990; Tooley and Jelgersma 1992; Warrick *et al.* 1993).

There are many problems involved in the analysis of tidal gauge records, including the fact that the data are incomplete and the coverage is biased in favour of Europe and North America. The records are affected by vertical movements of the land, changes in water level caused by a variety of oceanographic and atmospheric factors, and changes resulting from the building of dams, irrigation, and land clearance for agriculture.

There has been a recent trend towards more conservative predictions of the rise in temperature and particularly sea level during the next century. A working group for the intergovernmental panel on climatic change (IPCC) concluded that sea level has probably been rising in the last hundred years, at an average rate of from 1 to 2 mm yr^{-1}. Although there is no convincing evidence that the rate has been increasing over the last one hundred years (Douglas 1992), it may have risen faster in this century than in the previous two. The revised IPCC 'best guess' estimate suggests that mean sea level will be about 48 cm higher by the year 2100 (Wigley and Raper 1992) (Fig. 1.3). Most of this will probably result from thermal expansion of the oceans and increased melting of small ice caps and mountain glaciers. The possible melting of polar ice sheets remains a source of much uncertainty. Because of increased snow accumulation with global warming, Antarctica is expected to make a zero or negative contribution to sea level, and rapid disintegration of the West Antarctic ice sheet is considered unlikely in the next century (Warrick and Oerlemans 1990).

Changes in sea level will be distributed unevenly around the world owing to geoidal perturbations, and changes in the elevation of the Earth's solid surface (Clark and Primus 1987) (Chapter 2). The consequences of rising sea level will also vary around the world according to the characteristics of the coast, including its slope, wave climate, tidal regime, and susceptibility to erosion. Global warming and rising sea level will cause tidal flooding and the intrusion of salt water into rivers, estuaries, and groundwater, and it will affect tidal range, oceanic currents, upwelling patterns, salinity levels, storm intensity and frequency, biological processes, and runoff and landmass erosion patterns. Increasing rates of erosion will make cliffs more susceptible to falls, landslides, and other mass movements, exacerbating problems where loose or weak materials are already experiencing rapid recession (National Research Council 1987; Bardach 1989; Warrick and Farmer 1990; Gornitz 1991). Nevertheless, some workers have emphasized the uncertainties of global warming, and they have suggested that the predicted rise in sea level may only pose a hazard where the land is subsiding.

It has been estimated that about half the world's population lives in vulnerable coastal lowlands, subsiding deltaic areas, and river flood plains. The effects of climatic change will be particularly acute in these densely populated regions. It is often the rate of sea-level change rather than the absolute amount that determines whether natural systems,

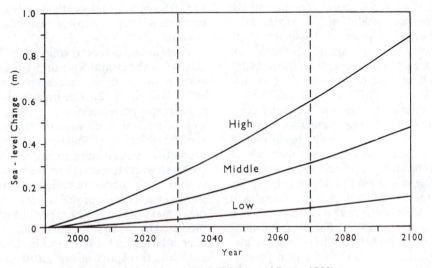

Fig. 1.3. Predicted global rise in sea level (Wigley and Raper 1992)

such as coastal marshes and coral reefs, can successfully adapt to changing conditions. Human and natural systems can adapt to slowly changing mean climatic conditions, but it is more difficult to accommodate changes in the occurrence of extreme events. It is not yet known whether higher sea temperatures will increase the frequency and intensity of tropical storms and spread their influence further polewards, and whether higher temperature gradients between land and sea will increase the intensity of monsoons and affect their timing (Warrick *et al.* 1993).

Human responses to the rise in sea level will depend upon available resources and the value of the land being threatened. High waterfront values will justify economic expenditure to combat rising sea level in cities, but less attention is likely to be paid to the deleterious effects on salt marshes, mangroves, coral reefs, lagoons, and ice-infested Arctic coasts. Coastal resorts and recreation represent very important economic activities for many tropical nations, and some are almost entirely dependent on this source of income. Recreational pressure on beaches will probably continue to increase in the future. Although higher sea levels reduce the width of beaches and result in increased crowding, the presence of existing structures or landscape patterns may preclude landward extension by pumping or the dumping of dredged sand (Bardach 1989). Beach erosion will also accelerate because: deeper water allows waves to get nearer the shore before breaking; wave refraction decreases, thereby increasing longshore transport; and waves and currents operate further up the beach (Leatherman 1990). Shifts in storm paths and the temporary rise in sea level associated with El Niño conditions in the equatorial Pacific, for example, resulted in accelerated erosion on the western coast of North America (Komar and Enfield 1987).

The gradient of rocky coasts largely determines the degree of shoreline displacement with rising sea level, but erosion plays a much greater role on sedimentary coasts. Several recession models have been used to predict shoreline erosion owing to rising sea level, although the Bruun rule and its variants continue to be the most widely employed (Bruun 1988; Komar *et al.* 1991; Dean *et al.* 1993; Roelvink and Brøker 1993). The Bruun Rule is based on the concept of the profile of equilibrium. Bruun found that the shape of beach profiles approximates the power law:

$$h_x = Ax^{2/3}$$

where h_x is the depth at a distance x offshore of the mean water line, and A is a scale parameter that is largely determined by grain size or fall velocity. Bruun proposed that a rise in sea level causes a concordant upward and landward translation of the equilibrium beach profile—the rise in the water surface being equal to the rise in the nearshore bottom. The sediment needed to raise the floor and maintain the original bottom profile is obtained through erosion on the upper part of the beach.

Other expressions have been suggested for the geometry of the equilibrium profile (Dean 1991; Bodge 1992; Komar and McDougal 1994; Lee 1994). Hands (1983) modified Bruun's equation, using field data describing beach and nearshore profile adjustment to rising water levels in Lake Michigan (Fig. 1.4):

$$x = \frac{ab}{c}(R_a) = \frac{1}{\tan \beta} a(R_a)$$

where x is the amount of shoreline recession for a given rise in the water level (a); b is the width of the bottom extending from the upper point of profile adjustment to the point of closure, the limiting depth of profile movement; c is the vertical distance from the top of the shore to the depth of closure; β is the slope of the active profile; and R_a is the percentage of eroded material, by volume, that is not stable on the active profile—this fine-grained material will be carried as suspended load beyond the closure depth, and deposited offshore. The model suggests that gentle beaches will retreat more rapidly than steep beaches, for a given rise in sea level.

There are a number of conceptual difficulties in the use of the Bruun Rule to predict coastal erosion, as well as difficulties in application (Healy 1991; Dubois 1992). The profile-of-equilibrium concept has been questioned, as has the use of a single equilibrium equation to represent all beach profiles (Pilkey *et al.* 1993). Deviations from the theoretical equilibrium profile are particularly common where there has been significant sediment sorting, with coarser material making the shoreward portion of the profile steeper, and finer sediment making the seaward portion gentler than predicted by the equilibrium equation. The Bruun Rule is based on the assumption that cross-shore sediment transport is the result of oscillatory flows, and it ignores the important contribution of

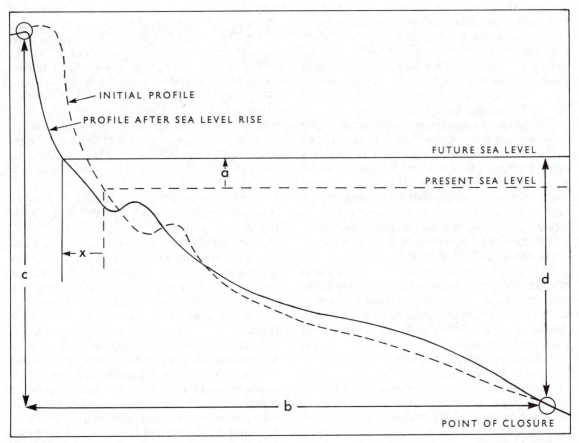

Fig. 1.4. Profile adjustment to rising sea level (Hands 1983)

undertows, rip currents, tidal-, wave-, and wind-induced changes in water level, and other factors (Chapter 3). The rule is also basically two-dimensional, and is not applicable to situations where sediment is carried alongshore or landwards into washover, inlet, or flood-tidal deltas, rather than into the offshore zone.

Coastal responses to rising sea level will be more complex than is predicted by the Bruun Rule. Shorelines could migrate landwards or seawards with slowly rising sea level, depending on the amount of available sediment (Dolotov 1992). Sorting processes tend to concentrate coarser grains in the littoral zone, while finer grains are carried offshore. Whether the amount of sediment in the nearshore increases or decreases may therefore depend upon the proportion of fine and coarse sediment in the eroded material (Komar *et al.* 1991). In addition to the erosional adjustment of

beaches to rising sea level, higher water levels encourage dune line breaching and overwashing by storm waves. This increases the tendency for coastal barriers to rollover and migrate landwards, or they may drown in place (Chapter 5). Rates of coastal erosion could be very low while the gradients of offshore profiles were adjusting to rising sea level. Barrier islands are slowly eroding and becoming narrower in several areas on the mid-Atlantic coast of the USA, for example, but when a critical width has been attained, they may begin to migrate landwards at rates that are from five to eight times higher than at present (Everts 1987).

The decision-making process associated with coastal erosion and flooding is complex, because of constraints imposed by financial considerations and a myriad of physical, social, economic, legal, political, and aesthetic factors. There is public and political pressure on coastal planners and managers to be seen to be doing something about the problem, and this can result in engineering projects that provide only short-term benefits, or which may even exacerbate the original problem. Several managerial options are available, however, ranging from the 'do nothing' approach to the construction of a completely artificial coast (Nordstrom and Allen 1980; Jolliffe 1983) (Fig. 1.5).

The attitude of most coastal geomorphologists to beach and dune erosion is quite different from that of coastal engineers. Engineers have traditionally recommended the hard stabilization of shorelines, including the building of structures parallel and perpendicular to the coast. Geomorphologists and coastal geologists usually emphasize the naturally dynamic nature of coasts, and advocate the use of other protectional measures. Not only are sea walls, riprap, and other structures unsightly, but they do nothing to remedy the causes of coastal erosion, which include rising sea level and negative sediment budgets. Sea walls and other hard structures prevent the exchange of sediment between the beach and foredune, and wave reflection from these structures scours the beach in front. Erosion and coastal recession of the unprotected shore on either side of structures eventually cause them to protrude into the surf zone, where they interfere with longshore sediment transport. An alternative approach is to place large amounts of compatible sand into the coastal zone to compensate for negative sediment budgets. Beach nourishment has been found to be technically feasible in many areas, although the cost may prevent its use in the poorer regions of the world. In sparsely populated regions it may be possible simply to relocate homes and other structures further inland, although public resistance to relocation often makes this a difficult political option (Pilkey and Wright 1988; Clayton 1990; Paskoff and Kelletat 1991).

The smallest amount of global warming is expected to occur in the tropics. An increase of 1–2 degrees in water temperature may increase coral growth rates in marginal areas, although prolonged periods of higher temperature and light conditions have bleached and killed corals in warmer regions (Chapter 10). Records extending back to 1870 suggest that recent bleaching events are unprecedented in scale, frequency, and severity (Glynn 1993). Three increasingly severe cycles of global bleaching have been attributed to the general warming trend through the 1980s (Bunkley-Williams and Williams 1990). Global warming

may initially cause a drop in sea level, because of the ability of a warmer atmosphere to hold more water. This could desiccate reef flats, causing mass coral mortality and making it impossible for the reef to recover and keep pace with the rapid submergence expected to occur a few decades later (Hoyt 1991). Assuming that reefs are not first exposed by lower sea levels, however, rising levels may actually be beneficial to their growth in the short term. Large areas of reef flat could be recolonized by corals, improving their appearance and increasing the rate of calcium carbonate production (Kinsey and Hopley 1991). Healthy reef frameworks generally appear capable of growing upwards sufficiently rapidly to match the expected rise in sea level up to the year 2100 (Spencer 1995). Widespread mortality would only take place when the reefs had fallen more than 20 m below the water surface, and this is not expected to occur within the next century. In many parts of the world, however, reefs are experiencing stress from anthropogenically induced siltation, the pollution and high turbidity of coastal waters, nutrient loading, and numerous other factors. Reduced calcification rates of corals stressed by higher temperatures and the effects of human population growth and resource exploitation, may render them incapable of compensating for rising sea level. Furthermore, a high proportion of most reefs consists of areas, including back-reef and lagoonal habitats, that accrete more slowly than well-covered reef flats, and are therefore more likely to succumb to rising sea level.

Rising sea level threatens the continued existence and habitation of low coral islands which have maximum elevations only 2–3 m above sea level (Miller and Mackenzie 1988; Oberdorfer and Buddemeier 1988; Roy and Connell 1991). Seawater incursion into freshwater lenses beneath low-lying atolls and sandy cays will affect water quality, altering vegetation type and density, and the protection it affords against storms. Freshwater supplies would not only be threatened by more frequent storm overwash, but also by increased coastal erosion—a decrease in island area causing a dramatic reduction in the volume of the groundwater lenses. Recharge rates would also be modified by changes in precipitation and evaporation amounts and temporal distributions. Higher surface water temperatures may increase the frequency and intensity of hurricanes and alter their distribution. This would not necessarily have

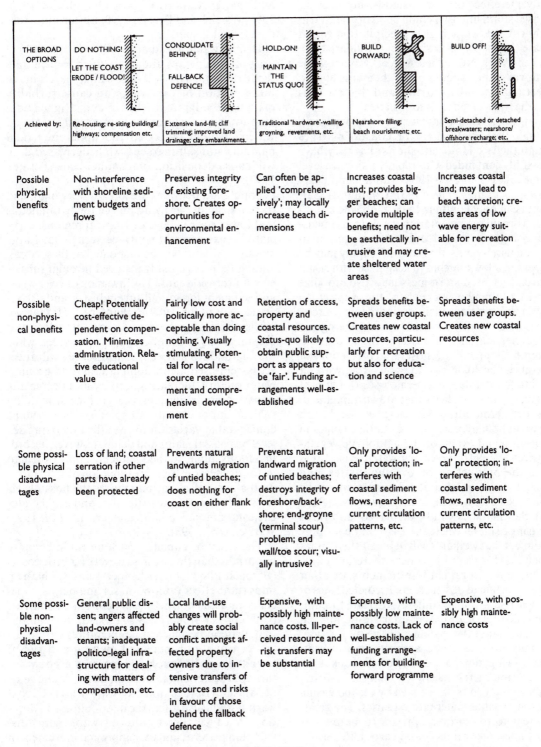

THE BROAD OPTIONS	DO NOTHING! LET THE COAST ERODE / FLOOD!	CONSOLIDATE BEHIND! FALL-BACK DEFENCE!	HOLD-ON! MAINTAIN THE STATUS QUO!	BUILD FORWARD!	BUILD OFF!
Achieved by:	Re-housing; re-siting buildings/ highways; compensation etc.	Extensive land-fill; cliff trimming; improved land drainage; clay embankments.	Traditional 'hardware'-walling, groyning, revetments, etc.	Nearshore filling; beach nourishment; etc.	Semi-detached or detached breakwaters; nearshore/ offshore recharge; etc.
Possible physical benefits	Non-interference with shoreline sediment budgets and flows	Preserves integrity of existing foreshore. Creates opportunities for environmental enhancement	Can often be applied 'comprehensively'; may locally increase beach dimensions	Increases coastal land; provides bigger beaches; can provide multiple benefits; need not be aesthetically intrusive and may create sheltered water areas	Increases coastal land; may lead to beach accretion; creates areas of low wave energy suitable for recreation
Possible non-physical benefits	Cheap! Potentially cost-effective dependent on compensation. Minimizes administration. Relative educational value	Fairly low cost and politically more acceptable than doing nothing. Visually stimulating. Potential for local resource reassessment and comprehensive development	Retention of access, property and coastal resources. Status-quo likely to obtain public support as appears to be 'fair'. Funding arrangements well-established	Spreads benefits between user groups. Creates new coastal resources, particularly for recreation but also for education and science	Spreads benefits between user groups. Creates new coastal resources
Some possible physical disadvantages	Loss of land; coastal serration if other parts have already been protected	Prevents natural landwards migration of untied beaches; does nothing for coast on either flank	Prevents natural landward migration of untied beaches; destroys integrity of foreshore/backshore; end-groyne (terminal scour) problem; end wall/toe scour; visually intrusive?	Only provides 'local' protection; interferes with coastal sediment flows, nearshore current circulation patterns, etc.	Only provides 'local' protection; interferes with coastal sediment flows, nearshore current circulation patterns, etc.
Some possible non-physical disadvantages	General public dissent; angers affected land-owners and tenants; inadequate politico-legal infrastructure for dealing with matters of compensation, etc.	Local land-use changes will probably create social conflict amongst affected property owners due to intensive transfers of resources and risks in favour of those behind the fallback defence	Expensive, with possibly high maintenance costs. Ill-perceived resource and risk transfers may be substantial	Expensive, with possibly low maintenance costs. Lack of well-established funding arrangements for building-forward programme	Expensive, with possibly high maintenance costs

Fig. 1.5. Managerial options to coastal erosion (Jolliffe 1983)

an adverse effect on coral islands, however, as increased storm intensity could build up the land in the short term, while higher precipitation would increase the supply of fresh water (Stoddart 1990; Fletcher 1992). Nevertheless, reef submergence and storm damage could make them less able to protect high islands, and any land that remains above the water surface on low islands.

Rising sea level also threatens coastal wetlands, although the development of coastal marshes through the late Holocene shows that they have successfully contended with rapidly rising sea level in some areas (Jacobson 1988; Funnell and Pearson 1989; Jennings *et al.* 1993) (Chapter 8). Salt marshes can only continue to develop where mineral and organic sedimentation is able to compensate for rising sea level and sediment compaction. In riverine environments, the sediment supply may be sufficient to allow marshes to keep pace with rising sea level. Increased storminess may also mobilize sediments, thereby increasing the supply to marshes that are dependent on inorganic sources. A mass balance model suggests that minerogenic backbarrier marshes in eastern England could accommodate all but the most rapid predictions of sea level rise up to at least the year 2100, without reverting to tidal flats. Moderate rises would produce accretionary deficits that are sustainable in the short term, although there would be an increase in tidal inundation and subtle changes in the location and extent of the halophytic vegetation (French 1993). Increasing water levels could reduce the production of organic matter, and marshes that rely on organic sedimentation may be unable to keep up with rapid rises in sea level (Reed 1995). Salt marshes will be displaced landwards with rising sea level where there is a limited supply of sediment, but wetlands will be lost if this cannot happen, as, for example, where there are cliffs, valuable agricultural land, settlements, or other natural or artificial obstructions. Predicted rates of sea-level rise are also unlikely to lead to catastrophic disruption of the largest, river-dominated or deltaic mangrove systems, or the major tide-dominated systems where large amounts of sediment are being transported. There will probably be some erosion of their seaward fringes, however, and they may also be threatened by the increasing intrusion of saline water into estuaries. The greatest threat to mangroves appears to be on low islands and on coral reefs and carbonate banks, where mangroves grow on carbonate sediments

and mangrove-derived peat (Woodroffe 1990, 1995; Warrick and Farmer 1990; Ellison and Stoddart 1991; Ellison 1993).

Deeper water in estuaries will reduce frictional drag on tidal waves and allow them to penetrate further inland. This will raise salinity levels in the inner portion of estuaries, and cause turbidity maxima to migrate further inland. Higher tidal amplitudes and current velocities will erode and widen channel banks in the upper intertidal mudflats and salt marshes, while decreased drag will cause deposition on channel beds (Pethick 1993).

Population densities are already very high, and they are rapidly growing on the low-lying deltas and lower river valleys of tropical regions, especially in eastern and south-eastern Asia. Large amounts of land will be inundated in these areas, adding to losses already incurred through subsidence and erosion caused by diversion of river water and sediment for agriculture, industry, and urbanization (Bardach 1989) (Chapter 9). Higher sea levels will also expose new areas of deltas and coastal plains to the effects of storm surges, while the migration of salt wedges up estuaries will affect ecosystems, sedimentation patterns, and groundwater quality. Increasing salinity and decreasing estuarine circulation could also reduce their ability to flush out pollutants (Chapter 7). There will be considerable variation in the effect of rising sea level on coastal plains and deltas, however, according, for example, to the sediment supply or length of the dry season. Simple estimation of land losses on the basis of elevation ignores the possible effects of sediment redistribution, and exaggerates the total amount of loss (Jelgersma *et al.* 1993; Warrick *et al.* 1993).

The effect of climatic warming could be more important than the rise in sea level on Arctic coasts (Bardach 1989). Higher temperatures will have an important effect on low coasts and deltas, especially in areas consisting of ice-rich permafrost (Chapter 12). It has been suggested, for example, that a rise in permafrost temperatures within the last century, and increased melting and runoff from the nearby Brooks Range, may account for the rapid accretion of deltas along the Arctic coast of Alaska (Barnes and Rollyson 1991). Global warming will also affect the distribution and duration of sea ice, thereby altering wave energy distribution patterns and the supply and movement of sediment in the littoral zone.

MODELS AND THE COASTAL SYSTEM

Models provide one of the best ways of investigating the structure, organization, and functioning of the complicated and poorly understood components of a coastal system. Models are abstractions or simplifications of systems. They can be used to generate hypotheses, and are valuable tools for simplification, reduction, experimentation, explanation, prediction, and communication. They provide insights into the interrelationships between and among variables, and they are indispensable in enhancing our efforts to monitor, manage, control, and develop the coastal system and its associated resources (Dalrymple 1985; Dyke *et al*. 1985; Fox 1985; Lakhan and Trenhaile 1989).

Physical models are simplified and scaled representations of prototypes. Scaling can be up or down, and in space, material properties, or time. Physical models can be used to provide a qualitative insight into phenomena not yet described or understood, to provide measurements to test theoretical results, and to measure complicated phenomena that cannot be theoretically analysed. Coastal engineers have constructed a wide variety of fixed-bed hydraulic scale models to study the action of waves, tides, and currents, and to assist in the design of coastal structures. Geologists and geomorphologists have used movable bed models to examine sediment transport, and the dynamics and formation of bars, barriers, and beaches. Although variables can be controlled and isolated on hardware scale models, however, they cannot duplicate aspects of coastal systems that are governed by uncontrolled entities and attributes which are unpredictable in magnitude and duration. Furthermore, unlike natural oceanic waves, the shallow water waves generated in most wave tanks have no orbital kinetic energy and are nearly pure solitons (Smith 1994). Physical models have therefore not been able accurately to describe the hydrodynamics and sedimentary processes operating in coastal systems, and the results obtained from them always have to be verified or corroborated.

The scale of physical models has to be carefully considered for the results to be applicable to their full-scale counterparts. The concept of similitude is used to determine scaling parameters. This refers to the amount by which some quantity in the model is multiplied to obtain the corresponding quantity in the prototype. Three basic laws are applicable to the study of fluids (Fox 1985):

a) The law of geometric similitude, or similarity of shape, requires that the flow field and boundary geometry of the model and prototype be the same. The ratio of model lengths to prototype lengths must therefore be the same, and equal to the scaling factor.

b) Kinematic similitude, or similarity of motion, requires that the ratio of the velocities and accelerations must be the same in the model and prototype.

c) There must also be dynamic similitude, or similarity of the forces acting on the fluids, in order to maintain kinematic similitude between geometrically similar models and prototypes.

The Froude number, which is the ratio of the inertial to the gravitational forces, is the main hydrodynamic scaling parameter for free liquid flows. It is the most important scaling parameter for coastal studies, although Reynolds numbers must also be considered if bottom friction is important. The Froude numbers must be equal to attain similitude in a model and prototype. Therefore:

$$\frac{u_{mo}^2}{gl_{mo}} = \frac{u_p^2}{gl_p}$$

where u_{mo} and u_p are the model and prototype velocity, respectively, g is gravitational acceleration, and l_{mo} and l_p are characteristic lengths or depths in the model and prototype respectively.

Other types of model concerned with symbolic or formal assertions, or a verbal or mathematical kind in logical terms, can be described as being theoretical, symbolic, conceptual, or mental. Mathematical models are used most widely because of their generality, versatility, and flexibility. Unfortunately, however, our lack of knowledge of coastal processes and the frequent reliance on laboratory data to determine the value of coefficients, casts doubt on the applicability of many mathematical models to the real world (Smith 1994).

Mathematical models can be classified according to a variety of criteria. The most widely used are: deterministic or probabilistic; optimization or simulation; static or dynamic; lumped parameter or distributed parameter; continuous or discrete; and analytical or numerical (Lakhan and Trenhaile 1989). Deterministic models, which are based on the principles of fluid mechanics, seem to work best in conjunction with laboratory experiments that allow parameters to be held constant while one is varied at a time. Simulation models involve the manipulation of process-response models on a computer. Re-running simulation experiments with varying input parameters allows the behaviour of a system to be determined under a variety of situations and conditions, and to test the sensitivity of the system to changing input parameters (Lakhan 1989). Simulation models and fast computer analysis can compress years of coastal development in the prototype into minutes. Static simulation models portray the processes operating at an instant in time, whereas the processes in dynamic models move ahead through time, with feedback loops representing the environmental response. The order of the steps and the relative frequency and intensity of different processes can be changed in probabilistic simulation models, using a random-number generator to express natural variability. This can help to identify the relative importance of the various processes in building up the prototype, and it can assist in the prediction of future prototype development. Statistical models can be used to study the relationships within a set of variables, and to verify possible relationships identified by theoretical models. To use equations derived from one area for predictive purposes in another, however, often requires the determination of a different set of coefficients (Fox 1985).

2 Waves, Tides, and Sea Level

Waves result from the deformation of the water surface by wind, submarine mass movements, and other agencies. Most are progressive, but standing waves can form in estuaries, bays, and along shorelines. Waves can be classified according to their frequency and magnitude (Fig. 2.1). Capillary waves are very small waves that are resisted by surface tension, whereas gravity is the restoring or resisting force for larger waves. The most important gravity waves for coastal scientists are those generated by the wind blowing over a water surface. Long, infragravity waves are of considerable importance to coastal processes, but they are often difficult to detect because of their low heights and long periods.

WIND WAVES

Wind waves are formed through the transfer of energy and momentum from the air to the water. Water movements are set up within the waves and there is a continuous, two-way transfer of energy between the kinetic and potential forms.

Wave Generation

Most explanations for wave generation are now based on the theories of Phillips (1957) and Miles (1957). Phillips attributed the formation and initial growth of small, wave-like deformations to the fluctuating pressures exerted on a water surface by

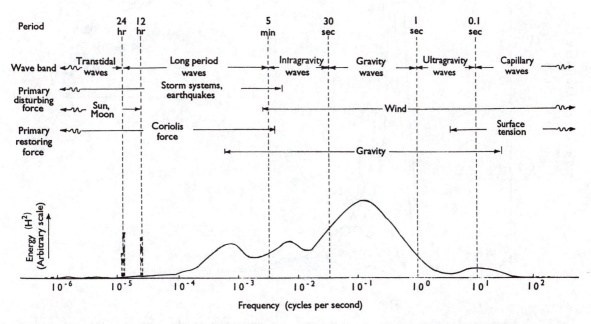

Fig. 2.1. Schematic estimate of the energy in surface waves. Most energy is contained within 4–12 s gravity waves. The 12 and 24 hr peaks represent tidal waves (Kinsman 1965)

a turbulent wind. Miles proposed that pressure perturbations associated with disturbance of the air flow over a small, pre-existing wave, cause it to increase in energy. His mechanism provides a possible explanation for the continued growth of waves that are initiated by Phillips's mechanism. Additional mechanisms have been developed to account for the rate of wave growth (Longuet-Higgins 1969*a,b*; Garrett and Smith 1976).

The height and period of waves generated in deep water are determined by the velocity and duration of the wind, and the fetch or distance of open water over which it blows. Wave spectra develop, through time and with distance of travel, as each component increases in energy. At equilibrium, the gain in wave energy is matched by the loss owing to breaking, and to a lesser extent to molecular and eddy viscosity. Higher-frequency waves

attain equilibrium before those of lower frequency. The peak of a growing wave spectrum therefore moves towards lower frequencies through time. The term 'fully arisen sea' (FAS) describes an equilibrium state that is independent of wind duration and fetch. Its development may take several days and hundreds of kilometres of travel under strong wind conditions. Instability of the highest waves in a FAS produces whitecaps, formed by water spilling down the advancing crests.

The Joint North Sea Wave Project (JONSWAP) is the most extensive investigation into the growth of waves yet conducted (Hasselmann *et al.* 1973; Ewing 1983). The main portion of this international study involved wave and air flow measurements along a line extending for about 160 km from the German island of Sylt. It was found that non-linear interactions between waves of different

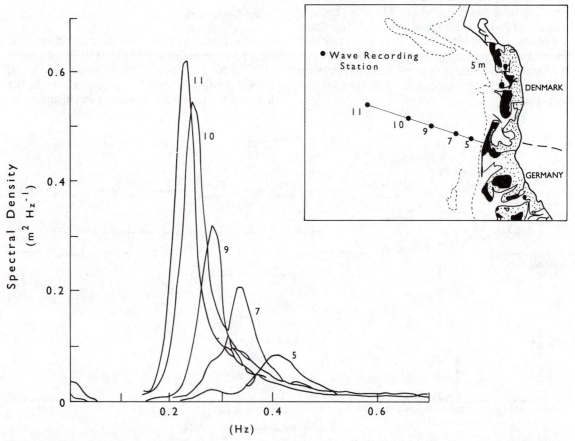

Fig. 2.2. Jonswap wave recording stations (inset) and corresponding stages in the development of the wave spectrum with fetch, for offshore winds. The graph shows an increase in wave energy and a progressive shift to lower frequencies in the downwind direction (Barnett and Kenyon 1975; Ewing 1983)

frequencies assume an important role in the down-wind development of a wave spectrum, particularly in transferring energy from the peak of the spectrum towards lower frequencies (Fig. 2.2). Non-linear interactions may also account for the 'overshoot' phenomenon, whereby wave energy and height temporarily attain values that are greater than under equilibrium conditions (Barnett and Sutherland 1968) (Fig. 2.3).

Wave Prediction

Wave forecasting, and hindcasting using historical wind fields, are often of great importance for the analysis of coastal processes and evolution. Without a satisfactory understanding of the mechanisms of wave generation, however, wave prediction has had to be partly based on theory, and partly on

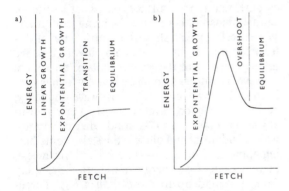

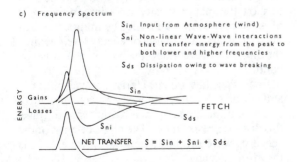

Fig. 2.3. (a) The growth of selected frequency component of the wave spectrum to energy saturation with infinite fetch, based on the theories of wave generation of Phillips and Miles; and (b) the overshoot effect owing to non-linear wave-to-wave interactions (Barnett and Sutherland 1968). Figure (c) shows contributions of various factors to the energy budget with increasing fetch (Hasselmann *et al.* 1976; Ewing 1983)

the measurement of waves generated under different wind conditions.

The SMB method (Sverdrup, Munk, and Bretschneider) uses empirical relationships, based upon a large body of wave observations, to predict, in a graphical format, significant wave height ($H_{1/3}$) and period ($T_{1/3}$) from known wind speed, duration, and fetch. The PNJ method (Pierson, Neumann, and James) predicts wave spectra, from which one can obtain the significant height or the complete distribution of wave heights. Other spectral models have been proposed, including the popular JONSWAP spectrum (Hasselmann *et al.* 1973, 1976; Bishop and Donelan 1989). Most models are empirical, however, and major theoretical advances are hindered by the lack of precise wave field data and the complexity of the physics of air-sea interaction (Huang *et al.* 1990). Furthermore, although spectral methods provide more realistic representations of actual wave conditions, they are more difficult to use, and many workers therefore continue to employ the SMB method.

Numerical models, which can be used to predict the sea state in deep or shallow water, are based on the transport or energy balance equation (Fig. 2.3*c*):

$$S = S_{\text{in}} + S_{\text{nl}} + S_{\text{ds}}$$

where S is the net source function, S_{in} is the input by the wind, S_{nl} is the non-linear transfer by resonant wave–wave interactions, and S_{ds} is the dissipation by wave breaking and whitecapping. Numerical models are concerned with the interplay of these factors in producing the spectral shape of waves, and the space–time evolution of the characteristic features of the spectrum, including the total energy and peak frequency (SWAMP group 1985; SWIM group 1985). In the first-generation models, the input function (S_{in}) was usually represented by terms based on Phillips's (1957) resonance and Miles's (1957) shear instability mechanisms. Second-generation models resulted from a fundamental reassessment of the wave spectral energy balance in the 1970s, and particularly, growing recognition of the dominant role of non-linear interactions in controlling wave growth. A third-generation model has now been developed which uses the spectral transport equation to compute the wave spectrum from first principles, rather than forcing it to comply with prior restrictions on spectral shape (WAMDI group 1988).

Wave Theories

Wave theories are used to describe wave motion and to contribute to an understanding of wave dynamics (Sleath 1984; Hardisty 1989, 1990a; Fenton 1990). A number of criteria are used to determine the optimum range of application of each theory, including the complexity of the equations, the depth of the water, and the required precision of the solution. Five theories are often employed:

a) Airy wave theory uses only first order functions, omitting terms involving wave height to the second (H^2) and higher orders. This is not a problem for low waves in deep water, but the theory is less suitable for shallow water. It is the mathematically simplest of the theories, however, and, because it seems to work quite well for a variety of purposes, many workers use it as a good alternative to Stokes wave theory, except in extreme situations.

b) Stokes wave theory was originally developed to the second order, but has since been extended to the fifth order. It is therefore applicable to waves of finite amplitude in all depths of water. This is the most widely applicable theory, and although some equations are very complex, tabulated solutions are available.

c) Gerstner trochoidal wave theory is limited to waves in water of infinite depth. The equations are quite simple and, although Stokes theory agrees better with observations, and is based on firmer theoretical grounds, the differences with trochoidal theory are fairly small.

d) Solitary wave theory is based on the fact that waves in shallow water resemble a series of solitary waves isolated from each other by broad, flat troughs. Because of its simplicity, and similarity to real waves, solitary wave theory has been widely applied to shallow water conditions. Nevertheless, small bottom slopes can cause the properties of solitary waves to differ quite markedly from theoretical predictions, and this, together with the difficulty of applying the theory to periodic oscillatory waves, casts some doubt on its suitability for nearshore studies.

e) Cnoidal wave theory describes a periodic wave that may have sharp crests separated by wide troughs—the type of wave characteristic of the area just outside the breaker zone. The cnoidal theory should be used for waves in shallow water, and although the mathematics are difficult, graphs can be employed to determine more easily its main properties (Wiegel 1964).

Predictions of small amplitude (Airy and Stokes) and shallow water (cnoidal and solitary) wave theories become quite poor near the breakpoint. Although numerical methods provide more accurate results for near-breaking waves, they are complex and laborious to use (Sleath 1984).

Deep Water Waves

Winds generate a confused mixture of waves of different sizes and shapes. Ripples and small waves are superimposed on larger waves, and wave crests are short, sharp-edged, discontinuous, and often the short-term result of several smaller waves coming together. This complexity is greatly increased by addition of wave trains generated by other storms. The constituent wave forms can be separated using harmonic or spectral analysis, which assumes that the sea surface consists of a series of simple sinusoidal waves of varying amplitude. Wave spectra, which show the contribution of different sources to the total energy involved in the undular motion of the sea, are plots of the square of wave height per frequency band, which is proportional to wave energy, against wave frequency, period, number, or wavelength. Most energy usually occurs in the 4- to 12-second range of periods, with semidiurnal and diurnal tides also contributing appreciable amounts (Fig. 2.1). Nevertheless, there is some wave energy at all periods, ranging from a second up to more than a day. Three-dimensional directional spectra, which can be used to show the energy density according to the angle of wave approach, can either be mapped as isolines, or by graphical plots of energy levels against the direction of wave approach for each frequency band.

Each wave has a characteristic wavelength (L) and period (T)—the distance between consecutive crests, and the time required by the wave to travel this distance, respectively. The phase celerity, or the speed of wave propagation in deep water (C), is therefore given by:

$$C = \frac{L}{T}.$$

More useful, however, are equations that describe the relationship between pairs of variables. Airy theory provides the following relationship between wavelength and period:

$$L = \frac{(g\,T^2/2\pi)}{r}$$

where $r = \tanh(kh)$, k is the wave number $(2\pi/L)$, h is the still-water depth, and g is the acceleration due to gravity. In deep water, where the depth is greater than one-fourth the wavelength, r is approximately equal to 1. In SI units, the expression then becomes:

$$L = 1.56T^2.$$

This expression allows wavelength, which is difficult to determine in the field, to be calculated from the period, which is measured using only a stop-watch. Also, since $C = L/T$, then:

$$C = 1.56T.$$

These expressions show that longer waves travel faster than shorter waves. Longer waves emerge from confused sea surfaces in storm areas before shorter waves, which also lose much more of their energy through air resistance, turbulent friction, and breaking. Wave spectra therefore change with increasing distance from the generating areas, their peaks moving towards lower frequencies as longer, faster waves become increasingly dominant. The term 'sea' is used for gravity waves that are receiving energy from the wind and are therefore in the process of being generated. 'Swell' refers to waves that have moved out and spread out from the area of generation. Long and short sea waves arrive almost simultaneously at the coast where there is a short wave fetch, creating choppy conditions. Coasts facing long oceanic fetches, however, receive a high proportion of long, regular swell waves with rounded crests, the shorter waves having been left far behind. They also sometimes experience the effects of swell waves originating from different storm sources, with frequencies slightly out of phase. The addition of two such waves produces regular alternations of groups of high and low waves. The interval between two groups of high waves is typically between two and four minutes, and it is often the sixth, seventh, or eighth wave that is the highest. This phenomenon (surf beat) creates rhythmic, infragravity vibrations in the nearshore zone, which have profound effects on surf zone processes, and important morphological implications (Chapters 3 and 4).

Variations in wind direction cause waves to move out of their generating area in slightly different directions, so that they spread out over a widening area. Waves are extremely efficient transporters of energy, and losses take place very slowly with the distance of travel. Swell waves travel enormous distances across oceans along great circle routes, although they can be refracted, reflected, and even trapped within oceanic currents (Barnett and Kenyon 1975). Waves transport energy and momentum over thousands of kilometres, but there is little net movement of the water itself. This is because water particles in deep-water waves rotate in essentially closed, circular orbits, without any net forward displacement. The diameter of the orbits, which for a particle on the water surface is equal to the height of the wave, diminishes exponentially with depth below the surface. According to Airy wave theory, the diameter of the orbits beneath the surface (O_d) in deep water is given by:

$$O_d = He^{kh_o}$$

where H is wave height, the vertical distance from crest to trough, and h_o is the depth of water to the centre of the orbit. This expression suggests that orbital diameter decreases most rapidly with depth for waves with short wavelengths and periods. As water particles must complete one orbit in one wave period, the orbital velocity must also decline with depth, becoming almost negligible at depths of about one wavelength.

Wave height can be expressed in several ways. The significant wave height ($H_{1/3}$), which is the mean of the highest one-third of the waves (usually recorded over a 20- to 30-minute period), roughly corresponds to the wave height visually estimated by an observer. Other possibilities include the average wave height, root-mean square wave height, the average of the highest 10 per cent of the waves ($H_{1/10}$), and the maximum wave height over a measurement period. Wave steepness, the ratio of wave height to wavelength, is also of considerable significance, the steepness of most waves falling between 0.1 and 0.056 (King 1972).

Waves have potential and kinetic energy associated with the deformation of the water surface and the orbital motion of the water particles, respectively. According to Airy theory, the two energy forms are equal, and the total amount is related to the square of the wave height:

$$E = 1/8\,\rho g\,H^2$$

where E is the energy per unit of wave crest, and ρ is the density of the water. The power or energy flux (P) is the rate at which wave energy is transmitted in the direction of propagation. Although this is determined by wave velocity, single waves in deep water move through wave trains at twice the group celerity, becoming smaller and eventually disappearing when they reach the front of the group. These waves are replaced by new waves that form at the rear of the group. Therefore, while single waves only exist for a short time, wave groups may travel across entire oceans. It is therefore the group celerity (Cn) that is used in the expression for wave energy flux:

$$P = EC_n$$

where:

$$n = 0.5\left[1 + \frac{2kh}{\sinh(2kh)}\right].$$

As waves advance, there is also a flux or transmission of momentum. The radiation stress has been defined as the excess flow of momentum owing to the presence of waves. It has a number of useful applications, having been used, for example, to calculate set-up and set-down changes in mean water level in the nearshore zone, and to investigate the origin of longshore currents and the generation of infragravity frequency motions (Komar 1976). The radiation stress or momentum flux (S) in the direction of wave advance is given by:

$$S = E(2n - 0.5)$$

and parallel to the waves crests by:

$$S = E(n - 0.5).$$

In deep water (where $n = 0.5$), the radiation stress is equal to $E/2$ in the direction of wave advance, and 0 parallel to the wave crest; the corresponding values in shallow water (where $n = 1$) are $3E/2$ and $E/2$, respectively.

Shallow-Water Waves

The orbital motion of a wave propagating through increasingly shallow water eventually extends to the bottom. Although very weak wave motions occur at depths as great as $0.84L$, it is generally accepted that waves begin to 'feel' the bottom when the depth is about half the deep-water wavelength ($h/L = 0.5$). The effects become really significant, however, at depths of one-quarter of the wavelength. In shallow water, the frictional resistance of the bed causes energy to be lost to the heat sink, and the wave then undergoes gradual changes

known as shoaling transformations. One or more secondary waves, which are harmonics of the primary wave frequency, often develop on the trailing edge of shoaling waves as energy is transferred to higher frequencies. As the waves shoal and break, each harmonic peaks and then declines as they move forward at different speeds (Fig. 2.4).

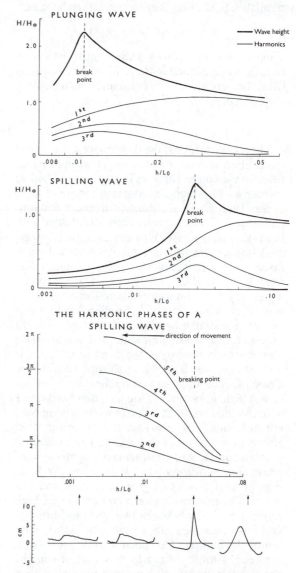

Fig. 2.4. Harmonic decoupling in plunging and spilling breakers in laboratory flumes (upper two diagrams). The lower two diagrams show the phases of the harmonics, relative to the primary wave, of a spilling breaker and its bore, and the associated, asymmetric shape of the wave (Flick *et al.* 1981)

Harmonic decoupling ultimately leads to the formation of a surf bore (Guza and Thornton 1980; Flick *et al.* 1981).

Simplified versions of Airy's equations are applicable when the depth of the water is more than one-quarter the deep-water wavelength. The full Airy expression must be used for wavelength, however, in intermediate depths ($0.25 > h/L > 0.05$):

$$L = \frac{gT^2}{2\pi} \tanh{(kh)}.$$

Once the depth becomes less than one-twentieth of the wavelength ($h/L < 0.05$), tanh kh becomes equal to kh and:

$$L_s = T(gh)^{0.5}$$

and as $C = L/T$:

$$C_s = (gh)^{0.5}$$

where the subscript *s* refers to shallow water. These two expressions show that shoaling wave phase celerity and wavelength progressively decrease. Wave period, however, remains constant. Shoaling transformations for Airy waves have been tabulated and plotted as functions of the ratio of water depth to the deep-water wavelength (Coastal Engineering Research Center 1973; Komar 1976).

Wave height is independent of the other wave properties in deep water, but changes in wave height in shallow water are related to changes to the other wave characteristics. To maintain deep-water wave power in shallow water, wave energy must increase to compensate for slower movement. As wave energy is related to the square of the wave height, the height of the waves must also increase in shallow water.

As waves enter shallow water, the orbits become more elliptical and forward movement under the high, narrow wave crests becomes increasingly stronger, although of shorter duration, than seaward movement under the long, flat troughs. In relative depths of less than 1/20, the ellipses flatten out into straight, horizontal lines that are the same length throughout the entire water column. The passage of each wave therefore causes the flow of water near the bottom to accelerate, decelerate, and reverse.

Wave Refraction and Diffraction

Wave crests bend or refract, and the direction of propagation changes as a portion of a wave, travelling through shallower water, slows down more rapidly than another portion in deeper water. From Snell's Law:

$$\frac{\sin\alpha_1}{C_1} = \frac{\sin\alpha_2}{C_2}$$

where α_1 and α_2 are the angles between adjacent wave crests and their respective bottom contours, and C_1 and C_2 are the corresponding phase celerities. For a straight coast with parallel offshore contours, the angle at a given depth between the wave crest and the bottom contour can be related to the angle of approach of the wave in deep water α_o:

$$\mathrm{Sin}\,\alpha = \frac{c}{c_0} \cdot \sin\alpha_0$$

where subscript o refers to deep water characteristics. The expression shows that as the phase celerity decreases in shallow water, the angle made by the wave with the bottom contour also decreases. Waves therefore tend to assume the shape of submarine contours, which may, in turn, resemble the shape of the coast.

Wave refraction causes energy to be concentrated or focused on some parts of a coast and dissipated on others. The concentration of wave energy on headlands, for example, can produce waves that are several times higher than in adjacent bays. Wave energy is also concentrated to the lee of submarine bars and other shallower areas, and weakened or dissipated behind deeper areas (Fig. 2.5). Because of the depths at which they begin to feel the bottom, long swell waves are refracted much better than shorter, locally generated storm waves, and they usually reach the shore parallel or at very low angles to it. Theory and procedures for the construction of wave refraction diagrams have been described in a number of places, and a variety of computer programmes have been written to provide automatic calculation and plotting (King 1972; Coastal Engineering Research Center 1973).

In some situations, waves are refracted back on themselves along a locus or caustic, a smooth surface, curve, or envelope tangent to a family of rays (Peregrine and Smith 1979). This can occur, for example, when reflected waves travel back into deeper water, or when waves pass an island. A caustic usually defines a zone of trapped wave energy. There is an illuminated zone on the wave side of a caustic, and a non-wave, or shadow zone on the other side. Although, in theory, the shadow zone should receive no wave energy, significant amounts are diffracted into this zone.

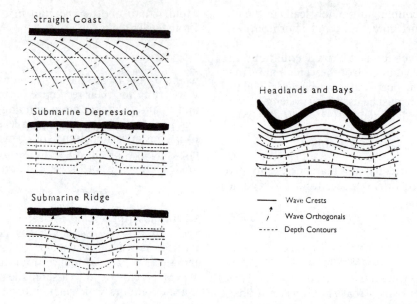

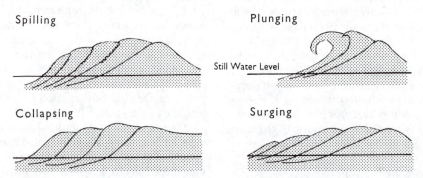

Fig. 2.5. Wave refraction over variable submarine topography (orthogonals show the direction of travel of the wave crests), and types of breaking wave (Trenhaile 1990)

Wave diffraction laterally transfers energy along the crest of waves, from areas where it is higher into areas where it is lower. This situation often occurs behind breakwaters, headlands, islands, and in other shadow zones. Diffraction and lateral transfer of energy by strong currents can also occur to the lee of islands, and in other places where wave rays cross at caustics.

Wave Reflection and Trapping

Waves can be reflected from vertical cliffs and breakwaters, sloping beaches, submarine bars, and other structures. If the reflected and incident waves are of approximately equal magnitude, a standing wave or clapotis develops, consisting of a vertical jet of water alternately rising and falling back on itself (Trenhaile 1987). Most natural features are not perfectly vertical, however, and they generally have fairly shallow water at their base—reflected waves are therefore usually smaller than incident waves. This produces a partial clapotis, and collapse of the vertical jet creates travelling waves on either side that collide and form other clapotis (Fig. 2.6).

Variations in topography can lead to the trapping of gravity waves and the development of edge

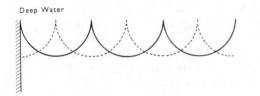

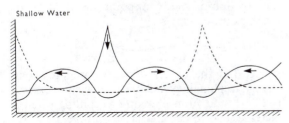

Fig. 2.6. True standing waves or clapotis in deep water and partial clapotis in shallow water (Trenhaile 1987)

waves, or large scale, long period, wind or baroclinic induced Rossby-type waves. Refraction of obliquely incident gravity waves from a cliff, for example, can cause the wave crests first to become perpendicular to the shore and then turn inward, to be reflected again at the shore. This process can be continued many times, the wave paths describing a series of arcs between the cliff and the wave caustic.

Pronounced wave breaking does not occur on steep, reflective beaches, where incident waves are strongly reflected at the shore, and resonant energy is transferred to subharmonic edge waves. Plunging and spilling breakers occur on gentle, dissipative beaches, and resonant energy is transferred to higher-than-incident wave frequencies. Reflectivity also varies with wave steepness and frequency, however, and tends to be higher for low- than for high-frequency waves (Tatavarti *et al.* 1988).

The surf similarity or surf scaling parameters, and a variety of other expressions, have been used to distinguish highly reflective and dissipative beach states. They have formed the basis for the study of rhythmic topography, longshore sediment transport, rates of energy dissipation, generation of secondary circulation, and many other nearshore phenomena, and they have been incorporated into general models of nearshore morphodynamics (Battjes 1974; Guza and Inman 1975; Guza and Bowen 1975; Wright *et al.* 1979). Nevertheless, the fact that these parameters describe nearshore conditions fairly well does not necessarily imply functional determinism or

predictive ability, and they should not be applied haphazardly to the surf zone, especially when the underlying assumptions are ignored (Bauer and Greenwood 1988).

Breaking Waves

Wave transformation as it moves through shallow water eventually causes it to break (Peregrine 1983; Basco 1985). Three explanations have been proposed to account for the breaking wave phenomenon:

a) Stokes proposed that waves break when the velocity of the water particles at the crest becomes greater than the wave velocity. This occurs when the angle at the crest reaches 120°, corresponding to a limiting wave steepness for deep water waves of 1/7. Breaking may therefore be a response to the rapid increase in steepness as waves enter shallow water.

b) The diameter and velocity of the orbits increase as wave height increases, but the velocity of the wave decreases. Eventually the orbital velocity of the water particles becomes greater than the wave velocity, causing the water to break through the wave form.

c) Wave velocity in shallow water is equal to $(gh)^{0.5}$. The velocity at the crest of a solitary wave is therefore:

$$g(h + H)^{0.5}$$

and at the trough:

$$g(h - H)^{0.5}.$$

The crest therefore moves faster than the trough, causing waves to become asymmetrical, with a steep leading edge and a gentle backslope. Instability stemming from this asymmetry may eventually cause the wave to break.

Each of these theories suggests that waves break when the water depth reaches some critical value in relation to wave height. The critical ratio (γ) between wave height and depth varies from about 0.6 to 1.2, with a mean of about 0.78. Values are higher for steep than flat beaches, and waves therefore break much closer to the shoreline than the same waves on gentle beaches. Waves also break differently on barred and non-barred planar beaches (Smith and Kraus 1992). The position of the breakpoint is influenced by currents off the mouths of rivers and by strong nearshore currents (Horikawa 1988). The wind also influences the location, geometry, and type of breaking wave. Waves tend to break earlier, in deeper water, and to spill when winds are blowing onshore, whereas offshore winds cause waves to break later, in shallower water, and to plunge. A change in the wind direction from offshore to onshore also increases the width of the surf zone (Douglass and Weggel 1988).

Several workers have provided expressions that allow breaking wave height to be estimated from deep water wave conditions (Horikawa 1988). Komar and Gaughan (1972), for example, obtained:

$$H_b = 0.39 g^{1/5} (T H_o^2)^{2/5}$$

where subscripts *o* and *b* refer to conditions in deep water and at the breakpoint, respectively.

The characteristics of breaking waves have important implications for current generation and sediment transport. Four types have been identified, although they belong to a continuum of forms that actually grade into each other, and some breakers do not really fit into any of these categories (Fig. 2.5):

a) Surging breakers are the result of strong reflection of low steepness waves from steep beaches. These waves peak up as they get close to the shore, but before they can plunge forward, their base surges up the beach with a smooth, sliding motion, causing their crests to collapse and disappear. Surging breakers resemble non-breaking standing waves, and they lose little of their energy at the 'breakpoint'.

b) Collapsing breakers are intermediate between the plunging and surging types. Collapsing occurs when the lower, landward portion of the wave front steepens until it is vertical, and then curls over as an abbreviated plunging wave. The irregular, turbulent water then slides up the beach without developing a bore-like front.

c) Plunging breakers tend to occur on steeper beaches than spilling breakers, with waves of intermediate steepness. The crest of plunging breakers curls over and plunges downwards as an intact mass of water, and there is a sudden loss of energy close to the breakpoint.

d) Spilling breakers are associated with steep waves and flat beaches. They are initiated a long way from shore when a small mass of water moves faster than the wave itself, so that the crest becomes unstable and cascades down the wave front. Entrainment of air bubbles makes the turbulent water lighter than the water below, and the foam on the front of the breakers retains its identify as it rides on top of the sloping sea surface. Spilling breakers eventually develop into bores or undulating bores, and they continue to lose energy from the breakpoint up to the water's edge.

Galvin's (1972) breaker coefficient (B_o) predicts the position of the breakpoint and the type of breaking wave:

$$B_o = \frac{H_o}{L_o \tan^2 \beta} \text{ or } B_b = \frac{H_b}{g T^2 \tan \beta}$$

where β is the beach slope. Surging breakers change to plunging when the coefficient increases to 0.003; and from plunging to spilling when it increases to 0.068.

There have been other attempts to quantify the relationship between breaker type, beach slope, and wave properties. The surf scaling parameter (ε) is similar to Galvin's breaker coefficient (Guza and Inman 1975; Guza and Bowen 1975):

$$\varepsilon = \frac{H_b \omega^2}{2g \tan^2 \beta}$$

where ω is the incident wave radian frequency ($2\pi/T$). Surging breakers occur when $2 < \varepsilon < 2.5$, and much of the incident wave energy is reflected from the beach into the oncoming waves. Plunging breakers appear when $\varepsilon > 2.5$, and spilling breakers when $\varepsilon > 20$, most wave energy being dissipated within wide surf zones.

Battjes (1974) used the surf similarity parameter (ξ) to predict the occurrence of breaker types:

$$\xi = \frac{\tan \beta}{(H_o/L_o)^{0.5}}$$

or the breaking equivalent (ξ_b):

$$\xi_b = \frac{\tan \beta}{(H_b/L_o)^{0.5}}$$

which is known as the Iribarren number. Using Galvin's data, and wave height at the breakpoint, Battjes determined that surging or collapsing breakers occur when $\xi_b > 2$; plunging breakers when $0.4 < \xi_b < 2$; and spilling breakers when $\xi_b < 0.4$.

Breaker type may also be identified by comparing the duration of onshore water movement—from the moment the wave breaks until it has run to its furthest point up the beach—with the wave period. Uprush times of surging breakers are about half the wave period, whereas spilling breakers have uprush times of several wave periods (Kemp 1975).

The Surf and Swash Zones

The surf zone extends from the breakers to the point where waves begin to run up the beach. Waves rapidly decay in the outer surf or transition zone, as they change from unbroken oscillatory waves into highly rotational or turbulent bores. Changes occur much more slowly in the inner, bore-like surf zone, where the waves are said to be saturated—their height being locally depth controlled and largely independent of their height outside the breakers (Thornton and Kim 1993). The fairly constant mean water level of the outer surf zone is replaced by a rapid rise or set-up in the inner region (Basco 1985; Basco and Yamashita 1986). The inner surf zone may not exist on steeply sloping beaches, and waves can therefore plunge, collapse, or surge directly up the foreshore.

We still lack an acceptable theory to describe water movement in the surf zone (Kobayashi *et al.* 1989; Hamm *et al.* 1993; George *et al.* 1994). Plunging and spilling breakers have similar breaking motions, although at very different scales (Basco 1985; Raichlen and Papanicolaou 1988), and they both develop into turbulent bores, with properties that are largely determined by the local depth of the water. Wave breaking usually occurs when a portion of a wave becomes vertical; a part of the surface then overturns and is projected forwards as a jet of water. Spilling breakers develop when the velocity of the jet is lower than the velocity of the wave, so that the splash is directed down the front of the wave. Overturning wave fronts are best developed on plunging breakers, particularly on steep beaches. The jet may be thrown beyond the base of the wave and the distance from the breakpoint to the plunge point can constitute a very significant portion of the transition zone.

In some cases, the jet lands in a small amount of water and is redirected up the beach as uprush (swash, runup). Usually, however, the wave front curls over, forming an air cavity as the jet closes onto the lower water surface. The cavity rapidly collapses, the trapped air sometimes venting through the surface as spray, and the entrapped air mixes with water. The jet either rebounds on the water surface or it pushes up a wedge-shaped mass of water, forming another jet over another cavity of air. Several jet-splash cycles are produced in this way by a single breaker, creating a series of discrete, large-scale vortices with a high concentration of air bubbles. Each successive vortex is weaker than the preceding one, and as they move more slowly than the wave, they drift to the back, expanding, slowing down, and generating foam as the bubbles rise to the surface behind the wave crest (Fig. 2.7). The vortices in plunging breakers are large relative to water depth, and the second and third may extend to the bottom before completely decaying. Vortices in spilling breakers are much smaller and are limited to regions near the water surface, usually above the still water level. Vortical motion contributes to rapid decay and transformation of breaking waves, but it has not yet been quantitatively modelled (Peregrine 1983; Basco 1985; Battjes 1988; Matsunaga *et al.* 1988; Tallent *et al.* 1990).

Jet-splashing ceases and vortex motions degenerate into turbulence as the wave proceeds into the inner surf zone. The breaking wave has now been transformed into a bore, consisting of a steep, turbulent front and a surface roller with recirculating flow between the toe of the front and a point of separation near the crest (Madsen and Svendsen 1983; Svendsen and Madsen 1984). The progressive reduction in water depth is usually sufficient to sustain turbulence in the bore, but if it is not, the bore becomes undular and its front develops into a smooth succession of long waves. In addition to rollers and vortices, which are perpendicular to the direction of wave propagation, secondary flows may occur in bores as a series of longitudinal vortices or helixes, oriented parallel to the main flow (Brenninkmeyer *et al.* 1977).

In the swash zone, thin sheets of water are propelled up and down the beachface by breaking waves and bores. Undular and fully developed bores behave differently as they approach the shoreline. The front of weak, undular bores overturn directly onto the beach surface, and the uprush consists of thin, smooth-surfaced layers of water. Fully developed bores first decelerate at the shoreline as the wave form compresses, until the front face becomes almost vertical. They then suddenly accelerate, and the turbulence on their front face appears to be advected on to the beach. The transition from bore to uprush is characterized by two motions, the turbulent uprush produced by a small, wedge-shaped body of water pushed up by the advancing bore, followed by the original incident wave motion (Yeh *et al.* 1989).

The uprush becomes progressively thinner, and its velocity decreases exponentially as it rises up the beach. It comes to rest when the leading edge

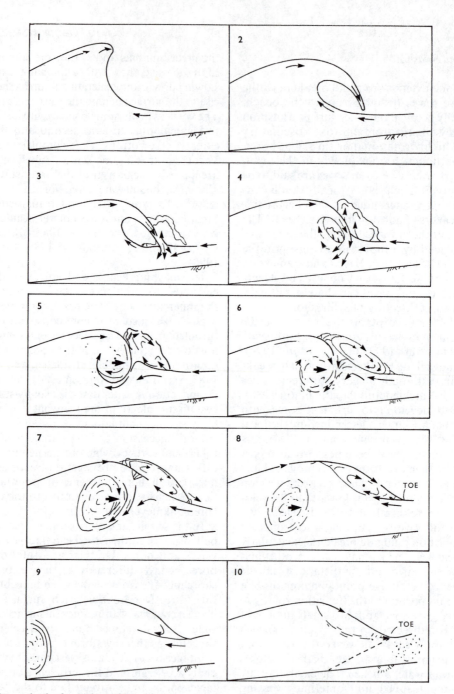

Fig. 2.7. Stages in the formation of breaking wave vortices. (1), (2), and (3), breaking occurs, and the overturning jet hits the oncoming trough and forms splash. (4) The jet penetrates the trough and is deflected downwards and backwards to form a vortex. (5) The trapped air is compressed and forms bubbles in the water. (6) A surface roller develops as the splash falls forwards. (7) The plunger vortex translates horizontally, pushing on the trough to create a secondary disturbance, and increase the size and strength of the surface roller. (8) Sliding of the toe of the roller down the front of the oncoming wave generates more vorticity as it grows in size. (9) Vortex translation slows down and stops as it rotates, enlarges, and drifts downwards, although the secondary disturbance continues to propagate forwards. (10) The inner breaker region begins where the surface roller reaches stable equilibrium, and horizontal translation of the plunger vortex stops generating the secondary disturbance (Basco 1985)

reaches its maximum position, and backrush (backwash) then takes over, the water running back down the beach as a gravity-induced flow. High backrush velocities retard the incoming wave front, promote earlier breaking, and influence the motion of the wave behind (Peregrine 1983; Battjes 1988). Kirk (1975) recorded swash zone velocities on 5° to 12° foreshore slopes, with breaker heights ranging from 0.3 to 2.44 m. The average uprush velocity was 1.68 m s^{-1}, with a maximum of 2.50 m s^{-1}, and an average duration of 2.98 s at the mid-swash zone position. Backrush velocities averaged 1.40 m s^{-1}, with a mean duration of 4.25 s. Points in the swash zone therefore experienced an initially rapid increase in water depth during the uprush phase, and a more gradual decrease during the backrush phase.

Swash zone processes or mechanisms are modified by collisions that occur when the period of the incoming waves is less than the uprush duration. The phase difference, the ratio of the time of uprush of a wave (t_s) to the wave period, has been used to classify swash zone regimes (Kemp and Plinston 1974). For low phase differences of $t_s/T < 0.3$, uprush cycles are completed before the next wave arrives. This condition is typical of surging waves on steep beaches. Transitional conditions have t_s/T ratios of between about 0.3 and 1.0, and the uprush arrives before the backrush of the previous swash cycle is completed, forcing the backrush to flow laterally. This condition is associated with plunging breakers. On high-phase beaches, with t_s/T ratios greater than 1.0, each uprush overlaps with the previous uprush, the backrush is completely absent, and water circulation occurs through percolation, lateral flow, or other means. This situation is typical of spilling breakers on flattish beaches.

The height that a wave runs up a beach depends upon such factors as its height and period, the slope and roughness of the beach, the angle of wave approach, and wave interference in the inshore zone. Whereas incident waves dominate uprush cycles and elevations on steep, reflective beaches, where waves break close to the shore and immediately send uprush up the beachface, they may be related to long-period, infragravity wave components on gentle beaches (Guza and Thornton 1982). A number of workers have derived equations and analysed statistical distributions to predict uprush elevations (Ahrens *et al.* 1985; Horikawa 1988; Holland and Holman 1993),

and others have attempted to correlate them with the surf similarity or surf scaling parameters (Battjes 1974; Guza and Inman 1975; Nielsen and Hanslow 1991). At present, however, the most accurate predictions are based upon empirical curves designed for specific conditions (Hallermeier *et al.* 1990).

Wave-Generated Currents

Numerous wave-generated motions coexist and interact in a complex way in the surf and nearshore zones. To simplify discussion, however, the distinction is made between those operating essentially normal to the shore, and those acting parallel, or along, the shore.

Currents normal to the shore When waves enter shallow water, the velocity of the landward movement under the wave crests becomes increasingly stronger, although of shorter duration, than the seaward movement under the troughs. Peak onshore and offshore flow velocities measured in wave flumes and in shallow water in the field, confirm the asymmetrical nature of near-bed flows (Horikawa 1988; Hardisty 1990a; Kuo and Chen 1990). These alternating, asymmetric currents and shear stresses have important implications for sediment movement and the evolution of beach profiles (Chapters 3 and 4).

Wave current asymmetry varies with water depth and wave characteristics, and therefore with beach gradient. Strong backrush off steep beaches may reduce the ratio of the forward-to-backward velocity close to the bed, just outside the breaker zone. The hydrodynamics of steep and gently sloping beaches have been compared in south-western England (Huntley and Bowen 1975). As spilling breakers crossed the wide surf zone off the gentle beach, they became lower and changed from a breaking, solitary wave-like form to a low breaking wave with a steep front. This was accompanied by a gradual increase in onshore current velocity and offshore current duration—the oscillations in the velocity-time curves becoming saw-toothed as the wave front steepened and became bore-like. There were plunging breakers in the 5–6 m-wide surf zone on the steep beach. Waves steepened rapidly towards the breakpoint, and there was some corresponding increase in onshore current velocity and offshore current duration. Horizontal velocities were high at the breakpoint. Inside the breakpoint, currents consisted of a strong, sustained offshore flow and a shorter onshore flow, as the collapsed breaker surged forwards. Areas close to the shore and well inside the breakpoint experienced strong uprush and backrush.

According to Airy wave theory, maximum orbital velocity at the sea floor (u_m) is given by:

$$u_m = \frac{\phi H}{T \sinh (kh)}.$$

Airy theory, however, also predicts that maximum velocity under the trough is the same as under the crest, although opposite in direction. Other wave theories, such as the Stokes or Cnoidal, must therefore be used to represent current asymmetry, and to accommodate higher waves and shallower water. Clifton (1976) defined the velocity asymmetry (Δu_m) as the absolute difference in the peak orbital velocities under the wave crest and trough. Using Stokes second order theory, he proposed that:

$$\Delta u_m = \frac{14.8 \, H^2}{LT \sinh^4 (kh)}.$$

Hardisty (1986) used the ratio of the maximum onshore (u_{in}) and offshore (u_{ex}) velocities to describe velocity asymmetry. His equations are also based on Stokes wave theory:

$$u_{in} = \frac{\gamma}{2}(1 + F) (gh)^{0.5}$$

$$u_{ex} = \frac{\gamma}{2}(1 - F) (gh)^{0.5}$$

where:

$$F = \frac{0.01\gamma^3}{(H_b/L_b)^2}.$$

The velocity ratio (V_r) is therefore given by:

$$V_r = \frac{u_{in}}{u_{ex}} = \frac{1 + F}{1 - F}.$$

The corresponding onshore (t_{in}) and offshore (t_{ex}) flow durations are represented by:

$$t_{in} = TA$$

and

$$t_{ex} = T(1 - A)$$

where:

$$A = \frac{1}{\pi} \cos^{-1} \frac{1 + (1 + 8F^2)^{0.5}}{4F}.$$

These equations suggest that the degree of wave asymmetry increases as waves become more surging. Stokes theory is of uncertain validity in the shallow depths near the breaker zone, however, where wave flow asymmetry is most important (Clifton and Dingler 1984). Hardisty (1989), for example, found that peak onshore and offshore flows near the bed in a wave flume were 0.63 and 0.89 times the values predicted by Stokes theory, respectively.

There is also some net movement of water towards the coast in shallow water. Slow onshore mass transport is predicted by Stokes but not by Airy theory. The Stokes model suggests that the orbits are no longer closed in shallow water, and each particle of water moves a little in the direction of wave advance with the passage of each wave. Longuet-Higgins (1953) provided an expression for the net shoreward velocity ($\bar{u}$) generated near the bottom by a Stokes wave:

$$\bar{u} = \frac{5}{4} \frac{(\pi H)^2}{(L)^2} \frac{C}{\sinh^2 (kh)}.$$

Although only small amounts of water are involved, and the velocities are low, mass transport is a unidirectional, shore-normal current, and it may therefore be important for the transport of fine, suspended sand.

The movement of water towards the coast must be compensated by an equal return flow. One possibility is that net onshore flow takes place at the surface and near the bottom, while the offshore return flow occurs at mid-depth. Measurement and modelling of cross-shore flows have led to growing recognition that fairly strong, seaward flowing, near-bottom undertows play an important role in the shore-normal movement of sediment, the formation and maintenance of longshore bars, and the evolution of beach profiles (Svendsen and Hansen 1988; Greenwood and Osborne 1991; Putrevu and Svendsen 1993). Undertows of up to 0.2 m s^{-1} have been measured in a barred surf zone during moderate storm events. This current, which generally decreased in velocity with increasing height above the bed, was strongly correlated with spatial and temporal variations in wave set-up. Similar topographically induced variations in the set-up and undertow on barred beaches have been reported from several areas (Hazen *et al.* 1990).

Longshore currents   Oblique waves drive longshore currents parallel to the shoreline. Incident angles for swell waves, which begin to feel the bottom a long way from shore, are rarely greater than

10°, but they can be much greater for shorter, less well refracted waves. Winds with a substantial longshore vector component can also generate longshore currents (Sherman and Greenwood 1985; Hubertz 1986).

There have been many attempts to account for longshore currents and to predict their velocity. Most recent work has related current generation to the longshore component of the radiation stress (momentum flux) of the waves. Komar (1983a) derived an equation for the mean longshore current velocity in the mid-surf position (V_1):

$$V_1 = 2.7 \, u_m \sin \alpha_b \cos \alpha_b$$

or

$$V_1 = 1.17 \, (gH_b)^{0.5} \sin \alpha_b \cos \alpha_b$$

where u_m, the maximum horizontal wave orbital velocity at the breaker zone, is equal to $(2E_b/\rho h_b)^{0.5}$; E_b is breaking wave energy; h_b is the depth at breaking; and α_b is the breaker angle. Cos α_b is approximately equal to 1 for angles less than 10°, and it can therefore be discounted. The expression shows that longshore current velocity depends upon the angle of wave approach and the orbital velocity at the breaker zone, which largely depends upon breaker height.

Longshore currents generated by oblique waves can coexist with longshore currents in circulation cells, which are the result of alongshore variations in wave height (Fig. 2.8). The two longshore currents can operate in the same direction, or they can oppose each other, and could, if equal, cancel each other out (Komar 1975).

Maximum longshore current velocities are generally between 0.3 and 0.6 m s^{-1}, although they tend to be temporally unsteady. In addition to velocity fluctuations corresponding to the incident wave frequency, variations with periods ranging up to several minutes may reflect the occurrence of infragravity waves, although a number of other explanations have been proposed (Guza and Thornton 1978; Komar 1983a). Longshore currents may also vary at tidal frequencies, possibly because of the effect of changing water depth on the height of saturated breaking waves (Thornton and Kim 1993).

Several workers have modelled spatial variations in longshore current strength across the surf zone, but few measurements have been made in the field. Most studies, however, suggest that velocities are quasi-parabolically distributed across the surf

zone, increasing seawards from the shoreline and usually reaching a maximum at, or just beyond, the midsurf position, before rapidly declining to zero outside the breaker zone (Komar and Holman 1986; Zampol and Inman 1989).

Analytical solutions for wave-induced longshore currents were developed for planar beaches, but the profiles of many natural beaches are concave upwards. McDougal and Hudspeth's (1983) solution for a concave beach profile suggests that the maximum longshore current is slower, and the current profile is significantly different from that predicted for planar beaches. The presence of submarine bars and other irregular topography can also have a marked effect on the strength and distribution of longshore currents. Several theoretical models suggest that longshore current velocities

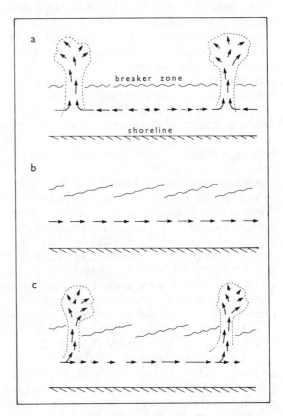

Fig. 2.8. (a) Rip currents, longshore currents, and nearshore cell circulation associated with incident waves approaching normally to the coast. (b) Longshore currents generated by oblique waves. (c) Cell circulation and currents generated by oblique waves (Komar and Inman 1970)

are highest over the crest of submarine bars, and lowest in the troughs, where waves reform after breaking (Symonds and Huntley 1980), although in the field, currents have sometimes been found to be strongest in bar troughs (Smith *et al.* 1993).

Rip currents, edge waves, and cell circulation   Shore normal and longshore currents often operate together in a circulatory current system. Cell circulation consists of the slow onshore mass transport of water through the breaker zone, longshore currents, and rip currents—strong narrow flows of up to 2 to 4 m s^{-1} that move the water seawards across the breaker zone and then broaden into a rip head. When waves have a normal approach to the coast, longshore currents diverge and become progressively stronger away from the midway point between adjacent rips. Circulation cells become asymmetric when waves approach the coast at an angle, and longshore currents flow in only one direction, their velocity increasing from zero on the immediately updrift side of a rip, to a maximum at the next rip current (Fig. 2.8). Asymmetric cells and rip currents may therefore migrate slowly alongshore. Rip currents and their longshore feeder systems can be strengthened by water draining from the beach during ebb tides. This is particularly significant when the water flows alongshore in troughs, before turning seawards into rip channels cut through bars.

Rip currents scour out channels that, in turn, act to stabilize their position. Cell circulation then becomes strongly influenced by beach and bottom topography, and it is not completely free to react to changes in wave conditions, except during high-energy storm events. Rip currents can exist where surf zones consist of alternating bars and troughs. In Florida, water flowing alongshore from bars, where there were spilling breakers and correspondingly higher set-ups, fed rip currents in intervening, seaward trending channels, where plunging breakers generated lower set-ups (Sonu 1972). The question remains, however, whether the topography generates the circulation, or the circulation generates the topography. Certainly, it is difficult to explain how the topography was created if not by cell circulation, and in any case, the existence of cell circulation is not dependent upon the presence of bottom irregularities.

Several workers have considered the factors responsible for fairly regular spacing between rip currents. Hino (1974) proposed that the spacing between adjacent rips is equal to four times the width of the surf zone, but others have found that there is considerable variation in this relationship. Huntley and Short (1992) found that rip current spacing is proportional to $H_b^{3/2}/w_s^2$, where w_s is the settling velocity of the sediment.

The occurrence of rip currents was first attributed to onshore mass transport, but a more successful explanation is now based upon wave-induced variations in the water surface. The radiation stress generates wave set-ups and set-downs, resulting in elevation of the mean water level in the surf and swash zones, and its depression within and just outside the breaker zone, respectively. Although set-up gradients are essentially independent of the size of the waves, larger waves produce higher set-ups than smaller waves because they break further from the shore. Maximum set-up is also generally inversely related to beach slope, and the shape of the water surface is influenced by the form of the beach profile. The set-up is confined to the swash zone on the beachface on steep reflected profiles, but it begins near the breakpoint on gentle dissipative beaches, and steepens its gradient at the inner breakpoint, and, still further, in the swash zone (Gourlay 1992).

Wave set-up was first measured in the laboratory, and it is only fairly recently that it has been confirmed by field measurement. In California, the maximum set-up on a gently sloping beach with a wide surf zone was found to be about 0.17 times the significant height of the deep water incident waves (Guza and Thornton 1981). A similar relationship was recorded from Georgian Bay, Ontario (Greenwood and Osborne 1990), but a higher value, of about 0.4 times the root mean square deep water wave height, has been reported from Australia (Hanslow and Nielsen 1993). In Britain, field measurement of wave set-ups, which ranged up to 0.4 m in height, also showed that there is a fairly good linear relationship between mean set-up and incident wave height (King *et al.* 1990). The relationship between wave height and set-ups, ranging up to 1.6 m in height, was quite variable on a moderately steep beach in North Carolina, although when scaled by deep water wave height, it was found to vary linearly with the surf similarity parameter (Holman and Sallenger 1985). Although set-ups at the mouth of rivers are minimal in some areas, even when breaking takes place, they can help to drive sand into estuaries (Tanaka and Shuto 1992).

Longshore variations in wave height and set-up provide pressure gradients to drive cell circulation, with longshore currents flowing from areas of higher waves and then turning seawards into rip currents where the waves are lower. Although set-up inside breakers is matched by set-down outside, any tendency for water to flow from positions of low to high breakers is countered by the gradient in radiation stress parallel to the wave crests. There is therefore no net force to drive circulation outside the breaker zone. Within the breaker zone, however, the longshore gradient in radiation stress acts in the same direction as the longshore variation in wave set-up. These two forces therefore combine within the surf zone to generate water flows away from high wave positions to areas of low waves. Longshore pressure gradients can be produced by wave refraction along coasts with irregular offshore topography. Longshore currents flow from areas of wave convergence and feed rip currents located in areas of divergence, where the waves and set-ups are lower. Refraction cannot, however, account for rip currents that are spaced at fairly regular intervals along long, straight beaches with a regular offshore.

Rip currents may be created by intersecting wave trains of the same frequency, periodically reinforcing and cancelling each other out as they propagate towards the shore. Rip currents develop in the surf zone where lines of cancellation, or nodal lines, intersect the beach. Suitable wave trains could be formed by reflectance, different storm systems, wave diffraction caused by islands, or refraction over a submarine shoal (Dalrymple 1975). Refraction of incident waves around pre-existing rip currents could also facilitate the growth of cell circulation systems. An alternative explanation is concerned with the effect of a small initial disturbance to an otherwise constant longshore wave set-up (Hino 1974). Others have also suggested that rip currents can be produced by hydrodynamic instability in the surf zone (Miller and Barcilon 1978).

Regular longshore variations in wave height and set-up can be produced by resonant interaction of incident waves with edge waves trapped by refraction between a shoaling beach and a caustic (Holman 1983; Schaffer and Jonsson 1992). Edge waves are free surface gravity waves with crests perpendicular to the incoming waves. Their amplitude varies sinusoidally in the longshore direction, but it can vary seawards in a number of ways, and

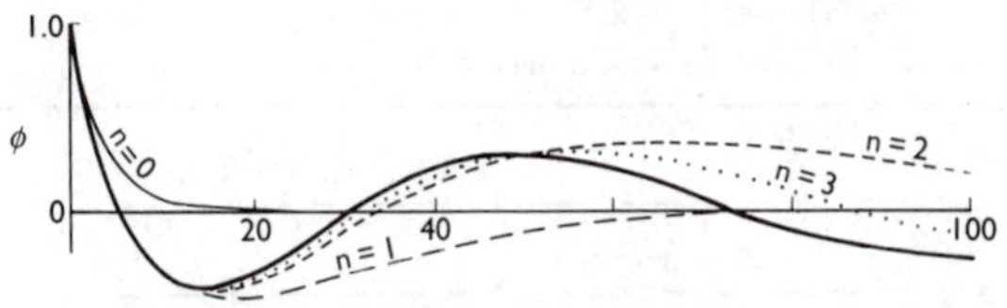

Fig. 2.9. Plot of amplitude variations of reflected wave (thick solid line) and edge wave modes 0 to 3 plotted against the dimensionless offshore distance (Huntley 1980)

cross-shore profiles are therefore more complicated than longshore profiles. Edge wave amplitude is at a maximum at the shoreline, and it rapidly decreases seawards, becoming almost negligible just outside the breakpoint (Fig. 2.9). Changes in tidal level, longshore currents, and non-planar or complex beach and bottom profiles, however, can modify this pattern (Evans 1988; Oltman-Shay and Howd 1993).

Edge waves can either travel along the coast as progressive waves, or, where two progressive edge waves of the same frequency travel in opposite directions, remain in a fixed position as standing waves. Stationary or progressive edge waves with the same period as the incident waves produce regular fluctuations in breaker height, the high and low points re-occurring at the same places along the beach with each breaking wave. Rip currents therefore develop where the breaking waves are lowest, at every other antinode (Fig. 2.10), and at intervals along the beach equal to the edge wave length (L_e):

$$L_e = \frac{g}{2\pi} T_e^2 \sin\left[(2n+1)\,\beta\right]$$

where T_e is the edge wave period—which for the generation of rip currents must be the same as for the incoming waves; and n is an integer wave mode number (0, 1, 2, 3...) which defines the number of shore-normal zero-crossings of the mean water level. The spacing of rip currents is therefore largely determined by the wave period, and to a lesser extent by the slope of the beach. Although different modal numbers provide a variety of possible wavelengths for each combination of period and beach slope, low modes are usually most easily excited and one mode generally becomes dominant.

The frequency of an edge wave (ω_e) is given by the dispersion relation:

$$\omega_e^2 = g k_e \sin\left[(2n+1)\,\beta\right]$$

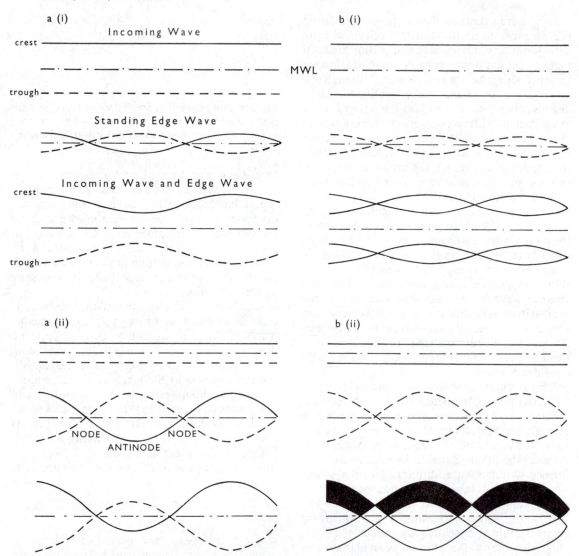

Fig. 2.10. Incident wave interaction with edge wave of (a) the same frequency, and (b) different frequency. In case (i), the incident waves are larger than the edge waves, and in case (ii), the edge waves are larger (Bowen 1972). The shaded area in b(ii) represents the perturbation of the maximum edge wave incursion due to the incoming waves

where k_e is the longshore edge wave number. The dispersion relation and the cross-shore shape of edge waves are sensitive to longshore currents. Longshore wave numbers increase when edge waves oppose the current flow, and decrease when they are coincident with the flow. The nodal structure of edge waves also shifts landwards as the longshore wave number increases, and seawards as it decreases (Howd *et al.* 1992).

Free, resonant edge waves satisfy the dispersion relation for values of $n = 1, 2, 3$, etc. There are no

trapped waves, however, if $(2n + 1)\beta > \pi/2$, and leaky waves with these frequencies and longshore wave numbers, including normal incident waves, may propagate to, or from, deep water. Forced, non-resonant waves have combinations of wave frequency and number that do not satisfy the dispersion relation (Bowen and Huntley 1984) (Fig. 2.11).

A wide variety of alongshore energy modulations can result from phase coupling of edge wave modes that coexist at the same frequency.

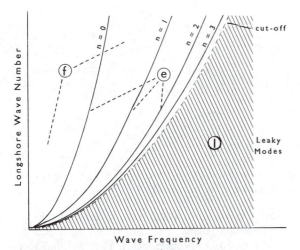

Fig. 2.11. The dispersion relationship between wave frequency and longshore wave number for a beach slope of 0.12. n is the modal edge wave number (e), and f and l are the regions of forced modes and leaky modes, respectively (Bowen and Huntley 1984)

Interaction of any two modes of the same frequency produces regular longshore variations in flow patterns or drift velocity, whereas coupling of three or more modes causes very irregular alongshore modulations (Holman and Bowen 1982). The predominant phase coupling is between a low mode (0 or 1) edge wave in the longshore current, and a higher mode edge wave in the cross-shore direction (Huntley 1988a).

Variations in the height of incoming waves owing to edge waves are usually too small to be seen. Their presence, however, is suggested by scalloped uprush on beaches, and rhythmic sedimentary features and nearshore circulation patterns. Edge waves have also been identified in the laboratory and field through the analysis of velocity spectra and variations in water surface elevation in the surf zone (Huntley *et al.* 1981; Oltman-Shay and Guza 1987).

Edge waves can be generated in a number of ways, including fast-moving atmospheric disturbances involving changes in wind or atmospheric pressure; interaction of two wave trains; reflection at the shoreline; and excitation by swash zone interactions (Huntley and Bowen 1975). It has also been proposed that standing edge waves may be generated by a longshore current passing over a localized protrusion on an otherwise uniformly sloping beach (Evans 1988). The instability of a single incident wave component reflected at the coast can generate short gravity edge waves on steep beaches. Low modes are most easily excited on reflective beaches, and synchronous ($T_e = T$) and especially subharmonic ($T_e = 2T$) edge waves have been observed. Guza and Davis (1974) showed how reflection of normally incident waves from a sloping beach generates subharmonic edge waves, and the theory was extended by Guza and Bowen (1975) to include incident waves approaching straight, planar beaches at an angle.

Theory suggests that in comparison with edge waves on steep, narrow beaches, those on wide, flat, and highly dissipative beaches should have lower frequencies, higher modes, and longer wavelengths. Water movement on dissipative beaches is usually dominated by oscillations with periods much greater than the incident waves, including infragravity edge waves with periods ranging up to 200 s or more (Holman and Bowen 1984, Kim and Huntley 1985). Between 42 and 88 percent of the longshore current variance on Californian beaches, for example, and as much as 50 percent of the variance in the swash spectrum have been attributed to low mode ($n \leq 2$) infragravity edge waves (Oltman-Shay and Guza 1987). Infragravity frequencies may be particularly dominant during storms, when breaking limits the height of the incident waves (Komar and Holman 1986; Guza and Thornton 1989; Russell *et al.* 1991). A dramatic increase in infragravity energy has been recorded during a storm off the eastern coast of Canada. Infragravity waves accounted for 84 percent of the spectral energy during the height of the storm, and a narrow 100 s peak dominated the spectrum throughout the storm (Holman *et al.* 1978).

Because of similar offshore profiles, it is difficult to distinguish infragravity energy associated with higher edge wave modes and standing incident waves. Mean water level is depressed beneath high waves and elevated under low waves. Groupy, normally incident waves, consisting of alternations of groups of high and low waves, therefore carry a forced wave component at the frequency of the wave groups (Fig. 2.12). When the incident waves break, the forced wave is reflected and escapes out to sea as a free, 'leaky' wave. The combination of the forced landward and free seaward propagating waves could then produce a standing component to the wave pattern (Guza and Thornton 1985; Kim and Huntley 1985).

The origin, exact nature, and relative importance of low-frequency motions remain to be

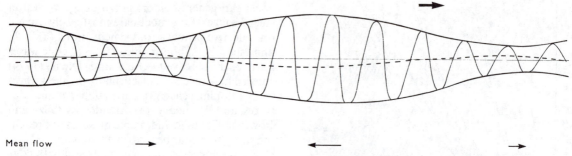

Mean flow

Fig. 2.12. The forced wave component associated with groupy incident waves. The mean water level is depressed under groups of high waves and elevated under low waves (Longuet-Higgins and Stewart 1962)

determined (Watson and Peregrine 1992; List 1992; Elgar *et al.* 1992). Infragravity frequencies could reflect the presence of forced (bound) waves coupled to groups of incident waves, or free waves that are decoupled from the incident wave groups, and therefore do not respond immediately to them. Free waves may take a number of forms:

a) Edge waves at infragravity frequencies are probably resonantly excited by groupy incident waves. Two groups of swells with a narrow frequency band can produce edge waves, which grow resonantly into large amplitude waves trapped to the shoreline (Gallagher 1971). Resonance continues to be important when the incident waves are breaking (Bowen and Guza 1978).

b) Leaky free waves are able to propagate towards or away from the coast. Free waves can also be generated within the nearshore zone. Variations in the position of the breakpoint and the set-up of groupy incident waves generate a standing wave in the surf zone, and a seaward propagating free wave seawards of the breakpoint (Symonds *et al.* 1982).

Free infragravity waves in the surf zone have therefore been considered to consist of progressive edge wave modes trapped to the nearshore, and leaky wave modes that escape out to sea. Surface gravity waves are not the only components, however, of the low frequency velocity field (< 0.05 Hz) in the nearshore. Shear or vorticity waves propagate in the same direction as the longshore current (Holman *et al.* 1990; Howd *et al.* 1990; Dodd 1994). They have been attributed to shear instability of a steady, longshore current (Bowen and Holman 1989; Putrevu and Svendsen 1992; Dodd *et al.* 1922; Falqués and Iranzo 1994), although it has also been suggested that they can be produced by

slowly migrating rip currents generated by two incident wave fields of slightly different frequency (Fowler and Dalrymple 1990). Typical longshore wavelengths are of the order of 10^2 m, an order of magnitude smaller than the lowest order edge waves of the same frequency. As frequencies are generally in the 10^{-2} to 10^{-3} Hz range, below the traditional limit of infragravity waves, this band has been termed the far infragravity band (FIG).

TIDES

Tides are periodic changes in the elevation of the ocean surface generated by the gravitational pull of the Moon and, to a lesser extent, the Sun. They move enormous amounts of water, creating strong currents and expending huge amounts of energy at the coast. Tides assume an important or dominant role in the formation and modification of mudflats, salt-marshes, and mangroves (mangals) in estuaries, bays, and other coastal inlets where there is a high tidal range and shelter from strong wave action (Fig. 2.13). They are also important, however, in wave-dominated environments, where they determine the degree of concentration and the vertical range of elevations over which waves and other marine processes can operate. Tides can generate longshore currents and they are responsible for variations in such factors as the width of the surf zone, the elevation of water tables in beaches and coastal rocks, and the position of the breaker zone.

High and low tidal levels correspond, respectively, to the crest and trough of tidal waves that are hundreds of kilometres in length. Tidal range, the difference in the height of consecutive high and low tidal levels, varies on a daily basis, and is therefore usually expressed as an average of the spring

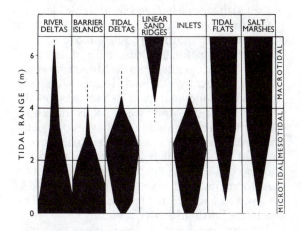

Fig. 2.13. The effect of tidal range on the morphology of coastal plain shorelines (Hayes 1975)

or neap tides. The terms macrotidal, mesotidal, and microtidal have been adopted to refer to spring tidal ranges of more than 4 m, from 4 to 2 m, and less than 2 m, respectively. Mean tidal range in the open ocean is between 0 and 1 m, but it can greatly increase towards the coast—the spring tidal range being more than 10 m in some estuaries and bays fronted by wide continental shelves, and more than 15 m in the Bay of Fundy in eastern Canada (Fig. 2.14).

Origin

Isaac Newton discovered the relationship between tides and the movement of the Moon and Sun in 1687, but it was almost a century later that a dynamic theory, encompassing the inertia of the water and the gyration of the Earth, was developed

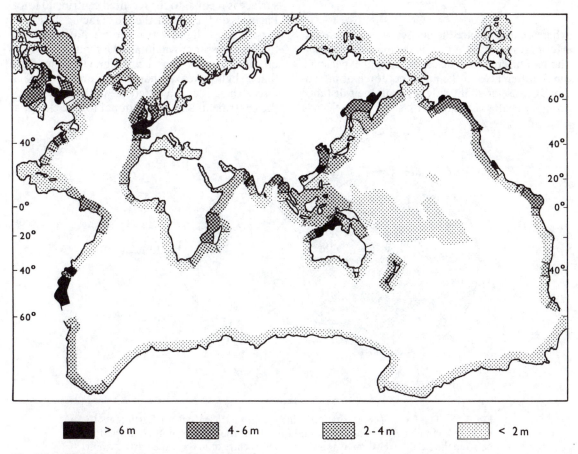

Fig. 2.14. Global spring tidal range (Davies 1972)

by the French mathematician Laplace. In this theory, the tides are considered to be long waves moving over an Earth covered in water of uniform depth. His theory continues to provide the basis for all modern work on tides.

The attractive force between any two elements of mass in the universe (F) is given by Newton's Law of Gravitation:

$$F = G \frac{m_1 m_2}{d^2}$$

where m_1 and m_2 are the masses of the two objects involved, d is the distance between them, and G is the universal gravitational constant ($6.67.10^{-11}$ N $m^2 kg^{-2}$). To calculate the total gravitational attraction between large bodies, such as the Earth and Moon, it can be assumed that their masses are concentrated at a single point at their centres. The net attractive force between the Moon and Earth is therefore:

$$F = G \frac{m_e m_m}{R^2}$$

where m_e and m_m are the masses of the Earth and Moon, respectively, and R is the distance between them. Different particles on Earth experience small differences in the mean net force of attraction because of slight differences in their distance from the centre of the Moon.

It is convenient for a moment to ignore the orbit of the Earth around its own axis, and consider only the Earth–Moon system. The Earth and Moon are attracted to each other, and they move about a common centre of mass. Because the Earth's mass is about 81 times greater than the Moon's, however, the common centre is actually located within the Earth, about one-quarter of its radius below the surface. While the Moon freely orbits the centre, the Earth wobbles around it with a fixed orientation, and all particles of mass on the Earth move in circles of the same radius. The centripetal acceleration needed to make the particles move in circles must therefore be of exactly the same magnitude everywhere, and directed towards the Moon parallel to the Earth–Moon axis (Fig. 2.15).

The centripetal acceleration is, on average, provided by the gravitational attraction of the Moon. The attraction varies with the distance of a particle to the Moon, however, and nowhere on the Earth's surface is it equal to the required centripetal force. Particles on the side of the Earth facing the Moon experience a slightly stronger gravitational force than is necessary to maintain the orbit, while those on the other side experience a gravitational force that is slightly weaker than the required centripetal force. There is also a difference in the direction of the centripetal and gravitational forces, except at

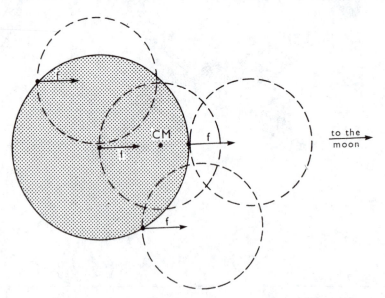

Fig. 2.15. The Earth–Moon rotation, which causes each particle of Earth mass to move in a circular path. The centripetal force (*f*) is the force of mutual attraction between the Earth and Moon. *CM* is the common centre of mass of the Earth–Moon system (Komar 1976)

the point directly under the Moon, and at the corresponding position on the opposite side of the Earth. At all other places, and in contrast to the centripetal force, the gravitational force is not parallel to the line joining the centres of the Earth and Moon (Fig. 2.16). These differences in magnitude and direction between the gravitational and centripetal forces generate tides on the Earth's surface.

For a particle of mass (m) in a sublunar position on the Earth's surface, the force towards the Moon is given by:

$$\frac{G(m\,m_{m})}{(R-r)^2}$$

where r is the radius of the Earth. The force required for particle revolution is the same as for a particle at the centre of the Earth:

$$\frac{G(m\,m_{m})}{R^2}.$$

The net lunar tide generating force at the sublunar position (T_{sl}) is equal to the difference between these two forces:

$$T_{sl} = G\,m\,m_{m}\left[\frac{1}{(R-r)^2} - \frac{1}{R^2}\right]$$

which is approximately equal to:

$$\frac{2G\,m\,m_{m}\,r}{R^3}.$$

This equation can be simplified. Newton's law of gravitation states that the gravitational force on a particle (mass m) on the Earth's surface is:

$$mg = \frac{G\,m\,m_{e}}{r^2}.$$

Therefore

$$G = \frac{gr^2}{m_{e}}.$$

Substituting for G, the equation for T_{sl} becomes:

$$T_{sl} = 2\,mg\frac{m_{m}}{m_{e}}\frac{r^3}{R^3}.$$

The gravitational force is therefore greater than the necessary centripetal force at the sublunar position.

Similarly, the net force on the opposite side of the Earth is:

$$-\frac{2G\,m\,m_{m}\,r}{R^3}$$

or:

$$-2\,mg\frac{m_{m}}{m_{e}}\frac{r^3}{R^3}.$$

The gravitational force at this point is therefore less than the necessary centripetal force.

The Earth exerts an attractive force on its surface that is about 300,000 times stronger than the Moon's. High tides therefore do not result from the Moon pulling water out towards itself. The radial or normal component of the tide-producing forces only slightly affects the magnitude of the gravitational force, but the horizontal component causes a slight change in its direction. Although the horizontal or tractive forces are small, they cause water to flow to the sublunar position, and in the corresponding location on the other side of the Earth, forming a second bulge that is only about 5 per cent smaller than the sublunar bulge.

The horizontal component of the Moon's differential force (T) at any point on the Earth's surface is given by:

$$T = \frac{3}{2}\,g\,\frac{m_{m}}{m_{e}}\frac{r^3}{R^3}\sin 2\theta$$

where θ is the angular distance of this point, measured at the centre of the Earth, from the line joining the centres of the Earth and Moon. This expression shows that the tractive force is zero where θ is 0°, or 180°, and on the great circle 90° distant from these points, and at a maximum on circles 45° from these points (Fig. 2.16). It also demonstrates that whereas the gravitational force exerted on a particle varies inversely with the square of the distance, the tide-raising force varies inversely with the cube of the distance.

While the Earth is rotating around its axis, the Moon is moving in the same direction around the Earth. As a result, a point on the Earth's surface occupies the same position, relative to the position of the Moon, on average every 24 hours 50.47 seconds. Two high tides and two low tides would therefore occur during this period, at each point on a water-covered Earth. The Sun's mass is much greater than the Moon's, but as it is much further away, its attractive force is only, on average, about 0.46 times as great. As attractive forces produce high tidal bulges on opposite sides of the Earth, the tide-raising force is a maximum when the Earth, Moon, and Sun are approximately aligned, the Moon either being in opposition (Full Moon) or conjunction (New Moon). Conversely, the tide-raising force is at a minimum when the Sun and Moon are in quadrature, with the tide-raising forces 90° out of phase. There are therefore two periods of maximum or spring (Anglo-Saxon

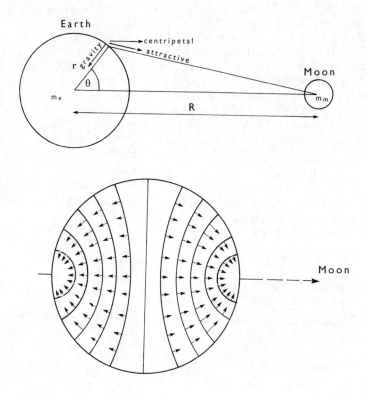

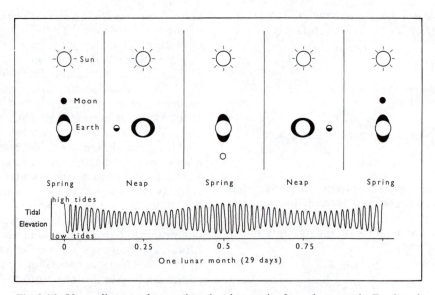

Fig. 2.16. Upper diagram, the centripetal and attractive forces between the Earth and Moon. Middle diagram, the reactive forces exerted on the Earth's surface by the Moon (Macmillan 1966). Lower diagram, tidal cycles during a lunar month (Pethick 1984)

springan—rising or upwelling of water) tides and two periods of minimum or neap tides about every 29.5 days (a synodic month)—the time the Moon takes to complete its orbit around the Earth (Fig. 2.16). Because of inertial and frictional effects, however, there is actually a lag of up to thirty-six hours between the occurrence of spring tides and the times of conjunction and opposition.

Tidal amplitude varies in a complex manner as a result of changes in the distance and declination of the tide-generating bodies, and in the relative position of the Moon and Sun. The Moon's orbit is elliptical rather than circular. The tidal generating forces are highest at the lunar perigee, when the Moon is about 357,000 km from Earth, and lowest at the lunar apogee, when it is about 407,000 km from Earth. This variation has a periodicity of 27.6 days (an anomalistic month). The tide-generating forces also vary as a result of the eccentricity of the Earth's orbit around the Sun, which has a periodicity of a little more than a true or sidereal year (366.5 days—an anomalistic year). The Earth's equator, and the orbits of the Moon around the Earth and the Earth around the Sun, do not lie on the same plane, but are tilted with respect to each other. Because of the tilt of the Earth's axis, the equatorial plane is at an angle of 23.5° with the plane of its orbit around the Sun. As the Earth moves around the Sun, with its axis maintaining a constant direction in space, the line of the solar tidal force varies between 23.5° north and 23.5° south on the Earth's surface. The Sun is overhead at the equator at the equinoxes, when the Earth, Moon, and Sun are almost aligned, and the highest

spring tides therefore occur around 21 March, and especially on 21 September when the Earth is closer to the Sun. Conversely, the Sun is high and the Earth, Moon, and Sun are not aligned at the solstices, and the lowest spring tides therefore occur around 21 June and 21 December. The declination of the Moon, which refers to its position relative to the Earth, varies over a period of 27.3 days (a tropical month), as it moves from the equator into the northern and southern hemispheres. The tidal bulges are therefore usually at an angle to the Earth's axis, creating an inequality in the two daily tides that increases with the declination. Slight variations in the tilt of the Moon's orbit cause the maximum lunar monthly declination to vary from 18.3 to 28.6° north and south of the equator, over an 18.61-year nodal period. Therefore all tidal variations associated with the lunar declination change in a regular manner over this period. Most of the world has semidiurnal tides, but diurnal tides, with only one high and low tide every twenty-four hours, occur where the inequality in the daily tides is so great that one of the two tides is almost negligible. Other areas have mixed tides that are semidiurnal at some times and diurnal, especially when the declination is very high, at others. Variations in the position of the Earth, Moon, and Sun are therefore responsible for semi-diurnal, diurnal, fortnightly, monthly, seasonal, annual, and even longer-period variations in the tide-producing forces.

Periodicities in tidal records can be expressed as the sum of the simple harmonic constituents, each with its own characteristic period, phase, and amplitude (Table 2.1). The relative importance of

Table 2.1. Principal tidal constituents or partial tides

Species	Notation	Period (solar hours)	Relative amplitude (%)	Description and Origin
Semi-diurnal	M_2	12.42	100.0	Main lunar semidiurnal constituent
	S_2	12.00	46.5	Main solar semidiurnal constituent
	N_2	12.66	19.1	Monthly variation in Moon's distance
	K_2	11.97	12.7	Changes in declination of Moon and Sun
Diurnal	K_1	23.93	58.4	Solar–lunar constituent
	O_1	25.82	41.5	Main lunar diurnal constituent
	P_1	24.07	19.3	Main solar diurnal constituent
Fortnightly	M_f	327.90	17.2	Moon's fortnightly constituent
Monthly	M_m	661.00	9.1	Lunar Monthly Phase
Solar semi-annual	S_{Sa}	4,385.00	8.0	Sun's semi-diurnal declinational constituent, vernal to autumnal equinox
Solar annual	S_a	8,759.00	1.3	Sun's annual distance, perigee to perigee
Nodal		163,024.00		Slight variations in tilt of Moon's orbit

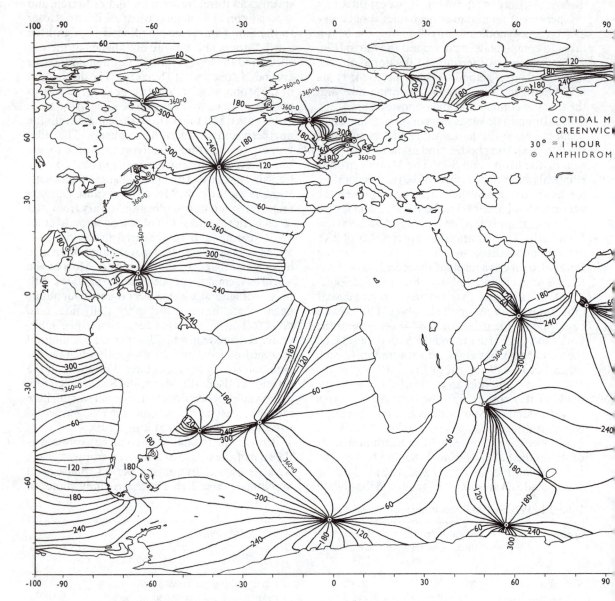

Fig. 2.17. Global amphidromic points for the M_2 tide based on a numerical model. Lines radiating from the points join places where high water occurs simultaneously (Schwiderski 1979)

each constituent can be determined through harmonic analysis, which provides the basis for tidal prediction. The first seven constituents (M_2, S_2, N_2, K_2, K_1, O_1, and P_1) can often be used to predict tides to within about 10 percent of their true figure, but up to sixty-five constituents may be needed when greater accuracy is required, or in estuaries and other environments where the influence of

bottom topography and non-linear effects are significant.

The shape, size, and other aspects of the morphology of water bodies have an enormous influence on the tides that are actually experienced at the coast. Tidal bulges travel across the oceans as tidal waves, moving from east to west. Reflection of the tidal wave from land masses introduces complex-

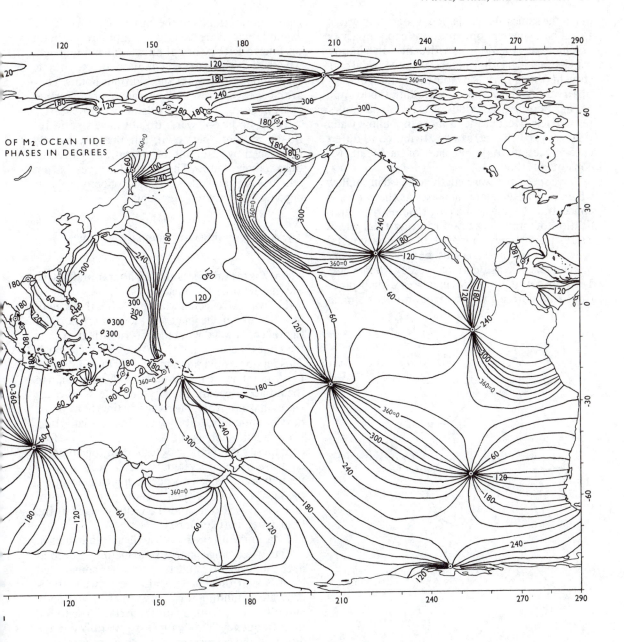

OF M₂ OCEAN TIDE
PHASES IN DEGREES

ities to the range and timing of the tides in coastal regions, as the reflected and oncoming wave crests interact and alter their relative positions through time. Because of the Coriolis Force, however, reflection does not create a standing wave about a nodal line, but rather a wave crest that rotates, usually anticlockwise in the northern hemisphere, about a nodal or amphidromic point of constant

sea level. The position of amphidromic points are determined by the geometry of the ocean and sea basins (Schwiderski 1979) (Fig. 2.17).

Tides in Shallow Water

Because of their length, tidal waves are in 'shallow water' in even the deepest oceans, and they obey much the same rules as shallow-water wind waves. Local bathymetric configurations can amplify or resonate certain tidal constituents. In shallow water, exact higher harmonics of the M_2 tidal constituent (the subscripts denote the number of daily observations) may appear, including the M_4 and M_8 harmonics with periods of 6.21 and 3.1 hours, respectively. Most of the variance in shallow-water tidal curves, however, can be attributed to differences in the amplitude and phase of the M_2 and M_4 constituents (Fig. 2.18).

Wave speed and wavelength decrease in shallow water, and wave height increases as the energy is concentrated into a smaller area. Similarly, amplification of progressive and rotating tidal waves causes tidal range to increase in shallow coastal waters, and as more water must be moved to accommodate this increase, current speeds also increase. The range of rotating tides increases with distance from amphidromic points. Tidal range is almost

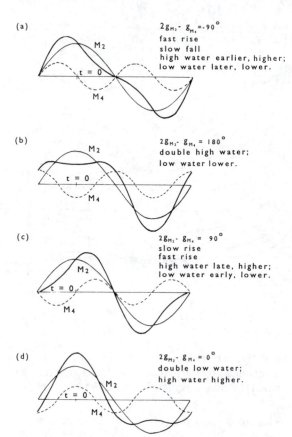

Fig. 2.18. The effect of M_2/M_4 tidal interactions on the shape of the composite sea level curve (Pugh 1987)

zero around islands in the eastern Caribbean, for example, which are close to an amphidromic point on Puerto Rico. Cotidal lines, connecting places where high tides occur simultaneously, radiate from and rotate around amphidromic points. They are close together in shallow water, but are further apart in deep water where tidal waves move faster.

Small bodies of water do not have significant tidal waves of their own, but they are greatly influenced by resonant amplification of oceanic tides at their mouths. The small tidal waves that are sent into bays and inlets travel at a velocity (C_s) that depends only on the water depth:

$$C_s = (gh)^{0.5}.$$

The corresponding wavelength is given by:

$$L_s = CT = T(gh)^{0.5}$$

where T is the tidal period (12.4 hours), and h is the average depth of the inlet.

The wave is reflected when it reaches the head of the inlet, and if the length of the inlet is just right, the reflected wave will reach its mouth at exactly the same time as the next high tide is about to enter. This reinforces the wave, increasing the tidal range to the point where it is several times greater than in the adjacent ocean. In very long inlets, reflected waves may not reach the mouth in time to meet the next oceanic tide, but they may meet a later tide. Long inlets, such as the Amazon, may therefore contain several tidal wave crests at any one time. In contrast to progressive tidal waves in oceans, where the maximum velocity occurs at high and low tide, the maximum velocity of standing tidal waves in inlets is at half-tide, and the water is slack at the tidal extremes (Chapter 7).

Tidal current velocity is usually less than 0.05 m s^{-1}, but it can range up to several metres per second in the mouths of estuaries and other constricted locations. Strong currents also occur where tides generate a standing wave, as in the Bay of Fundy and other inlets with a natural period close to the tidal frequency. Tidal currents generally change direction and rotate in a clockwise or anticlockwise ellipsoidal pattern during a tidal period, and ebb and flood currents therefore follow different paths.

SURGES, SEICHES, AND TSUNAMIS

Tidal levels are continuously modified by non-periodic meteorological influences. Forcing by weather systems generates surges that raise the water surface to extreme elevations, especially when they coincide with high spring tides (Heaps 1983; Murty 1984). In 1969 Hurricane Camille attained a surge peak of 7.4 m in the northern Gulf of Mexico one of the highest ever recorded. The piling up of water on the windward coasts of enclosed seas and lakes may be matched by a marked depression, or set-down, on leeward coasts. The most severe flooding occurs along low-lying coasts fronted by areas of shallow water and affected by the passage of storms. Storm surges develop in the North and Baltic Seas, along the Japanese Pacific coast and elsewhere in eastern Asia, where they are associated with typhoons, and along the Gulf and Atlantic coasts of the USA, where they are frequently generated by hurricanes. Low-lying islands in the Bay of Bengal are particularly vulnerable owing to the shallowness of the water body, the triangular shape of the basin, large tidal range, and favourable cyclonic tracks. A 7 m-high storm surge killed about 200,000 people at the northern end of the Bay in 1970. A surge in April 1991 was estimated to have been even larger, but although more than 125,000 people were killed in Bangladesh, the number of fatalities was less than in 1970 owing to the installation of an early-warning system and construction of storm shelters.

Elevation of the water level by storm surges and generation of surge currents can result from a number of factors, including direct wind shear, differences in atmospheric pressure, precipitation and runoff, and wave set-up (Carter 1988). Water levels can also be raised when winds force long trapped waves, with periods of from 60 to 600 hours, over the continental shelf. The effects of wind stress are usually most important for tropical storms, whereas barometric effects may be equally important for extra-tropical storms, which are generally larger and slower moving.

Storm surges appear to depend on resonant phenomena, although they can be generated by passing storms when the resonance is not exact. Two mechanisms have been proposed:

a) Lower pressure in the centre of a hurricane could raise the water surface about a metre above surrounding areas. Coupling resonance occurs if the hurricane moves at the same speed as the shallow-water wave generated by the differences in pressure. Only rarely, however, do storms move as fast as shallow-water waves in the deep ocean or even on the continental shelf.

b) Resonant coupling can also occur when the wind shifts in concert with the natural seiche period of the water. The wind direction can change by 180° as a hurricane or another type of cyclonic storm passes over a point. Winds therefore blow towards one end of a bay, and then, as the eye of the storm passes, turn round and blow, for a similar amount of time, towards the opposite bank. The seiche generated by the hurricane would therefore be enhanced if the period of the wind shift matched the natural period of the bay.

Water slowly oscillates back and fore in enclosed bodies of water after the wind stress has declined. For a seiche in a channel, open at one end to the ocean, the boundary conditions require a node at the mouth of the channel and an antinode at its head. As the length of the channel (l) is therefore a quarter of a wavelength, the velocity of the wave is:

$$C_s = L/T = 4l/T = (gh)^{0.5}$$

and the natural period of the channel is:

$$T = 4l/(gh)^{0.5}$$

or, if the channel has more than one node:

$$T = 4l/n(gh)^{0.5}$$

where n is an odd-number integer.

The water in an inlet or bay may oscillate almost continuously if its natural period is close to the period of the incoming swell, tidal oscillations in the ocean, or other natural forcing functions. The lack of a second reflecting boundary makes it difficult to sustain the oscillations, however, and the phenomenon is therefore less important for seiches than for tidal resonance and amplification in estuaries and over continental shelves.

Tsunami (Japanese for harbour wave) are long gravity waves that are usually generated by earthquakes, earthquake-induced landslides, or volcanic explosions (Murty 1977; Loomis 1978; Kajiura and Shuto 1990). Tsunamis usually result from shallow-focus earthquakes that involve vertical crustal movements, rather than horizontal movements that do not displace water vertically.

Tsunamis propagating in deep water typically have periods of ten minutes or more, and their wavelengths are long in comparison with the water depth. Using the relationship for shallow water waves ($C_s = (gh)^{0.5}$), a tsunami with a period of 10 minutes, travelling in water 4,000 m deep, has a velocity of 720 km hr^{-1} and a wavelength of 120 km. Their height in deep water is probably little more than 1–2 m, and they are therefore undetectable to the eye at sea. Tsunami gain in height and steepen as they reach the continental shelf, arriving at the coast as a series of bore-like waves. The first sign of an approaching tsunami is often a sudden drop in sea level, and the first wave is usually not the largest. Tsunami may be up to 10 m or more in height in coastal waters, and the uprush can extend several tens of metres above sea level. Their occurrence and height are difficult to predict, however, not only because they can sometimes be generated by fairly small earthquakes, but also because of their amplification in shallow water and the effects of refraction and reflection.

Tsunamis can be devastating to coastal populations, which may be situated great distances from the point of origin. The Pacific coasts are particularly vulnerable because of the seismically active subduction zones around its margins. In 1896, for example, an earthquake about 200 km off north-eastern Japan generated a tsunami consisting of three waves up to 15.2 m in height, which killed more than 17,000 people on the Japanese coast. Another 36,000 people were drowned by a tsunami, up to 30 m in height, generated in the Sunda Strait between Sumatra and Java by the 1883 explosion of the island of Krakatoa. A possibly even more powerful explosion took place in 1450 BC, on the volcanic island of Santorin in the Aegean Sea. The resulting tsunami, which inundated the coasts of the eastern Mediterranean, is thought to have been over 100 m high.

In many areas, the largest storms cannot account for sedimentary and erosional features at considerable elevations above present sea level, and there is therefore increasing recognition of the possible effects of tsunamis, near their source and in tectonically more stable regions further away (Atwater 1987; Minoura *et al.* 1994; Darienzo *et al.* 1994). Possible tsunami-deposited features along the south-eastern Australian coast include boulder masses tossed on to rock platforms and backshores or jammed into crevices; highly bimodal mixtures of sand and boulders; and dump deposits of coarse, well-sorted debris. There is also evidence of catastrophic wave erosion on rock ramps up to at least 15 m above present sea level (Bryant *et al.* 1992). A tsunami generated by a submarine slide on the continental slope off western Norway may have been responsible for the deposition of a thin but extensive layer of sand, several metres above contemporary high water level, in eastern Scotland, about 7,000 years ago (Dawson *et al.* 1988).

CHANGES IN RELATIVE SEA LEVEL

Coastal morphology has developed, and continues to change, in response to variations in sea level and vertical movements of the land. The most conspicuous vestiges of former relative sea levels are found along tectonically active collision coasts, where there are flights of raised erosional and depositional terraces, or fringing coral reefs. These surfaces were originally formed at about present sea level during Pleistocene interglacials, and were tectonically raised to their present elevation. Wide marine surfaces on more stable trailing edge coasts, however, were formed during periods of higher sea level during the Tertiary.

The contribution of relative sea levels that were similar to today's is generally less clear on stable than on tectonically active coasts. As the sea returned to its present position during the Holocene, it inherited and reoccupied coasts whose gross form had already been sculptured during periods of high interglacial sea level. In many places, for example, contemporary shore platforms were cut into interglacial platforms that were only a few metres higher. The stratigraphic record also shows that modern coastal elements, including barriers and coral reefs, have frequently been superimposed on older, but otherwise similar formations. Older depositional landforms have been cannibalized to provide sediment for the construction of modern coastal features. Because sea level was slightly higher during the last interglacial stage, however, modern depositional features have frequently formed a little seaward of, and at slightly lower elevations than, their ancient counterparts.

Holocene changes in sea level had an enormous influence on coastal development. It caused tidal range to change in some areas, particularly in semi-enclosed seas and where there are wide, shallow shelves (Gehrels *et al.* 1995). Rising sea level also provided abundant sediment for the formation of coastal structures, facilitated erosion of consolidated and unconsolidated coasts, caused barriers to migrate landwards over backbarrier deposits, drowned or forced the upward growth of coral reefs, drowned river mouths to form estuaries, and possibly triggered or accelerated the development of sand dunes. In high- and mid-latitude areas that were under thick ice, however, rapid isostatic recovery has caused relative sea level to fall during the Holocene, creating suites of elevated deltas, gravel ridges, beaches, wave-cut terraces, and other features of abandoned shorelines.

Until fairly recently it was assumed that changes in sea level are synchronous and of uniform magnitude on stable coasts around the world, and that abandoned shoreline features and sediments can therefore be correlated according to their elevation. Sea levels, however, may have changed around the world at different times, by different amounts, and even in different directions. The introduction to coastal studies of radiocarbon dating in the 1960s, and uranium series dating and amino acid analysis in the 1970s, has been particularly valuable in determining changes in relative sea level. Uranium series disequilibrium dating has helped to identify periods of high sea level during the last 150,000 years. Fossil corals have been dated using ^{230}Th, with a half-life of 75,000 years, and ^{234}U, with a half-life of 250,000 years, in the ^{238}U series, and ^{231}P, with a half-life of 32,000 years, in the ^{235}U series. Dates obtained from fossil shells are generally less reliable. Oxygen isotopic analysis of deep sea cores, speleothems, and glacial ice has provided independent corroboration of the palaeo-sea level evidence from dated shorelines.

The Cretaceous and Tertiary ✓

Even if all the ice on Earth melted, sea level would only rise by between 60 and 80 m, and in any case, this did not happen during the Pleistocene. Therefore marine planation surfaces up to 200 m or more above sea level on fairly stable trailing edge coasts cannot be reconciled with the palaeo-climatic and palaeo-sea level record of the Pleistocene. Many studies have suggested that sea level has been falling since the Cretaceous, although the general trend is obscured by the oscillations that are superimposed on it. It is also clear that sea level ranged from well above to well below today's level in the Cretaceous–Tertiary periods.

There were several major transgressions and regressions in the Cretaceous, but we need to learn much more about their occurrence and order of magnitude. Miall (1992) criticized the implied precision of the so-called Exxon or Vail curve, and he doubted whether more than a few of the depicted fluctuations in sea level were eustatic or global in nature (Fig. 2.19). According to the latest version of this scheme (Haq *et al.* 1987), the sea was between 200 and 250 m above today's level in the Upper Cretaceous—the curve may, however, be biased towards the high side (Kerr 1987). It may overestimate the amplitude of some fluctuations, and there is evidence to suggest that, about 30 million years ago, the height of the sea above its present level was only about half the amount shown. The Exxon curve suggests that the sea was not close to its present level for any significant period during the Tertiary.

Although some fluctuations in sea level may have been tectonically induced, ^{18}O isotopic data suggest that glacio-eustatic fluctuations could have occurred throughout much of the Cretaceous. The growth of ice sheets has affected sea levels since at least 10 million to 15 million years ago, when the east Antarctic ice sheet had attained or even surpassed its present size. It may have been growing for at least the previous 10 million years, however, and the west Antarctic ice sheet had reached its present size by 7 million years ago. The fall in sea level during the later part of the Tertiary, therefore, probably reflects the progressive growth of ice sheets, which was not reversed during the Pleistocene.

The Pleistocene ✓

The most detailed resolution of sea-level changes has been obtained through the dating of stairways

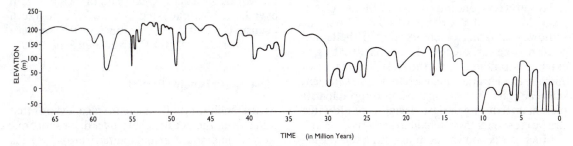

Fig. 2.19. Tertiary changes in sea level (Haq *et al.* 1987)

of elevated reefs in Barbados, New Guinea, the Ryukyu Islands, and tectonically similar areas elsewhere (Chappell 1974; Pirazzoli *et al.* 1993). As much of the middle and all of the early Pleistocene lies beyond the limits of uranium series disequilibrium dating methods, however, we are dependent upon less-reliable morphological and sedimentary evidence for the interpretation of ancient shorelines from periods preceding the last interglacial. Several shorelines have been dated at around 400,000 years ago, but there is no globally consistent evidence of high sea level stands before that time. There is also evidence of a higher sea level between about 350,000 and 300,000 years ago, which was similar to today in New Guinea. Uranium series dating and other evidence suggest that there was a high sea level between 220,000 and 180,000 years ago, although difficulties in separating the effects of tectonic and eustatic influences make it difficult to determine its height. Some workers suggested that sea level in the penultimate interglacial was between 10 to 30 m above present sea level, but most recent studies show that it did not exceed its present level.

Many studies have demonstrated that the sea was between 3 and 10 m above today's level between about 135,000 and 120,000 years ago, during the last interglacial stage (oxygen isotope substage 5e). Indeed, strandlines dating from this period are so widespread and conspicuous that they are used as an important reference level around the world. The interglacial maximum may actually have been composed of two similar peaks, separated by a minor regression (C. E. Sherman *et al.* 1993), although the evidence is lacking in some areas (Muhs and Szabo 1994). There were also two later sea level maxima at about 103,000 and 82,000 years ago, although there is some disagreement over their elevation, with estimates ranging from a few metres to more than 16 m below the interglacial maximum.

It was once thought that sea level was similar to today's in at least one interstadial of the last glacial stage. Much of the evidence was based upon the radiometric dating of contaminated shells and other marine carbonates. More-reliable dating of emerged strandlines, and the lack of evidence of climatic amelioration and ice retreat on a sufficient scale, suggest that the sea did not approach its present level during the last glacial stage, although there were several lower maxima during this period.

Much less is known about low sea levels during the Pleistocene. This is because the evidence is generally below water, and the alteration of sediments makes it difficult to obtain reliable dates. Most estimates are therefore based upon the presence of terraces, the occurrence of terrestrial or nearshore sediments on the continental shelf, and estimates based on the water equivalence of ice masses during glacial stages. Although levels of between about −80 and −130 m are generally accepted as typical of the amount of lowering in the last glacial stage, much greater amounts, up to −200 m, may have occurred in some parts of the world (Pratt and Dill 1974).

The Postglacial Period

Markedly different opinions have been expressed on the way that the sea reached its present level. It has been suggested that:

a) it rose at a diminishing rate, reaching today's level asymptotically;
b) it attained its present level between 5,000 and 3,000 years ago, and has been largely stable ever since; and
c) after reaching its present level between 5,000 and 3,000 years ago, the sea either oscillated above and below present sea level, or it rose above today's level and then gradually declined.

Even when sufficient data are obtained from single areas, it is difficult to determine the elevation and age of former shorelines with enough accuracy to determine the precise manner in which the sea reached its present level. Unless the elevation and age of shorelines can be determined very precisely, the errors only allow a broad sea level band to be identified, making it impossible to resolve whether changes were oscillatory or smooth. Nevertheless, there is a strong geographical component to variations in Holocene sea level curves from around the world. In the last 6,000 years the sea has risen over large non-glaciated areas of the northern hemisphere, and fallen over much of the southern hemisphere (Pirazzoli 1977; Newman *et al.* 1980).

The Oxygen Isotopic Record

At the beginning of a glacial stage, evaporated water from the oceans contains a disproportionately large amount of the lighter isotope ^{16}O. As this water is used for glacial growth, oceanic water becomes slightly enriched with the heavier residual isotope ^{18}O. Variations in the oxygen isotopic composition and the temperature of the water are recorded in the shells of foraminifera and other marine organisms contained in deep sea cores. These data provide an approximation of global continental ice volume and therefore changes in sea level (Figs. 2.20 and 2.21)

The isotopic record is broadly similar for most deep sea cores obtained from around the world, although there are some important differences (Ruddiman *et al.* 1986, 1989; Shackleton *et al.* 1990). A comparison of eleven of the most detailed records show that isotopic values were lower in interglacial stages 7, 13, 15, 17, and 19 than in the Holocene, either because of colder deep ocean water, or because more ice remained in the northern hemisphere—with the result that sea level would not have attained its present level. Sea levels were similar to today's during stages 5e, 9, and 11—which was probably the warmest in the last 1 million years. Stage 25, about 0.9 million years ago, is the only interglacial younger than stage 11 which was isotopically similar to the present (Shackleton 1987). It therefore appears unlikely that sea level could have been glacio-eustatically more than a few metres higher than today during the last 2.5 million years.

Despite some persistent discrepancies, there is general agreement between the major elements of sea level curves based upon radiometrically dated strandlines, and those inferred from the oxygen isotopic record (Chappell and Shackleton 1986). On Barbados, and possibly in Taranaki, New Zealand, each isotopic interglacial stage back to between 640,000 and 680,000 years ago is represented by at least one reef or emerged platform (Bender *et al.* 1979; Pillans 1983). Deep sea records are further substantiated by variations in the microfaunal composition of deep sea cores, the isotopic composition of Arctic and Antarctic ice, and radiometric dating and oxygen isotopic analysis of speleothems (Andrews 1985). Nevertheless, glacio-hydro-isostatic effects make it difficult to correlate shoreline elevations with ocean volume. Because of isostasy, the age of shorelines from the last interglacial vary according to their position relative to the former ice sheets, and the assumption that ocean volume was larger during the last interglacial than in the late Holocene has been disputed (Lambeck and Nakada 1992).

Mechanisms of Sea Level Change

The results of the International Geological Correlation Programme (IGCP Project 274) confirm that no sea level curve is of global applicability (Pirazzoli 1991). All curves record changes in relative sea level, and they are therefore relevant only to the areas from which the data were gathered (Fig. 2.22).

Glacio-eustasy involves the effect of the alternate growth and decay of ice sheets on ocean water volume, whereas geoidal-eustasy is concerned with

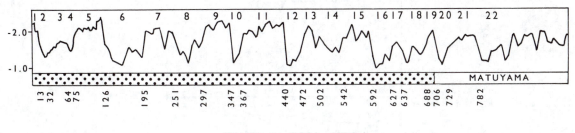

TIME (1,000's YRS.)

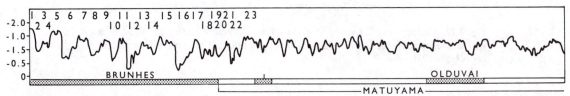

Fig. 2.20. Implied changes in sea level obtained from deep sea cores V28–238 (upper) and V28–239 (lower) from the Pacific Solomon Rise (Shackleton and Opdyke 1973, 1976)

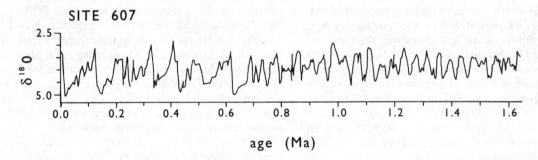

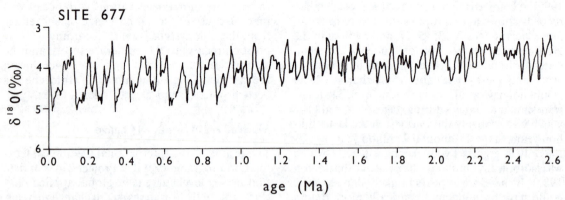

Fig. 2.21. Deep sea isotopic data from the Atlantic Ocean. DSDP Site 607 is at 41°N, 32°58′W and site ODP677 is at 1°12′N, 83°44′W (Ruddiman *et al.* 1989; Shackleton *et al.* 1990). The data suggest that there was a gradual change, between about 0.7 million and 0.6 million years ago, from dominance of the 41,000-year period of orbital obliquity, to dominance of the 100,000-year period of orbital eccentricity

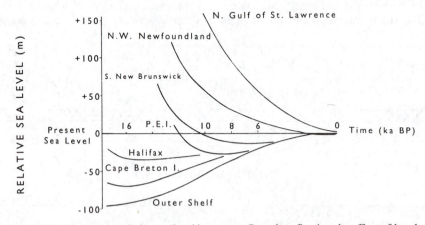

Fig. 2.22. Changes in relative sea level in eastern Canada reflecting the effect of local ice thickness and the timing of deglaciation on the rate and amount of isostatic recovery (Grant 1980)

the distribution of ocean water mass and elevation. Measurements from satellites orbiting the Earth have shown that the ocean surface consists of a series of humps and depressions. In extreme cases, changes in the configuration of the geoid or geodetic sea level could cause large changes in sea level. There would be a 200 m change in level, for example, if the New Guinea hump (+76 m) interchanged with a depression near the Maldive Islands (–104 m). Changes in the geoid induced by eustatic changes in sea level could therefore generate marked differences in the direction and magnitude of the changes in level actually experienced around the world.

Many mechanisms could induce changes in the geoid (Morner 1987). They could be associated with changes in the Earth's core–mantle coupling, for example, possibly involving movement of a topographically irregular interface. The gross form of the geoid could also be deformed by polar shift and changes in the tilt of the Earth's axis. A change in the rate of rotation would affect the centrifugal forces that cause the geoid to bulge out at the equator, resulting in circumferential variations in sea level around the Earth. Other changes could result from the shift in mass associated with the formation of ice sheets, glacially induced isostatic movements, and eustatic changes in sea level.

Sedimento- and tectono-eustasy involve changes in the level of the ocean surface as a result of variations in the volume of ocean basins. The effect of sedimentation on ocean basin volume is probably quite small, but tectonic activity can cause major changes in their shape and capacity (Worsley *et al.* 1984). Epeironesis, orogenesis, and ocean floor spreading are closely related within the framework of plate tectonics (Hallam 1981), and the growth or decay of mid-oceanic ridges could cause sea level to rise or fall by hundreds of metres.

Changes in the water load resulting from the growth and decay of ice sheets induce hydro-isostatic adjustments of the sea floor, particularly on wide continental shelves (Johnston 1995). It has been suggested that crustal subsidence may cause sea level to rise by only about two-thirds of the amount expected on the basis of the volume of meltwater returned to the oceans. This is only true, however, if the ocean floor responds simultaneously to the rise in water level. With a lag in the response, the altitudinal range of the strandlines on a stable continental coast could be as much as one-third greater than the eustatic change in sea level. Hydro-isostasy may therefore explain differences in relative sea level within shelf areas. Systematic variations in sea level perpendicular to the coast in the Australian Great Barrier Reef Province, for example, have been attributed to the effects of hydro-isostasy.

Models of Sea Level Change

There have been several attempts to model changes in sea level that are consistent with the decay of ice sheets. These models make assumptions regarding the Earth's rheology and the size and distribution of the ice sheets during and following the maximum of the last glacial stage. The model employed by Clark and co-workers considers deformation of the ocean floor by glacial and water loading; geoidal perturbations resulting from gravitational attraction of the water surface to large ice sheets; and redistribution of matter within the Earth. It is assumed that mantle viscosity is constant, sea level rose by 75 m between 18,000 and 5,000 years ago, and there has been no eustatic change in sea level since that time. The model distinguished six sea level zones (Clark 1980) (Figs. 2.23 and 2.24):

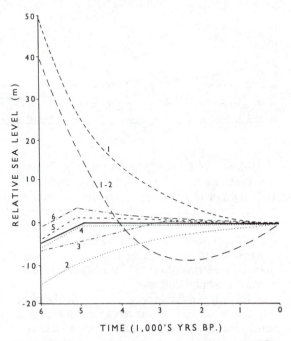

Fig. 2.23. Representative changes in relative sea level for the last 6,000 years. The solid line is the assumed eustatic change in sea level (Clark and Lingle 1979)

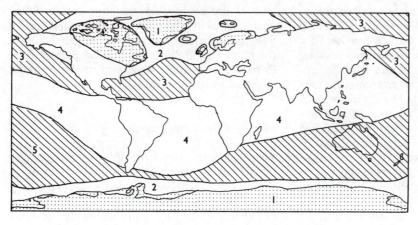

Fig. 2.24. Distribution of sea level zones. The numbers refer to the changes in relative sea level plotted in Fig. 2.21. The upper diagram is based on the assumption that the disintegration of northern ice sheets was competed by 5,000 years ago, and that eustatic sea level subsequently rose by 0.7 m because of Antarctic melting. The lower diagram assumes that ice disintegration was completed by 5,000 years ago in the northern and southern hemispheres. Stippled areas in both diagrams experienced continuous emergence, whereas Holocene sea levels have been higher than today in the shaded areas (Clark and Lingle 1979)

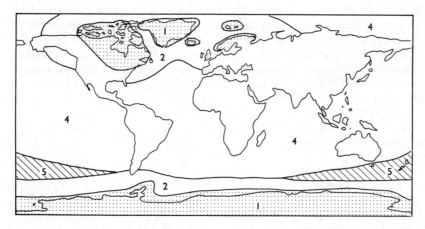

a) Zone 1 consists of regions that were beneath ice sheets. Immediate emergence occurred in this zone because of elastic uplift and reduction in the gravitational attraction of the ice on the ocean surface. Emergence has continued because of the viscous flow of material beneath the mantle.

b) Submergence in zone II was the result of mantle material flowing into uplifted areas, causing forebulge collapse.

c) Initially rapid submergence in zone III gradually slowed down and was replaced, several thousand years ago, by slight emergence of less than one metre.

d) There was continuous submergence of between 1 and 2 m in the last 5,000 years in zone IV,

despite the assumption that ocean volume was constant during that period.

e) Initial submergence in zone V gave way to slight emergence of between 1 and 2 m when meltwater was no longer being added to the oceans.

f) All the continental margins, other than those adjacent to zone II, are within zone VI. Increased water load on the ocean floor forced mantle material to flow into areas beneath the continents, causing them to emerge during the mid-Holocene, although the model assumed that ocean volume was not increasing at that time.

Morner (1981) believed that glacial loading models cannot explain global variations in relative sea level. He argued that glacial loading in Fenno-scandia was mainly compensated locally, and that differences in sea level curves around the world are the result of geoidal eustatic changes. Nevertheless, differential loading and geoidal models show that changes in sea level caused by the return of melt-water to the oceans, cannot be uniform around the world. Therefore, although the responsible mechanisms remain to be determined, the evidence for Holocene sea levels higher than today's in some parts of the world can now be reconciled with concurrent levels that were below today's in other areas.

3 Beach Sediment: Sources, Characteristics, and Transport

Although beaches can consist of any available material of suitable size, ranging from microscopic particles up to boulders, this chapter is primarily concerned with sandy sediments. The transport of silts and clays is discussed in Chapters 7 and 8, and coarse clastic material in Chapter 4.

SOURCES, SINKS, AND SEDIMENT BUDGETS

Beach material is derived from a wide variety of sources. Most sediment is local, although it is sometimes transported, generally by longshore transport, from distant locations. In addition to primary sources, sediment is also obtained through erosion or cannibalization of material stored in depositional features that were formed in the past, when sea level was similar to today's. Sediment re-cycling, in response to changes in sea level, climate, and other environmental factors, can make it very difficult to identify the original source.

Cliffs in resistant bedrock are eroded too slowly to provide significant amounts of beach material, and fine-grained sediment released by the rapid erosion of cohesive clay cliffs is carried offshore as suspended load. Cliff erosion is locally important, however, where easily eroded sand and gravel is delivered to beaches by coastal landslides or falls. Protection of rapidly retreating cliffs will there-fore have adverse consequences for beaches if it prevents replenishment of sediment transported alongshore. Erosion of stratified and unstratified glacial sediments supplies large amounts of sand and gravel to the nearshore zone in Pleistocene glaciated regions. Active glaciers and meltwater streams are still the primary source of beach mater-ial in some areas, and ice-rafting and ice-push are important transport mechanisms in high- and some cool mid-latitude regions (Chapter 12). Biogenic sources, including shell, coral, and calcareous algae, are especially significant in low latitudes where there is high biological productivity. There are also carbonate-rich, shell-dominated beaches in higher latitudes, where there is a large supply of shells, little terrigenous sediment or longshore transport to dilute the shell content, and aeolian deflation, waves, or some other sorting mechanism to concentrate it (Gell 1978; Leonard and Cameron 1981). High carbonate content in western Britain, and in other places along the exposed western coasts of the mid-latitudes, may also reflect the ability of large, energetic waves to bring shells and other carbonate material onshore (Davies 1972). Whether sediment delivered to coasts by rivers is suitable for beach nourishment depends upon such factors as the climate, vegetation, rock type, and elevation of drainage basins. It has been estimated that the rivers of the world have an annual sus-pended sediment load of about 13.5×10^9 tons (Milliman and Meade 1983). The monsoonal and tropical rivers of Asia are responsible for more than 75 per cent of this, and two-thirds of that is carried by the Huanghe (Yellow, Hwang-Ho) and Ganges/Brahmaputra. Hot, wet climates are con-ducive to chemical weathering and the production of fine-grained sediments, however, which are generally unsuitable for beach development. In any case, much of it is trapped in deltas, estuaries, or behind dams, or it is advected seawards. Small, steep rivers draining sandy, high relief basins therefore tend to deliver much higher proportions of beach-sized material to the coast than much larger rivers draining areas of low relief (Griggs 1987). Volcanic lava and ash are deposited directly on coasts, or they are carried there by rivers. Black sand beaches on volcanic islands often consist entirely of sediment derived from basaltic or andesitic lavas. Volcanic glass may also occur near the site of recent eruptions, and green beach sands in areas with a high olivine content.

Several other sources are of local significance. Offshore winds transport sandy terrestrial materials into the coastal zone, particularly where there are coastal deserts. Beaches on the lagoonal side of barriers are also supplied with large amounts of windblown sand. Although they are rare, beaches on the outer margins of the Bahamian Platform, and in a few other areas, consist of oolites, quasi-spherical, sand-sized grains formed by non-biogenic precipitation of calcium carbonate in warm, shallow water. The amount of sediment transported from the continental shelf to the littoral zone is difficult to determine, although much of the sediment presently being delivered to the coast from offshore appears to be material eroded from beaches during storms (Chapter 4). Some human activities also supply material to beaches, including the dumping of waste from coastal quarries, coal-mines, and steel-works. Sand is also pumped or dumped on to beaches artificially to nourish or replenish them. Suitably sized sand can be dredged from coastal inlets or lagoons, or quarried from dunes.

Much of the sediment on beaches was driven onshore when sea level was rising during the early and mid-Holocene. Progressive inundation of coastal plains and the acquisition of new sediment sources created a sediment economy that was strongly in surplus. By the later Holocene, when sea level was no longer rapidly rising, sediment had become trapped or locked up in the littoral zone, in prograding beaches, barriers, estuaries, lagoons, dunes, and other coastal structures. There is now a slight disequilibrium in the supply and depletion of sediment—rivers, coastal erosion, and other sources apparently being unable fully to compensate for normal losses of sand. The sediment economy of abundance has therefore been replaced by one of scarcity, and this has resulted, in many cases during only the last few centuries, in the world-wide erosion of coastal deposits. It has been estimated that more than 70 percent of the world's sandy coasts have experienced net erosion over the last few decades, while only 10 percent have prograded (Bird 1985*a*). Although the role of other mechanisms must also be considered (Fenster and Dolan 1994), coastal erosion has probably been too great simply to attribute to the effects of a recent rise in sea level, and too widespread for human influences or an increase in storminess to be responsible.

Beaches represent temporary storage sites for sediment, before it is deposited in sinks, which retain or trap sediment for periods ranging from a few years to thousands of years. The most common are re-entrant sinks, formed where a headland or other obstruction juts out seawards from a coast. This interrupts longshore transport of sediment and usually produces bayhead beaches, although large barrier systems and dune complexes can also develop. Salient traps occur where a sudden change in the direction of the coast produces barrier or comet tail spits. Equilibrium sinks, which occur where there is an approximate balance between opposing forces, may be occupied by forelands, angled spits, or estuarine deposits (Davies 1972).

Sediment carried into deep water is permanently lost to the coastal system while it remains at its present level. This can occur off the ends of salient sinks and through submarine canyons. Rip currents move coarse sediment offshore, beyond the zone of wave transformation (Gruszczynski *et al.* 1993), and it can then be channelled into submarine canyons and transported into deeper water. Nevertheless, although this has been well documented in southern California (Everts *et al.* 1987), and is also known to occur off modern deltas (Chapter 9), the general contribution of submarine canyons to coastal budgets remains to be determined.

Corrosion and abrasion gradually reduce beach sediment volume, although they are insignificant on quartz grains. Shells and other calcium carbonate sediments are more susceptible, although it is still unclear whether beaches lose significant quantities of material in this way. Human removal of beach material has been going on since ancient times, and it continues in some areas today, although legislation has been enacted to discourage it in many areas. The mining or quarrying of beach material can have very deleterious effects on local communities where natural replenishment is very slow. A classical and frequently cited example is provided by the former fishing village of Hallsands in southern Devon, England (Tanner and Walsh 1984). Most of the coarse clasts on the beach consists of flints that originated from chalk outcrops on the sea floor over 40 km away. Deep, tidally scoured channels now separate the source from this coast, and most of the original beach material must therefore have come on to the coast when sea level was rising. As the original source is now inaccessible, any material lost to the coast cannot be replenished by the sea. Between 1897 and 1902 the intitidal clastic material near Hallsands was dredged to provide aggregate for an extension of the Devonport naval dockyard. The ledge on which Hallsands was constructed was, for the first time, left unprotected by a clastic bank during the winter of 1900. Buildings were almost immediately demolished, and despite subsequent construction of sea walls, all of the remaining houses were severely damaged or destroyed by the great storm of January 1917.

Sediment budgets are used to evaluate the contribution of active sources and sinks in increasing or decreasing the amount of sand in a selected control volume (Fig. 3.1). Budgets help to determine the reasons for erosion or accretion in a coastal unit, which is usually measured by comparing a series of beach profiles. The boundaries of a coastal unit depend upon the time scale of interest, and the purpose and area of the study. The unit could be a section of beach, a dunefield, or, best of all, a coherent sand budget compartment or cell, with boundaries created by submarine canyons, headlands, or other features that significantly change the littoral system. Often, however, there are no obvious natural boundaries, and those selected are therefore somewhat arbitrary.

Budgets are calculated by determining all the gains and losses of sediment occurring in a unit over a specified period of time (Coastal Engineering Research Center 1984). Point sources and sinks add and remove sediment from a limited part of a boundary of a coastal unit, whereas line sources and sinks operate across an extended portion of a boundary. Rivers and eroding cliffs, for example, are often important point and line sources respectively. Convecting processes, including longshore transport and tidal flows through inlets, function as both sources and sinks. They add and subtract material at the same rate, and therefore have no net effect on beach volume. Longshore transport is generally evaluated indirectly because of the difficulty of making direct measurements, and the need for long-term net transport data. Suitable estimates can be based on: progressive dilution of heavy minerals with distance of travel from an identifiable source, as other sands are added to the beach; the rate of accumulation or bypassing of sediment at a breakwater or other obstruction; and equations that relate rates of longshore transport to wave conditions.

One of the greatest sources of error in sediment budget analysis is the assessment of sediment gained or lost to the offshore. There is increasing evidence of sediment exchanges between beaches and surf zones and the innermost portion of the continental shelf, which forms a connecting pathway to the mid- and outer shelf (Wright 1987). Considerable amounts of sediment may be involved, although the quantities are very difficult to determine. In a littoral cell in southern California, for example, about 35 percent of the volume of sand in the foreshore was recently transported from the inner continental shelf. In another cell, 42 percent of the sand on the inner shelf migrated onto the foreshore in the year following a severe storm (Osborne *et al.* 1991).

SEDIMENT CHARACTERISTICS

A variety of techniques is used to determine the grain size of coastal sediments, including callipers and sieves for coarse beach material, and Coulter

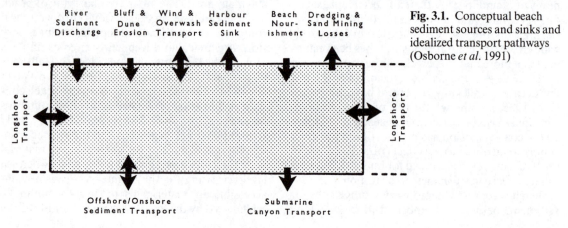

River Sediment Discharge   Bluff & Dune Erosion   Wind & Overwash Transport   Harbour Sediment Sink   Beach Nourishment   Dredging & Sand Mining Losses

Longshore Transport   Longshore Transport

Offshore/Onshore Sediment Transport   Submarine Canyon Transport

Fig. 3.1. Conceptual beach sediment sources and sinks and idealized transport pathways (Osborne *et al.* 1991)

Counters, pipettes, hydrometers, optical settling instruments, and electron microscopes for finer sediments. Grain size can be expressed in a number of ways. The Wentworth scale is based on classes that are separated by factors of 2, so that each is twice the size of the one below. Integers for each of the Wentworth class limits are obtained through the use of a $\log_2$ transform. The phi (ϕ) grain size is therefore:

$$D_\phi = -\log_2(D_{mm})$$

where D_ϕ is the grain diameter in phi units and D_{mm} is the corresponding diameter in millimetres (Table 3.1). Sand, ranging from very coarse to very fine, has ϕ values between -1 and $+4$, respectively.

Table 3.1. Sediment grain size classification

Type	ϕ units	Wentworth (mm)
Boulder	> –8	>256
Cobble	–8 to –6	256 to 64
Pebble	–6 to –2	64 to 4
Granule	–2 to –1	4 to 2
Sand		
Very Coarse	–1 to 0	2 to 1
Coarse	0 to 1	1 to 0.5
Medium	1 to 2	0.5 to 0.25
Fine	2 to 3	0.25 to 0.125
Very Fine	3 to 4	0.125 to 0.0625
Silt		
Coarse	4 to 5	0.0625 to 0.0312
Medium	5 to 6	0.0312 to 0.0156
Fine	6 to 7	0.0156 to 0.0078
Very Fine	7 to 8	0.0078 to 0.0039
Clay		
Coarse	8 to 9	0.0039 to 0.00195
Medium	9 to 10	0.00195 to 0.00098

Source: King 1972.

Grain diameter is an ambiguous term which can refer to (Sleath 1984):

a) mesh size of the sieve through which the grains are just able to pass;

b) the diameter of a sphere of the same volume;

c) the length of the long, short, or intermediate axes of the grain, or some combination of these lengths; or

d) the diameter of a smooth sphere of the same density and settling velocity as the grains.

Grain size distributions can be represented as histograms or frequency curves by plotting weight-percentages against the phi values. The method most frequently used, however, is to plot the grain size data on a probability, cumulative percentage ordinate, and the phi scale on an arithmetic abscissa. Graphical approximations of simple descriptive statistical measures, including the mean size, standard deviation, skewness, and kurtosis, can then be determined from the percentiles on the cumulative size distribution, although the calculations can also be made by computer (King 1972; Dyer 1986). The median or geometric mean grain size, or the size of the grain coarser than some percentage of the sample can be used to represent sediment samples for comparative purposes.

Many workers have attempted to use sediment size distributions to identify recent and ancient depositional environments, but they have generally been found to be unreliable. Others have therefore analysed and interpreted them according to the processes and physics of sediment movement. Although normal distributions plotted on a Gaussian probability axis produce straight lines, beach sediment distributions often consist of three straight-line segments. This has been attributed to the presence of subpopulations representing different modes of transport and deposition, corresponding to the coarse bed load, the fine suspended load, and material of intermediate size that moves in intermittent suspension (Middleton 1976). An alternative explanation is concerned with the packing controls on a grain matrix, the larger grains being viewed as a lag deposit, with the finest grains resting in the spaces between grains of median size (Moss 1972). A further possibility is that the sample consists of grain populations derived from a number of laminae, representing several depositional episodes.

Segmentation of grain size distributions on lognormal cumulative probability paper may reflect the use of an inappropriate probability model. Distributions may conform much better to a hyperbolic probability function than to the lognormal model, which poorly represents the extremes of natural grain size distributions. Some workers have found the four parameters needed to specify a logarithmic hyperbolic distribution are more sensitive to sedimentary environments and dynamics than the statistical moments of the normal probability function. Others have concluded that there is little difference in their ability to distinguish depositional environments (Lund-Hansen and Oehmig 1992; Hartmann and

Bowman 1993; Sutherland and Lee 1994). Grain sizes may also be fitted to a skew log-Laplace model, which is a limiting form of the log-hyperbolic distribution. This distribution is essentially described by two straight lines, and is defined by three parameters (Fieller *et al.* 1984).

Many indices have been developed to describe the overall form or some specific aspect of the shape of coastal sediments. Grain roundness, which is a measure of the smoothness of its surface, can be defined as the ratio of the radius of curvature at its corners, to the radius of curvature of the largest inscribed circle. Grains are also frequently described by their sphericity, a measure of the degree to which their shape approaches a sphere with three equal orthogonal axes. The shape of grains, which can range from spherical, to plate, to rod-like forms, can be defined and measured in many ways, using a variety of indices (Whalley 1981; Willetts and Rice 1983; Illenberger 1991). They include the E shape factor (ESF):

$$\text{ESF} = D_s \left[\frac{D_s^2 + D_1^2 + D_1^2}{3} \right]^{-0.5}$$

and the Corey shape factor (CSF):

$$\text{CSF} = \frac{D_s}{(D_i \, D_1)^{0.5}}$$

where D_1, D_s, and D_i are the long, short, and intermediate axes of the grain, respectively. The relationship between the three axes can also be used to determine grain shape, using a ternary diagram.

Although the sphericity and other aspects of the shape of coarse clasts can be determined fairly easily, direct measurement is generally too time-consuming for sand and other small grains. The roundness of sand grains has usually been estimated visually, by comparing them with a set of standard grain images of known roundness, although Fourier analysis is increasingly being used (Powers 1953; Lee and Osborne 1995; Thomas *et*

al. 1995). Winkelmolen (1971) determined the 'rollability' of sand grains, the time taken to roll down the inside of a slightly inclined, revolving cylinder. Rollability is easier to measure than other shape parameters, and it is more closely related to the physical behaviour of the grains. Shape distribution factors, obtained by plotting relative grain rollability against grain size, may therefore be characteristic and indicative of the mode of origin of coastal sediments.

The density of beach sediments partly depends on the density of the grains themselves, which reflects their mineralogy (Table 3.2). Most beach material in temperate regions was derived from continental granitic rocks, and it largely consists of quartz, and to a much lesser extent feldspar, grains. The sediments of enclosed pocket beaches, however, tend to be derived from restricted source areas, and they can strongly reflect the mineralogical character of the local geological outcrops, or the accumulation of shelly carbonate material. Most continental beach sediments contain small amounts of heavy minerals, such as magnetite, horneblende, and garnet, which can help to identify the source rocks, their relative importance, and the direction of longshore transport. Beaches can consist almost entirely of heavy minerals in volcanic areas, and carbonates may dominate in the tropics, especially where there are coral reefs.

The density of beach sediments also partly depends upon bulk density, which reflects the way that grains are arranged or packed together. There are basically four ways of systematically packing spherical grains (Fig. 3.2). With unstable cubic packing, the position of the grain centres describes the corners of a cube. Packing density—the ratio of the volume of the solids to the total volume including voids, is only 0.52, and the porosity is therefore 48 percent. A tetragonal arrangement, with a packing density of 0.70, is formed by moving the upper layer of grains until they occupy

Table 3.2. The mean density of some minerals found in beach sands (kg m^{-3})

Mineral	Density	Mineral	Density
Aragonite	2930	Microcline	2560
Augite	3400	Muscovite	2850
Calcite	2710	Orthoclase	2550
Foraminifera shells	1500	Plagioclase	2690
Garnet	3950	Quartz	2650
Hornblende	3200	Rutile	4400
Magnetite	5200	Zircon	4600

the hollows between the grains below. In ortho-rhombic packing, which has a packing density of 0.60, the centres of the lower layer of grains form a diamond pattern, and the grain centres of the upper layer are directly above. A rhombohedral arrangement is created by moving the upper layer of grains into the hollows created by the lower layer. This arrangement has the highest packing density of 0.74, and the lowest porosity.

A number of factors influence packing geo-metry, and sediment stability, porosity, permeabil-ity, and other bulk properties. The shape of the grains is particularly important. Bits of shell or other flat, flaky, or plate-like particles can create small cavities in a deposit and greatly increase its porosity. Variations in grain size also have a marked effect on packing density and porosity. When grains are non-uniform or poorly sorted, smaller grains occupy the spaces between the larger grains, increasing packing density and decreasing porosity. Grains less than about 1/7th the size of the larger grains can also pass down through voids between the larger grains. Coarse, poorly sorted sand, almost uniform fine-grained, well-sorted sand, and coral sand consisting of a mixture of coral and shell fragments, have porosities of about 40, 45, and from 50 to 65 percent, respectively (van Rijn 1989). Packing is also determined by depos-ition rates. Rhombohedral packing can develop when slow deposition allows grains to settle into their optimum positions, but when depositional rates are high, grain collisions and other inter-actions result in cubic arrangements, voids, and increased porosity. Jostling or vibration of the underlying layers by settling grains with high fall velocities tend to increase packing density and reduce the porosity. Grains deposited by waves and currents are also more densely packed than those that settle out from suspension in still water.

Grain transport or depositional processes can orient non-spherical grains in a preferred direc-tion. Gentle bed load transport, with grains sliding or rolling along the bed, can produce imbrication—the grains lying with their longest axes normal to the flow, and their short–intermediate planes dip-ping at a small angle upstream. When grains saltate in stronger flows, however, their long axes tend to become oriented parallel to the flow.

The steepest surface gradient of loose sediment is referred to as its yield angle, the angle of repose or rest, or the angle of internal friction (ϕ). The mass angle of repose ranges from about 32° to 35°, generally increasing with decreasing grain size and sphericity. Individual grains, however, can have much higher angles of rest.

Sorting usually occurs as a result of a number of variables working together, or independently, to separate grains according to their shape, size, and

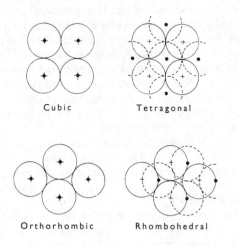

Cubic   Tetragonal

Orthorhombic   Rhombohedral

+ Centres of upper layer of spheres, full circles
• Centres of lower layer, dashed circles
✛ Centres on top of each other

Fig. 3.2. The packing of spherical grains of equal size (Dyer 1986)

Table 3.3. Factors controlling sediment sorting

1. Rate of sediment accumulation:
   (a) slow—allows reworking of grains
   (b) rapid—allows little or no reworking of grains
   (c) none—scour
2. Nature of the sediment surface:
   (a) size distribution of grains
   (b) packing and arrangement of grains
   (c) type of bedforms present
3. Style of grain motion:
   (a) traction, including sliding and rolling
   (b) saltation
   (c) suspension
4. Fluid characteristics:
   (a) velocity or shear velocity
   (b) turbulence
   (c) depth
5. Grain characteristics:
   (a) size
   (b) shape
   (c) density

Source: Steidtmann 1982.

density (Table 3.3). Beach sediments tend to be better sorted than river sediments, but less well than dunes. They are often negatively skewed, with a tail of coarse sediment, although a few cases of positive skew have also been noted. This can be attributed to the removal of fines, or addition of coarse clasts or shells. Skewness may also arise from a single sedimentary event, however, and is not necessarily symptomatic of the mixing of two or more sedimentary populations (McLaren 1981).

Spatial changes in beach sediment characteristics can be attributed to several possible mechanisms, including mechanical and chemical breakdown, more rapid transport of grains of one size than another, longshore variations in wave energy, the addition or loss of sediment, and mixing of two or more distinct sediment populations. Sorting occurs through selection, breaking, and mixing (Carter 1988). Rejection and acceptance phenomena play an important role in the selection process, and in perpetuating sorted grain distributions on beaches. Shielding impedes the movement of fine grains over coarser material, whereas 'rejection' of foreign material accelerates movement of coarse grains over finer sediment. There is, however, a high probability of the assimilation or acceptance of grains moving over material with similar textural characteristics. If the source sediment is eroded, the erosional lag must be coarser, better sorted, and more positively skewed than the original sediment. If the transported sediment is then completely deposited, the deposit must be finer, better sorted, and more negatively skewed than the source. If the transported sediment is only selectively deposited, the sorting will be better and the skew more positive than the source. The deposit will be finer than the source if erosion only removes material finer than the mean size of the source, but it may be coarser if sediment larger than the mean size is removed from the original deposit (McLaren 1981).

The mean size of beach material is determined by the sedimentary processes and the characteristics of the source. There are, therefore, often differences between the exposed and sheltered sides of islands, longshore variations in grain size along a beach, and variations in grain size with beach stage, as a storm-eroded beach gradually recovers to its fully accreted state (Richmond and Sallenger 1984; Medina *et al.* 1994) (Chapter 4). The coarsest sediments are generally found at the plunge point of breaking waves. Grains tend to become finer seawards and landwards, although coarser sediments are often found on the higher parts of the beach, particularly on top of the berm. This coarse sediment could be stranded over the berm crest by large swash events, or it could be a deflation lag deposit, resulting from finer material being blown inland to form dunes. There is less agreement over changes in the degree of grain-size sorting normal to the beach. Some workers have found that the poorest sorting occurs in the breaker and surf zones and the best in the swash zone, whereas others have found that the degree of sorting declines on either side of the breaker zone.

It is not known whether larger or smaller grains are more easily transported alongshore, and therefore whether examples of beach sediments becoming coarser downdrift represent anomalous or normal situations. In any case, although longshore grading often develops on beaches in closed or almost closed embayments, it is generally lacking or poorly developed where large amounts of sediment are moving alongshore, and where active sediment throughput does not allow enough time for it to develop (Carter 1988).

Sediment sorting is also sensitive to variations in grain density, and particularly to the abundance and mineralogy of the heavy mineral component. Small, dense heavy mineral grains occupy spaces between larger and less dense quartz and feldspar grains, shielding them from the flow so that they are less easily entrained. Heavy mineral grains are carried alongshore less rapidly than lighter quartz grains, even when they have the same settling velocity—presumably because their smaller size inhibits entrainment during each brief suspension episode (Slingerland 1977; Trask and Hand 1985).

Steidtmann (1982) found that grains that are either smaller or much larger than the bed roughness grains move continuously, and the transport rates for heavy and light mineral grains are essentially the same. Grains that are equal or slightly larger than the roughness elements, however, move intermittently, and the heavy particles move more slowly than light mineral grains of the same size. The size–density sorting of sand grains depends upon whether transport and deposition take place over plane or rippled beds. Over a plane bed, grains move fairly continuously, though with brief halts, and the size and proportion of heavy minerals decrease with transport distance. With a rippled bed, however, grain motion is discontinuous and deposition occurs through grains avalanching down the lee slope of ripples. There is a great deal of variation in the size–density relation of these grains, and the size relationship between light and heavy minerals remains essentially constant with distance of travel.

Heavy minerals tend to be locally concentrated on beaches, often forming bands or streaks near the high tide or upper swash zones, in the troughs of ripples, or where there are shells, coarse clasts, or other flow obstructions (Li and Komar 1992*a*). Viewed in cross-section, the upper swash zone often consists of layers of fine sediment grading upwards into layers of coarser sediment. The alternating layers are between about 1 and 25 mm in thickness, and they typically extend along the beach for a few tens of metres (Clifton 1969). They are most apparent when the fine layers are rich in dark, heavy minerals, while the coarse sands consist of light-coloured quartz-feldspar grains. Swash laminae are thought to develop as a result of the shearing of dense concentrations of coarse and fine grains by backrush. Shear sorting causes the coarser, or lighter, grains to migrate upwards into the zone of lower shear, while finer, or heavier, grains move downwards, into the zone of maximum shear at the bed. Alternateively, smaller particles may tend to fall into the spaces between the larger grains, thereby displacing coarser grains towards the surface. Selective longshore transport of lower density minerals may concentrate heavy minerals in erosive lag deposits (Frihy *et al.* 1995). Heavy mineral concentrations in the cross-shore direction have been attributed to wave asymmetry—heavy minerals being carried onshore by high current velocities, but not by the weaker offshore flows, or beach erosion and offshore transport of quartz-feldspar grains (Komar and Wang 1984). There may be poor separation under vigorous wave conditions, however, when heavy and light minerals can be entrained and transported together.

There have been few investigations of the effect of sand grain shape on longshore and cross-shore sorting patterns. The proportion of angular grains increases in the direction of longshore transport between Delaware and Chesapeake Bays. This may reflect their lower settling velocities, which allow them to remain in suspension longer, so that they are carried further and at higher rates than more rounded grains. On the other hand, there is increasing grain rounding with longshore transport on Long Island (Williams and Morgan 1988). It has been found that similarly sized quartz grains were differentially transported and sorted within the swash zone in the laboratory and field—the more spherical and rounded grains being deposited near the top of the uprush (Trenhaile *et al.*, in press).

SETTLING VELOCITY

The settling (fall) velocity of grains in a motionless fluid is an important element in the theory of sediment transport. Despite the fact that sediment seldom settles through still water in coastal regions, it is generally assumed that grains with high settling velocities are less easily transported than those with low velocities.

The nature of the flow around a falling spherical grain depends upon its settling rate. Viscosity is important if the grain falls slowly, but inertia is dominant if it settles rapidly. The settling Reynolds number (Re_s) is a measure of these effects:

$$Re_s = \frac{w_s D}{v}$$

where w_s is the settling velocity of a spherical grain, D is grain diameter, and v, the kinematic viscosity of the fluid, is given by:

$$v = \frac{\mu}{\rho}$$

where μ is the molecular viscosity and ρ is the density of the fluid. When Re_s is less than about one, laminar flow occurs and grains distort the flow as they settle through the fluid, leaving no wake. When Re_s is greater than one, separation occurs in the lee of the grains and a turbulent boundary layer develops around the particles. According to Stokes's Law, the settling velocity at low Reynolds numbers (< 1) is:

$$w_s = \frac{D^2}{18} g \left[\frac{\rho_s - \rho}{\mu} \right] = cD^2$$

where g is the acceleration due to gravity, ρ_s is the density of the grain, and c is a constant. For sea water, with salinity of 35‰, c for quartz grains is about 8,526 at 20°C and 6,536 at 10°C. The settling velocity cannot be calculated at higher Reynolds numbers and has to be determined experimentally. Empirical equations have been derived for the settling rate of heavy minerals and grains of approximately the density of quartz (Komar 1981). Roux (1992) has devised a new way of determining the relationship between sphere size and settling velocity based on Newton's Impact Law. This technique is applicable over a wide range of Reynolds numbers, and can be used for grains ranging up to several centimetres in diameter.

The settling velocity for natural sand grains, which spin and oscillate as they fall, is less than for perfect spheres. Small differences in the shape of angular grains may have far greater hydraulic significance than small differences between nearly spherical grains. The log of the settling velocity, for differently shaped grains, has been plotted against the log of the diameter of a sphere with the same volume and weight as the non-spherical grain (Komar and Reimers 1978) (Fig. 3.3). This study was extended to consider the settling velocity of cylindrical grains, which are characteristic, for example, of several types of heavy mineral (Komar 1980a).

A number of relationships have been found between the settling velocity of natural grains (w_n) and the settling velocity of spheres with diameters equal to the intermediate diameter of the natural grains (w_s) (Baba and Komar 1981a). When the intermediate diameter is determined with a microscope:

$$w_n = 0.761\, w_s.$$

Although less suitable, sieve diameters can also be used to determine settling velocities from:

$$w_n = 1.473\, w_s^{0.842}.$$

These simple equations, however, represent average approximations based on mean grain shape. A more precise determination of settling rates, using the Corey (CSF) or E shape (ESF) factors, requires measurement of the largest, intermediate, and short diameters:

$$\frac{w_n}{w_s} = 0.808\,(\mathrm{CSF}) + 0.192$$

$$\frac{w_n}{w_s} = 0.992\,(\mathrm{ESF}) + 0.008.$$

These equations can also be used to determine the settling rates of heavy minerals and other non-quartz grains. The settling velocity of coral sand may be considerably lower than quartz sand, largely because of its more angular shape, although its density may also be a little lower than quartz (van Rijn 1989).

Equations have been derived for the settling velocity of irregularly shaped grains (Baba and Komar 1981b). The asymmetry of irregular particles is important, as imbalances cause grains to oscillate, spin, or tumble through water. At low Reynolds numbers, irregular grains settle faster than symmetrical grains of the same weight and general shape, probably because they have smaller projection areas perpendicular to the fall direction. The effect of grain roundness on settling velocities is much less, however, than the non-spherical shape of the grains. Using published data on terminal settling velocity, Hallermeier (1981) developed three empirical equations which provide estimates of the settling velocity of commonly occurring sands.

The settling velocity of a grain is determined not only by its shape, size, and density, but also by the presence of other grains. The settling velocity of grains contained within a small cloud of sediment, in otherwise clear water, is greater than that of a similar, isolated grain in clear water. The upflow of displaced water induced by the fall of one grain, however, retards the settling of other grains, and

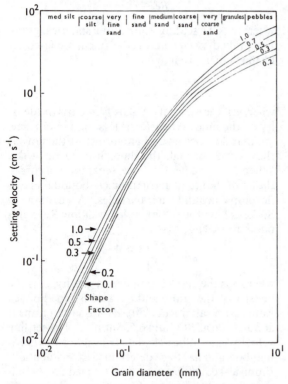

Fig. 3.3. The settling velocity of grains in water at 20°C as a function of grain diameter and shape (Komar and Reimers 1978)

the settling velocities of grains in uniform suspensions are therefore lower than that of the particles in isolation; this effect is known as hindered settling (Sleath 1984). The settling velocity in clear water (w_o) and in water with a uniform sediment concentration (w_c) is given by (Richardson and Jerónimo 1979):

$$\frac{w_c}{w_o} = (1 - C_o)^n$$

where C_o is the fractional volumetric concentration, and the exponent n varies from 2.3 at high Reynolds numbers to 4.6 at low Reynolds numbers.

BOUNDARY CONDITIONS AND FLOW REGIMES

Friction affects the velocity of flowing water in the boundary layer near the bed. Mean velocity is zero in a thin layer extending up to the focus, at a height z_o from the bed. Above the focus, flow velocity increases asymptotically with distance from the bed, and the boundary layer is generally considered to terminate at the point at which it is within 99 percent of the free stream velocity.

Boundary layer structure is different for laminar, smooth turbulent, and rough turbulent flow regimes, and the transition zones between them (Fig. 3.4). In laminar flows, successive layers of fluid flow over each other at a rate determined by the molecular viscosity, whereas in turbulent flows, water moves in random eddies throughout the liquid. Under smooth turbulent flow, there is a viscous sublayer, a few millimetres in thickness, next to the bed. Velocity increases linearly with height in this sublayer, although wave-like disturbances can be generated in it by eddies in the fully turbulent portion of the boundary layer above. These disturbances are of an entirely viscous nature, with stress being transmitted to the surface by molecular viscous forces. Velocity increases logarithmically with height in the fully turbulent layer, which is separated from the viscous sublayer by a transitional or buffer layer. The viscous sublayer and transitional zones are absent in rough turbulent flows, and the fully turbulent layer extends all the way down to the top of the roughness elements.

The hydraulic roughness of the bed, which determines its resistance to applied stresses, can be represented by the ratio of grain diameter, or the length scale of bed elements (bedforms or biological

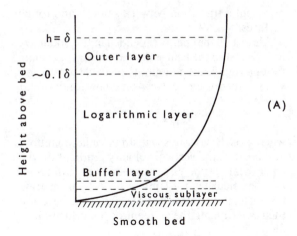

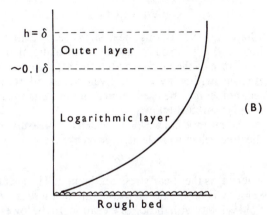

Fig. 3.4. Velocity profiles for (A) smooth turbulent, and (B) rough turbulent flow. The thickness of the various layers is not drawn to scale (Dyer 1986)

microtopography), to the thickness of the viscous sublayer (δ_v). An alternative measure is provided by the grain or boundary Reynolds number (Re_g), the ratio of the destabilizing inertial forces to the stabilizing viscous forces:

$$Re_g = \frac{u_* D}{v}$$

where u_* is the shear or drag velocity. Smooth surfaces, in which the grains are entirely enclosed within the viscous sublayer, have D/δ_v values of < 1 and Re_g < 5. Rough surfaces, which occur at higher flow speeds or with coarser sediments, have D/δ_v values > 12 and Re_g > 70. The flow over rough surfaces is turbulent, and small eddies are shed off from grains that project up through the viscous sublayer. Intermediate D/δ_v and Re_g values represent a transition between smooth and rough surfaces.

Almost all the energy expended in the boundary layer is lost to the heat sink, and only a very small, though significant, amount is available to perform work. According to Newton's law for laminar flows:

$$\tau_l = \mu \, du/dz$$

where τ_l is the viscous shear stress for laminar flow, and du/dz is the vertical velocity gradient or viscous shear strain. Viscous shear stress can therefore be calculated for laminar flows by measuring flow velocity at two elevations above the bed, and then determining the gradient of the velocity profile.

When the flow is turbulent:

$$\tau_t = (\mu + \eta) \, du/dz$$

where τ_t is the viscous shear stress for turbulent flow, and η is the eddy viscosity, a measure of the resistance of the fluid to eddy-generated shear. As η varies with eddy size and velocity, a different technique must be used to measure viscous shear stress in turbulent flows.

The shear or drag velocity, which is a measure of the fluid velocity gradient, is given by:

$$u_* = (\tau_o/\rho)^{0.5}$$

where τ_o is the shear stress on the bed. The shear velocity is proportional to the rate of increase in fluid velocity with the log of elevation. In the lowest 10 to 20 percent of the boundary layer above the viscous sublayer, the shear velocity is equal to the fluid velocity gradient divided by 5.75. Knowing the velocity at two or more points above the bed allows the shear velocity to be determined, and the equation above can then be used to calculate the shear stress exerted on the bed (Leeder 1982). The fluid velocity (u_z) at any height z above the bed can then be determined using the von-Karman–Prandtl ('law of the wall') equation:

$$u_z = \frac{u_*}{\kappa_n} \log_e \frac{z}{z_o}$$

or:

$$u_z = 5.75 \, (\tau_o/\rho)^{0.5} \log_{10} \frac{z}{z_o}$$

where κ_n is von-Karman's constant. Although values used in the oceanographic literature range from 0.33 to 0.45, recent work has generally supported the original estimate of 0.41 for this constant (Nowell 1983).

The von-Karman–Prandtl equation can only be used to calculate shear stress if velocity measurements are made very close to the boundary. An alternative method involves the use of the quadratic stress law:

$$\tau_o = \rho \, C_{Dz} \, u_z^2$$

or in modified form:

$$\tau_o = 0.5 f_w \rho \, u_m^2$$

where C_{Dz} is the drag coefficient at height z above the bed; f_w is the friction factor, and u_m is the maximum orbital velocity near the bed. Characteristic drag coefficients for typical types of sea bed have been provided by Soulsby (1983) (Table 3.4), and appropriate values for the friction factor have been calculated by Jonsson (1966) and Kamphuis (1975). The inertial dissipation method provides another means of estimating the bottom shear stress from measurements of turbulent velocity spectra. This technique has the advantage of requiring only one flow metre to be placed at a specified distance above the bed (Huntley 1988b; Xu *et al.* 1994).

Wave-induced boundary layers are layers of intense shear that grow and decay under oscillating waves every half-wave period. They are therefore both intermittent and discontinuous, although they begin to overlap as the relative depth decreases. Steady flows in the sea are almost always turbulent. Wave-induced boundary layers in shallow water are also rough turbulent under vigorous wave conditions, but they are not necessarily turbulent during the long, fairly calm intervals between storms. Boundary layers may change from laminar to turbulent with flow acceleration

Table 3.4. Typical roughness lengths and drag coefficients for different sea bottoms

Bottom type	z_o (cm)	C_{Dz}
Mud	0.02	0.0022
Mud/sand	0.07	0.0030
Silt/sand	0.005	0.0016
Sand (unrippled)	0.04	0.0026
Sand (rippled)	0.6	0.0061
Sand/shell	0.03	0.0024
Sand/gravel	0.03	0.0024
Mud/sand/gravel	0.03	0.0024
Gravel	0.3	0.0047

Note: C_{Dz} is based on velocities measured 1 m above the bed.

Source: Soulsby (1983).

and deceleration during a half-wave cycle. They are only a few millimetres thick in non-turbulent flows under swell waves, but they can be up to a few centimetres if the flow is turbulent, and up to about 50 cm over sharp crested ripples (Grant and Madsen 1986; Wright 1989; Nielsen 1992). This compares with tidal and other current induced boundary layers that can attain thicknesses of 100 m or more on continental shelves. The phase and amplitude of the velocity oscillations vary through wave-induced boundary layers, with the phase lead increasing towards the bed. For laminar flows, the phase of the maximum velocity at the bed is 45° ahead of that in the free stream (Sleath 1984; Dyer 1986) (Fig. 3.5), but with turbulent flows it appears to vary with the wave Reynolds number and the relative bed roughness, and may approach zero as waves near the shore (Conley and Inman 1992).

Whether flow near the sea bed is laminar or turbulent has an important effect on velocity distributions, sediment movement, wave-induced currents, and wave attenuation. The transition from laminar to turbulent flow occurs as increasing bed roughness or orbital velocity causes the viscous sublayer to become progressively thinner, until it is replaced by a turbulent boundary layer. There is considerable uncertainty, however, about the conditions governing this transition in thin wave boundary layers. Laminar flow can persist in wave boundary layers at Reynolds numbers that correspond to turbulent conditions in steady flows. At these high Reynolds numbers and with rough bottoms, the formation of vortices during each wave half-cycle causes intense mixing close to the bed, similar to that usually attributed to turbulence. At high Reynolds numbers but low bottom roughness, flow is unstable for only a small portion of each half-cycle. This brief instability takes the form of sudden jets of water moving in opposite directions at the crest and trough of the roughness elements, causing vigorous mixing close to the bed (Sleath 1974a).

The wave or amplitude Reynolds number (Re_w) and the relative bed roughness (B_r) determine the type of boundary layers beneath waves (A. G. Davies 1983; Sleath 1984, 1990). The wave Reynolds number is given by:

$$Re_w = \frac{u_m A_b}{v}$$

where u_m, the maximum horizontal water particle velocity at the upper limit of the boundary layer, is equal to $\pi H/(T \sinh(kh))$; and A_b, the amplitude or half the length of the horizontal water motion at the bottom (the excursion amplitude), is equal to $H/(2 \sinh(kh))$.

The relative bed roughness is given by:

$$B_r = \frac{A_b}{k_s}$$

where k_s, the Nikuradse roughness length, or the equivalent roughness of the bed surface, is determined by grain size and the presence and growth of ripples.

With a smooth turbulent boundary layer, the viscous sublayer is thicker than the bed roughness elements, and its properties are determined only by the wave Reynolds number. With a rough turbulent boundary layer, the roughness elements extend up into the turbulent flow and its properties are determined only by the relative bed roughness. Several workers have classified flow regimes by plotting relative bed roughness against the wave Reynolds number (Jonsson 1966; Kamphuis 1975; A. G. Davies 1983) (Fig. 3.6). Nevertheless, formulae for the transition between flow types derived from laboratory measurements should be treated with caution when applied to field conditions, where the random nature of the waves may generate extreme bursts of turbulence rather than regular periodic fluctuations (Sleath 1990).

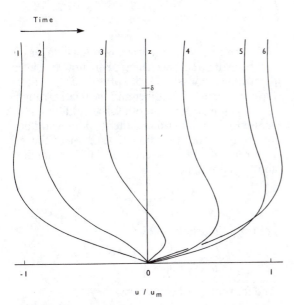

Fig. 3.5. Velocity profiles and phase shifts at equal time intervals during a half-wave cycle in a laminar wave boundary (Dyer 1986)

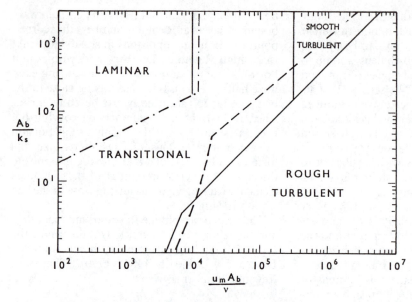

Fig. 3.6. Classification graph of flow regions (A. G. Davies 1983). The dashed and dash-dot lines represent the flow boundaries of Jonsson (1966), and the solid lines are the boundaries as determined by Jonsson (1980). The two curves delineating the rough turbulent regime are from Sleath (1974*b*) for very rough walls, and Kajiura (1968) for less rough walls

Grain diameter can be used to represent the equivalent roughness of a flat, smooth bed of sand. For a less well-smoothed bed, it has been suggested that $k_s = AD_n$, where A is some value between 1 and about 5, and n ranges between 35 and 90 percent. Sleath (1984) tabulated the experimental results of a number of workers. Critical Re_w values for the transition from laminar to turbulent oscillatory flow over smooth beds ranged between 1.3×10^4 and 6×10^5. It is more difficult to delineate flow regimes over rough, rippled beds, when one needs to know the relative bed roughness. Vongvisessomjai (1984) suggested:

$$k_s = 12.5\, H_r \frac{H_r}{L_r}$$

where L_r and H_r are ripple wavelength and height, respectively, whereas according to P. Nielsen (1979):

$$\frac{k_s}{A_b} = 25 \frac{(H_r)^2}{L_r A_b}.$$

Other workers have obtained values for the constant in the above equation of 16, 20, and 27.7 (Raudkivi 1988; van Rijn 1989). Vitale's (1979) equation considers both ripple and grain size:

$$\frac{k_s}{A_b} = \frac{2D_{90} + 0.01\, H_r}{A_b}.$$

Jonsson (1980) proposed that the lower limits for rough turbulent flow are approximately:

$$Re_w = 10^3 \frac{A_b}{k_s}, \qquad \text{when } 10 \le \frac{A_b}{k_s} \le 10^3$$

or, if the bed is very rough:

$$Re_w = 10^4, \qquad \text{when } 1 \le \frac{A_b}{k_s} \le 10.$$

Kajiura's (1968) work suggests that the lower limit for less rough beds is:

$$Re_w = 2000 \frac{A_b}{k_s}.$$

The thickness of the wave boundary layer (δ) can be estimated once the type of flow has been determined. Jonsson (1966) plotted δ/A_b against the wave Reynolds number and the relative roughness. In laminar and smooth turbulent flows, δ/A_b is essentially a function of the Reynolds number, and in rough turbulent flows, a function of the relative roughness. For purely sinusoidal laminar flow above a smooth bed:

$$\frac{\delta}{A_b} = \frac{\pi}{(2Re_w)^{0.5}}.$$

For smooth turbulent flow:

$$\frac{\delta}{A_b} = \frac{0.0465}{\sqrt[10]{Re_w}}.$$

And for rough turbulent flow:

$$\frac{30\delta}{k_s} \log_{10} \frac{30\delta}{k_s} = 1.2 \frac{A_b}{k_s}.$$

The friction factor is also a function only of the Reynolds number in laminar and smooth turbulent flows, the relevant expressions being:

$$f_w = \frac{2}{(Re_w)^{0.5}}$$

for laminar flows; and:

$$\frac{1}{4(f_w)^{0.5}} + 2\log_{10}\frac{1}{4(f_w)^{0.5}} = \log_{10} Re_w - 1.55$$

for smooth turbulent flows.

The friction factor in the rough turbulent regime is a function only of the relative roughness:

$$\frac{1}{4(f_w)^{0.5}} + \log_{10}\frac{1}{4(f_w)^{0.5}} = -0.08 + \log_{10}\frac{A_b}{k_s}.$$

The friction factor has been plotted against the wave Reynolds number and the relative roughness of the boundary (Jonsson 1966; Kamphuis 1975; Justesen 1988) (Fig. 3.7). The friction factor varies, however, under the crests and troughs of asymmetrical waves in shallow water. Measurements of bottom shear stress in a wave flume with a smooth bottom showed that friction factors under wave crests, corresponding to high bottom velocities, can be considerably lower, depending on the Reynolds number, than Jonsson's predictions, whereas under wave troughs, where velocities are lower, friction factors can be much higher than Jonsson's values (Kuo and Chen 1990). It has also been demonstrated that friction factors are strictly applicable only to fixed bed roughness, and considerably under-represent actual friction under sheet flow conditions, which increases in direct proportion to the shear stress (Wilson 1989). Wave friction factors are further enhanced by interaction with currents (Bijker 1966; O'Connor and Yoo 1988).

A video camera has been used to record the response of a sandy bottom to near-breaking waves in the field (Conley and Inman 1992). The boundary layer under the wave crest consisted of up to three distinct regimes, depending on the intensity of wave motion (Fig. 3.8):

a) Streaking occurs when the flow, accelerating from rest after orbital reversal, becomes strong enough for grain motion. Grains begin to roll over the bed and quickly develop into a series of rather long and regularly spaced lines, which may be visually enhanced by separation of small, dark heavy mineral grains from larger quartz grains. Streaking appears to occur under laminar oscillatory flow conditions, and if the motion is weak, it can persist for the entire wave half-cycle.

b) Roiling begins, apparently with the onset of turbulence, with the abrupt transition of the

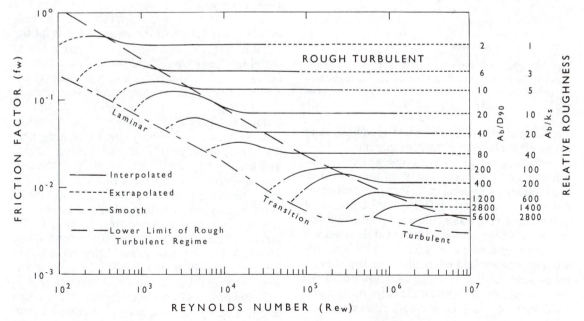

Fig. 3.7. Relationship between the wave friction factor, the Reynolds number, and the Relative Roughness (Kamphuis 1975)

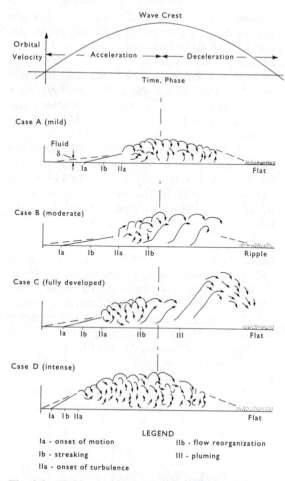

Fig. 3.8. Fluid–granular boundary layer development beneath the crest of near-breaking waves. The sequences are shown in order of increasing intensity of motion at the interface (Conley and Inman 1992)

streaking regime to thicker sheet-type flow, sand lifting off the bottom in large, roiling tufts.

c) Pluming occurs under fairly low, long period waves, at or just after the passage of the crest when the flow begins to decelerate. Explosive lifting of the roiled layer from the bed injects great plumes of sediment into the water column. It is then carried by the free stream until it settles out at the cessation of motion. Pluming may result from the spatially varying pressure field imposed on the bed over an entire wavelength, with ventilation flow out of the bed and into the fluid under the wave trough, and suction, or flow into the bed, under the crest.

Flow under the wave crest is characterized by more rapid and distinctive turbulent boundary layer development than under the trough, and it is less homogeneous, more structured, and better able to mobilize sediment. Despite turbulence under intense wave conditions, roiling and pluming regimes did not develop under the trough. As roiling and pluming move more sand onshore than streaking moves it offshore, there is pronounced asymmetry in the instantaneous sand transport rate between the wave crest and trough. This is independent of the asymmetry in near-bed orbital velocity and acceleration.

Turbulent Bursting

Short, intermittent suspension of grains projecting above mean bed level, or on ripple crests, has been attributed to eddies penetrating the viscous sublayer from the turbulent boundary layer. In turbulent flows, water is generally carried upwards into faster-moving flows, and downwards into slower flows. The local shear stresses exerted by these exchanges of momentum, or 'bursts', are known as Reynolds stresses. Upward movements of low-velocity fluid are referred to as ejections, and downward movements of high-velocity fluid towards the bed as sweeps. Sweeps are presumably most effective in moving bed load, whereas ejections promote suspended load. This reasoning suggests that sweeps give way to ejections as the prime agent of sediment movement some distance from the bed, possibly near the top of the mean trajectory of the saltating particles. Outward interactions, motions with high longitudinal and vertical fluctuations, could be even more important suspension mechanisms than ejections, but they occur less frequently (Dyer 1986).

The viscous sublayer consists of alternations of wavy high and low velocity streaks, running parallel to the direction of flow. The spacing, height, and length of these streaks increase with the kinematic viscosity, and decrease with the shear velocity of the fluid (Weedman and Slingerland 1985). Bursting may be associated with the formation of a transverse vortex, initiated from a low-velocity streak in the viscous sublayer (Offen and Kline 1975). A low-velocity streak lifted by a three-dimensional disturbance causes a local recirculation cell to develop beneath it. As the cell rotates with the flow, its forward edge creates the sweep and its trailing edge the ejection. In rough turbulent boundary layers, which lack viscous sublayers, vortices could be formed by instantaneous separation to the lee of obstacles. Vortices advected downstream become horseshoe-shaped, transverse to the flow, and inclined at 16° to 20° to the horizontal. They travel a little slower than the mean flow velocity, and gradually break down as they are stretched by the velocity gradient.

Although the effect of rough beds and mobile sediment on turbulent structure remains to be determined, because of high instantaneous shears bursting processes are potentially very important for sediment movement. Most sediment is moved in the sea by stresses only a little above the threshold value. Temporal and spatial variations in the intensity and rate of bursting could therefore have an important effect on the rate of sediment transport. Bursting has been recorded in slowly oscillating tidal currents in shallow seas (Soulsby 1983), and it may account for vigorous upward flow and sediment entrainment in kolks and surface 'boils' in estuaries (Kostaschuk *et al.* 1991). The movement of gravel by tidal currents is thought to be largely associated with sweep-type events, but ejections are important for the suspension of sand (Thorne *et al.* 1989; Williams *et al.* 1989*a*). It has been proposed that high suspended sediment concentrations are the result of burst-induced helixes or vortices touching the bottom (Brenninkmeyer *et al.* 1977; Clarke *et al.* 1982). Nevertheless, Reynolds stresses were negligible more than 6 mm from the bed in a flume, and could not have helped to keep sediment in suspension (McDowell 1983). Although it is questionable whether burst cycles, involving the lift up of low-speed streaks in the viscous sublayer, can occur in the rapidly reversing flows beneath shoaling waves, they could be responsible for parting lineations, low parallel ridges running down the swash zone, a few grain diameters in height (Sleath 1984). There are explanations, other than burst cycles, however, for current-parallel sedimentary features, including secondary helicoidal flows generated by differences in bottom roughness (McLean 1981).

THRESHOLDS OF SEDIMENT MOVEMENT

Grains begin to move when sufficient stress is applied to the bed. The relationship is complicated, however, and particles of the same size may not begin to move at the same time. As flow speeds increase, grains first begin to vibrate, and then, when the critical flow is exceeded, individual grains

start to move downstream, followed by flurries, and finally by the whole surface. Although the threshold for sand grain movement tends to occur fairly close to the transition from laminar to turbulent flow, actual values vary according to such factors as the shape, size, and density of the grains, and their position and orientation.

A spherical grain, lying on a bed consisting of similar grains, is subjected to drag parallel to the flow, lift acting upwards, and gravity directed downwards. The drag acts at a distance of about 0.29 of the grain diameter below the top of the grain (Fig. 3.9). Fluid drag is composed of drag on the area of the grain normal to the flow, and, because it projects above the general bed surface, much of the drag exerted on the area in its wake. The drag on the grain (F_D) is therefore:

$$F_D = \pi \frac{D^2}{4} \frac{\tau_o}{N}$$

where N, the ratio of the mean drag and lift forces over the whole bed to the drag and lift on the protruding grains, varies between 0.2 and 0.3.

Lift results from distortion and acceleration of the flow, and consequent reduction in pressure resulting, according to Bernoulli's equation, from the deflection of stream lines over the top of protruding grains. Backspin rotation of the grain can increase the relative velocity of the upper surface, and increase the lift through the Magnus effect (Bagnold 1974). The lift force (F_L) is given by:

$$F_L = 0.5 \rho C_L A u^2$$

where A is the protruding area of the grain, and C_L is a coefficient which depends upon grain protrusion and the reference height for the measurement

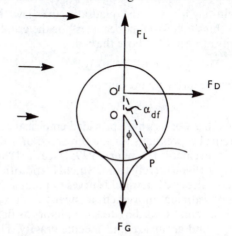

Fig. 3.9. The forces acting on a static grain

of flow velocity. For spherical particles, the lift force is approximately equal to 0.85 of the drag force.

The gravitational force (F_G) on the particle is equal to its immersed weight acting through the centre of the sphere:

$$F_G = \frac{\pi}{6}(\rho_s - \rho)\, g\, D^3.$$

The three forces act through the point of contact (P) with the grains at the critical moment when the grain is about to rotate and become dislodged (denoted by subscript cr). At this moment therefore:

$$F_D = (F_G - F_L)\tan\alpha_{df}$$

where α_{df} is the angle of dynamic friction, the angle made with the vertical by a line drawn from the point in the grain through which the drag operates, to the centre of the grains below.

Substituting for the drag, gravitational, and lift forces:

$$\tau_{cr} = \rho u_{*cr}^2 = \frac{0.66\, Dg\, (\rho_s - \rho)\, N \tan\alpha_{df}}{1 + 0.85\tan\alpha_{df}}.$$

Using Chepil's (1959) values of $\alpha_{df} = 24°$ and $N = 0.3$, and the density of quartz sand and sea water:

$$\tau_{cr} = 103.5\, D.$$

As it is the maximum bed shear stress rather than the mean, however, that initiates grain movement, threshold values should be reduced by a turbulence factor, which is, according to Chepil, about 2.5. Therefore, taking turbulent fluctuations into account:

$$\tau_{cr} = 41.4\, D.$$

Substituting for τ_{cr} in the quadratic stress law, the mean threshold formula becomes, for the critical velocity at a height z above the bed:

$$u_{zcr} = (103.5\, D/\rho\, C_{Dz})^{0.5}$$

and for the turbulent maximum:

$$u_{zcr} = (41.4\, D/\rho\, C_{Dz})^{0.5}.$$

Many other workers have provided empirical and theoretical expressions for the threshold of sediment movement (Dingler 1979; Sleath 1984; Hardisty 1990a). Huljstrom, Shields, and other workers calculated threshold curves to predict the onset of sediment movement in steady unidirectional currents, according to such factors as flow velocity, and grain size and specific gravity. The Shields parameter (Θ), which has a threshold value

Θ_{cr}, is the ratio of the entraining to the stabilizing forces on a grain:

$$\Theta = \frac{\tau_o}{(\rho_s - \rho)gD} = \mathbf{f}(Re_g) = \mathbf{f}\frac{u_* D}{v}.$$

Alternatively, it can be expressed in terms of the friction factor (Jonsson 1966):

$$\Theta = \frac{0.5 f_w \rho u_m^2}{(\rho_s - \rho)\, gD}.$$

The Shields parameter plotted against the grain Reynolds number has provided adequate prediction of grain thresholds under oscillatory flow conditions in the laboratory and field (Komar and Miller 1974; Larsen *et al.* 1981). A more practical approach, however, is to plot the Shields parameter against a dimensionless parameter (S) that depends only on the properties of the sediment and fluid (Madsen and Grant 1976):

$$S = \frac{D}{4v}\cdot [gD(\rho_s - \rho)/\rho]^{0.5}.$$

Plotting these variables on a modified Shields diagram shows that the critical value of the Shields parameter for quartz grains with a specific gravity (ρ_s/ρ) of 2.65 g cm^3 and diameter 0.2 mm to 2 mm, is from 0.03 to 0.05.

Komar and Miller (1973) proposed that the threshold for grains of less than 0.5 mm in diameter, which is attained while the boundary layer flow is still laminar, is best determined using:

$$\frac{\rho u_m^2}{(\rho_s - \rho)\, gD} = a(d_o/D)^{0.5}$$

where d_o is the length of the horizontal water motion at the bottom ($2A_b$), and a is a proportionality coefficient that has values ranging between 0.13 and 0.46, according to the authority. The left hand side of the equation, which is a modified form of the Shields parameter based upon velocity rather than shear stress, has been termed the Mobility number (**M**).

The threshold for coarse sands and other grains more than 0.5 mm in diameter is reached when the boundary layer is turbulent. In this case, the threshold can be determined using the equation:

$$\frac{\rho u_m^2}{(\rho_s - \rho)\, gD} = 0.46\pi(d_o/D)^{0.25}.$$

The effect of bed slope on grain thresholds also has to be considered on sloping beaches. Allen (1982) provided an expression for the ratio of the threshold velocity on a slope (u_{crb}) to the threshold on a horizontal bed:

$$\frac{u_{crb}}{u_{cr}} = 1.3204 \, [\sin (\phi - \beta)]^{0.5}.$$

Dyer (1986) proposed that:

$$\frac{u_{crb}}{u_{cr}} = \frac{\tan \phi - \tan \beta}{\tan \alpha_{df}} \cos \beta.$$

Whitehouse and Hardisty (1988) tested these slope correction factors and found that both are satisfactory, although Allen's fitted the data more consistently. They also found that while there was a significant decrease in grain thresholds when flow was downslope, thresholds did not increase as rapidly as predicted by both models when flow was upslope.

Most investigations of sediment thresholds and transport have been conducted in the laboratory, using single grain sizes, short, monochromatic, sinusoidal waves, shallow water, and initially smooth, flat beds. The turbulence generated in small wave flumes is also generally too weak to suspend sediment. These conditions are quite different in the field, where there are very irregular waves, a variety of grain sizes, and rippled beds. Nevertheless, reasonable agreement has been found between Bagnold's (1963) threshold Shields curve for a rippled bed and sediment thresholds in the sea (Sternberg 1971). Because of changing form drag, however, thresholds vary during a tidal cycle and with the tidal range, and considerable care must therefore be taken when using laboratory curves to determine thresholds of sediment movement over rippled beds (Dyer 1980).

Standard threshold curves must be modified to accommodate the effects of mixed grain sizes, shapes, and imbrication on the selective entrainment of grains in heterogeneous sediments (Komar and Li 1986). Only 5–9 percent clay content is enough to begin to bind sand grains together, and as the proportion increases, so does the threshold (Dyer 1986; van Rijn 1989). Threshold shear velocity is also higher when the water contains low concentrations of suspended clay (Best and Leeder 1993). Because of lower pivoting angles and protrusion into faster flows, the larger grains in a mixed population have lower threshold entrainment stresses than the same grains on more homogeneous beds (Fig. 3.10), and the finer grains have higher threshold stresses than the same grains in a uniform deposit. The importance of grain protrusion and sheltering has been demonstrated by Hammond *et al.* (1984), who found that the

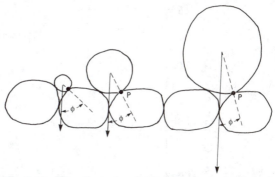

Fig. 3.10. The effect of grain size, relative to the size of the underlying grains, on the pivoting angle (ø) about the point of contact (*P*) (Li and Komar 1986)

threshold for loosely packed clasts ($D \leq 5$ cm) under the influence of tidal currents in the sea, is represented by the relation $u_{*cr} = 7 \, D^{0.2}$, where D is the diameter of an equivalent sphere. Differences in mineral density are also responsible for preferential entrainment and transportation of light mineral sand grains, and the concentration of less-easily entrained heavy minerals (Li and Komar 1992*b*).

The concept of critical threshold conditions, below which no single grain is moving, has been questioned (Lavelle and Mofjeld 1987). The magnitude of instantaneous stresses in turbulent flow fluctuates about the mean stress, and any mean stress level, even very low ones, will intermittently provide instantaneous stresses that exceed threshold levels for some particles. Therefore some transport can occur under turbulent conditions at time-mean stresses below those that are often considered to be threshold values. A. G. Davies (1985) studied the relationship between wave generated near-bed velocities in shallow water, and sand grain thresholds on flat and rippled beds. There was a clear relationship between grain motion and orbital velocity, with movement generally occurring with high velocities but not with low. There were fairly wide transitional zones with intermediate velocities between these extremes, however, in which some waves moved sediment, while others, of the same velocity, did not. These data demonstrate that there is a transitional range of velocities for the initiation of sediment movement, rather than a single threshold value. Threshold velocity amplitudes measured in unperturbed free stream flow 1 m above the bed were at least twice as high for flat as for rippled bottoms. This is probably

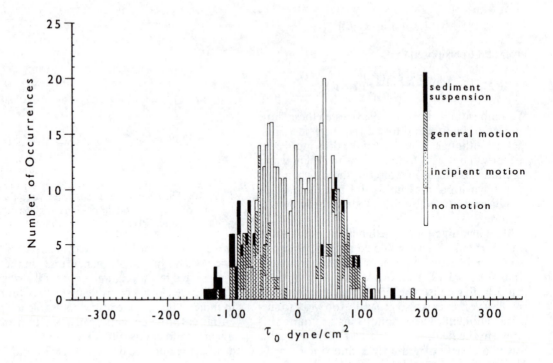

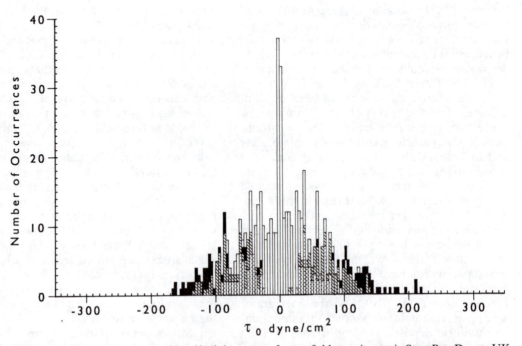

Fig. 3.11. Sediment movement and total bed shear stress for two field experiments in Start Bay, Devon, UK (Davies 1985)

because near-bed velocities over ripple crests were significantly greater than free stream flows. Bed shear stress, however, had better predictive capabilities than free stream velocity. The use of bottom stress instead of free stream velocity reduced the width of the transitional range by 19 per cent, although it was still quite wide. For non-separating flow, almost all waves moved sediment when stresses were greater than about 10 dynes cm^{-2}, while in the transitional zone, corresponding to stresses from about 5 to 10 dynes cm^{-2}, only some waves moved sediment. For flows that separated over ripples every half-wave period, the total stress at the threshold of bed load movement was about 30 dynes cm^{-2} (Fig. 3.11).

Several workers have considered sediment thresholds under the combined action of waves and currents. When waves operate alone, the critical velocity is lower for shorter waves, possibly because of their greater flow accelerations. In a laboratory investigation, however, it was found that when waves were combined with a unidirectional current, there was an inverse relationship between threshold velocity and wave length—possibly because the probability of turbulent bursts caused by steady flows coinciding with near maximum orbital velocities increases with increasing wave period (Hammond and Collins 1979). Lee-Young and Sleath (1988) measured thresholds in a flume with steady currents operating at right angles to the oscillatory flow. The results suggest that the Shields curve can be used to predict thresholds of sand movement in combined flows. A similar study indicated that there is a tendency for threshold shear stress to be higher for combined flows than for steady currents (Katori *et al.* 1984).

BEDFORMS

The distortion of mobile beds by flowing water has an important effect on bed roughness, friction factors, near-bed currents, wave height attenuation, and sediment transport. Bedforms begin to develop with the initial movement of sediment. As the magnitude of the orbital diameter increases, flat beds are replaced by rolling grain ripples, and then by vortex ripples. As orbital velocities increase even further, there is a decline in ripple height and sediment begins to move over the bed as sheet flow—highly concentrated layers of grains

shearing over each other, with an intrusive depth of several millimetres (Kraus and Horikawa 1990; Dibajnia and Watanabe 1992; Asano 1992). Oscillatory flow velocities are generally too low to form dunes and other higher stage bedforms in the sea, although they can appear as temporary forms in the backrush of sandy beaches, and where there are strong tides or other quasi-steady currents.

Vortex ripples tend to develop when there are low to moderate rates of sediment transport. Flow separation, whereby the zone of high shear leaves the bed and rejoins it a short distance downstream, causes vortices to develop to the lee of ripples every half-wave cycle. The vortices propel grains towards ripple crests until flow reverses, when they are swept towards the crests and up into the flow. Although most grains are deposited at the crests, finer grains can be placed into suspension by energetic waves. At the end of the return stroke, a second vortex develops on the other side of the ripple, and this too is lifted into the flow (Fig. 3.12). The effect of ripples on the formation and movement of suspension clouds can result in net sediment transport being in the opposite direction to that of the strongest flows (Nielsen 1988).

Flow resistance or form drag resulting from the physical obstruction of the bedform, varies with the shear stress. It reduces the capacity of the flow

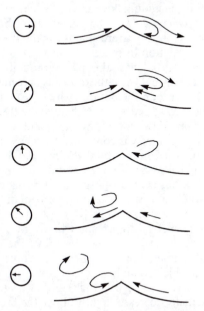

Fig. 3.12. Vortex generation, separation, and advection over ripples during a wave oscillation. The rotating vector on the left shows the wave phase (Dyer 1986)

to carry sediment and, depending in part on the steepness of the bedform and whether flow separation occurs, it can contribute more to the total stress on a rippled bed than surface shear stress or skin friction (Davies 1985; Dyer 1986). Maximum velocities in ripple vortices are two to three times wave-induced velocities near and parallel to the bed. Formation and ejection of vortices is therefore very effective in placing sediment into suspension, even though turbulence usually extends for only about twice ripple height above the bed. Although greater shear stress is required to carry grains up the upstream slope of bedforms, this is partly offset by flow acceleration over the undulations. The effect of bedforms on the direction and amount of sediment movement, however, depends upon their height, asymmetry, and other aspects of their shape and size.

Wave-induced ripples have symmetrical and rounded profiles, although strong nearshore currents tend to produce asymmetric ripples in shallow water. There have been numerous investigations of ripple geometry and sediment and water movement over rippled beds (Hay and Bowen 1993; Li 1994; Wiberg and Harris 1994; Mogridge *et al.* 1994). Based upon detailed measurements of the flow over ripples in the laboratory, Sleath (1975) determined that ripples first introduce random instability, including flow separation above the lee slopes, when $A_b/L_r \geq 0.5$. Although the flow becomes more and more chaotic as the frequency of the oscillation increases, the transition to fully developed turbulence takes place gradually, and it is only completed when the ratio has become many times larger than the point at which instability or chaotic flow appears. Nevertheless, the flow over active ripples is almost always turbulent.

Ripple geometry is controlled by many factors, including maximum orbital velocity and diameter near the bed, grain diameter, and possibly wave period. It has been proposed that up to a critical break-off point, which varies approximately with $\sqrt{D}$, ripple wavelength is roughly equal to the orbital diameter (d_o), irrespective of grain size or density. Above the break-off point, increasing orbital diameter has no effect on ripple wavelength, which remains constant. More-recent work, however, suggests that ripple wavelengths may actually decrease slightly above the break-off point, before reaching a constant, intermediate value (Miller and Komar 1980*a*, *b*; Clifton and Dingler 1984).

Under moderate flow conditions, ripple wavelengths are between about 0.65 and 0.8 times the orbital diameter (Miller and Komar 1980*a*, *b*; Clifton and Dingler 1984). These conditions are typical of shallow lakes and bays, as well as laboratory wave tanks and oscillating wave tunnels. Under oceanic conditions, however, where there is deeper water and the waves have much longer periods, orbital diameters are usually many times larger than ripple spacing (Miller and Komar 1981). Vortex ejection could account for the breakdown of the relationship between ripple wavelength and orbital diameter under more vigorous flow conditions. Wavelengths would then be determined by the factors that influence the way that sediment is lifted from the bed and distributed over adjacent ripples (Sleath 1984).

There have been many attempts to predict bedform occurrence. None is completely satisfactory, however, probably because of the inability of a small number of variables to represent the wide range of flow and sedimentological conditions that occur in nature. Some workers have used the wave form of the Shields criterion, or Mobility number (**M**), to predict the inception of sheet flow over flat beds (Dingler and Inman 1976). Ribberink and Al-Salem (1990), for example, found that the transition occurs when **M** > 100 to 200. Ripple occurrence and transition to sheet flow have also been related to the Shields parameter. According to Komar and Miller (1975), the transition occurs when:

$$\Theta = 0.413\, D^{-2/5}$$

where the grain diameter is in centimetres. Bedforms have been found to persist in the field under high energy conditions with Shield parameter values greater than 1, and ripples may even be able to develop when values are higher than two (Kawata *et al.* 1992; Osborne and Vincent 1993). The non-dimensional shear stress (Θ') can also be used to predict when ripples are replaced by sheet flow over a flat bed. This occurs when (P. Nielsen 1979):

$$\Theta' = \frac{\tau_o'}{(\rho_s - \rho)\, g D} \geq 1.0$$

where τ_o' is the peak bed shear stress owing to skin friction. Although Ribberink and Al-Salem's (1990) experimental data supported Nielsen's prediction for the transition from ripples to plane beds, they found that sheet flow can occur over bedforms without flow separation in the high-velocity regime. Sunamura (1981) proposed that bedform demarcation and stages of sediment motion can be represented by a combination of the Ursell parameter (U_r), which is a measure of the asymmetry of the velocity field, and a dimensionless parameter (F). The relation between F and U_r is given by:

$$F = k\, U_r^{0.25}$$

where:

$$F = \frac{\rho\, u_m^2}{g\,(\rho_s - \rho)\,(D\, d_o)^{0.5}}$$

and

$$U_r = \frac{H\, L^2}{h^3}.$$

Values for k were less than 0.1 for a flat bed with no movement, 0.2 to 0.52 for a rippled bed, and greater than 0.69 for a flat bed with sheet flow. A similar relationship between F and U_r, but with different power and k values, has been found to predict grain movement thresholds in the laboratory (Moutzouris 1990).

MODES OF SEDIMENT TRANSPORT

Most sediment moves in water as bed load or suspended load. Saltation is much less important than in air, and the maximum height of the trajectories is usually equivalent to only a few grain diameters. Bed load remains close to the bottom, rolling or even sliding along the bed or moving in low trajectories, whereas suspended sediment is supported in the water by an excess in the upward flux of turbulent fluid momentum over the downward flux. Suspension results from the intense agitation of the bed by breaking and broken waves, and particularly by the intermittent shedding of sand-laden vortices from the lee side of ripple crests. A variety of instruments has been employed to measure suspended sediment in the nearshore, including sound and light scattering and transmission devices, and self-siphoning or diver-operated samplers. Bed load transport is more difficult to measure, although an estimate can be obtained through the migration of ripples. Sediment can be continuously exchanged between bed load and suspended load during a wave oscillation. Moving sediment concentrations, the height and length of trajectories,

and the velocity of movement are functions of the excess stress. This can be expressed in terms of the transport stage, the ratio of the ambient to the threshold shear velocities (u_*/u_{*cr}).

The occurrence and relative importance of transport modes vary spatially and temporally, with differences in wave and breaker types, beach profiles, sediments, and other factors (Horikawa 1988; Kraus and Horikawa 1990). In general, however, sediment movement in the offshore zone is dominated by wave-induced orbital motion; in the surf zone by wave breaking and wave-induced currents; and in the swash zone by alternating uprush and downrush. Sand moves predominantly by bed load outside the breaker zone, but suspension above ripples becomes more important as bottom velocities increase in the shallow water near the breakpoint. In the surf zone, strong agitation of the water by wave breaking can result in intense suspension, although bed load may also be significant. Large amounts of sediment are also thrown into suspension where the backrush meets the incoming surf. In the swash zone, sediment is moved by sheet flow and in dense suspended concentrations (Yu *et al.* 1990).

Shibayama and Horikawa (1982;—see also Horikawa 1988) distinguished the following transport modes:

a) Bed load over an essentially flat surface with no suspended sand clouds.

b) Bed load and suspended load occurring together, with clouds of suspended sediment forming over ripples. There are two subtypes:

(i) If the orbital diameter is less than three to five times the ripple wavelength, strong vortices keep the clouds of suspended sediment within a distance equal to one ripple wavelength. Some of the sediment suspended during the landward flow is confined within the vortices, carried seawards as the flow reverses, and then deposited on the bottom. Flow reversal does not occur for bed load, however, which is carried landwards. Sediment that started moving in the first half of the wave period can therefore be carried landwards or seawards, depending on whether it is carried as bed load or suspended load.

(ii) If the orbital diameter is more than three to five times the ripple wavelength, vortices are too weak to keep the sediment suspended within them, and the suspension clouds are not confined to within one ripple wavelength. Suspended sediment is moved landwards by the vortices during the first half-wave period, and then released and deposited on the bottom. Sediment carried as suspended load and bed load are therefore moved landwards.

c) Suspended sediment transport is predominant, and consists of the two subgroups described in b(i) and b(ii) above, but with little or no bed load.

d) Sheet flow occurs as a layer over a flat, ripple-less bed with high bottom shear stresses. Grains are transported landwards.

There is continuing debate over the relative importance of suspended load and bed load to the total-load transport. The ratio of suspended load to bed load is likely to vary, however, according to the type of breakers, wave intensity, and other surf conditions (Zampol and Inman 1989). For example, Kana and Ward (1980) found that sediment suspended 5 cm or more above the bed accounted for all the material carried by waves during storms, but only 30 per cent under post-storm, swell wave conditions. There may therefore be a relationship between the relative importance of suspended sediment transport and beach state (Chapter 4). Suspended sediment accounted for all the predicted longshore transport on a narrow reflective beach, for example, but only 47 per cent of the total on a wide dissipative beach (Downing 1984; Sternberg *et al.* 1989).

Komar (1978) derived an expression to determine the contribution of suspension to the total-load longshore transport rate:

$$\frac{I_s}{I_t} = \frac{7.0 \, C_o}{\gamma \tan \beta} \frac{\rho_s - \rho}{\rho}$$

where I_t and I_s are the total and suspended immersed weight longshore transport rates, respectively, and γ is the ratio of breaker height to breaker depth. Walton and Chiu (1979) expressed the ratio of bed load to suspended load in terms of grain characteristics, and the breaker size (H_b) and angle (α_b):

$$\frac{\text{bed load}}{\text{suspended load}} = 9.4 \frac{w_s \sin 2\alpha_b}{H_b^{0.5}}.$$

This equation suggests that bed load is dominant for high breaker angles, coarse sediment, and low breakers.

Sawaragi and Deguchi (1980) proposed that the ratio of bed load to suspended load can be expressed in terms of the parameter N_{sr}, where:

$$N_{sr} = \frac{g \, T \, D_{50}}{2 \, \pi \, v}.$$

The time-averaged bed load is greater than the time-averaged suspended load when $N_{sr} > 1200$, whereas the suspended load is greater when N_{sr} is < 640. They also found that suspended load is greater than bed load when $u_*/w_s > 0.8$. Bowen (1980) proposed that the relative importance of suspended and bed load depends on the value of u_m/w_s. This implies that the proportion of suspended to bed load increases in the surf zone with distance offshore, and then begins to decrease seawards of the breakpoint.

The lack of precise definitions for modes of transport are partly responsible for different estimates of their relative importance. The distinction between suspended and bed load is arbitrary, and there are marked differences in the elevation above which sediment is considered to be in suspension. Other problems involve the lack of field data on bed load, and the lack of samples from the higher concentrations near the bed, which result in considerable underestimation of the transport mode to which they are assigned.

Suspension

The asymmetry of flows and ripples plays an important role in the formation of sand clouds. The rate of sediment entrainment is largely determined by the strength of the lee vortices, ripple spacing, and the velocity and shear stress just upstream of the ripple crests. Once suspended, sediment entrained in vortices disperses and settles as it is swept back and fore by the reversing currents.

Suspended sediment concentrations vary with maximum oscillation velocity, the phase within each oscillation, and height above the bed. There are often very rapid variations in concentration through time, with high concentrations generally corresponding to periods of high fluid velocity. Hysteresis complicates the relationship between orbital velocity and sediment concentrations. As higher velocities are needed to resuspend sediment than to keep it in suspension, for the same orbital velocity, concentrations tend to be higher when velocities are decreasing than when they are increasing. The response of ripple geometry to wave groupiness can have the opposite effect. Steep ripples develop during the passage of a group of small waves, and less steep ripples with larger waves. There is a lag in the response of the bed to changing wave size, however, and ripple steepness is usually out of equilibrium with the waves. Therefore, high waves pass over steep ripples that developed under the preceding group of small waves, and low waves over flat ripples formed during the preceding period of high waves. Despite similar wave velocity, more sand is suspended by the ejection of sand laden vortices from steep ripples when wave size is increasing, than when the ripples are flat and wave size is decreasing (Vincent *et al.* 1991). Theory, supported by some flume data, also suggests that suspended sediment concentrations increase through a wave cycle much faster than they decline. This is because increases in concentration reflect increasing turbulent intensity and boundary layer thicknesses, whereas decreases take place at lower flow velocities, when grains fall at almost their settling velocity (Fredsøe *et al.* 1985). There are also often phase lags between maximum sediment concentrations at any elevation, and the maximum velocity or stress near the bed. These lags, which increase with height above the bed, and decrease with increasing velocity, may cause net transport reversals within short distances from the bed (Greenwood *et al.* 1990*a*, *b*).

Larger material tends to be carried fairly close to the bed, but turbulence can carry finer sediment high into the water column. With nonbreaking waves, sediment concentrations are generally very low more than three ripple wavelengths above the bed, but suspended sediment can be found throughout the entire water column with breaking waves. Instantaneous sediment concentrations, within a few tenths of a metre of the bed, vary considerably during wave cycles as a result of varying turbulent intensity and the shedding of vortices from rippled beds. Instantaneous profiles may therefore reveal decreasing, constant, or even increasing concentration levels with elevation above the bed. Nevertheless, intermittent resuspension of coarse material tends to result in a decrease, usually exponential, in time-averaged concentrations with distance above the bottom, despite the presence of a fairly constant wash-load of fine-grained sediment at all elevations within the surf zone (Sternberg *et al.* 1989; Greenwood *et al.* 1990*a*, *b*; Hay and Sheng 1992; Nishi *et al.* 1992). In the SUPERTANK project, however, mean suspended sediment concentrations only decreased exponentially with height just outside the breaking region. They decayed roughly linearly with distance from the bed in the region of initial breaking, and increased

with height just shorewards of this region (Barkaszi and Dally 1992).

Sediment resuspension is extremely episodic, and it takes place at a variety of time scales, including that of the incident waves. Several peaks in concentration may be identified during each wave period, although they vary in height according to differences in the maximum onshore and offshore velocities (P. Nielsen 1979). There are often four peaks within a wave cycle, but the number can range between one and five, depending on how far the bursts are swept back and fore in relation to the wavelength of the ripples, and the location of the sampling position relative to the ripple crests. High sediment concentrations also occur at frequencies that are much lower than that of the incident waves, in some cases because of long period variations in the envelope of the incident waves (Greenwood *et al.* 1991; Hanes 1991; Davidson *et al.* 1992; Shibayama *et al.* 1992; Jaffe and Sallenger 1992). Groups of large waves are more effective than individual waves in carrying sediment high into the water column, but individual waves are most influential near the bed. Suspensions may also respond more slowly to individual waves with increasing elevation because of a relative increase in fine sediment, which can remain in suspension for more than one wave period (Osborne and Greenwood 1993).

Suspended sediment concentrations are especially high in the turbulent water near the breakpoint, and in the inner surf region where breaker vortices and obliquely descending eddies bring highly intermittent turbulence to the bottom (Nadaoka *et al.* 1989; Mocke and Smith 1992; Pedersen *et al.* 1992). Sediment is suspended under plunging waves by the large vortex touching the bottom and entraining sediment immediately behind the advancing wave crest. Some sediment settles to the bottom as the suspended sediment cloud grows larger and is carried seawards as the current reverses. Any sediment remaining in suspension is then swept landwards by the next wave crest. Plunging breakers suspend more material than spilling breakers, but the effect of breaker height is less clear. Kana (1978) found that concentrations with plunging waves reached a maximum just inside the breakpoint, and then declined gradually landwards. With spilling waves, however, concentrations increased rapidly inside the breakpoint, and then stayed fairly constant under the shoreward propagating bores. Local increases in concentration can also occur in the surf zone over the crest of bars.

Combined Flows

Surface waves and currents operate together in the coastal zone, interacting with bottom sediments in the nearshore zone and on the continental shelf down to depths as great as 200 m. Wave and current fields interact in several ways (Soulsby *et al.* 1993):

a) wave refraction by horizontally sheared currents;

b) modification of wave kinematics by currents;

c) wave generation of mass transport currents;

d) wave generation of radiation stresses which give rise to longshore and other currents in the surf zone;

e) enhancement of the bottom friction felt by currents owing to interaction with the wave boundary layer; and

f) enhancement of bed shear stresses and energy dissipation of the waves owing to interaction with the current boundary layer.

Sediment suspended by waves can be carried by steady tidal or other currents, even though they may not, by themselves, exceed the threshold for movement. Prediction of sediment transport rates therefore depends upon understanding the interaction of currents and waves over rough bottoms, with particular regard to velocity distributions and bottom shear stresses (Soulsby *et al.* 1993; Arnskov *et al.* 1993; Bender and Wong 1993). Several workers have provided expressions to predict suspended sediment concentrations and transport by combined flows (Glenn and Grant 1987; Watanabe and Isobe 1990; Soulsby 1991). Some data have been collected on combined flow transport in the field (Whitehouse 1992). For example, it has been estimated that wave–current interactions on a dissipative beach increased the suspended sediment load and longshore sediment transport by 50–60 per cent over transport owing to waves alone (Beach and Sternberg 1992). Much more data are required, however, before predictive transport rate equations can be confidently incorporated into mathematical models.

Vortex formation and shedding over asymmetric, combined flow ripples exert a strong influence over suspended sediment transport. Wave and current motions act in the same direction on the forward stroke, creating a very strong vortex (A) on the lee slope of ripples (Fig. 3.13). As the current reverses, these vortices are carried upwards from the crest high into the flow. The combined velocity is less on the backward stroke, but smaller vortices may develop on the stoss slope (B) if the orbital velocity of the waves is greater than the velocity of the current. These vortices are then lifted into the flow by the next forward stroke (Dyer 1986). As the lee vortex is carried higher into the flow than the smaller stoss vortex, it can travel further downstream; the sand suspended in it, however, may be deposited during the weaker backstroke. The sedimentary effects of wave–current interactions are therefore quite complex, and in some cases, fine suspended sediment is transported offshore while coarser bed load is carried onshore. There is an important difference in the effect of waves that are able to reverse current flows and those that cannot. When currents are reversed, vortices can be generated on both strokes of the wave oscillation, whereas in the absence of flow reversal, a single strong vortex develops that cannot be ejected into the flow in the manner previously described.

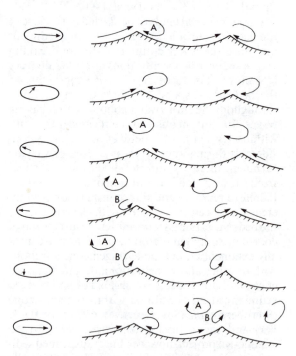

Fig. 3.13. Vortex generation, separation, and advection over ripples under the combined effect of waves and currents. The rotating vector on the left shows the wave phase (Dyer 1986)

LONGSHORE SEDIMENT TRANSPORT

Shore-parallel movement of sand and other beach material (longshore or alongshore transport, littoral drift, littoral transport) creates barrier spits, barrier beaches, and other coastal formations; it helps to determine beach configuration and the character of tidal inlets; and it is an important component of beach sediment budgets. It causes sand to accumulate on the upstream side of headlands, estuary mouths, groynes, and other natural and human obstructions, and this often leads to severe erosion on the downdrift side. The amount of sediment moving alongshore can be enormous, ranging up to more than 1 million m³ yr⁻¹ in areas of high transport.

The longshore transport rate varies with the type of breaking wave, although the relationship is still not clear (Ozhan 1983):

a) Longshore transport by spilling breakers tends to occur when orbital motion stirs up the sediment and places it in suspension, so that longshore currents can move it.

b) Plunging breakers play a similar role, except that the sand is raised from the bottom by turbulence and vortex motion of the breaking waves, which extend down to the bed.

c) Surging breakers, which generate strong uprush and downrush, probably move sand most effectively alongshore in the swash zone. This occurs when the waves break at an angle to the shoreline, causing the uprush to travel up the beach at an angle to the shore, whereas the downrush returns along a path that is more normal to the shore. Sediment therefore travels along a saw-tooth path in the swash zone.

The velocity of longshore sediment transport depends upon the mode of transport. Suspended sediment moves at about the same velocity as the longshore current, but bed load velocities are far lower. Many expressions have been developed to predict longshore transport, but the methodology is still fairly crude and is probably capable only of providing order-of-magnitude estimates (Sleath 1984; Davies 1984; Horikawa 1988; Hardisty 1990*a*). The expressions and models are largely based on field and laboratory data that are of poor quality, and the lack of reliable data inhibits the development of improved methods of prediction. Uncertainty over the thickness of the mobile layer is one of the greatest problems arising from the use of radioactive or fluorescent tracers to determine longshore transport rates, and the use of these data to verify the reliability of predictive equations (White and Inman 1989*b*). The volume transport rate (Q_s) can be estimated using:

$$Q_s = u_1 X_b b$$

where u_1 is the mean tracer advection velocity, X_b is the width of the surf zone, and b is the thickness of the layer of sand moving along the beach.

Estimates of the thickness of the moving sediment layer have generally fallen between 0 m and 0.24 m although values up to about 0.56 m have been reported (Drapeau *et al.* 1990; Jackson and Nordstrom 1993). The depth of mixing has been related to wave height. In a low-energy, reflective beach environment, for example, the ratio, averaged across the surf zone, varied between 0.16 and 0.30 (Sherman *et al.* 1994). The average depth of mixing in the surf zone has also been related to the Shields parameter (Sunamura and Kraus 1985):

$$b = 81.4D\,(\Theta_b - \Theta_{cr})$$

where Θ_b is the Shields parameter at the breakpoint.

Discrepancies in reported mixing depths partly reflect the different methods that have been used to determine the thickness of the moving layer. These range from simple observation of the depth of penetration of a tracer, to objective, semi-empirical estimators. Discrepancies can also reflect differences in the depth of tracer burial with its position within the breaker, surf, and swash zones. Depths of disturbance may be greater for steep than gentle beaches (Jackson and Nordstrom 1993). Much greater depths of mixing also occur as a result of deep scour associated with the migration of megaripples (D. J. Sherman *et al.* 1993*a*).

Komar and Inman (1970) used sand tracers to measure longshore transport rates, and to assess the applicability of two predictive models. The first model assumes that the instantaneous immersed weight longshore transport rate is proportional to the longshore component of the wave energy flux:

$$I_t = K\,(ECn)_b \sin\alpha_b \cos\alpha_b$$

where I_t is the underwater weight of material moved alongshore per second, E is the wave energy, Cn is the wave group celerity, and K is a dimensionless coefficient. The subscript b denotes conditions at the breakpoint. The immersed weight longshore transport rate is related to the volume transport rate by:

$$I_t = (\rho_s - \rho) \, ga'Q_s$$

where a' is a correction factor for pore space, such that $a'Q_s$ is the volume of solid sand without any pore spaces. The value of a' is usually assumed to be 0.6, although it can range between ±17 percent of this value (Bodge and Kraus 1991). The model can be used to predict the saw-toothed movement of sediment by oblique uprush in the swash zone, as well as by longshore currents. It successfully predicted longshore sand transport rates using values of $K = 0.77$, but the average value was found to be 1.71 in a field study in southern California (Sternberg *et al.* 1989).

The second model is based on the assumption that longshore currents are responsible for the net transport of sediment initially suspended and placed into motion by waves:

$$I_t = K' \, (ECn)_b \cos \alpha_b \frac{V_1}{u_m}$$

where K' is a dimensionless proportionality factor, V_1 is the average longshore current velocity in the surf zone (Chapter 2), and u_m is the maximum horizontal wave orbital velocity near the bottom under the breaking waves. The best fit between model predictions and longshore transport rate data was obtained with $K' = 0.28$, although other studies have found values ranging between 0.18 and 0.32 (Kraus *et al.* 1982; Sternberg *et al.* 1989).

Theory suggests that K' is not a constant, but varies with u_m/w_s, beach slope, longshore current strength, and breaker angle. It also suggests that K is a complex function of beach slope, incident wave conditions, and sediment size. Although $K = 0.77$ may be satisfactory for most sandy oceanic beaches, lower values may be more appropriate in lower energy environments, and on coarse clastic beaches (Komar 1980b). Experimental evidence of a relationship between the value of K and the surf similarity parameter (Chapter 2), or a similar coefficient, has been confirmed in the field (White and Inman 1989a; Bodge and Kraus 1991). This suggests that longshore transport is greater for collapsing and plunging breakers than for spilling breakers, for a given level of longshore energy flux.

Few studies have been made of the effect of grain size on longshore transport rates. Several workers have indicated, however, that there is an inverse relationship between K and grain size, which suggests that the finest grains are transported most rapidly alongshore (Nicholls and

Wright 1991; Walker *et al.* 1991; Watanabe 1992). Laboratory investigations have also indicated that longshore sediment transport increases with decreasing grain size, although the effect is quite slight (Kamphuis 1990a). On the other hand, mathematical modelling suggested that the longshore transport rate increases with sand grain diameter (Hallermeier 1982a).

The relative longshore transport rates of fine- and coarse-grained sediment may depend on the mode of transport and the morphology of the beach, although the effect of cross-shore variations in grain size must also be considered. Several studies have indicated that the finest grains move most rapidly alongshore on beaches with wide surf zones (Duane and James 1980). Larger grains ($D_{mm} = 1.19$) moved more rapidly alongshore on a swash-dominated beach, however, possibly because they remained near the breaker zone, where there were strong longshore currents, whereas finer grains ($D_{mm} = 0.3$) were swept up the beachface (Komar 1977). Maximum longshore transport rates were also significantly higher for coarse than fine sand in another study. Nevertheless, total sand transport rates were similar because the zone of faster, coarse sand transport near the breakers was much narrower than the zone of slower, fine sand transport higher up the beach (Quick and Ametepe 1991). If longshore transport rates increase with beach slope, the tendency for coarse grains to form steep beaches could result in their being carried alongshore more rapidly than finer grains on more gentle beaches. The cross-shore processes responsible for beach slope may therefore have a strong influence on rates of longshore transport.

Although theory suggested that beach gradient has no influence on the longshore transport rate (Komar and Inman 1970), there is increasing evidence to the contrary. The tendency for longshore transport to be greatest with collapsing and plunging breakers, which are associated with steep slopes and bars, suggests that transport rates increase with beach gradient. This is supported by theoretical modelling and by laboratory and field data (Bodge 1989; Kamphuis 1990a; Quick and Ametepe 1991). More-efficient longshore transport on steep beaches with narrow surf zones could reflect the greater stress and turbulence generated on the bed by plunging breakers, and greater bottom dissipation per unit area of the surf zone. An equation for the longshore sediment transport rate, based upon laboratory and field data, considers the effect of beach slope and particle size (Kamphuis 1991*a*):

$$Q = 2.27 \, H_{sb}^2 \, T_p^{1.5} \, \beta^{0.75} \, D_{50}^{-0.25} \sin^{0.6}(2\alpha_b)$$

where Q is the alongshore sediment transport rate in kilograms of immersed mass per second, H_{sb} is the breaking significant wave height, T_p is the peak period of the offshore wave spectrum, and β is the beach slope in the breaker zone. The expression suggests that the transport rate increases with beach slope, and that although there is an inverse relationship with grain size, the effect is quite small.

At least fifteen models have been proposed to describe the distribution of longshore sediment transport across the surf zone (Bodge 1989). Most are based on the assumption that sediment is locally mobilized by bed shear stresses generated by peak wave orbital velocities; energy dissipated from breaking waves; or by a combination of orbital velocities and longshore currents. It is then assumed that longshore currents move the mobilized sediment downdrift. Most theoretical models predict maximum longshore transport rates between the breakers and the middle of the surf zone on planar beaches (Bodge 1989; Watanabe 1992). The profiles of few natural beaches are planar, however, and many tend to be concave-upwards. Theory suggests that the maximum transport rate and the total longshore transport of sediment are lower on concave than on planar beaches. Transport maxima migrate towards the shoreline as beaches steepen, and towards the breakpoint as slopes decrease. Longshore currents and sediment transport maxima are located closer to shore than to the breaker zone on concave beaches, in contrast to most predictions for planar beaches (McDougal and Hudspeth 1984; Bodge 1988).

Field and laboratory studies suggest that the distribution of longshore sediment transport is much more complex than predicted by theoretical models. Most models predict no movement seawards of the breakers or at the shoreline, and few consider longshore transport in the swash zone. It has been estimated, however, that at least 5 percent to over 60 percent of the total longshore transport occurs within the swash zone—with the higher values corresponding to the occurrence of collapsing breakers, and the lower values to spilling breakers; and about 10 percent to 30 percent outside the breaker line (Bodge 1989). There are great variations in the shape of measured longshore

transport distribution profiles, and maximum transport is as likely to occur shorewards as seawards of the mid-surf position (Downing 1984; Fulford 1987; Sternberg *et al.* 1989; Rosati *et al.* 1991). Experiments conducted on a beach over a period of only three days recorded unimodal distributions with a prominent peak in the outer surf zone; bimodal distributions with peaks in the swash and outer surf zones; flat distributions with no peaks; and distributions with a broad peak in the inner surf zone (Kraus and Dean 1987).

Maximum longshore transport rates do not necessarily coincide with the location of peak longshore current velocities. Bimodal distributions, usually with transport maxima near the breakpoint and in the inner surf and swash zones, have been predicted in mathematical models, and identified in the laboratory and field (Katoh *et al.* 1984; White and Inman 1989*a*; Morfett 1991). Bodge and Dean (1987) also found that longshore transport was generally bimodally distributed, with maxima near the breakers and at the shoreline—their relative significance changing from dominance of the former to the latter with the transition from spilling to plunging to collapsing breakers. Bimodal distributions have been observed in wave flumes, although the peaks tend to merge together with decreasing wave height and grain size (Kamphuis 1991*b*).

Rates of longshore transport can also increase across the crest of bars, where the water is shallower and more turbulent, and there are strong longshore currents, and diminish in the deeper waters of the adjacent troughs (Deigaard *et al.* 1986; Bodge 1989). Nevertheless, longshore transport can be greater in the troughs if they contain strong, tidally induced longshore currents, and water deep enough over bar crests to prevent breaking.

CROSS-SHORE TRANSPORT

If wave-induced flows were sinusoidal in shallow water, the amount of sediment carried onshore over a horizontal bed during one half of the cycle would be equal to the amount moved offshore during the other half. Waves move sediment in the nearshore zone, however, because they are not purely sinusoidal, nor are the beds usually horizontal.

Cornish (1898) recognized that sand can be moved back and fore by onshore and offshore oscillatory currents, whereas only the higher onshore velocities may be capable of moving coarse clasts. Cornaglia developed a similar hypothesis in a number of papers published in the late nineteenth century, although, in addition to wave asymmetry, he also emphasized the effect of gravity on threshold velocities and the movement of bed load over sloping bottoms (Cornaglia 1889; Zenkovitch 1967; Komar 1976). Two zones were distinguished. The first is located some distance from the shoreline, where the water is shallow enough for the movement of fine bottom material by waves, but deep enough for essentially symmetrical bottom velocities. The second zone is in the shallower water nearer the coast, where onshore flows are stronger but of shorter duration than seaward flows (Chapter 2). In the first zone, the effect of gravity on grain thresholds causes them to be carried further downslope in one half-wave cycle than they are moved upslope during the other half of the cycle (Fig. 3.14a). Grains on sloping bottoms will therefore be carried seawards in this deeper water zone. Closer to shore the thresholds for some coarse material may be attained only by strong onshore flows, and it can therefore only be carried landwards. Less coarse material can be swept back

and fore by the onshore and offshore currents, but despite lower thresholds for downslope movement, stronger landward flows result in net onshore transport. Net transport of still finer material, however, which has lower thresholds of movement, may be offshore (Fig. 3.14b). For each size of wave and each size of particle, there must be some point, seaward of the breakers, where the gravitational and wave-induced forces are in balance. When a grain is at this point, which Cornaglia called the 'neutral line' ('neutral point'), it moves an equal distance shorewards and landwards under passing waves, and retains the same general position. He proposed that the position of the neutral line moves offshore with increasing wave size and increasing bottom slope, and with decreasing size and specific weight of the sediment. Although the presence of a neutral line or null point outside the breakers has been confirmed in the laboratory (Ippen and Eagleson 1955), it has not been located in the field. Assuming that the neutral line concept is valid, the difficulty in locating it in the field could reflect the lack of regularity in the incoming waves, which would cause it to vary considerably in position (Dyer 1986).

Classical null point models consider the effect of wave asymmetry and gravity on the movement of bed load. Asymmetrical threshold models are not

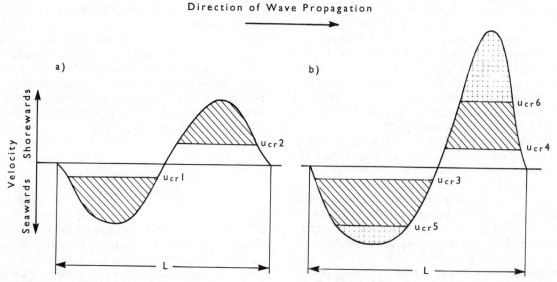

Fig. 3.14. Cornaglia's theory showing the net movement of sediment according to grain size. In (a), u_{cr1} and u_{cr2} are the velocity thresholds for seaward and landward grain movement, respectively. u_{cr1} is lower than u_{cr2} because of the effects of gravity on a sloping bed. In (b), u_{cr3} and u_{cr4} are the seaward and landward thresholds for fine grains, and u_{cr5} and u_{cr6} are the corresponding thresholds for coarser grains

concerned with the gravitational effect of sloping beds, however, which is considered to be negligible on very gentle slopes, especially if they are rippled. Bowen's (1980) null point model indicates that gravity has a more important effect on suspended load than bed load. It suggests that there is a strong tendency for onshore movement of bed load, together with suspended load coarser than the equilibrium size, and offshore transport of finer material. This conclusion is supported by the results of other models, and by experimental data (Chappell 1980a; Bailard 1981; Sawaragi and Deguchi 1980; Leont'ev 1985; Horn 1992; Liu and Zarillo 1993). The tidal version of Horn's (1992) asymmetrical threshold bed load model suggests that grain sizes range between the largest and smallest threshold diameters on the landward portion of beaches, whereas only grains with diameters less than the smallest threshold diameter occur on the seaward portion. Field observations support her conclusion that the degree of sorting increases seawards—macrotidal beaches tending to have fine, well-sorted sediments on the low tidal terrace, and coarser, more poorly sorted sediments landwards. Although some null point models suggest that fine grains are transported landwards, and coarse grains seawards (Ippen and Eagleson 1955), most field evidence shows that the opposite usually occurs, including reports of iron, chalk ballast, and other heavy material washing ashore from wrecks in fairly deep water.

Net cross-shore transport rates must be known reliably to predict changes in beach topography, and more fully to understand longshore transport—which may be related to cross-shore transport through the effects of sediment sorting and entraining mechanisms. Unlike longshore transport, however, cross-shore movement of sediment is driven by reversing, oscillatory flows, and it cannot be easily measured using tracers or traps. Net sediment transport also results from small differences in the sometimes very large amounts of sediment moving on- and offshore. Reliable prediction of net transport rates must therefore be based upon accurate prediction of transport directions and instantaneous transport rates, as well as the effect of various bedforms upon them. The mode of sediment movement, which has an important influence on cross-shore transport rates, must also be identified. Factors responsible for net cross-shore sediment transport include: the asymmetric profile of shoaling waves, with resulting differences in onshore–offshore directed velocities, accelerations, and shear stresses; the asymmetrical shape of combined flow sand ripples; crest–trough asymmetry in fluid–granular boundary layer development (Conley and Inman 1992); bottom slope; mass transport, undertow, rip currents, and other wave induced nearshore currents; tidal-, wave-, and wind-induced changes in water level; and standing and long-period waves. Forced, long waves of infragravity frequency are an effective mechanism for the offshore transport of bed load and suspended load. This is because higher wave sets in long wave troughs move more sand than lower wave sets on the crests, and the net seaward drift in the troughs of long waves carries it offshore (Shi and Larsen 1983/4) (Fig. 2.12). Numerical beach change models do not generally consider the effect of long waves, however, because of the difficulty in selecting suitable forms of long wave motion, and the lack of data with which to test its influence.

Numerous attempts have been made to derive algebraic and graphical criteria to predict beach erosion or accretion under given wave conditions (Kobayashi 1988; Kraus *et al.* 1991; Sunamura and Takeda 1993). For example, Sunamura (1982—see Horikawa 1988) plotted the Ursell parameter (U_r) against a flow intensity parameter I_p, where:

$$I_p = \frac{\rho \, (d_o \, \omega)^2}{gD \, (\rho_s - \rho)}.$$

No motion occurs if I_p is less than 17. Sediment is carried offshore if I_p is greater than 17 and greater or equal to $0.048 U_r^{1.5}$, and onshore if I_p is greater than 17 and less than $0.048 U_r^{1.5}$.

The predictive value of criteria expressed in terms of simple wave and sediment parameters has been questioned, partly because most were formulated from laboratory data using monochromatic waves. They are also only concerned with the effect of short-period wind waves, and they fail to consider the effect of other factors, including infragravity waves, mean currents, and changes in grain size. Seymour and Castel (1989) tested the ability of six models to predict the direction of wave-driven, cross-shore transport, using field data from southern California and the Atlantic coast of the USA (Table 3.5). They found that the models only correctly predicted the point at which the direction of net cross-shore transport changes between about two-thirds and one-half of the time. It was therefore concluded that the models have only 'barely useful predictive capabilities'. Kraus et al. (1991), however, came to a different conclusion after evaluating eight criteria. They found that the predictive capabilities of criteria based upon two parameters are good, and they are still reasonable for some criteria based on only one parameter. The parameters $N_o = H_o/w_n T$ and $F_o = w_n/(gH_o)^{0.5}$, which are related to the mobilization and suspension of sediment, were suitable alone, or in combination, to predict the cross-shore direction of sediment movement.

Many equations predict rates of cross-shore sediment transport (Sleath 1984; Allen 1988; Horikawa 1988; van Rijn 1989; Hardisty 1990a; Fredsøe 1993) (Table 3.6). Some apply only to bed load, some to suspended load, and some to total load; some consider the occurrence and effect of bedforms, while others assume a plane bed. Many equations are basically similar, however, and the lack of reliable field data on transport rates, particularly of bed load, makes it difficult to determine which is the most reliable. Three basic groups of equations can be distinguished, although it is questionable whether any provide more than a rough estimate of transport rates under a wide range of conditions (Dyer 1986; King and Seymour 1989):

a) Empirical relationships have been determined between transport rates and a characteristic flow variable based upon data obtained from flumes (Shibayama and Horikawa 1982; Hallermeier 1982b).

b) Theoretical equations are based on the physics of grain movement, although they contain constants that must be determined from data. Many of these equations were derived by modifying and adapting equations for steady, undirectional stream flow to coastal conditions (Bowen 1980; Bailard 1981; Nairn and Southgate 1993).

c) The third group of equations is the result of grouping the flow and sedimentary variables together as dimensionless numbers, and using flume data to determine the resulting constants and coefficients (Yalin 1963).

In Bagnold's (1963, 1966) oscillatory flow model, which is one of the most frequently used, the transport rate is proportional to the energy dissipated per unit of bottom area. No allowance was made for threshold conditions, however, and it was assumed that movement takes place at all flow velocities. The total immersed weight sediment transport rate across unit width of the bed (i_t) for zero slope is given by:

$$i_t = W(K_b + K_s)$$

where:

$$W = \tau_o u_m = \rho\, C_D\, u_m^3,$$

Table 3.5. Model Predictors of onshore–offshore transport

Model	Predictor	Onshore	Offshore
Dean (1973)	$2R\dfrac{H_b}{w_s T}$	< 1	> 1
Short (1978)	H	< 120 cm	> 120 cm
Short (1978)	$\left[\dfrac{\rho g^2}{16\pi}\right] H^2 T$	< 30 Kw m^{-1}	> 30 Kw m^{-1}
Hattori and Kawamata (1980)	$2\dfrac{H\tan\beta}{w_s T}$	< 0.5	> 0.5
Quick and Har (1985)	$\left[\dfrac{H}{w_s T}\right]_{initial}$	$>\left[\dfrac{H}{w_s T}\right]_{final}$	$<\left[\dfrac{H}{w_s T}\right]_{final}$
Sunamura and Horikawa (1974)	$\dfrac{1.845}{g^{0.33}}\dfrac{(\tan\beta)^{0.27}}{(TD)^{0.67}} H$	< 4	> 8

Note: R is an arbitrary constant.

Source: Seymour and Castel 1989.

Table 3.6. Some sand transport equations (cm^3 cm^{-1} s^{-1})

Source	Equation
Madsen and Grant (1976)	$\dfrac{\bar{q}}{w_s D} = 12.5\Theta^3$
Sleath (1978)	$\dfrac{\bar{q}}{\omega D^2} = 47(\Theta - \Theta_{cr})^{1.5}$
Vincent et al. (1981), Hardisty (1983)*	$\bar{q}_b = A(u^2 - u_{cr}^2)u$
Williams et al. (1989b)*	$\bar{q}_b = A(\iota_0^2 - \iota_{cr}^2)u$
Bowen (1980)*	$q_s(t) = \dfrac{A\,C_D\,\rho\,u^3\,\lvert u\rvert}{w_s - u\tan\beta}$
	$q_b(t) = \dfrac{B\,C_D\,\rho\,u^3}{\tan\phi - (u\tan\beta/\lvert u\rvert)}$
Hallermeir (1982b)	$\dfrac{q_{net}b}{\omega D^2} = A\Theta'^{1.5}$,
where	$\Theta' = \dfrac{\rho u_m^2}{(\rho_s - \rho)Dg}$
Watanabe et al. (1980)	$\dfrac{q_{net}}{w_s D} = A(\Theta - \Theta_{cr})$
Shibayama and Horikawa (1982)	$\dfrac{q_{net}}{w_s D} = 19\,\Theta^3$ with no ripples
	$\dfrac{q_{net}}{w_s D} = 19A\,\Theta^3$ with ripples
Yamashita et al. (1984—see Horikawa 1988)	$\dfrac{q_{net}}{w_s D} = 2.2\,(u_*/w_n)^3$ under sheet flow conditions
Dyer (1986)*	$i_s = \dfrac{k\,A\,\rho_s\,u_*^3\,z_0}{(k\,u_* - w_s)^2}\dfrac{\rho u_*^2 - \tau_{cr}}{\tau_{cr}}$

Note: * = dimensionless equations.

q refers to the volume transport rate and i to the immersed weight; q_{net} is the net transport rate over a wave cycle; $\bar{q}$ is the average sediment transport rate in a half cycle; $q(t)$ is the instantaneous transport rate; A and B are empirical coefficients; and subscripts s and b refer to suspended and bed load, respectively.

Source: Partly from Horikawa 1988.

$$K_b = \frac{e_b}{\tan\phi - \tan\beta},$$

and:

$$K_s = \frac{e_s(1 - e_b)}{(w_n/u) - \beta}$$

where u is the mean flow velocity; W is the rate of energy dissipation by friction; and e_b and e_s are efficiency factors with values less than 1. The efficiency values depend upon such factors as grain size, and whether wave conditions are, for example, laminar or turbulent, or steady or unsteady. Bagnold (1966) suggested that $e_b = 0.15\tan\phi$ and $e_s = 0.01$ for stream flow, whereas Bailard (1982) proposed $e_b = 0.21$ and $e_s = 0.025$ for the long-shore current, and $e_b = 0.10$ and $e_s = 0.020$ for cross-shore flow. Nairn (1991) found that these constants provide reasonable estimates of cross-shore sediment transport rates under a wide range of conditions.

For the movement of sediment over a sloping bottom, Bagnold's equations can be represented by (Hardisty *et al.* 1984):

$$i_{net}\,(in) = \frac{C_1\,u_{in}^3\,Tf_{in}}{\tan\phi + \tan\beta}$$

$$i_{net}\,(ex) = \frac{C_2\,u_{ex}^3\,Tf_{ex}}{\tan\phi - \tan\beta}$$

where i_{net} is the net transport, Tf is the duration of the seaward and landward flows, C_1 and C_2 are empirical constants that vary with grain size, and subscripts 'in' and 'ex' refer to shorewards and sea-wards flow, respectively. Hardisty *et al.* calibrated Bagnold's beach equations in the field, using sand traps to measure the movement of bed load by uprush and backrush in the swash zone. They found that there is no significant difference between C_1 and C_2, and that the mean value of 12.78 kg m^{-4} s^{-2} is similar to that obtained in the laboratory. King (1991) proposed that any bed load transport model for a flat bed can be used to predict movement on a sloping bed, according to:

Transport on a slope =

(transport over flat bed) $\times \dfrac{\tan 30°}{\tan 30° \pm \tan\beta}$

where the plus and minus signs refer to upslope and downslope movement, respectively.

Most transport models suggest that the rate of sediment transport is a function of the fluid velocity, although there is considerable disagreement about the precise nature of this relationship. The transport rate is related to the sixth power of the velocity in the models of Madsen and Grant (1976) and Shibayama and Horikawa (1982), for example; the fourth power in the models of Sleath (1978) and Hallermeier (1982b); the third power in the model of Kobayashi (1982); and between the third and fourth power in Bagnold's (1963) model.

Most models are also based upon data obtained from experiments using regular or monochromatic waves in flumes, or from field measurement under calm wave conditions. There is some evidence, however, to suggest that irregular waves are less efficient in moving sediment in the field than regular waves (Horikawa 1988). Furthermore, large waves and strong currents move most sediment in shallow water during storms, when there are rapid changes to nearshore bottom topography. High concentrations also extended further seawards and higher in the water column during storms.

Estimates of the suspended sediment transport rate require precise information on the vertical distribution of horizontal velocities and sediment concentrations. The direction and strength of sand movement varies throughout the water column, and the total or resultant movement must be determined by measuring along the whole profile, rather than at a few levels. There must therefore be some reservation about transport models that use depth averaged terms, rather than considering the contribution of each distinct component of fluid motion. This is particularly true of sediment transport over rippled beds. For example, Vincent and Green (1990) found that onshore-skewed, peak wave-induced flows were responsible for strong shoreward transport of suspended sediment near the bed, in about 1.7 m of water, seaward of the surf zone. Although there was also weak, onshore movement at greater heights above the bed, the shedding of sand-laden vortices carried sediment offshore at elevations between 50 and 150 mm above the bed.

Few transport equations are applicable to the swash zone, although sediment transport rates are known to be high. Ogawa and Shuto (1982, see Horikawa 1988) identified three types of cross-shore sediment transport in the swash zone—by bore-like waves moving up a dry bed (I) or through shallow water (II), and transport by receding waves (III). Field measurements showed that the relationship between the nondimensional sediment transport rate (S_t) and the Mobility number is in the form:

$$S_t = A\,\mathbf{M}^n$$

for:

$$S_t = \frac{a'\,q}{w_s\,D}$$

where q is the volumetric sediment transport rate, and A is an empirical coefficient. For transport types I and III, $n = 1.5$, and for type II, $n = 1.0$.

The formulation of reliable cross-shore transport models must await the acquisition of more, high-quality field data, but the situation is improving as a result of a number of well instrumented, interdisciplinary, and inter-institutional programmes. They include NSTS, DUCK82, DUCK85, SUPERDUCK, and DELILAH in the USA, C²S² and C-COAST in Canada, NERC in Japan, and B-BAND in Britain. These programmes are demonstrating the frequency dependent nature of cross-shore sediment transport in the nearshore zone, and they are helping to identify spatial (horizontal and vertical) and temporal variations in the absolute and relative contributions of quasi-steady mean and oscillatory currents.

A number of studies have been conducted on barred and non-barred coasts in eastern Canada and in the Great Lakes under the C²S² and C-COAST programmes (Greenwood et al. 1990a, b). They have reported strong onshore transport of suspended sediment by incident waves, and weaker offshore transport by long period, infragravity motion (Huntley and Hanes 1987; Doering and Bowen 1988). Greenwood et al. (1991) also found that net oscillatory transport is usually directed onshore outside the surf zone, although long wave domination can drive sediment offshore. At the breakers, wind waves continue to drive sediment landwards when vortex ripples are present, but offshore transport can take place if increasing velocities near the bed cause ripple steepness to decrease and decouple vortex-shedding. Mean currents in this study were the result of a set-up driven undertow. They were directed offshore and increased in magnitude at the breakers, where there was also an increase in the average suspended sediment concentration. Net cross-shore transport of suspended sediment therefore depends upon the relative contribution of wind waves, group-bound long waves, and time-averaged mean currents. Net transport can have a distinct vertical structure as a result of variations in the contribution of these transport components with distance above the bed.

Further evidence of the contribution of temporally and spatially variable oscillatory and quasi-steady components to cross-shore suspended sediment transport has been obtained from a non-barred shoreface in Nova Scotia, during a storm event. Sediment was transported by offshore-directed mean currents or undertow, and by oscillatory currents at wind and swell wave frequencies and group bound long waves at low frequencies (Osborne *et al.* 1990; Osborne and Greenwood 1992*a*). As noted by Huntley and Hanes (1987), primarily onshore transport, related to the passage of skewed, asymmetrical incident waves, was much stronger than offshore transport by oscillatory currents generated by wave groups at lower frequencies. There was a distinct vertical structure in the near-bed water column under shoaling waves, with oscillatory currents driving net onshore transport near the bed and higher in the water column, and mean currents causing net offshore transport in the middle portions of the lower water column. There were also spatial variations in suspended sediment transport. Net onshore transport at wind wave frequencies was at a maximum just seaward of the breakers and it decreased onshore, while net offshore transport occurred in the seaward portion of the shoaling zone and landward of the point of initial wave breaking, where there were fairly strong mean currents and weak oscillatory currents. Erosion and accretion patterns, and local time-averaged total and net sediment volume flux, suggested that there was almost a balance in sediment transport for the complete storm event.

A similar study has been conducted on a barred shoreface on Lake Huron (Osborne and Greenwood 1992*b*). Undertows and oscillatory currents at various frequencies were again of great importance, although sediment transport was also strongly influenced by other types of low-frequency wave, which may have been edge waves or reflected long waves. On the lakeside slope of the bar, offshore sediment transport by group bound, low frequency oscillatory currents (< 0.05 Hz) was often equal to the high rates of transport associated with oscillatory currents at wind wave frequencies. The additional contribution of mean currents near the bed therefore resulted in net offshore transport across the slope. On the crest of the bar, there was a rough balance between onshore net oscillatory transport resulting from the interaction of high-and low-frequency waves, and offshore directed mean current transport. Sediment transport by wind waves decreased in the deeper water on the landward and lakeward side of the bar crest, where net sediment transport may have been dominated by either long waves or undertows.

Net suspended sediment transport was spatially and temporally very variable across a barred surf zone during a storm on Prince Edward Island. The amount of sediment transported by mean and oscillatory currents was almost equal when the total surf zone was averaged over the storm event. Near infragravity waves were more important during the waning period of the storm, however, whereas incident waves were important during the early stages, when the waves were not saturated. Incident waves were also important throughout the storm on the seaward slope of the bar, where wave saturation occurred for only a very short period. Mean currents carried sediment offshore in the inner surf zone, but onshore transport, probably related to mass transport induced by surf bores, or drift velocities induced by the infragravity wave boundary layer, occurred further seawards. Oscillatory currents at wind wave frequencies usually carried sediment onshore during the storm, but transport direction and magnitude varied at low frequencies, partly in response to changes in the infragravity wave field (Aagaard and Greenwood 1994).

The relative importance of the various transport components have also been investigated on macrotidal beaches in Britain, under the B-BAND programme (Russell *et al.* 1991; Davidson *et al.* 1992, 1993). In Wales, net offshore suspended sediment transport resulted from mean offshore transport coupled with the undertow, and infragravity oscillatory flows, which dominated the hydrodynamics of the inner surf zone. Offshore directed infragravity components also dominated transport processes in the inner surf zone in Yorkshire, but onshore directed, incident wave transport dominated seawards of the surf zone. Although these studies showed that hydrodynamic processes on macrotidal beaches are similar in many ways to those on meso- and microtidal beaches, they also identified some important differences. For example, it was found that the amount of sediment suspended on the ebb tide is an order of magnitude greater than on the flood, probably because of beach de-watering during the ebb, and a time lag in the response of bedforms to changing hydrodynamic conditions.

There was strong onshore sediment transport by incident wind waves, and almost equally strong offshore movement by infragravity waves, during a period of low energy conditions in the outer surf zone of a swell-dominated beach in North Carolina (Beach and Sternberg 1987, 1991). Low-frequency motions, however, dominated the inner surf zone of a high-energy, dissipative beach in Oregon. Although net transport was directed onshore throughout the incident frequency band, it was at much lower levels than in the infragravity band. Much greater amounts of sediment were transported in the swash zone of a high-energy, dissipative beach in Washington. Transport in the incident band was directed onshore, but the quantities were much less than in the infragravity band. The direction of sediment movement in the infragravity band was both frequency and depth dependent. At 5 cm off the bed, the steady component moved a little more sediment seawards than the fluctuating components moved it landwards, and net transport was therefore slightly seawards. Both components were directed seawards 2.8 cm from the bed, however, resulting in strong net offshore movement. Sediment transport on the three beaches indicated that wind waves primarily move sediment onshore, whereas long waves can move it onshore or offshore, depending upon the relative phase between flow velocities and sediment concentrations. Net sediment flux in the swash and inner surf zones is therefore determined by the relative strength of the fluctuating (wind waves and infragravity) and steady (mean) components, which, although of comparable magnitude, may act in different directions.

Cross-shore transport modes have been studied on the inner shelf, in depths of 7 to 17 m, off the mid-Atlantic coast of the United States (Wright *et al.* 1991; Wright 1993). Although mean flows, incident and infragravity waves, and gravity help to move sediment, the transport direction was often contrary to expectations. Wind-driven downwelling flows dominated during a storm, and wind shear and tidally induced mean flows also made a significant contribution to onshore and offshore sediment transport during fairweather and moderate energy conditions. Oscillatory flows associated with surface gravity waves were the most important reasons for bed agitation and sediment suspension, but they were also responsible for net landward and seaward sediment advection. Low-frequency waves made secondary contributions to cross-shore transport, which was directed shorewards as often as seawards.

4 The Beach and Nearshore Zones

About one-fifth of the world's coasts consists of sandy beaches backed by beach ridges, dunes, or barriers (Bird 1985b). Beaches develop wherever there is available sediment and a suitable site for its accumulation. They are the most widely distributed coastal depositional features, and one of the most dynamic.

Although there is no universally accepted definition of the term 'beach', it is generally considered to be an accumulation of unconsolidated sand, coarse clastic material, or other sediment extending from the uppermost limit of wave action down to the low tidal level. The backshore, which is usually subaerial, extends from the normal high tidal level to the landward margin of the beach. Although it is only affected by waves during severe storms or exceptionally high tides, it may include the lower dune or cliff under extreme wave conditions. The intertidal foreshore or beachface extends from the crest of the berm, or the upper limit of the uprush at the high tidal level, to the ordinary low tidal level (Fig. 4.1). The nearshore is a zone of undefined width extending seawards from the shoreline to, according to the authority, just beyond, or well beyond, the breaker zone. The term 'beach' is sometimes used in a loose sense to refer to the entire area extending from the landward limit of the backshore to the seaward limit of the nearshore.

BEACH PLAN FORM

There are two main components of the plan form of beaches, their orientation and shape.

Beach Orientation

The type of wave that is responsible for beach orientation depends upon sediment size and the degree of exposure (Lewis 1938; Schou 1952). On the western coast of Ireland, the orientation of wide sandy beaches is closely related to the offshore relief and refraction of the long ocean waves coming from the Atlantic Ocean. The orientation of coarse clastic beaches and ridges, however, reflects the dominant role of storm waves, which are often generated by local winds. These waves are short and little refracted in fairly sheltered areas, and coarse-grained beaches therefore face in the direction of the approaching storm waves (King 1972). In Australia, wide beaches are oriented essentially parallel to the crests of the refracted swell waves (Jennings 1955; Davies 1960). These long waves begin to be refracted well offshore, in considerable depths of water, and the offshore relief is therefore one of the most important factors in determining beach alignment. The effect of wave refraction explains why beaches usually have a curved rather than straight plan form, and greater refraction at the head of bays could account for the relationship between beach curvature and the length and width of inlets. Constructional features are not aligned completely parallel to hardrock coasts, and there is fairly close correspondence between beach orientation and global wave generation patterns. For any coast which has a tendency for beaches to be aligned in a certain direction, there is therefore a marked tendency for them to occur most frequently in locations that allow this alignment to be achieved (Davies 1972).

A distinction has been made between swash- and drift-aligned equilibrium beach forms. A swash-aligned beach is constructed parallel to wave crests, and net longshore transport is at a minimum. Swash-aligned beaches are therefore associated with impeded longshore transport, as on irregular coasts where the important wave trains reach the shoreline almost normally. Drift-aligned beaches

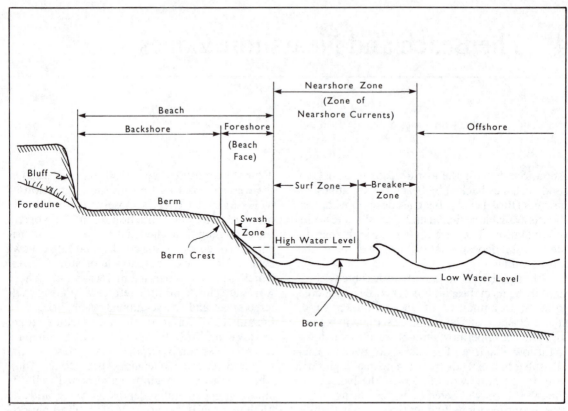

Fig. 4.1. Beach profile nomenclature, using Komar's (1976) definition of the nearshore

are parallel to the line of maximum drift—usually between about 40° to 50° to the direction of wave approach. These beaches occur on coasts of free transport, where sediments can be carried great distances in one direction. Drift-aligned beaches, therefore, develop where the initial coastal outline is fairly regular, or important sediment-moving waves come in at an angle to the coast.

Dynamic beach equilibrium requires a through-put of sediment, as, for example, when a coastal cell is coupled to the mouth of an estuary. Static equilibrium can be attained in several ways (Carter 1988):

a) through swash-alignment, when sediment movement is restricted to cross-shore trans-port;

b) by the longshore current velocity becoming equal to zero owing to cancellation of the longshore currents. This can occur where there is a strong wave height gradient, or where two wave trains interact; and

c) by the alongshore grading of beach sediments in such a way that the critical entrainment threshold is never exceeded.

All three situations are common, and in some cases equilibrium is attained through a combination of options.

Plan Form

Long stretches of some coasts are straight, and a few, especially at the mouths of rivers, have convex shapes. The plan form of most beaches is concave seawards, however, especially where they are bounded by headlands. The planimetric shape of an embayment between two headlands largely depends upon the angle of wave approach. Beaches develop a circular arc plan where wave approach is normal to the coast, and an asymmetrical shape where the waves are oblique (McLean 1967). A variety of terms have been used to describe the plan shape of asymmetrical beaches and bays, including headland-bay, logarithmic spiral, long spiral, half-heart, crenulate, hook-shaped, and zetaform.

The plan form of beaches lying to the lee of headlands and other natural and artificial obstructions is the result of interruptions to longshore transport, dissipation of wave energy by induced turbulence or reflection, and redistribution of wave energy by refraction and diffraction of wave trains. Beaches in the lee of headlands usually have a concave-seawards plan shape, with the radius of curvature progressively decreasing towards the headlands. It has been suggested that beach plan curvature follows a logarithmic spiral law, with the distance from the beach to the centre of the spiral (*r*) increasing with the angle *θ* according to (Yasso 1982):

$$r = e^{\theta \cot \alpha}$$

where *θ* is the angle of rotation, or spiral angle, which determines the tightness of the spiral; and *α*, the logarithmic spiral constant, is the angle between a radius vector and the tangent to the curve at the point—this is a constant for a given log spiral (Fig. 4.2). Although log spirals generally fit the plan form of beaches between headlands quite well, however, there are some exceptions. Log spirals are also more difficult to fit to beaches than parabolic equations, and they fail to describe the fairly straight reach tangential to the downcoast headland (Hsu *et al.* 1989*a*).

Bays that develop between model headlands under oblique wave approach consist of three sections: the curved, nearly circular, section in the lee of the headland, which may be absent in some areas; a logarithmic spiral section; and a nearly linear to curvilinear reach tangential to the downcoast headland. When the beach is in equilibrium, the middle segment is parallel to the crests of the approaching waves, and there is no nourishment from upcoast. The value of the logarithmic spiral constant changes as a straight coast gradually assumes a more curved form. When a beach has attained its final stable or equilibrium outline, the spiral constant is related in a consistent manner to the angle between the two headlands and the incident wave crests—or the tangent to the shoreline, which is parallel to the wave crests. Graphing the headland alignment angle against the logarithmic spiral constant can therefore provide a measure of bay stability, and a further indication is given by

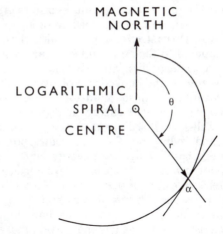

Fig. 4.2. Log spiral nomenclature

the ratio of the depth of the indentation to the distance between the headlands. The time required for bays to attain equilibrium decreases with increasing wave energy levels, although their final equilibrium shape is independent of their size, or the length and period of the waves.

In the field, log spiral beaches have usually been examined after they have attained a state of static equilibrium. The development of a log spiral bay has been recorded on Lake Michigan, however, following construction of a marina (Terpstra and Chrzastowski 1992). As erosion proceeded, the spiral angle decreased from 47° to 31° and the spiral centre moved closer to the origin. The angular extent of the log spiral portion of the shoreline also increased through time, from 126° to 196°. There was a consistently good fit between log spiral curves and measured plan forms throughout the period of development. This suggests that a developing log spiral plan form is in a state of dynamic equilibrium as it expands landwards and downcoast, with successive plan forms fluctuating within a family of log spiral curves that are gradually approaching a static equilibrium configuration.

Changes in wave height and energy along log spiral beaches may cause systematic variations in beach sediments and morphology—except along the tangential section, which is aligned roughly parallel to the refracted wave fronts of the dominant swell. Beaches are finer-grained and more gently sloping in the sheltered portions of log spiral bays in California, and the tendency for beach slope to increase with decreasing wave energy is therefore subordinate to the grain size–beach slope relationship in this area. In south-eastern Australia, however, the sheltered portions of log spiral beaches tend to be steep or reflective, while the portions exposed to the dominant swell and storm waves are gentle or dissipative (Wright 1980). The shadow zone at the hooked end of five Alaskan bays, consisting of mixed sand and coarse clasts, generally had low wave energy, small grain sizes, fairly well-sorted sediment, gentle beachface slopes, and eroding shorefaces. The central portion of the beaches had high wave energy, large grain sizes, poor sorting, moderate beachface slopes, and shorelines transitional between erosion and deposition. Sediment transport was mainly to the tangential end of the bays, which had similar characteristics to the shadow ends, except that the beachface slopes were steep and the shorelines were depositional (Finkelstein 1982) (Fig. 4.3).

Wave refraction can influence the location of stream outlets in bays. Damming of stream mouths by storm waves creates elongated lagoons, and as the water level rises, the streams eventually break out over the lowest part of the berms, cutting channels through them. Therefore, in contrast to drift-aligned beaches, where stream outlets are usually

deflected downdrift, on swash-aligned beaches they are generally where the berms are lowest, in the sheltered lee of headlands. Although outlet deflection to the protected zones of log spiral beaches is most common in swell wave environments, where there are high berms, it also occurs on coarse clastic beaches in storm wave environments (Davies 1972).

The relationship between the direction of beach skew and dominant longshore drift has been used to identify global systems of coastal sediment movement (Silvester 1962). Coasts consisting of a series of log spiral bays have a characteristic appearance, each successive bay being recessed behind its neighbour. Offset coasts are especially common in low latitudes, where depositional features are more continuously distributed than in higher latitudes, and constructional waves have very consistent directions of approach. Although the headlands are usually rocky, offset coasts can be formed entirely in sediments, the headlands usually being located at the mouths of rivers, where fairly large amounts of sediment are being discharged (Davies 1972). Log spiral bays and offset coasts may also develop entirely in rock, the

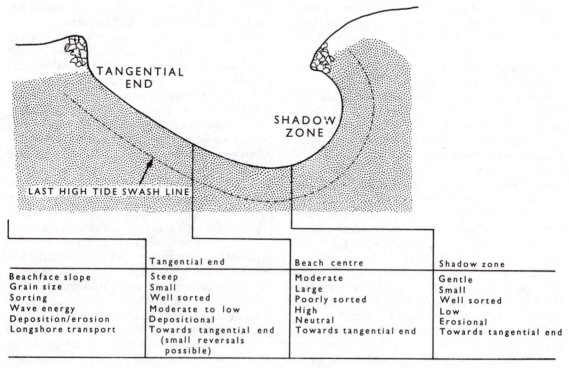

	Tangential end	Beach centre	Shadow zone
Beachface slope	Steep	Moderate	Gentle
Grain size	Small	Large	Small
Sorting	Well sorted	Poorly sorted	Well sorted
Wave energy	Moderate to low	High	Low
Deposition/erosion	Depositional	Neutral	Erosional
Longshore transport	Towards tangential end (small reversals possible)	Towards tangential end	Towards tangential end

Fig. 4.3. Variations in beach texture, morphology, and dynamics along a log spiral beach (Finkelstein 1982)

headlands being cut back in such a way as to fit in with the general run of the coast (Jennings 1955).

Gradual decrease in beach curvature downdrift of headlands is generally assumed to reflect increasing exposure to wave action. Shorelines attempt to attain an equilibrium configuration that is determined by the pattern of offshore wave refraction and diffraction, and the distribution of wave energy flux (Wind 1994). The log spiral is an expression of the balance between the effects of headlands and offshore bathymetry on wave refraction and diffraction, and the relationships between beach slope, wave energy, and grain size (LeBlond 1979). Phillips (1985), however, suggested that the lack of consistent, systematic longshore variations in beach slope, grain size, and wave energy along a log spiral beach to the lee of an artificial protrusion shows that the headland effect on wave energy distributions is not the controlling factor in the evolution of crenulate embayments. Although asymmetrical beach plan forms may represent distance-decay effects from headlands, which tend to be curvilinear and exponential or logarithmic in nature, wave energy is but one of a number of factors that must be considered. Log spiral bays are common in estuarine environments, for example, where wave energy is low, and steep, short period waves undergo limited refraction. Factors other than the shadow effect of marsh headlands are therefore likely to control the development of bay configurations in these environments—one possibility being the decline in sediment supply with distance from estuarine tributaries.

Crenulate-shaped bays formed under oblique wave attack are generally considered to be the most stable in nature. These bays may be in dynamic equilibrium with sediment supply, or, when no further littoral drift is taking place, in static equilibrium. When in static equilibrium, the tangential and nearly straight downcoast section is parallel to the wave crests approaching from offshore, and as waves diffract into the bay, they break simultaneously along the whole periphery. There is therefore no longshore component of breaking wave energy and no littoral drift, and the plan shape, local beach slope, and sediment size distribution remain constant through time. The beach has therefore attained a state of complete harmony with the prevailing wave field (LeBlond 1979; Hsu *et al.* 1989*a*).

The stability of existing bays can be evaluated by comparing their plan form with the shape for static equilibrium (Hsu *et al.* 1989*b*). The crenulate bay concept also has important implications for the provision and maintenance of stable beach forms, and it provides an alternative means of stabilizing coasts against erosion. Headland control, which involves the insertion of fixed points along the shoreline, can be used to reduce littoral drift, and to inhibit or prevent long-term erosion (Silvester 1976; Bishop 1983).

Coastal Cells and the 'a–b–c . . .' Model

A simple, ideal coastal cell is a coherent sand budget compartment, consisting of an area of erosion, an area of deposition, and a connecting transport path between. Fixed cell boundaries, including headlands and river mouths, are morphological features that affect the refraction pattern or inhibit longshore transport. Free boundaries are determined by the wave field, and their occurrence and location vary with changes in wave power and breaker angle. Cell boundaries may separate converging or diverging longshore transport, or they may serve to transmit sediment between adjacent cells. Some cells are well isolated from each other, with little or no leakage of sediment across their common boundary. Others are less well isolated, and changing wave conditions, which cause boundaries to shift and switch, allow sediment to pass from one cell to another as it is transported alongshore (Carter 1988).

The 'a–b–c . . .' model considers the longshore transport of sediment from a headland or the updrift end of a cell, to the head of an adjacent bay or the downdrift end of the cell (May and Tanner 1973; Tanner 1977) (Fig. 4.4). Because of wave refraction, total wave energy density (E) decreases systematically from a maximum at the drift divide at point a to a minimum at point e. The longshore component of wave power (P_L) is at a minimum on the headland and at the head of the bay, and at a maximum at point c, where the refracted waves reach the coast at a small angle. The rate of change in the amount of sediment in motion with distance along the beach (dQ_s/dx), attains its largest positive value at point b, and its largest negative value at point d. The greatest amount of erosion therefore occurs around point b, while the greatest amount of deposition is around point d. There is no change in the amount of material being carried at points a, c, and e, where dQ_s/dx is close to zero.

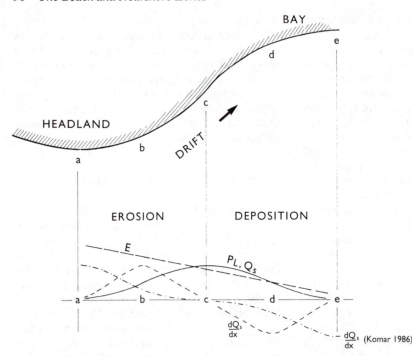

Fig. 4.4. The 'a-b-c . . .' model, showing variations along half a re-entrant in the energy density (E), the longshore component of wave power (P_L), the volume transport rate (Q_s), and the rate of change in the amount of sediment in motion along the beach (dQ_s/dx), Erosion occurs where dQ_s/dx is positive, or lies above the baseline, and deposition where it is negative (May and Tanner 1973). The dQ_s/dx line of Komar (1985) is also shown

Komar (1985) also modelled headland erosion and sediment transport into an adjacent embayment. He suggested that although the breaker angle, and therefore the littoral drift, become zero at the apex of a headland, the local rate of erosion depends upon dQ_s/dx, the gradient of the littoral drift. Although Q_s, the volume of sediment being transported past a point per unit time, is zero at the headland apex, dQ_s/dx is at a maximum at this point. Similarly, dQ_s/dx attains its largest negative value in the centre of the embayment. Therefore, in contrast to the 'a–b–c . . .' model, Komar's numerical model predicts maximum erosion at the headland apex, rather than at point b, at some unspecified distance downdrift, while maximum deposition occurs in the centre of the bay, rather than at point d, some distance updrift.

Komar's model suggests that maximum erosion at the apex of headlands causes them to develop a blunt nose, whereas in May and Tanner's model, maximum erosion on the flanks of headlands produces narrow promontories. Carter (1988) attempted to reconcile the two views. He noted that as many headlands function as divides between diverging longshore currents, there must be some point at which there is no longshore transport, and consequently no erosion. This would, under ideal conditions, produce a needle-like headland, but

because of the variable nature of the incident wave field, the point of no erosion sweeps back and fore across the face of the headland, producing a more rounded form. The argument has been developed further, using data collected from two eroding drumlin headlands in Nova Scotia (Chapter 5) (Fig. 5.1). Because of wave refraction, wave height and uprush are greatest on headlands, which are therefore more susceptible to erosion than embayments. Continued erosion also requires efficient removal of the debris, however, which cannot occur where there are divergent longshore currents. Although this would tend to produce narrow promontories, the effect of changing wave height and direction on rates of erosion and transportation tends to blunt their ends (Carter *et al.* 1990a).

Rhythmic Features

There are frequently crescentic projections on beaches and wave-formed bars, spaced at quasi-regular intervals, and oriented normal, or at high angles, to the shoreline. Rhythmic features occur at a wide variety of scales, ranging from forms less than 1 m apart on small lakes, up to features more than 1 km apart on oceanic coasts (Fig. 4.5). Smaller rhythmic features are frequently superimposed upon the larger elements.

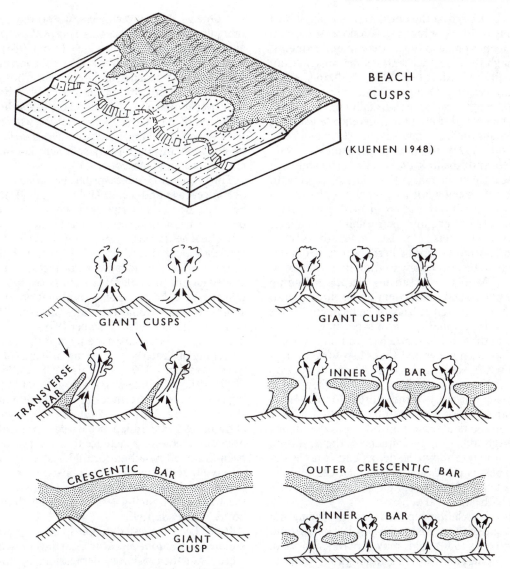

Fig. 4.5. Some types of coastal rhythmic feature

Beach cusps Beach cusps are arcuate scallops, generally considered to be less than about 25 m apart, that develop in the swash zone in sediments ranging from boulders and cobbles to fine sand and shell. They are particularly prominent on reflective, coarse clastic beaches, but they are also found on the reflective faces of otherwise highly dissipative sandy beaches with wide surf zones. On coarse clastic beaches, the submarine topography tends to mirror the shape of the shoreline, with re-entrants offshore from cusps, and small, submarine deltas off embayments. Relief is lower on the gentler slopes of sandy beaches, and deltas tend to merge into a uniform step at the base of the swash zone.

Beach cusp morphology largely depends upon the sediment, beach slope, and tidal range (Komar 1983b). They exhibit enormous differences in form, and although they appear to be regularly spaced, careful measurement may reveal considerable variation (Sato *et al.* 1992). Typical cusps consist of a series of triangularly shaped horns or ridges of clasts or coarse sand extending down the beachface, and separated from each other by shallow, curved embayments. Their amplitude, the vertical distance from the trough to the horn, ranges from almost nothing to 1–2 m or more. Beach slopes are generally steeper on cusp horns than in embayments, although the difference is sometimes insignificant (Russell and McIntire 1965; Williams 1973; Entsminger 1977; Dean and Maurmeyer 1980).

Sediment in cusp horns is usually coarser than in embayments, although the difference tends to be very slight on fine sandy beaches (Komar 1973*a*; Williams 1973; Dean and Maurmeyer 1980; Chafetz and Kocurek 1981). The permeability of coarse cusp horns makes them less susceptible to erosion than embayments, and although changes in wave conditions can form or destroy cusps within a few hours, high permeability sometimes allows them to persist for days under adverse conditions. Sorting according to shape may be dominant where there is little variation in grain size or density. More slowly settling shapes, including discs, plates, and blades, tend to be thrown onto cusp horns and to the back of embayments, while more-mobile spherical grains become concentrated at the foot of the beach (Flemming 1964).

It has been suggested that cusp formation is favoured by oblique waves, but most workers believe that it is most likely to occur when waves approach normally to the shore (Gorycki 1973; Komar 1976; Carter and Orford 1993). Cusps develop in tidal and non-tidal environments, although tidal range influences the resulting morphology. Submarine deltas and troughs develop best in tideless or microtidal areas, and the horns tend to become simple elongated ridges in areas with large tidal ranges.

There is a relationship between cusp spacing and the length of the uprush (Williams 1973; Dean and Maurmeyer 1980). This may explain changes in cusp spacing along beaches, although variations in beach gradient and grain size are among other factors that must be considered (Williams 1973; Sallenger 1979; Carter and Orford 1993). Furthermore, when there are two or three levels of cusps on a beach, the shortest spacing is invariably on the series that is furthest down the beach, and therefore formed by the lowest waves (Kirk 1980; Antia 1989; D. J. Sherman *et al.* 1993*b*; Carter and Orford 1993).

Most workers have found that cusp horns divide the uprush into streams that turn and flow into embayments, thereby interfering with the surge of water flowing directly up into the embayments (Komar 1973*a*; Dean and Maurmeyer 1980; Inman

and Guza 1982): nevertheless, water may sometimes flow from embayments to cusps (Dalrymple and Lanan 1976; Williams 1973). Dyer (1986) suggested that flow is from horns into embayments on steep, reflective beaches, and from bays to horns on beaches of lower slope. In the former case, breaking waves shorten cusp horns and deposit sediment seawards, thereby reducing beach slope, whereas in the latter case, spilling, refracted waves travel faster up embayments, depositing sediment landwards and increasing beach slope.

The origin of beach cusps has been debated for over a century-and-a-half (Johnson 1919). It has been suggested that cusps are primarily the result of: depositional processes, or that they develop when beaches are accreting (Russell and McIntire 1965; Schwartz 1972; Takeda 1984); erosional processes, or that they form when beaches are being eroded (Johnson 1919; Dolan and Ferm 1968); or that they are the result of a combination of erosional and accretional processes (Gorycki 1973; Guza and Inman 1975; Sallenger 1979). It has also been proposed that cusps are formed through the breaching of beach ridges, berms, or banks of seaweed (Williams 1973; Dubois 1978; Sallenger 1979), although some workers doubt that breaching is a necessary precursor to cusp development (Komar 1973*a*; Inman and Guza 1982; Seymour and Aubrey 1985). In Japan, cusp development has been observed on a beachface that had previously been flattened by a bulldozer. These cusps were purely accretionary features consisting of sand carried landwards from the lower foreshore, with the bays experiencing less deposition than the horns (Sato *et al.* 1992).

It has been suggested that cusp development is initiated by random depressions already present on a beach, or the erosion of depressions by uprush (Russell and McIntire 1965). Cusps have also been attributed to breaking waves separating into roughly equal units (Cloud 1966), instability of the surf zone bed to perturbations by longshore currents (Schwartz 1972), the tendency for breaking waves and uprush to become structured into salients (Gorycki 1973), wind-driven Langmuir circulation—a type of helical flow parallel to the direction of wave propagation (Campbell *et al.* 1977), and intersecting wave trains (Dalrymple and Lanan 1976; Entsminger 1977).

Escher (1937) attributed cusp formation in a wave tank to the presence of standing cross waves at right angles to the incoming waves, but of the same period. The identification of edge waves on beaches has spurred renewed interest in the possible role of standing waves in beach cusp development. Many workers now accept that beach cusps result from the interaction of standing edge waves and incident waves, which produces regular longshore variations in the uprush. Cusp wavelengths could be equal to either the wavelength of synchronous edge waves ($T_e = T$), or to half the wavelength of subharmonic edge waves ($T_e = 2T$). Standing, subharmonic, and synchronous edge waves have produced beach cusps in laboratory wave tanks, with embayments and horns corresponding to the position of the maximum and minimum, edge-wave controlled uprush, respectively (Guza and Inman 1975; Kaneko 1985; Takeda *et al.* 1986). Cusp spacing in the field has also been found to be generally consistent with the presence of subharmonic edge waves, and possibly on occasion with smaller synchronous edge waves (Sallenger 1979; Inman and Guza 1982; Seymour and Aubrey 1985; Aagaard 1985; Komar and Holman 1986; D. J. Sherman *et al.* 1993*b*).

The longshore wavelength of an edge wave (L_e) is given by (Chapter 2):

$$L_e = \frac{g}{2\pi} T_e^2 \sin\left[(2n + 1)\beta\right]$$

where T_e is the edge wave period, n is the wave mode number, and β is the beach slope angle. The equation suggests that cusp wavelength should increase with beach steepness and wave period, or with beach slope and the vertical excursion of the uprush (Inman and Guza 1982). The tendency for cusp spacing to increase with beach slope is therefore consistent with the edge wave mechanism (Sallenger 1979). The effect of beach slope also suggests that cusp spacing increases with grain size, although the opposite relationship has been observed in northern California (Dean and Maurmeyer 1980). As cusps grow, the effects of strong bedform feedback cause the amplitude of the edge waves to diminish (Guza and Inman 1975). Although they may eventually disappear, circulation patterns generated by incident waves force the cusps to continue growing, until they attain maturity. The contribution of edge waves to cusp development may therefore be limited to initiating slight topographic perturbations on otherwise uniform beaches (Inman and Guza 1982).

Despite general acceptance of the role of standing, subharmonic edge waves in cusp formation, they have not been definitely observed in the field while cusps are developing. Indeed, the edge wave hypothesis has recently been challenged in favour of earlier theories that cusps develop through the erosion of incipient beach depressions. Werner and Fink (1993) simulated the evolution of uniformly spaced cusps by modelling water flow and sediment transport in the swash zone. Beach cusps developed as a result of velocity-dependent erosion and deposition, and positive feedback between swash flow and beach morphology. Erosion occurred in random topographic lows that deflected and accelerated the backrush, whereas deposition occurred on topographic highs that decelerated the uprush. The lows developed into cusp bays and the highs into cusp horns, with smoothing and the interaction between water particles helping to produce a regularly spaced pattern. Although the edge wave and Werner and Fink's self-organization models are based on quite different mechanisms, they provide similar predictions of cusp spacing and occurrence. The validity of these incompatible models can therefore be determined only as a result of the detailed measurement of beach morphology and swash flow during cusp formation.

Rhythmic topography   Rhythmic topography develops in coarse clasts as well as sand, and it has been described from a wide variety of coastal environments (Sallenger *et al.* 1985; Shaw 1985; Jagger *et al.* 1991). Two types can be distinguished (Sonu 1973; Komar 1976) (Fig. 4.5):

a) Ephemeral beach and inner bar topography is associated with rip currents and cell circulation. Inner bars are more sensitive to changes in wave regime than outer bars, and they often migrate towards and on to the shore. Although crescentic inner bars can exist off straight shorelines, their horns often extend across surf zone shoals into very large shoreline cusps (sand waves, shoreline rhythms, or mega cusps). These projections are up to hundreds of metres apart, and they can project several tens of metres seawards (Inman and Guza 1982; Aagaard 1988*a*). With oblique wave approach, rhythmic topography on the inner bar and shoreline becomes skewed, as the shoals or bars align with the wave crests in a similar manner to rip currents: high longshore transport rates, however, can straighten cuspate shorelines.

b) Outer crescentic bars may be completely detached from shore. They extend into deeper water than inner bars, and are generally much more stable features. They either occur in isolation off smooth or non-rhythmic shorelines, or they coexist with rhythmic inner bar and beach topography, although in the latter case, they are often out of phase with the inner bars and remain essentially independent of nearshore rhythms.

Beach profiles, grain size, water depth, and current velocity vary alongshore in the presence of rhythmic topography. This has an important effect on the wave energy reaching the shore, and it may account for the regular longshore distribution of zones of accelerated beach and dune erosion, and the location of barrier overwashes (Wright 1980; Dolan and Hayden 1983). The position of these zones changes through time in response to slow, alongshore migration of the topography. Giant cusp migration rates of between 0.5 to 2.5 m yr^{-1} have been measured in several areas, but the larger rhythms on outer bars may remain virtually unchanged from year to year.

Cell circulation creates rhythmic topography consisting of bars segmented by rip current channels, with giant cusps located on the shore midway between the rips. These cusps are therefore erosional remnants, formed by rip currents eroding the bays on either side, and transporting the sand into deeper water. This pattern has been observed on Lake Michigan, although there were only shoals between rips on the Gulf of Mexico coast. Cusps can also develop where eddies cause sediment to be deposited on the lee or shoreward side of rip currents. These cusps tend to develop on more gently sloping beaches than those that are situated between rip currents.

Crescentic or lunate bars are concave shorewards. Their wavelengths vary from about 100 m to more than 2,000 m, although those in the 200–500 m range are probably most common. The wavelengths of multiple crescentic systems tend to increase with increasing distance offshore. Crescentic bars are particularly well developed in bays situated between headlands, but they also develop off long, straight beaches. They appear to be restricted, however, to sandy areas with gentle offshore slopes. Although it has also been suggested that rhythmic topography is restricted to microtidal areas, it has developed where the tidal range is as high as 4 m (Sonu 1973).

It is now generally accepted that rhythmic topography can be produced by drift velocity fields or

cell circulation generated by standing edge waves. Sand moves to crescentically shaped zones where the drift currents associated with standing edge waves converge and are at a minimum. Small-scale crescentic bars, and other components of rhythmic topography, could be formed by subharmonic edge waves generated by long ocean waves with periods of about 15 to 20 seconds (Komar 1976), but the scale of most elements is closer to the wavelengths of infragravity edge waves. Field evidence supports the low-frequency, edge wave hypothesis for the formation of rhythmic bars (Bauer and Greenwood 1990). On the Atlantic coast of the United States, longshore variations in erosion rates, the position of inlets, and the length of barrier islands are compatible with the wavelengths of hypothetical standing waves trapped between shoals extending off the capes (Dolan *et al.* 1979). Although the possible origin of these long edge waves has not been determined, they may be generated by atmospheric pressure fluctuations, or internal gravity waves on the continental shelf.

Although the emphasis has been on regular longshore periodicities in flow patterns or drift velocity associated with standing edge waves, the same result can be obtained by the interaction of any two modes of the same frequency (Holman and Bowen 1982). If one of a pair of interacting edge waves has a fairly short offshore lengthscale, for example, periodic morphology is generated alongshore near the shore, and linear morphology offshore. The spatially varying morphology of a three-bar system in Denmark, however, is the result of temporal, rather than spatial, segregation of edge wave modes or frequencies. In contrast to the linear outer bar, which is active only during intense storms when the edge waves are progressive, the middle bar is also active during periods of more moderate energy when the edge waves are standing, and it is therefore crescentic. The inner bar has a more intricate or smaller-scale rhythmicity than the middle bar, probably because it is active under even lower energy conditions, when the standing edge waves have slightly higher frequency (Aagaard 1991).

BEACH PROFILES

Gradients

Beach gradient is determined by differences in the strength of the uprush and backrush, and consequently in the quantity and direction of the net cross-shore sediment transport. The gradient of equilibrium beachfaces largely depends upon the amount of water that is lost into the beach through percolation, which is primarily a function of grain size. A beachface is in a state of dynamic equilibrium when its gradient is able to compensate for the weaker backrush, resulting in the same amount of sediment being moved seawards as landwards. Coarse clastic beaches are particularly steep as a result of rapid percolation of the uprush into the beach, and the consequently very weak backrush. By comparison, only a small amount of water is lost into a fine sand beach, and the weak gravitational effect of a gently sloping beachface is therefore sufficient to compensate for the small difference between onshore and offshore transport rates. The relationship between grain size and beach slope is valid, however, only if the waves are able to transport all sizes of the available sediment. For example, as only the strongest waves are capable of moving boulders, these very coarse-grained beaches have anomalously gentle gradients, which are characteristic of high energy conditions (Oak 1984).

The degree of sorting also determines the rate of percolation—beaches of well-sorted sediment tending to be steeper than those consisting of poorly sorted materials. In New Zealand, for example, the gradients of poorly sorted, mixed sand and coarse clastic beaches are lower than those of beaches consisting of better sorted sediments with the same medium grain sizes (Kirk 1980). In general, the relationship between grain-size characteristics and foreshore gradient is more complex for mixed sand and gravel beaches than for sand beaches, and the relationship tends to be more polynomial than linear.

The gradient of a beach tends to decrease with increasing wave steepness, presumably because higher uprush velocity makes it easier to transport sediment up the slope. Variations in wave steepness can account for changes in beach slope through time, and for differences in the gradient of exposed and protected beaches, or portions of beaches, with the same grain size. There may also be a relationship between beach slope and wave height or energy level, which would explain why, for the same wave steepness, slopes are greater in low-energy environments than on exposed oceanic coasts. Heavy minerals increase the weight per unit volume of the sediment and, consequently, its resistance to

removal by backrush. The proportion of heavy minerals in beach sediments may therefore be significant in determining its gradient. Foreshore gradient increases with the proportion of fine-grained heavy minerals on Lake Michigan, for example, despite decreasing grain size and percolation (Dubois 1972).

A variety of descriptive, empirical, and mathematical models consider the effect of grain size and wave parameters on beach slope (Komar 1976; Hardisty 1990b; Quick 1991). It is not easy to determine when slopes are in equilibrium, however, or even how to define the slope of beaches that usually consist of more than one slope element. The slope that is usually considered is that in the swash zone, where it is essentially planar.

The first attempts to model beach gradient were largely concerned with correlations with wave and sedimentological parameters. Doornkamp and King (1971) found that a linear equation accounted for about 72 per cent of the variation in slope on twenty-seven beaches representing a wide range of conditions:

$$\text{Beach slope} = 407.71 + 4.2D_\phi - 0.71 \log E$$

where the beach slope is in the form of the log cotangent of the gradient. The equation confirms that grain size is the single most important variable affecting beach slope. Using published field data from Britain, Japan, and the Pacific coast of the United States, Sunamura (1989) concluded that beach slope in the field is approximately represented by the curve:

$$\tan \beta = \frac{0.12}{(H_b/g^{0.5}D^{0.5}T)^{0.5}} = 0.25 \, (D/H_o) \, (H_o/L_o)^{-0.15}.$$

Iterative models are based upon computer simulations that run process equations until an equilibrium slope has developed, whereas in the analytical approach, equations for beach gradient are solved by assuming that there is no net sediment transport at equilibrium. Inman and Bagnold (1963) derived an expression for the equilibrium slope by assuming zero net transport at equilibrium and a balance of forces induced by asymmetry in the onshore–offshore velocities. The equilibrium slope is given by:

$$\tan \beta = \tan \phi \, \frac{1 - E_r}{1 + E_r}$$

where E_r is the ratio of energy dissipation in the offshore flow to that in the onshore flow. The value of

E_r is approximated by (Inman and Frautschy 1966):

$$E_r = \frac{u_{m.ex}^3}{u_{m.in}^3}$$

where $u_{m.ex}$ and $u_{m.in}$ are the maximum offshore and onshore velocity components, respectively. Good agreement has been found between the seaward gradient of a submarine bar during a storm, when there was no net sediment transport, and the theoretical slope calculated from the asymmetry of the maximum orbital velocities (Greenwood and Mittler 1984).

In Bowen's (1980) model, dynamic equilibrium represents a balance between the effect of gravitational forces, which facilitate downslope sediment transport, and flow asymmetry, which drives sediment landwards. The flow velocity (u) consists of two components, the symmetrical orbital velocity (U_o) and a perturbation (U_1), which may be caused by a steady current, wave asymmetry, or a wave of a different, unrelated frequency. Therefore the flow asymmetry can be expressed in the general form:

$$u = U_o + U_1$$

where the orbital velocity is a sinusoid with an amplitude u_o:

$$U_o = u_o \cos ft.$$

If sediment transport is only by bed load, as on coarse clastic beaches, equilibrium can only be attained on steep slopes. Equilibrium on most sandy coasts, however, reflects the combined effects of bed load and suspended load. Bowen's total expression for the equilibrium slope with zero net transport is:

$$\tan \beta = \frac{\frac{u_o}{C}(2 + (\sinh kh)^{-2})(1.8\, e_s \frac{u_o}{w_s} + 1.325 \frac{e_b}{\tan \phi})}{0.8\, e_s \frac{u_o^2}{w_s^2} + \frac{e_b}{\tan^2 \phi}}$$

where e_b and e_s are efficiency factors for bed load and suspended load, respectively (Chapter 3), and w_s is the settling velocity and ϕ the angle of repose of the sediment.

Bailard's (1981) model differs from Bowen's in placing less emphasis on the effect of gravity on suspension transport. Nevertheless, his equation for equilibrium beach slopes is closely related to Bowen's, the main difference being the presence of the term e_s^2 in the dominator. This produces equilibrium beach slopes for strong suspension conditions that are significantly greater than in Bowen's

model. Both models imply that equilibrium beach slope increases with decreasing depth, however, and with increasing wave period and settling velocity (or grain size). Bailard's model also suggests that beach slope slightly increases with increasing wave angle, decreases with the increasing strength of strong longshore currents, and increases with the increasing strength of weak longshore currents.

Jago and Hardisty (1984) expressed beach gradient in terms of the flow asymmetry and the transport ratio ($i_{net}(r)$):

$$\tan \beta = \tan \phi \frac{V_r^2 - i_{net}(r)}{V_r^2 + i_{net}(r)}$$

for:

$$i_{net}(r) = \frac{i_{net.in}}{i_{net.ex}}$$

where $i_{net.in}$ and $i_{net.ex}$ are the shoreward and seaward sediment mass transport per wave, respectively (Hardisty 1986) (Chapter 3), and V_r is the velocity ratio, the ratio of the maximum onshore (u_{in}) and offshore (u_{ex}) velocities (Chapter 2). At equilibrium ($i_{net}(r) = 1$), the beach slope is given by:

$$\tan \beta = \tan \phi \frac{V_r^2 - 1}{V_r^2 + 1}.$$

The model suggests that the beach system is self-stabilizing. The slope of a beach that is steeper than the equilibrium gradient would be reduced through a simultaneous increase in seaward transport and a decrease in landward transport, whereas the reverse situation steepens beaches whose gradients were less than their equilibrium values. Although the original model was developed only for bed load transport, good agreement was found between model predications and measured beach slopes. The model has since been modified to include suspended load (Hardisty 1990b). The equilibrium gradient under total load transport is given by:

$$\tan \beta = \frac{(V_r^2 - 1)(e_b \tan \phi \frac{w_s}{|U|^2} + e_s \frac{w_s}{|U|} \tan \phi^2)}{(V_r^2 + 1)(e_b \frac{w_s^2}{|U|^2} + e_s \tan \phi^2)}$$

where $U = (u_{in} + |u_{ex}|)^2$. The model predicts beach gradients increasing exponentially shorewards, and grain size decreasing seawards. Also in accordance with field observations, predicted slopes decrease with increasing deep-water wave height or decreasing wave period.

Water Tables

Groundwater exerts an important influence on swash zone morphology and processes. Mathematical models of tidally induced fluctuations in beach water tables have been developed by several workers (Nielsen 1990; Turner 1995). Others have described the elevation and shape of the water table, and their effect on beach accretion and erosion. Groundwater levels continue to rise after the tide has started to fall, and continue to fall after the tide has started to rise. The water table therefore generally slopes seawards near the beachface when the tide is falling, and landward when the tide is rising. The lag in the response of the water table to tidal ebb and flow increases landwards, and it may be up to a few hours at stations 10 m or so from the water's edge. Water table profiles are also sensitive to the permeability of the beach material, and differences within a beach can produce a high degree of irregularity. Lags are likely to be much smaller in highly permeable beach materials, and the amplitude of tidally induced fluctuations decreases landwards more gradually than in less permeable materials.

The elevation of the water table, relative to the beach surface, plays a vital role in determining whether erosion or accretion takes place. When the water table is some distance beneath the surface, percolation rapidly reduces the depth of the uprush, causing an early transition from turbulent to laminar flow, and rapid deposition on the upper foreshore. Small ridges of sand or swash marks are formed by percolation of the uprush and deposition on the dry upper portion of beaches, at the point of greatest advance. The ability of the backrush to carry sediment back down dry beaches is also reduced by its smaller volume and lower velocity, which act to prolong the laminar flow. The effect is particularly marked on gently sloping beaches, where gravitational acceleration is weak. Deposition on the upper foreshore may be compensated by rill erosion further down the beach, however, as water flows out from the saturated effluent zone beneath the water table.

When the water table is essentially coincident with the surface, little water is lost by percolation into the beach, and there is little deposition on the upper foreshore. Outflow causes erosion in the effluent zone and supplements the backrush, however, thereby encouraging the onset of turbulent flow earlier in the cycle. High water tables can be the result of large waves propelling water on to the backshore, heavy rainfall, an impermeable underlayer, the addition of fine sand to coarse clastic beaches, or fresh water draining from the land. This latter factor helps to account for the erosion of beaches in front of the mouth of streams.

Previous work has been concerned with the effects of water tables on beach dynamics, particularly over tidal cycles. Turner (1993), however, emphasized the importance of the saturated capillary fringe zone above the water table. The fringe zone and the zone beneath the water table are saturated, but whereas pore water pressure below the water table is greater than atmospheric pressure, it is less than atmospheric pressure within the fringe zone. The thickness of the fringe varies according to the rate of water table rise and the rate of infiltration from above. Water tables rise through unsaturated sand when uprush infiltrates the beachface faster than interstitial air can be expelled. No evidence was found on two microtidal Australian beaches of sand reverting from completely dry to wet conditions during the flood half of a tidal cycle, the greatest change in saturation occurring at any one point being only 65 per cent.

Although groundwater response to tidal cycles has received the most attention, beach water tables fluctuate at a variety of scales, ranging from individual swash events to seasonal and longer oscillations (Waddell 1980; Eliot and Clarke 1986). Katoh and Yanagishima (1992) attributed berm erosion during storms to higher water tables resulting from infragravity waves running beyond the berm crest. In addition to water table fluctuations in the immediate vicinity of the beachface, Waddell (1976) found that standing waves are responsible for larger fluctuations that are transmitted further into the beach. He noted that as the uprush is quite shallow near the point of maximum runup, even a small loss of water by infiltration can have an important effect on the potential energy available to carry sediment downslope. Although Hegge and Masselink (1991) accepted that Waddell's pressure forces may be important on the lower portion of the beachface, they attributed the rise in beach water tables above the mean groundwater level to uprush infiltration and the reverse Wieringermeer effect. This latter factor refers to rapid and large rises in the water table that result from the addition of a thin film of water to a fully saturated capillary fringe which extends to the surface. Field measurements provide further support for the contention that rapid water table fluctuations occur when the top of the saturated capillary fringe is coincident with the beachface (Turner 1993). Very small amounts of water can cause pore pressures in the capillary fringe to equal or exceed atmospheric pressure, producing an almost instantaneous rise of the water table through the capillary fringe to the sand surface.

Water table elevations vary spatially as well as temporally. Saturation levels are affected, for example, by the occurrence of higher water levels near rip catchment divides and over inshore bars, and lower levels near rip current channels (Lanyon *et al.* 1982). Water tables first begin to increase with rising tides in the lower embayments landwards of the rip currents, and finally on the higher portions of the bar profile. The variable response to rising tides along the coast is reflected in the shape of the water table, which slopes landwards on steeper rip profiles, and seawards on bar profiles.

It has been found that the removal of water generates initially rapid and large amounts of accretion on the foreshores of laboratory beaches, and therefore provides a potentially practical way of constructing or replacing berms: the same conclusion has been reached through computer modelling (Machemehl *et al.* 1975; Kawata and Tsuchiya 1986; Urish 1989). The results of a field experiment strongly suggest that beach conservation and aggradation can be encouraged by maintaining the water table at a low level (Chappell *et al.* 1979). It was found that pumping induces greater infiltration in the upper and middle swash zone, as well as reducing sand liquefaction by pressure waves propagating into the water table from the shore. These high-frequency waves ranged up to 200 mm in amplitude beneath the upper swash zone under moderate energy conditions. During the rising phase, they increased the elevation of the water table at rates of at least 100 mm s^{-1}, generating a buoyancy force close to that required for the liquefaction of sand. Water table waves may therefore make a substantial contribution to beach front failure and slumping, which play an important role in the cutting back of beaches during storms. A one-year test on the western coast of Jutland, which is exposed to vigorous wave action, showed that a drainage system helps to build up and stabilize the beach (D.G.I. 1986). There has also been a full-scale demonstration of the beachface dewatering technique on the eastern coast of Florida. The treated portion of the beach was wider and more stable than the adjacent untreated portions, and it contained a greater volume of sand. Although it has been proposed that artificial beach drainage systems are only effective under mild wave conditions (Bruun 1989), the treated beach withstood three storm events better than untreated areas. The results therefore suggest that beach dewatering could provide an important and economical means of stabilizing commercially valuable beaches where groundwater is a problem (Parks 1989; Terchunian 1990).

Beach States

Beaches cause waves to break and dissipate their energy, thereby providing effective protection to the land behind. Beach morphology and sediments vary in space and over a variety of time scales in response to variations in the size and energy of the incoming waves. Two broad types of beach profile have been identified. Profiles with wide berms or swash bars and steep foreshore slopes, sometimes with steps at the breaker line, have been termed swell, step, berm, ordinary, or non-barred profiles. Profiles with gentler foreshore slopes and longshore submarine bars have been termed storm or barred profiles. The transition from a non-barred to a barred profile takes place during storm periods, when large waves cause the sediment to be shifted seawards, whereas the reverse occurs when smaller swell waves move sediment back onshore.

Strong annual periodicity in the wave climate of the western coast of the United States led to the terms 'winter' and 'summer' profiles becoming synonymous with barred and non-barred profiles, respectively. Seasonal terminology, however, is inappropriate for storm wave environments which have persistently high wave power throughout the year; east coast swell, Trade Wind, monsoonal, and low and high latitude protected sea coasts, where erosional profiles dominate or are significant during the summer; and west coast swell environments in the southern hemisphere, where there is little seasonal variation.

The transition from a barred to a non-barred profile may depend upon wave steepness, or the ratio of the time of uprush of a wave to the wave period. Dean (1973) suggested that the boundary between barred and non-barred profiles is given by:

$$\frac{H_o}{L_o} = 1.7 \frac{\pi w_s}{gT}.$$

The dimensionless fall velocity (H_b/w_sT) is a measure of whether a particle which is lifted from the bottom by breaking waves settles back to the bed before it can be acted on by offshore horizontal velocities. If this is the case, it will only be affected by onshore velocities and net sediment transport will be landwards or zero. The beach will therefore have, or will develop, a steep profile. Alternatively, grains tend to be moved offshore if settling velocities are low. The dimensionless fall velocity and related indices have been used to determine the cross-shore direction of sediment transport, and to predict foreshore gradients (Quick and Har 1985; Masselink 1993).

Beach morphology changes in a fairly predictable manner as it is gradually restored to its equilibrium state, following erosion during a storm (Sonu and van Beek 1971; Fox and Davis 1973). Wright, Short, and co-workers at the University of Sydney, as well as other Australian workers, have related beach states to wave power, and identified the process signatures associated with them. This work has shown that in south-eastern Australia there is a close relationship between beach morphology and such factors as the amplitude and frequency of standing waves, and the relative strength of rip and longshore currents. Near-bottom currents vary according to the beach state and the relative importance of motions associated with incident waves, infragravity and subharmonic oscillations, and mean longshore and rip currents (Short 1979; Wright *et al.* 1979; Chappell and Eliot 1979; Wright and Short 1984).

Beach morphology depends upon immediate and antecedent wave, tide, and wind conditions, sediment characteristics, and the antecedent beach state. Two types of profile represent the extremes of the range of forms that can be assumed by beaches as they adjust to the size or power of the waves (Fig. 4.6). Much of the incoming wave energy is reflected by steep beaches ($\beta > 6°$), which have well-developed berms, no nearshore submarine bars, and a pronounced step of coarser sediment where the incoming waves meet the backrush (Bauer and Allen 1995). Reflective domains, which are analogous to the 'summer' or 'fully accreted' profile, have surging to collapsing breakers. Infragravity oscillations are very weak, and most energy is at incident wave frequencies under low

wave or energy conditions, and at the first sub-harmonic (twice the incident wave period) with somewhat higher wave conditions. The subharmonic oscillations, which are invariably standing, are interpreted as edge waves standing alongshore. Highly rhythmic beach cusps, which develop in the swash zone when there is moderately energetic swell, may be spaced at one-half the predicted length of the most easily excited, zero-mode subharmonic edge waves. Reflective domains have surf scaling parameter or reflectivity values (Chapter 2) of less than 2.5, and dimensionless fall velocity values less than 1.

At the other end of the spectrum, gently sloping dissipative beaches absorb or consume much of the incoming wave energy in turbulence, before the waves reach the beach. Dissipative domains are analogous to 'storm' or 'winter' profiles. They have concave-upward nearshore zones and wide, flat surf zones, and they usually have spilling breakers. Although longshore rhythms are rarely present, dissipative profiles are normally more complex and varied than reflective systems, and there are generally one or more shore-parallel, submarine bars. Waves break up to a few hundred metres from dissipative beaches, resulting in low uprush at incident wave frequencies, and high set-ups. Shoreward decay of wave energy at incident wave frequencies is countered by an increase at infragravity frequencies. Standing infragravity waves are usually present, and they, rather than incident waves, dominate beachface uprush, bottom currents, and sediment transport in the inner surf zone. Infragravity oscillations encourage the formation of rip currents, although they may fail to develop on highly dissipative beaches if the mass balance is attained through the vertical division of the water column into seaward and landward flows. Dissipative domains have surf scaling parameter values ranging from 20 to more than 200, and dimensionless fall velocity values that are typically between 6 and 30.

Intermediate beach states have coexisting dissipative and reflective elements, and they are characterized by rip currents and the onshore migration of submarine bars (Fig. 4.6). Intermediate states represent stages in the gradual return of sediment to the beach, and subaerial exposure of material initially stored subaqueously in the nearshore, following a period of high wave energy. Intermediate beaches are more spatially and temporally variable than fully dissipative or reflective beaches, and

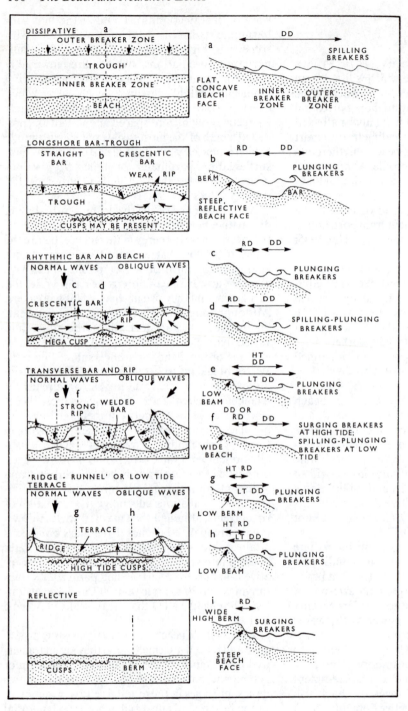

Fig. 4.6. Six major beach states. HT and LT are the high and low tidal levels, and RD and DD are the reflective and dissipative domains (Wright *et al.* 1985)

they can rapidly change in response to variations in wave height and directions of sediment transport. At least four intermediate types can be identified. The longshore bar-trough and rhythmic bar and beach states develop in an accretionary sequence from an antecedent dissipative profile. The relief of the bar-trough is much higher than in the dissipative state, and the beachface is much steeper. Waves of low steepness surge up the beachface, whereas steeper waves plunge near its base. The uprush is quite high and cusps often develop in the swash zone. The rhythmic bar and beach state is distinguished by rhythmic longshore undulations of the beach and bar, the horns of the crescentic bar or shoreline protrusions usually being between about 100 and 300 m apart. Rip current circulation is weak to moderate, though persistent in location. The transverse-bar and rip state tends to occur when the horns of crescentic bars become welded to the beach. This creates alternations of relatively dissipative transverse bars or megacusps, and relatively reflective and deeper embayments occupied by strong rip currents. The 'ridge and runnel' / low tide terrace state has a beachface which is reflective at the high tidal level, and a moderately dissipative, flat accumulation of sand at, or just below, the low tidal level. Any rips that are present tend to be small, weak, and irregularly spaced.

Sunamura (1989) devised a similar three-dimensional beach change model based upon field data from Japan, North America, Australia, and the eastern Mediterranean. His model consists of eight topographical, two limiting, and six transitory stages. It is applicable to beaches in moderate to high energy, microtidal environments, with moderate nearshore bottom slopes, sediment diameters of between 0.1 and 0.2 mm, and dominant shore-normal sediment transport. The transition from one stage to another is predicted using a dimensionless parameter (K_*):

$$\bar{K}_* = \bar{H}_b{}^2 / \bar{T}^2 Dg$$

where $\bar{H}_b$ and $\bar{T}$ are the daily average breaker heights and wave periods, respectively. Onshore bar migration occurs when $5 \leq K_* \leq 20$ and offshore migration when $K_* > 20$.

During a storm, increased wave power causes beach erosion, formation of a submarine bar, and transformation of the beach from a more reflective to a more dissipative state. Following the storm, lower wave power causes bars to migrate shorewards, and the system gradually changes from the dissipative to the reflective or fully accreted state. In nature, however, increases and decreases in wave power tend to occur in cycles that are shorter than that between the two extremes, thereby providing a variety of morphological sequences. The time needed to change from one beach state to another is proportional to the degree of disequilibrium induced by changing morphodynamic conditions, and inversely proportional to the energy available to enact the required changes. Fast responses and frequent changes in state probably involve fairly small scale exchanges of sediment over short distances between the beach and surf zone, whereas slower, less frequent responses involve larger exchanges of sediment between the surf zone and the inner continental shelf. The shift from reflective to intermediate or dissipative beach states is much faster during a high energy storm event than restoration of the reflective state under low energy conditions (Thom and Hall 1991). Equilibrium dimensionless fall velocity values have been estimated for each beach state, using a 6.5-year time series of quasi-daily data on wave and beach conditions from south-eastern Australia (Wright *et al.* 1985) (Table 4.1). Beach state equilibria and directions and rates of change were then calculated from the data (Fig. 4.7). The immediate dimensionless fall velocity was much less successful in predicting beach states than a weighted mean value that considered recently antecedent wave conditions. Nevertheless, because of overlap in the fall velocity values between proximal states, antecedent conditions could not reasonably predict individual beach states, although they were fairly successful in predicting the occurrence of broader categories obtained by merging similar pairs of states.

Higher mean wave groupiness values (Chapter 2) represent more-energetic surf conditions for a

Table 4.1. Equilibrium dimensionless fall velocity parameter values and beach states

Beach state	$H_b/w_s T$	Standard deviation
Reflective	< 1.50	—
Low tide terrace	2.40	0.19
Transverse bar and rip	3.15	0.64
Rhythmic bar and beach	3.50	0.76
Longshore bar and trough	4.70	0.93
Dissipative	> 5.50	—

Sorce: Wright *et al.* 1985.

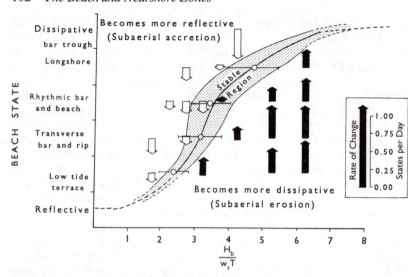

Fig. 4.7. The relationship between beach state and the dimensionless fall velocity (H_b/w_sT). The central curve represents the mean fall velocity for each beach state, and the shaded region is within one standard deviation of the mean. The length and colour (black or white) of the arrows illustrate the rate and direction of beach changes from an initial state, towards the mean (Wright *et al.* 1985)

given dimensionless fall velocity value. Greater incident wave groupiness therefore favours more dissipative beach states, assuming that wave conditions are also suitable. Thus, greater groupiness causes the transverse bar and rip state to replace the low tide terrace state, and assuming the dimensionless fall velocity parameter is suitable, it is the principal factor determining whether the rhythmic bar and beach, or the transverse bar and rip states develop (Wright *et al.* 1987).

The single bar, beach state model has been extended to include multiple bar systems (Short and Aagaard 1993). On the assumption that bars are the result of infragravity edge waves, their number can be estimated using a parameter (B^*):

$$B^* = \frac{X_s}{g \tan \beta \, T^2}$$

where X_s is the distance from the shoreline to the point at which the gradient of the bottom becomes zero, and T is the incident or edge wave period. The data suggest that no bar forms when B^* is less than 20, one bar when B^* ranges from about 20 to 50, two bars when B^* is from 50 to 100, three bars when B^* is from 100 to 400, and four bars when B^* is greater than 400. High waves, short periods, and fine sand favour more dissipative, multi-bar beaches, whereas moderate to low waves, long periods, and coarser sand favour more intermediate-reflective, single bar beaches. All bars respond to dominant infragravity processes during storms and other periods of high waves, but the outer bar(s) may be inactive when moderate or low waves break on the inner bars during accretionary periods. Sequential

wave breaking across a series of bars also results in a shoreward decrease in wave height, and a descending heirarchy of beach/bar types, with more intermediate-reflective types near the shore. High mode, long period progressive or standing edge waves, and/or leaky mode standing waves, tend to produce linear bars far from the shoreline. Decreasing wave height and increasing wave period and grain size (slope) constrict the surf zone and favour low mode standing edge waves nearer the shore, and the formation of more rhythmic bars. Bars become welded to the shore when low waves, long periods, and coarse sand restrict surf zone dissipation and infragravity wave growth.

The most frequently occurring or modal beach state in an area reflects the recurrent breaker and sediment characteristics. Beach state curves show their expected occurrence on open, microtidal, sandy coasts (Fig. 4.8). The relationship between wave power and beach states suggests that global wave environments generate characteristic beach state curves (Short 1978). Storm wave environments have persistently high wave power, and the beaches are usually in a highly dissipative state, especially if the sediment is fine-grained. More prolonged periods of calmer conditions may occasionally allow some transformation into more reflective intermediate states, however, especially in bays and other sheltered areas. Alternatively, beaches in low swell or sheltered environments are generally in reflective states, particularly if the sediments are coarse-grained, although infrequent storms can generate more dissipative states. All beach states occur in east coast swell environments,

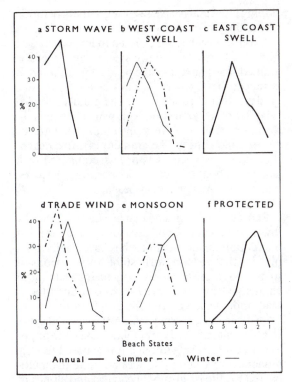

Fig. 4.8. Percent frequency of occurrence of each beach state for the major global wave environments for sandy beaches on open, microtidal coasts. The fully dissipative and reflective states are numbered 6 and 1, respectively (Short 1978)

although the modal state is the rhythmic bar and beach. Beach states in west coast swell, Trade Wind, and monsoonal environments vary according to seasonal changes in wave power. The transition to a more accretive state requires waves of sufficient energy to transport sediment landwards—this may not be available in low energy environments. On the northern coast of Zealand, Denmark, for example, the modal state is the rhythmic bar and beach, and the evolutionary sequence rarely attains the final reflective state. This is not because of persistent storms, however, but because the movement and morphological evolution of the bars are arrested during prolonged periods of calm, when energy levels are too low to move sand shorewards (Aagaard 1988b).

In addition to changes caused by fluctuations in wave intensity, beach morphology also varies over the long term in response to changes in sediment supply, and atmospheric and oceanographic processes. Long-term temporal and spatial fluctuations in the amount of sediment stored sub-aerially on a beach interact with, but are independent of, changes in the nearshore wave climate. Eliot and Clarke (1982) analysed beach profile data collected every two weeks over a five-year period in south-eastern Australia. Onshore–offshore sediment transport was dominated by biennial, seasonal, and six-monthly cycles. Temporal beach changes occurred as a result of long period oscillations in mean sea level, changes in coastal rainfall, long-term changes in wave climate, and water circulation within coastal embayments. The water table fluctuates in response to each of these factors, and through its effect on swash zone processes, beaches may erode and accrete without any change in the nearshore wave climate. Analysis of a long-term record of the high tidal position on a beach in New South Wales, confirmed that changes in rainfall and many other environmental variables play important roles in determining rates of coastal erosion, in addition to changes in storm frequency and intensity (Bryant 1988).

Macrotidal Beaches

Most work on beach states has been conducted in south-eastern Australia, and in other areas with low tidal ranges. The situation is more complex, however, in macrotidal environments. Beach states change with tidal level, and the upper portion of a beach profile may be morphodynamically different from the lower part. Gradients, breaker conditions, and reflectivity vary in surf and swash zones as they migrate back and fore across concave-upward profiles. Beach mobility and the relative role of sediment transporting mechanisms, including incident and standing waves, infragravity waves, and tidal currents, may also be expected to vary across the beach. Whereas the morphology of microtidal beaches is mainly the result of processes operating in the surf and swash zones, this is true only of the high tidal zone on macrotidal beaches, and even here, they act only intermittently. The dominance of tidal currents, and more especially shoaling, unbroken waves in the lower portions of macrotidal beaches, probably makes them dynamically similar to the nearshore and offshore zones of microtidal beaches. Because of constantly changing water levels, increasing tidal range has the effect of reducing rates of sediment movement and changes in beach morphology (Davis 1985). Tides also have an important influence on water circulation in the nearshore zone. Rip currents and the offshore directed bottom flow, or undertow, tend to be strongest when the tide is falling, and shore parallel tidal currents play a significant role in longshore sediment transport on the lower intertidal and subtidal zones of macrotidal beaches (Wright *et al.* 1982).

Macrotidal beaches are globally widespread, but we know far less about their dynamics than microtidal, and particularly swell-dominated, beaches. Wright *et al.* (1987) found that the effect of mean tidal range is most significant for intermediate states at the reflective end of the sequence. If wave conditions are suitable for either the low tide terrace or transverse bar and rip states, then spring or other high tidal ranges favour the occurrence of the low tide terrace/'ridge and runnel' state. Higher tidal ranges also discourage highly accentuated bar and trough topography, and if wave conditions are suitable for bar formation, the most accentuated topography therefore develops when tidal range is lowest. Short (1991) recognized three groups of macrotidal beach:

Group 1 beaches are exposed to fairly high waves, and they have a planar, flat, uniform surface. The upper foreshore or high tidal zone is exposed to the highest waves, and because of the asymmetrical flows of plunging and surging breakers, it has the coarsest sediment and the steepest gradients. This zone is dominated by swash zone and surf or breaker processes. Subharmonic edge waves may be responsible for the frequent occurrence of cusps in this reflective beach zone, although infragravity standing wave energy also increases shorewards. The intertidal zone is essentially uniform and flat, although its gradient slightly decreases seaward. The height of the breaking waves also decreases seawards with increasing wave attenuation, and swash zone oscillations become more symmetrical as the breakers change from plunging to spilling with the falling tide. Most of the work is performed in the intertidal zone by unbroken shoaling waves, however, rather than by wave breaking and surf zone processes, although tidal currents also increase in importance seawards (Wright *et al.* 1982; Jago and Hardisty 1984).

Group 2 beaches are exposed to waves of moderate height, and are characterized by multiple longshore bars or ridge and runnel. The steep, high tide beach is often cusped and reflective, whereas dissipate conditions probably dominate on the gently sloping intertidal beach, resulting in the generation of standing infragravity waves.

Group 3 beaches are exposed to low waves. The high tidal zone is usually steeper and coarser-grained than the rest of the beach, but although it is reflective, it generally lacks cusps. This zone grades rather abruptly into a fine-grained, rippled tidal flat of very low gradient, which may be uniform, have multiple, very low amplitude bars, or ridges and runnels. The gradient of these beach-tidal flats is typically 0.1° or less, compared with slopes of about 0.5° on moderate energy beaches, and 2° to 6° on high energy beaches.

Short (1991) suggested that because of generally lower wave energy and the lower frequency of inundation, macrotidal beaches have equilibrium profiles that remain fairly stable over periods of weeks to months. This conclusion has been challenged by Horn (1993), who found that a macrotidal beach on the Isle of Man can change from a swell to a storm profile over one tidal cycle. It is difficult to fit this very mobile beach into any of Short's categories. Although it is exposed to only moderate wave energy, it does not have the multi-barred profile of group 2 beaches, and slopes of 0.6° on the lower portion of the beach and almost 6° on the upper beach are greater than those suggested by Short. Group 2 beaches may therefore represent only one of a number of intermediate beach states between the group 1 and 3 extremes of a macrotidal morphodynamic continuum. Carter (1988), for example, identified five energy-dependent types of macrotidal beach. High energy beaches have a washed out or rippleless flat, uniform surface. With slightly lower energy levels, bedforms may survive tidal emergence on a uniform surface, and with declining energy, intertidal dunes may survive. Low energy, dissipative beaches have ridges and runnels or parallel bars, while cellular water motion generates transverse bars on very low, extremely dissipative beaches.

Beach morphology is determined by the relative effect of waves and tides, rather than by their absolute values. Coasts exhibiting tide-dominated characteristics can therefore occur in microtidal environments where there is low wave energy, whereas wave-dominated coasts can occur in macrotidal environments where there is high wave energy. Equilibrium beach profiles and beach states have been modelled as a function of the relative tidal range (RTR)—the ratio of tidal range to breaker height, and the dimensionless fall velocity (Masselink 1993; Masselink and Short 1993; Masselink and Hegge 1995) (Fig. 4.9). High RTR values represent tide-dominance, and low values wave-dominance. The model is based on the microtidal (RTR < 3) and macrotidal (RTR > 3) beach literature, and field data from the macrotidal coast of central Queensland. Swash and surf zone processes dominate the entire intertidal and upper subtidal zones when the RTR is less than 3. These beaches, which are usually in microtidal environments, can be classified according to the scheme of Wright and Short (1984). Swash and surf zone processes still dominate the upper part of the profile when the RTR is between 3 and 7, but shoaling waves dominate the lower part. These beaches are usually in high-to-moderate energy, macrotidal environments. Swash and surf zone processes are only significant on the uppermost portion of the beach when the RTR is greater than 15, and the transition begins into tidal flats. Beaches with RTR values between 7 and 15 essentially overlap with group 1 in Short's (1991) classification, and those with an RTR greater than 15 with group 3. RTR values between 3 and 7 were not considered in Short's classification, whereas group 2, or ridge and runnel systems, were not recognized as a separate beach type in this model.

Wave-formed Bars

The nearshore zones of most sandy coasts have one or more submarine bars, and they are generally

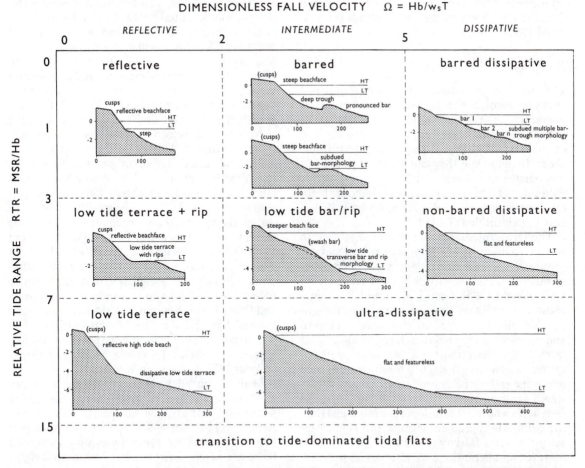

Fig. 4.9. Beach state as a function of the dimensionless fall velocity and the relative tidal range. MSR is the mean spring tidal range, and HT and LT are the high and low tidal levels, respectively (Masselink and Short 1993)

absent only from areas dominated by long swell waves or where steep nearshore slopes allow waves to break at the shoreline. Although they are almost always formed of sand, they are occasionally found in fine clastic material. Wave-formed bars, as opposed to those formed by tidal currents or other mechanisms, play an important role in the near-shore sediment budget, and they exert a strong influence on wave action and sediment movement.

Bars contain primary sedimentary structures produced by bedforms generated by symmetrical and asymmetrical oscillatory currents, and uni-directional longshore and rip currents. Avalanch-ing during the migration of very mobile bars can produce medium to large scale sets of tabular cross bedding. Trough sediments are generally less well sorted than those in bars, although they can be coarser or finer than bar sediments, depending upon whether they are erosional lags or suspension deposits laid down during calm periods (Green-wood 1982).

Bar Types

The morphology, sedimentology, and stability of bars vary according to the nature of the processes operating in the area. The ridges can be single or multiple, symmetrical or asymmetrical, and oriented parallel, oblique, or perpendicular to the shore. In plan view, they can form continuous or compartmented linear, sinuous, or crescentic patterns, and they often produce periodic or rhyth-mic longshore topography. Six main types of bar have been identified (Table 4.2).

The term 'ridge and runnel' was first used in Britain to describe low intertidal ridges and de-pressions, which develop on gently sloping beaches in macrotidal areas with a limited fetch (King 1972; Wright 1984). Gaps cut through the ridges allow water to escape seawards from the runnels, as the tide falls. The ridges are either stable in form and position, or they migrate slowly landwards, at rates of less than a metre a day, in response to spring–neap tidal cycles. It was originally proposed that ridge and runnel is produced by short period waves, creating equilibrium swash bar slopes on foreshores whose natural slopes are much less than the equilibrium gradient. Ridge and runnel de-velops under a fairly wide range of conditions, however, and definitions based upon its perman-ency, mobility, and sediment availability may be too rigorous. Although four to five ridges have

generally been observed in Britain, there are up to fourteen low, quasi-stable ridges in the lower por-tion of a 1,200 metre wide intertidal zone on the eastern side of Vancouver Island. These ridges, which developed over a flat surface cut in deltaic cobbles and pebbles, demonstrate that, in addition to areas with a surfeit of sediment, ridge and run-nel can also form on erosive platforms with a paucity of sediment (Hale and McCann 1982). Ridges have formed under conditions of limited and abundant sediment availability on the eastern coast of Ireland, and in contrast to those on Van-couver Island, they are destroyed or suppressed by storm wave conditions (Mulrennan 1992).

Cusp or bar type sand waves are frequently destroyed by storm waves, and then reformed at the toe of the swash zone by long swell waves during calmer periods. The bars are usually parallel to the coast, although they can become cusp-like under oblique wave action. They tend to migrate onshore, at rates ranging between 1 and 30 m day[-1], and eventually become welded onto the foreshore. Migrating bars act as mobile, flexible obstacles to waves and wave-induced currents, and interactions between the bars and wave-current fields change beach and bar morphology. Although they are genetically distinct, and much less stable, the morphology of welded bars can resemble classical ridge and runnel. This led to the use of the term 'ridge and runnel' in North America to refer to migratory bars in the swash zone, which are moved landwards by the lower energy waves following a storm (Davis *et al.* 1972). The modified American use of the term has also been adopted by some workers in Australia and Europe (van den Berg 1977; Dabrio and Polo 1981; Wright *et al.* 1985). In addition to the swash bar and welded bar explana-tions, it has also been suggested that ridges can be residual features formed by runnel erosion (Orme and Orme 1988). Nevertheless, the term 'ridge and runnel' has specific genetic implications, and should only be used in its original sense, to describe a series of essentially stable, swash bar slopes (Komar 1976; Orford and Wright 1978).

Multiple parallel bars occur on gentle slopes with moderate to low wave energy (restricted fetch) and a fairly low tidal range (Davidson-Arnott and Pember 1980; Guillen and Palanques 1993). The bars are generally symmetrical, and there are often slight increases in their height, spacing, and the depth of water over their crests, with increasing distance from the shore (Mei 1985;

Table 4.2. Types of wave-formed bar

Type	Approximate		Planform	Profile	Number seaward	Location	Wave energy	Wave processes	Tidal range	Slope	
	Height (m)	Length (m)								SN	I
Ridge and runnel	0.2–1.5	10^3	Straight, SP	AL	1–4	I	L–M	Surf-swash beach drainage	Ma–Me	< 0.01	
Cusp or bar type sand wave	0.2–1.5	10^2	Straight to spit-shaped, SP	AL	1–2	I and LT terrace	M	Breakers, surf-swash	Me–Mi		> 0.01
Multiple parallel	0.2–0.75	10^3	Straight to sinuous, SP	NeS	4–30 or more	I and NS	L–M	Spilling breakers, surf	Mi–Me	< 0.01	
Transverse	0.2–0.75	10^2	Straight, SN	SL and AL	1	I and NS	L	Surf-swash, spilling breakers	Mi	< 0.01	
Nearshore I	0.25–1	10^2 (?)	Straight, SP	AL	1–2	NS	M–H	Plunging breakers	Mi–Me	< 0.1	
Nearshore II	0.25–3	10^3	Straight, sinuous to crescentic, SP	AL	1–4	NS	M	Spilling breakers	Mi	< 0.01	

Notes: SP = shore parallel; SN = shore normal; AL = asymmetric landwards; SL = symmetric landwards; NeS = nearly symmetric; I = Intertidal; LT = low tide; NS = nearshore; L = low; M = moderate; H = high; Ma = macro; Me = meso; and Mi = micro.

Source: Greenwood and Davidson-Arnott 1979.

Davidson-Arnott and McDonald 1989). In eastern Canada and the Great Lakes, there are significant relationships between fetch length and the height of the outer bar, depth to its crest, and its offshore distance, while bar number increases with decreasing nearshore slope (Davidson-Arnott 1988).

Transverse bars, which are oriented perpendicularly or at high angles to the shoreline, usually develop under lower wave energy conditions than multiple parallel bars, although bed slopes may not be as low (Niedoroda 1973). Some irregularly spaced transverse bars are the coarse lag deposits of former barrier spits, tombolos, or islands (Taylor *et al.* 1985), but the origin of others is less clear. Transverse bars create a subtle nearshore circulation pattern which is strongly dependent on their length. Refraction concentrates wave energy over the crest of fairly short bars, generating narrow and fairly strong onshore currents. This carries sand onto the beach, and as it cannot be returned seawards by the fairly weak return currents between bars, cusp-like features develop where the bars meet the beachface. The circulation pattern is more complex where there are long transverse bars. Although the currents are also onshore over the seaward portion of the bars, the friction induced loss of wave energy over bar crests nearer the shore, increases the capacity for mass transport between the bars, resulting in offshore sediment transport.

Chevron beach ridges, consisting of coarse clasts or very coarse sand, are a type of transverse bar found in the Canadian Arctic (King 1972). Short barbs, running from the elongated ridges at fairly high angles, create arrow-like features that point to the coast and are oriented almost perpendicularly to it. The barbs of adjacent, subparallel ridges sometimes join together to form ridges parallel to the coast. Chevron beach ridges appear to develop where there are low gradients, and falling relative sea level. The barb pattern is related to refraction of the waves that form them. When the ridges are fairly close together, and the water depth between them is roughly uniform, little refraction occurs and the barbs form almost at right angles to the ridges. Greater refraction occurs in the shallower water near ridges that are spaced a kilometre or more apart, and the barbs are therefore formed at acute angles to the ridges.

Two types of nearshore bar were distinguished (Greenwood and Davidson-Arnott 1979). Narrow, low, and highly unstable type I bars are formed on fairly steep slopes, possibly by the vortices of high,

plunging breakers. Larger and more stable type II bars, which may be straight or crescentic, develop seawards of the low tidal mark. They are essentially permanent forms that are rarely destroyed and rebuilt, although they do move and are subjected to some modification.

Bars often occur in association with nearshore, rip cell circulation. This has been well documented in areas with bar and cusp type sand waves (Sonu 1973), transverse bars (Niedoroda 1973), and crescentic and straight nearshore type II bars (Greenwood and Davidson-Arnott 1979; Greenwood and Mittler 1984).

Origin

Barred profiles tend to develop where there are steep waves and high wave-height to grain-size ratios, although the way that wave energy is reflected or dissipated is also important. Some theories of bar development depend upon incident wave characteristics and their related cross-shore flow structures (Dally 1987; Zhang 1994), whereas others are concerned with the effect of waves of infragravity frequency (Bauer and Greenwood 1990; Holman and Sallenger 1993). Most theories only consider monochromatic or non-random wave fields, with a single wave height and period. The position of the breakpoint varies considerably in the field, however, and it is not known whether the integrated effect would create a bar, and if so, where it would develop. Formation of bars by infragravity waves also requires that there be a few dominant frequencies and modes. Most field studies have found that observed spectra in the nearshore zone are usually broad banded, with no dominant frequency (Oltman-Shay and Guza 1987; Holman and Sallenger 1993), although a few have identified significant spectral peaks during storms (Aagaard 1990; Bauer and Greenwood 1990; Aagaard *et al.* 1994).

Longshore bars have been attributed to the offshore transport of sediment excavated, during trough formation, by vortices under plunging breakers (Fig. 4.10). Formation of an initial perturbation in the profile would further concentrate wave breaking and facilitate bar growth. Multiple bars could develop where breakers reform and establish subsequent breakpoints, or where strong resonant reflection from pre-existing bars initiates bar formation further seawards (Heathershaw and Davies 1985; Mei 1985). Bars also develop under spilling breakers, even though their smaller vortices are confined to the upper portion of the water column (Dally 1987; Sunamura 1988). Spilling breakers generate vortices with oblique rotating axes that extend down through the bore. Those that reach the bottom act like small tornadoes, whirling the bed material and lifting it into suspension, so that it can be carried offshore by the mean flow field. Several types of horizontal vortices are generated by plunging breakers, and this may determine the size and number of bars that are produced (Zhang 1994). Other explanations for breakpoint bars involve onshore sediment transport seawards of the breaker zone, and either seaward movement in the surf zone or decreasing landward

movement because of breaking. Extensive and temporally persistent near-bed undertows play an important role in maintaining bar equilibrium (Davidson-Arnott and McDonald 1989; Osborne and Greenwood 1992b), and it has been proposed that they are formed by undertows carrying sediment seawards to the breakpoint, where the current disappears (Roelvink and Stive 1989). Bar formation has also been attributed to alongshore and offshore movement of sediment beneath meandering or cellular nearshore circulations, resulting from instability of longshore flows at the fluid-sediment interface (Hino 1974) (Fig. 4.10).

The results of several wave tank experiments provide support for the breakpoint bar hypotheses,

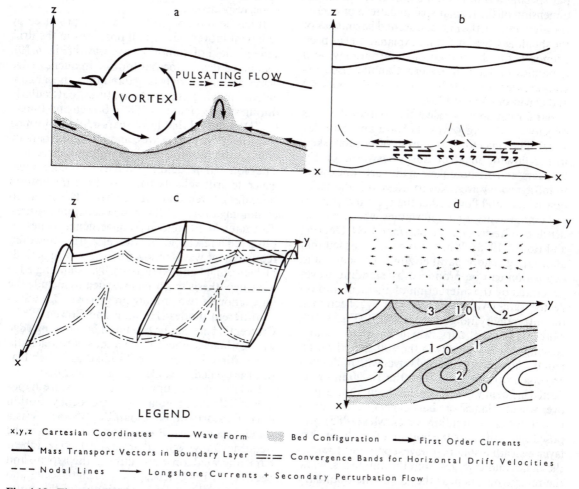

LEGEND

x,y,z Cartesian Coordinates   ——— Wave Form   ▒ Bed Configuration   ➡ First Order Currents

➡ Mass Transport Vectors in Boundary Layer   ═·═ Convergence Bands for Horizontal Drift Velocities

--- Nodal Lines   ➡ Longshore Currents + Secondary Perturbation Flow

Fig. 4.10. Theories for the origin of bars: (a) generation of a vortex by plunging breakers; (b) mass transport in the boundary layer of a strongly reflected wave; (c) drift velocities associated with standing edge waves; and (d) longshore current and secondary perturbation flow resulting from hydrodynamic instability (Greenwood 1982)

and possible explanations for the relationship between bar occurrence and wave steepness (Sunamura and Maruyama 1987; Dally 1987; Dean *et al.* 1992). Bar crests are often less consistently related to the breakpoint in the field (Sallenger and Howd 1989), however, and some bars are found in fairly deep water considerable distances from the shore, where they are clearly unrelated to wave breaking.

The interaction of regular progressive waves with accompanying lower, slower, secondary waves, or solitons, could influence the magnitude and direction of sand transport, in such a way as to produce bars. Multiple shore-parallel bars could be formed by interactions between wave harmonics and bottom sediment. In Lake Huron, for example, bar spacing was found to be similar to the length dimension of the regular spatial pattern of energy transfer between the first and second harmonics of the shoaling waves. A similar explanation has been proposed for the formation, morphology, and dynamics of bars in eastern Canada (Boczar-Karakiewicz *et al.* 1987*a*, *b*; Boczar-Karakiewicz and Davidson-Arnott 1987).

Bar formation may reflect the combined efforts of short and long waves (O'Hare and Huntley 1994). They may form as a result of onshore skew in the flow of incident waves seawards of the breakers, and offshore skew in the surf zone owing to infragravity waves. Nevertheless, it is the skewness of the total flow, rather than just the oscillatory components, that determines whether a bar develops (Greenwood and Sherman 1984; Doering and Bowen 1988). It has also been proposed that alternating, shore-parallel zones of scour and deposition can be produced by standing waves generated by the interaction of incident and reflected waves (Allen *et al.* 1991), or by reflection in the infragravity range (Short 1975) (Fig. 4.10). Multiple parallel bars, which have cross-shore spacing that is typically in the 100–200 m range, require infragravity reflected waves with nodes that are much further apart than the 10–20 m of reflected incident waves. There has been some debate over whether longshore bars develop at the nodes or antinodes of standing waves. Mei (1985) proposed that drift near the bottom of the boundary layer beneath a standing wave, converges towards the nodes and diverges from the antinodes, while the reverse occurs near the top of the boundary layer. Heavy particles rolling on the seabed therefore tend to drift towards the nodes, whereas light particles in suspension move towards the anti-

nodes. In the field, Aagaard and Greenwood (1994) found that suspended sediment was consistently transported by standing infragravity waves towards the antinodes in the water surface elevation, and most workers now concur with the view that standing long or short waves cause bars to form at the antinodes when suspension transport dominates, and at the nodes when bed load dominates. The offshore distance of the crest of a bar induced by a standing wave (x_c) is given by (Sallenger *et al.* 1985).

$$x_c = C \frac{g\,T^2 \tan \beta}{4\pi^2} = C \frac{g \tan \beta}{\omega^2}$$

where C is a constant equal to 3.5 or 6.5, depending on whether deposition occurs at a node or antinode, respectively.

It has been proposed that bars are produced by sediment movement to null positions in the drift velocity field of standing edge waves (Fig. 4.10). Estimated drift velocities for low-frequency edge waves are surprisingly large, and they appear capable of moving crescentic features, particularly during storms (Huntley 1980). Bowen and Inman (1971) proposed that linear bars could be formed by progressive edge waves, and crescentic bars by edge waves standing in the longshore direction. They found that standing subharmonic edge waves generate drift velocity patterns above the bottom boundary layer that are remarkably similar to commonly observed rhythmic nearshore features. To match drift velocity convergence zones to observed crescentic bar wavelengths, however, requires edge waves of low frequency, with periods of between 40 and 60 seconds. The standing edge wave hypothesis has been extended to include the interaction of two or more progressive edge waves of different modes (Holman and Bowen 1982). Crescentic bars can be produced by standing edge waves, whereas other combinations of wavelength, mode, direction, and amplitude create such features as welded, meandering, and linear bars.

Field evidence supports the edge wave hypothesis. The predominant low frequency motion during a storm on Georgian Bay, Ontario was a standing edge wave of mode three or greater, in the frequency range centred around 0.035 Hz. Linear bars migrated onshore during the storm and became rhythmic, reflecting the length scales and positions of the nodes and antinodes of the standing edge wave. The evidence suggested that bar accumulation took place under the antinodes of the standing wave (Bauer and Greenwood 1990).

In multiple bar systems, the inner bars are usually more dynamic and mobile than outer bars, and they often have a tendency for smaller scale rhythmicity. The spatially varying morphology of a three-bar system in Denmark has been attributed to temporal, rather than spatial, segregation of modes or frequencies of edge waves or leaky mode waves. Mode 3 edge waves, with a period of 50–55 seconds, dominate during storms, when the outer bars are active. The edge waves were progressive during a very intense storm and standing during a less intense storm. Single edge wave modes dominated the energy spectra in both cases, and their longshore and cross-shore length scales corresponded fairly well to bar patterns. The occurrence of progressive edge waves during the highest energy conditions may explain why the outermost bar tends to remain linear. In contrast to the outer bar, however, the middle bar is affected by standing edge waves during weaker storms, or as energy levels decline after a strong storm, and it therefore tends to assume a rhythmic pattern. Mode 1 standing edge waves of slightly higher frequency occur under more moderate energy conditions, when only the inner bar is active, and they are probably responsible for its smaller-scale rhythmicity (Aagaard 1991).

It remains to be determined whether infragravity waves actually produce bars, or whether it is the presence of bars that creates a dominant infragravity wave. The fairly constant form and position of multiple bars, for example, could reflect their ability to modify edge wave profiles (Holman and Bowen 1979). A theoretical model suggested that the resonant frequency for leaky waves should have an antinode at the crest of the bar, which is possibly maintained by the convergence of suspended load at this point (Symonds and Bowen 1984). Kirby et al. (1981) applied a numerical model to a beach with multiple bars corresponding to the antinodes of a standing wave generated by reflected infragravity wave energy. The results suggested that edge wave amplitudes on barred coasts are amplified away from the shoreline, and the antinodes of the resonant edge wave modes are attracted to the bar crests. This facilitates the maintenance or even growth of bars, despite shifts in infragravity wave frequency or changes in tidal level. It has also been proposed that bar formation and evolution may result from modification of edge wave drift velocit-

ies by strong longshore currents (Howd et al. 1991). There is some field evidence to support the hypothesis that bars amplify discrete frequencies, creating a dominant infragravity wave (Huntley 1980; Wright et al. 1986).

Storm Effects

There is a lack of data on changes to bar morphology during stormy periods, when the most rapid changes take place. In the eastern Mediterranean, non-rhythmic bars develop within seven to ten days after significant wave heights have fallen to below 1 m, single crested bars require fifteen days, and double crested bars twenty to thirty days (Bowman and Goldsmith 1983). The evidence from other areas, however, suggests that bars respond much faster to changing wave conditions than has previously been recognized (Birkemeier 1985). In particular, the conclusion that there is a significant lag between the occurrence of storms and the development of crescentic bars conflicts with the results of experiments conducted under the DUCK programmes.

Experiments designed to investigate the response of a nearshore bar system to a significant storm and the following recovery period, were conducted in the nearshore at Duck, North Carolina in October 1982, under the DUCK82 programme (Sallenger et al. 1985; Mason et al. 1985). There was a low bar, only about 13 m from the shoreline, prior to the onset of the storm. The high reflectivity of the beach, together with the proximity of the bar to the shoreline, suggest that it may have developed through convergence of sediment at the node or antinode of a standing wave of incident wave period. The bar responded to the increase in wave energy during the storm by becoming better developed, and migrating rapidly, at rates of up to 2.2 m hr^{-1}, to a position about 70 m from shore. The bar was within about one-tenth of the width of the surf zone from the shoreline during most of the storm and could not have been affected by breaker zone processes, and it was too far from shore to be attributed to incident standing wave motion. Bar morphology and offshore position suggested that it could have been generated by a mode 1 edge wave with an infragravity period of between 55 and 75 seconds. Surf zone wave data confirmed that there was significant energy at the appropriate infragravity frequencies, but no definite relationship was established with bar formation. There was rapid development of crescentic bar morphology after the storm had peaked, with portions migrating landwards at rates of more than 1 m hr^{-1}. A rough crescentic form appeared one day after the peak of the storm, although a couple more were required for the development of a classical crescentic configuration. The complete cycle, beginning with storm modification and offshore migration, and ending with recovery of the bar to almost its pre-storm configuration, therefore required only a few days.

The DUCK85 experiment, conducted in September 1985, provided further confirmation that rapid changes occur in the nearshore during the passage of even a moderate storm. The pre-storm morphology consisted of a terrace with incised rip channels. A linear sand bar developed with the onset of the storm, and it migrated seawards as it intensified. Morphological changes were similar to those reported in the DUCK82 experiments, including rapid transformation to a crescentic form as the incident wave height began to decline, and as the bar steepened and began to migrate landwards (Howd and Birkemeier 1987).

Spatial and temporal variability of the inner bar at Duck has been analysed using estimates of bar morphology based upon daily time exposure images of incident wave breaking over a two year period. Eight types of bar morphology were identified, using four classification criteria: the presence or absence of a bar; whether cross-shore, and where applicable, longshore bar scaling corresponds to incident or infragravity wavelengths; longshore variability, whether linear, rhythmic, or non-rhythmic; and whether the trough is continuous or discontinuous (Lippmann and Holman 1990). Although the classification is similar to Wright and Short's (1984), the latter's longshore bar and trough state was broken into two types, representing linear bars with no longshore variability, and bars with nonrhythmic longshore variability. Wright and Short's transverse bar and rip state was also subdivided into two bar types attached to the shore, which were distinguished by the presence or absence of dominant periodicity. All other bar types were essentially the same in the two classifications. All the bar types were infragravity scaled, with the exception of the two incident scaled classes at the reflective end of the spectrum. Longshore periodic or rhythmic bars were the most common type, having been observed in 68 per cent of the data. Linear bars occurred under the highest wave

conditions during stormy periods, although they were unstable, with mean residence times of only two days. Following the decline in wave energy, the bars rapidly evolved into offshore and attached rhythmic forms, the transition typically occurring within five to seven days of peak wave events. Shore attached rhythmic bars, which generally formed between five and sixteen days after peak wave events, were the most stable forms, having a mean residence time of about eleven days. An extended data set has also been used to examine the behaviour of the two-bar system at Duck, over a five-year period (Lippmann *et al.* 1993). The outer bar does not migrate seasonally in the on-offshore direction, although it is generally reduced in amplitude during prolonged, low wave periods. In general, the presence of an outer bar reduces the distance that the inner bar migrates seawards under storm conditions. During storms, when large waves break over the outer bar, the wave field over the inner surf zone is saturated, and wave height inside the breakpoint does not change significantly with increasing wave height outside the breakpoint. In the absence of an outer bar, however, the wave field over the inner bar is not saturated, and is therefore able to respond to increasing incident energy.

Horizontal, near-bed velocities have been measured during a storm across a lacustrine shoreface with two bars (Greenwood and Osborne 1991). An offshore directed mean flow or undertow, generated by a wave induced set-up, was spatially extensive and temporally persistent across the landward slope of the linear outer bar. In the inner bar system, persistent onshore directed skewness in the velocity distribution under propagating bores was only opposed by large offshore mean flows at the height of the storm. Local slopes were in a steady state over the storm cycle on the outer bar, although there was extensive sediment transport. Weak offshore mean flows and marked asymmetry in the near-bed velocity field, however, resulted in the inner bar migrating landwards through the storm decay period. Low frequency oscillations were possibly responsible for this bar assuming a sinuous three-dimensional form as it moved onshore.

Coarse Clastic Beaches

Although there is no consensus on the terminology of coarse-grained beach materials, 'coarse clastic' is used in this text, in preference to alternatives such as 'gravel', 'shingle', 'pebble', 'cobble', and 'boulder', to refer to sediments that are usually between about 2 and 2,000 mm in diameter (Carter and Orford 1993). Coarse clastic beaches are found throughout the world. They are rare in most tropical regions, however, although they occur on some dry coasts in desert and seasonally dry tropical climatic zones, and there are often ridges of coral fragments behind reefs (Chapter 10). Coarse clastic beaches are most common on the coasts and continental shelves of polar regions, in the storm wave environments of the mid-latitudes of the northern hemisphere, and on tectonic coasts where bed load is carried to the shore by high gradient streams (Fig. 4.11). The general association of coarse clastic beaches with high latitudes is probably largely a reflection of high energy levels in storm wave environments, and marine erosion of detrital material produced and transported to the coast by non-marine agencies, including weathering, mass movement, rivers, and glacial ice. In vigorous wave environments, angular joint blocks are quarried from cliffs and shore platforms and rounded by abrasion as they disperse along the coast, although splitting, spalling, or crushing of blocks also produce small, angular rock fragments. Wave erosion of terrestrial and marine tills, stratified, glacio-fluvial drift, and periglacial mass movement deposits are particularly important for

the supply of clasts to coasts bordering areas of Pleistocene glaciation.

Some steep beaches are composed entirely of coarse clasts, although they also accumulate as steep, narrow ridges at the rear of gentle sandy beaches, often in front of abandoned cliffs and ponds. The presence of a lower, dissipative sandy beach can have an important effect on the morphodynamics of a reflective clastic ridge, which may respond quite differently to changes in tidal level or set-up than a more homogeneous clastic beach. Coarse clastic ridges are often particularly pronounced across the mouths of rivers, and in front of lagoons and coastal flats, but they are generally absent or poorly developed where the material accumulates against the foot of cliffs.

There are important differences in the hydrodynamic behaviour of coarse clastic and sandy beaches, and in their response to storm wave action. Frictional drag and high rates of percolation increase the asymmetry of the uprush over the backrush on stony beaches, despite the inevitable presence of fine, interstitial material. The slopes of clastic beaches, which typically range between about 4° and 16°, and occasionally up to 25°, are therefore much higher than sandy beaches. Clasts are mainly transported by saltation, and by sliding, rolling, and other tractive processes. Bluck (1967) observed the collective movement of similar-sized clasts in a form of surface creep, and the shuffling

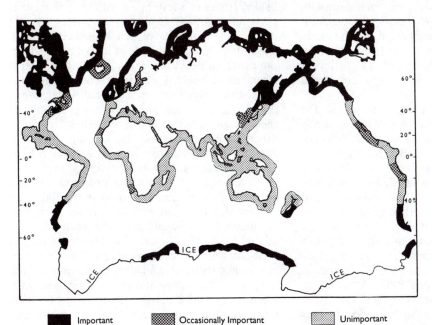

Fig. 4.11. The global distribution of pebble beaches (Davies 1972)

Important   Occasionally Important   Unimportant

movement of imbricated disc-shaped forms. This latter mechanism may be the result of the flipping over of dipping discs by strong backrush, rotation of underlying spherical clasts, or the orbital movement of passing waves. The selective transport and deposition of a mixed population of coarse clasts can also produce inverse grading and a stable, armoured or paved surface (Isla 1993). This develops when smaller clasts, which have higher pivoting angles, are trapped within the bed during decelerating oscillatory flow, while rounder, larger clasts continue to pass over them. A stable arrangement is created when these rounded, sorted clasts are finally deposited over the finer material.

Constructive waves create small, berm-like ridges at the limit of the uprush. The transition from spring to neap tides can therefore produce a series of ridges, each corresponding to a previous high tidal level. A larger and more permanent ridge is built above the high spring-tidal level by storm waves, and a series of these ridges may develop if there is an adequate supply of sediment. Because of rapid percolation, clasts deposited at the limit of the storm wave uprush cannot be carried back down the beach by the backrush, and they accumulate high on the beach crest, in positions well beyond that reached by normal waves. Orford (1977) challenged the traditional view that coarse clastic ridges are formed by plunging storm waves, which throw clasts beyond the uprush limit, to considerable heights above mean sea level. Although plunging breakers can throw clasts up a beach, it does not occur on a sufficient scale, and he therefore attributed storm ridges to constructive spilling breakers overtopping the beach crest during periods of very high tides, high waves, and storm surges.

Coarse clastic beaches usually fall within the reflective domain, and they are often confined between headlands. The consequent excitation of standing, sub-harmonic, usually zero mode, edge waves may produce a series of cusps that dominate the dynamics of the narrow swash zone—concentrating the uprush, modifying gravel facies, and occasionally directing uprush breaching of the ridge crest (Orford *et al.* 1991; D. J. Sherman *et al.* 1993*b*). Progressive edge waves usually inhibit cusp development, however, and they are therefore generally absent or poorly defined on drift-aligned beaches (Carter and Orford 1993). Destructive waves can rapidly remove clasts from the upper foreshore to just outside the breakpoint. This

material is then returned to the beach by constructive waves after the storm. Clastic beaches therefore tend to move towards more dissipative beach states with bar morphology under storm conditions, and seasonal and longer-term changes in beach profiles, gradients, and rates of profile change have been observed and inferred (Caldwell and Williams 1986; Powell *et al.* 1992). Because of their high entrainment thresholds, hydrodynamically rough surfaces, and strongly reflective profiles, however, clastic beaches tend to respond less readily to changes in tidal levels and wave conditions than sandy beaches, and their profiles therefore have long preservation potential. Features that develop under calm weather conditions can survive well into storm periods, when they can act to modify wave processes at the shoreline. Changes in sandy beach profiles are largely the result of cross-shore sediment transport, but the profiles of coarse clastic beaches fronted by sand can only change through the longshore redistribution of sediment. If there is a constant longshore supply of sediment, the loss of clasts along one profile is matched by a gain of similar magnitude along another. There is therefore generally no uniform longshore erosion or accretional trend along these beaches.

Coarse clasts often exhibit imbricate and vertical packing, with smaller clasts lodged between larger ones (Massari and Parea 1988; Isla and Bujalesky 1993). Vertical packing of flat disc or bladed stones of similar size, with their short axes perpendicular to the oncoming waves, greatly increases the stability of beach surfaces. Bladed clasts usually dip seawards at higher angles than the beach slope. Offshore dipping imbrication is strongest on beaches with weak backrush, and it may be destroyed or even replaced by landward dipping clasts where the backrush is strong, or there is an impermeable layer below the discs. Strong imbrication may be lacking, however, where the energy of the breaking waves is high enough to transport clasts by sliding.

Pronounced shape sorting produces distinct zones or frames, dominated by particular clast shapes and oriented parallel to the shoreline. Although it may be difficult to separate the interrelated effects of clast size, disc- and blade-shaped clasts are generally concentrated on the upper part of beaches, with rods and especially spheres on the lower part. The shape and size of clasts also vary alongshore and cross-shore according to their location within cusp systems, however, and this can produce quite complicated zonation patterns on the foreshore (D. J. Sherman *et al.* 1993*b*).

Bluck (1967) identified four distinct zones, based on grain shape and size, on coarse clastic beaches in southern Wales (Fig. 4.12). Large discs are typical of the ridge furthest shorewards, and they are followed seawards by an imbricated zone dominated by seaward dipping discs. There is sometimes a sand run seawards of the imbricated zone, which allows clasts to move very quickly over it, followed by a band on its seaward side where spherical and rod-shaped forms accumulate, with the long axes of the rods oriented normally to the beach. In the most seaward zone, spherical and rod-shaped clasts infill a framework of larger, mainly spherical stones, which form the outer margin of the storm ridge behind a flat, sandy berm. In addition to clast movement on the beach surface, movement also takes place within the beach during erosive periods, particularly beneath the surface of the large disc zone. This occurs when backrush infiltrates through the sieve-like frame and combs finer material seawards through the pore spaces. The evidence suggests that spherical particles move more quickly through the pores than other shapes. Shape sorting on a microtidal beach in Crete was generally similar to that on the macrotidal beaches of southern Wales, although there was no recognizable outer frame zone, and infilling appeared to take place over the entire beachface (Postma and Nemec 1990).

The shape sorting of clasts reflects their varying suspension and pivotability potential. The pivoting angle, which is the angle a clast makes about its contact point with an underlying clast, determines when it is tipped out of its rest position and entrained by the drag of a flowing fluid. The angle depends on the shape of the clast (rollability and angularity), the ratio of the size of the pivoting grain to those beneath, and on imbrication and other factors (Fig. 3.10). Pivoting angles decrease with increasing clast size, and they are also dependent on the ratio of the smallest to intermediate axial diameters, which controls the ability to roll and pivot. Clasts with small ratios tend to slide out of position, whereas true pivoting, with smaller angles, occurs with clasts that have ratios closer to unity. The pivoting angle increases approximately with the angle of imbrication, and flat clasts in an imbricated arrangement have much larger angles than when they are laid in a horizontal position. The interlocking of angular clasts also provides high pivoting angles. Isla and Bujalesky (1993) found that spherical clasts are more likely to move by saltation than other shapes, and the measurement of pivoting angles suggests that entrainment becomes increasingly difficult for clasts that are spherical, ellipsoidal, angular, and imbricated (Li and Komar 1986; Komar and Li 1986).

Discoidal clasts dominate the back of beaches because they are carried further landwards than clasts of other shapes, possibly reflecting the greater

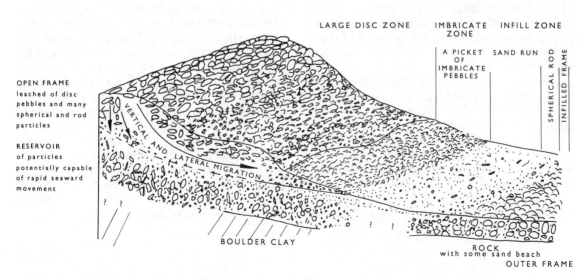

Fig. 4.12. The sedimentary structure of the storm beach at Sker Point, South Wales (Bluck 1967)

ease with which they are lifted from the sea floor by storm waves, and their generally lower settling velocity. They are also less pivotable than spheres and rods, and are therefore less likely to be entrained by the backrush, or to roll down the steep slopes. Orford (1975) suggested that the degree of shape zonation on a coarse clastic beach in western Wales varies with wave energy, and is greatest under swell wave conditions. The role of shape is not, however, simply related to the wave energy expended on a beach, but also to the wave phase and breaker type. Although he observed shape and size zonations similar to those of Bluck, Orford proposed that they are simply examples of the range of possible facies responses to changing wave energy. Williams and Caldwell (1988) also observed shape sorting on coarse clastic beaches in southern Wales, although the imbricate zone was generally absent. They concluded that when swash zone processes are at the critical transport threshold, the more easily suspended oblate material is thrown forward during the brief energy peak of the uprush. Spherical and prolate material is then winnowed down the beach by the longer, weaker backrush. This explanation for shape sorting, which is similar to Bluck's (1967), occurs during periods of berm buildup. Mass rather than shape, however, becomes the dominant factor in determining net cross-shore transport when energy levels are high, and entrainment forces are much greater than the threshold values. Shape sorting is therefore less pronounced on beaches that experience high energy levels. Grain size may also affect the degree of shape sorting. Massari and Parea (1988), for example, found that the degree of shape sorting increases with grain size, and is weakest in less-responsive, finer-grained clasts.

Most investigations of clast shape and size sorting have been conducted on beaches that are essentially swash-aligned, and therefore have a fairly finite volume of sediment normal to the shore. Although cross-shore shape sorting may be similar on swash- and drift-aligned beaches, the latter has a dynamic quality resulting from longshore movement and trading and exchange of clasts between adjacent zones, particularly in the transport corridor that is usually located in the upper foreshore

(Carter and Orford 1991). The degree of shape sorting on coarse clastic barriers is also influenced by their mobility. Reworking of the beachface can create discrete cross-beach sorting frames on barriers that are migrating landwards at rates of less than 1 m yr^{-1}. Shape sorting may be quite poor, however, on more mobile barriers that roll over their old backbarrier platforms, exhuming large discoidal clasts on the beachfaces too rapidly for the re-establishment of sorting frames (Orford *et al.* 1991).

Coarse clastic beaches are often well sorted alongshore according to grain size, although there is still considerable debate over the nature of the responsible mechanisms. Acceptance and rejection phenomena are especially important (see Chapters 3 and 5), although longshore variations in the size and shape of the clasts can also result from progressive attrition and impact breakage, rather than selective transport. Clasts increase from pea-sized on the western part of the Chesil Beach barrier in southern England, to diameters of about 6 cm at Portland, almost 29 km to the east. This increase in size, which is accompanied by an increase in beach height, may reflect an easterly increase in wave energy, or it may be the result of clast sorting by waves approaching from different directions. The strong grading of clasts on Chesil Beach is, to a large extent, a result of approximately swash-alignment to the dominant waves, which prevents vigorous longshore movement, the largely relict nature of the beach material, and longshore variations in wave energy (King 1972; Gleason and Hardcastle 1973). Shape sorting is apparently unimportant along Chesil Beach, possibly because of the effect of the clast traction carpet and the high wave energy, which enables clasts of all shapes to be readily transported. It was found that the largest tracer material travelled furthest on Chesil Beach, as well as on the Hurst Castle barrier spit in southern England. This may possibly occur because cross-shore sorting processes place the largest material in the lower portions of the beach, where they are most accessible for longshore transport. On the other hand, smaller material travels further on Slapton Beach in south-western England (Nicholls and Webber 1988).

5 Barriers and Related Accumulation Features

Accumulation forms develop where local conditions encourage deposition, and they are therefore common where obstacles interrupt longshore flow, there is a sudden change in the direction of the coast, and in sheltered, wave shadow zones between islands and the mainland. Primary accumulation features also create conditions that may cause secondary, or induced, forms to develop. A number of factors complicate their development, including erosion and retreat of the coast to which they are attached, and changes in relative sea level, wave energy and direction, and sediment supply (Zenkovitch 1967).

There have been many attempts to classify coastal accumulation forms, using a variety of criteria, but none is entirely satisfactory (Zenkovitch 1967; King 1972). The simplest type of classification is based upon how they are attached to the land, although this can result in the same term being used to describe features that were formed in different ways and in different situations (Table 5.1).

BARRIER COASTS

Barriers are separated from the mainland coast along much or all of their length, by lagoons, swamps, or other shallow water bodies. Barrier spits are attached to the land at one end. Although most tend to eliminate or reduce coastal irregularities, some project out towards the ocean, and then turn and trend approximately parallel to it. A

Table 5.1. Accumulation forms

A) Beaches attached to the land at one end
  1. Length greater than width
      a) Continuation of original coast, or parallel to the coast[1] (barrier spits)
      b) Extending out from the coast at very high angles[2] (arrows), or extending out from the lee side of an island (comet tail spits)
  2. Length less than width—forelands

B) Beaches attached to the land at two ends
  1. Looped forms extending out from the coast
      a) extending from lee side of an island (looped barriers)
      b) a barrier spit curving back onto the land[3] or two barrier spits or tombolos joining up[4] (cuspate barriers)
  2. Connecting islands with islands, or islands with the mainland (tombolos)
      a) single form (tombolos)
      b) single beach looped at one end (Y-tombolos)
      c) two beaches (double tombolos)
  3. Closing off a bay or estuary (barrier beaches)
      a) at the mouth (front) of a bay (baymouth barriers)
      b) between the head and mouth of a bay (midbay barriers)
      c) at the head (back) of a bay (bayhead barriers)

C) Forms completely detached from the land (barrier islands)

Notes: [1] A winged headland is a special case in which an eroding headland provides sediment to barrier spits extending out from each side. [2] A flying spit is a former tombolo connected to an island that has now disappeared. [3] Looped spits. [4] Double fringing spits.
Source: Trenhaile 1990.

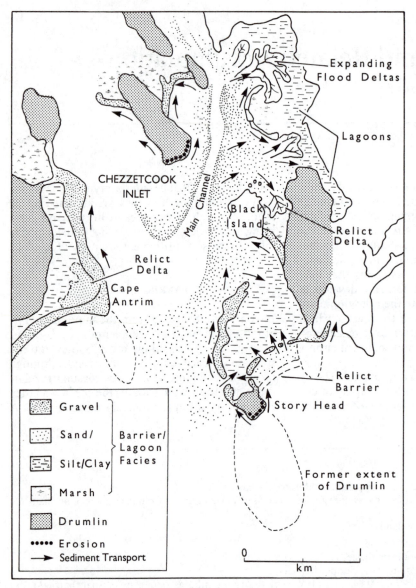

Fig. 5.1. Erosion of a paraglacial coast, Chezzetcook Inlet, Nova Scotia. Barriers are constructed of shingle derived from the erosion of drumlin promontaries. The finer sediment is transported landwards, via flood ramps and deltas, and eventually deposited in lagoons and in the upper reaches of the estuary (Carter *et al.* 1989)

barrier spit is described as recurved if its end curves away from the incoming waves, and compound recurved if there are a number of successively landward deflected termini along its inner side. The term 'serpentine' is used for barrier spits that have meandering axes caused by shifting currents. Longshore movement of material down each side of an island can produce comet tail spits extending from its rear. Single spits normally trail behind narrow islands, but they may extend from each side of wider islands. Arrows are spit-like features, fed by longshore movement of material from either

side, that grow out from a coast. Barrier spits extending out from the opposite banks of inlets may result from unidirectional growth and breaching of baymouth barriers. Spits also grow out to each other, however, as a result of opposing longshore currents generated by changes in coastal orientation or bottom topography.

A distinction has been made between subaerial spits and their submarine platform foundations. Experimental work suggested that the depth of the water above a platform is constant. Spit and platform growth generally occur in alternating cycles

a)   b)

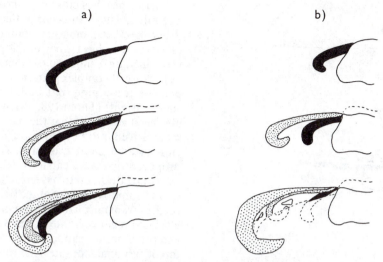

Fig. 5.2. Possible evolution of a barrier spit: (a) from curved spits, of similar length, developing seawards of their predecessors; or (b) from spits, of increasing length, extending a little further along the coast than their predecessors (Leatherman 1987)

and are inversely related, so that when one grows at a faster rate, the development of the other slows down (Meistrell 1966). Repeated inlet breaching and downdrift migration allow flood-tide deltas to coalesce within lagoons on the US mid-Atlantic coast, creating platforms that facilitate onshore barrier migration (Niedoroda *et al.* 1985). Platforms formed by onshore transport and trapping of fine material before an advancing erosional front also facilitate barrier migration or breakdown in eastern Canada (Forbes *et al.* 1991) (Fig. 5.1). In south-western Newfoundland, seaward progradation and alongshore extension of coarse clastic barriers over sandy, subaqueous platforms, provide support for the contention that platform and spit growth are inversely related (Shaw and Forbes 1992). Although platforms preceded spit formation in fairly deep water in eastern Denmark, however, spits were able to prograde over the bottom in depths of less than a few metres, without first forming platforms.

Barrier spits develop where sediment, moving alongshore, is deposited at the mouth of estuaries and other places where there is an abrupt change in the direction of the coast. Growth could be by curved spits, of similar length, forming seawards of their predecessors, with erosion of the sediment-supplying cliffs to which they are attached. Alternatively, each spit could extend a little further along

the coast than its predecessor, before curving back towards the land (Fig. 5.2). Spit elongation can sometimes take place where there are low rates of longshore transport. This could occur, for example, where ebb-tidal flow in an inlet channel is subparallel to the coast, causing a channel margin linear bar to develop at the distal end of a spit (Aubrey and Gaines 1982) (Fig. 5.3).

Three types of berm-induced beach growth have been identified along a recurved, actively migrating barrier spit on Cape Cod (Hine 1979) (Fig. 5.4). Where the beach is straight and longshore transport well developed, sand first accumulates as neap-tidal berms, and is later redistributed by uprush over the main berm during spring high water. Decreasing longshore transport with changing beach orientation produces excess sand along the middle portion of the spit, where it begins to curve landwards. Intertidal swash bars migrate onshore in this area and weld on to berms, forming new ridges immediately seawards. Berm-ridges develop in the strongly recurved, southernmost, distal portion of the spit, where beach progradation is most rapid. Berm-ridges are most common on rapidly accreting spits, and the resulting arcuate, vegetated, and sometimes dune-covered ridges and intervening wide, marsh-infilled swales are prominent features of barrier islands and spits (Nielsen *et al.* 1988).

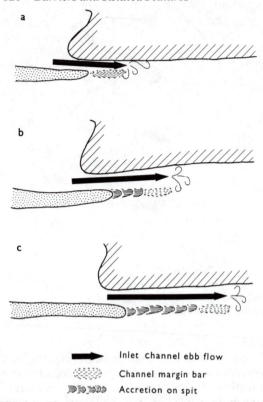

Inlet channel ebb flow
Channel margin bar
Accretion on spit

Fig. 5.3. The development of a barrier spit overlapping an adjacent headland: (a) formation of a channel margin bar by ebb-tidal flow; (b) accretion and subaerial growth; and (c) development of an inlet throat parallel to the coast (Aubrey and Gaines 1982)

Barrier beaches close or almost close off bays and inlets. They can occur at the mouth, close to the back or head, or at some point within the central portion of bays. They develop in a number of ways, including single spits extending across bays, or because of complex patterns of wave refraction, pairs of converging spits built by opposing longshore currents (Thom 1983; Melville 1984). It has also been suggested that they can be built by sediment driven into bays by wave action, independently of alongshore transport. Zenkovitch (1967) proposed that wave capacity declines as they refract and advance into bays, resulting in the formation of a midbay barrier at the point beyond which further transport cannot take place. In formerly glaciated regions, however, many midbay barriers and other accumulation forms have developed on top of terminal moraines that run across bays and other inlets (King 1972; Coakley 1989).

Barrier islands enclose estuaries, embayments, or narrow lagoons, which are connected to the open sea through channels or tidal inlets between the islands. Some portions of long barrier island chains may consist of large barrier spits or barrier beaches, that are still attached to the land at one or both ends. There is a strong relationship between the occurrence of barrier coasts and types of tectonic plate margin. About 49 percent of the world's barriers are on trailing margins, 24 percent on collision margins, and 27 percent on marginal sea coasts (Chapter 1). Barriers are therefore most

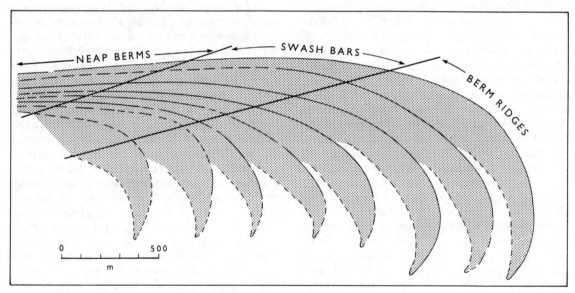

Fig. 5.4. Berm widening zones on an idealized barrier spit (Hine 1979)

numerous off coasts with gently sloping inner shelves, and adjoining coastal plains containing large amounts of detrital sediment. Only about 10 percent of the world's barriers are on coasts that have tidal ranges greater than 3 m (Glaeser 1978). Mesotidal barriers tend to have drumstick shapes as a result of wave refraction around large ebb-tidal deltas, which causes sediment transport reversal on the downdrift side of inlets. The reversed flow slows down sediment bypassing of inlets, and allows swash bars, and other sediment associated with the deltas, to be welded on to the beach. Nevertheless, despite some characteristic differences between micro- and mesotidal barriers (Hayes 1979) (Table 5.2), there are exceptions to the model, as in the microtidal Gulf of St Lawrence, where barriers have some mesotidal characteristics, including the number and importance of the tidal inlets (McCann 1979).

Barrier coasts contain a number of distinct depositional environments, including the mainland, lagoon, inlet and inlet deltas, barrier, barrier platform, and shoreface. Tidal inlets are the most dynamic of these environments, and they experience the most rapid morphological changes. There are two main types: those formed by storm-generated scour tend to be shallow and prone to rapid migration or closure, whereas those formed by the growth of barrier spits across the mouth of estuaries are generally deeper and more stable (Hayes 1991). There are marked differences in tide- and wave-dominated inlets along the US Atlantic coast. Cross-shore and longshore sediment transport gradually fills ephemeral, wave-dominated inlets, whereas ebb-delta bar bypassing concentrates tide-dominated inlet deposits at the updrift ends of barrier islands. Inlets are long and narrow along wave-dominated coasts, and they migrate in the direction of the prevailing longshore current, whereas the more stable forms in tide-dominated areas tend to be wider and shorter, and the lagoon and flood-tidal deltas of wave-dominated areas are replaced by salt marshes (Oertel 1979; Moslow and Tye 1985) (cf. Fig. 7.8). The characteristics of barrier coasts, however, are determined by the relative effects of tidal and wave processes, rather than by their absolute values (Davis and Hayes 1984). Therefore, despite much lower tidal range and wave energy, wave- or tide-dominated barriers along the coast of west-central Florida are similar to those in the higher energy environments off Georgia, and in the German Bight (Davis 1989).

Sandy Barrier Systems

Long, low sandy barriers may be supplied with sediment from alongshore or offshore sources. Both were important along the barrier coast of the Netherlands during the late Atlantic and sub-Boreal. Most of the barrier systems of North America appear to have resulted from the erosion and longshore transport of fluvial deltaic sand (Nummedal 1983), but most sand came from offshore sources on the embayed, compartmentalized coast of New South Wales (Roy *et al.* 1980; Thom *et al.* 1981).

Three basic mechanisms have been proposed for the origin of barrier islands (Fig. 5.5). Historical charts suggest that many small, and a few larger, islands have emerged from shoals since the eighteenth century in Louisiana, Alabama, and Mississippi. Barrier islands in this area overlie nearshore open marine deposits, and they have been

Table 5.2. Geomorphological differences between microtidal and mesotidal barrier islands with medium wave energy

Barrier type	Length (km)	Shape	Washover features	Tidal inlets	Flood-tidal deltas	Ebb-tidal deltas
Microtidal	30–100	elongated	Abundant; many wash over fans and terraces	Few	Large, often coupled with washovers	Small to absent
Mesotidal	3–30	drumstick	Minor beach ridges or washover terraces; washover fans are rare	Many	Moderate size to absent	Large, with strong wave refraction effects

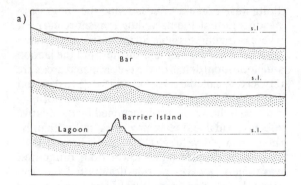

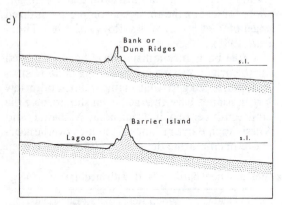

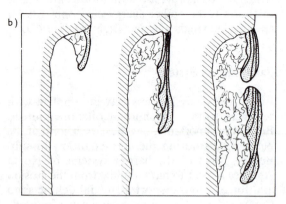

Fig. 5.5. Theories for the origin of barrier islands from (a) a submarine bar, (b) spit elongation, and (c) submergence of beach ridges or dunes by rising sea level (Hoyt 1967)

attributed to shoal or barrier spit progradation and subsequent segmentation (Otvos 1985). On the Texan coast, Matagorda Island may have formed from emergence of a landward migrating sand shoal during the Holocene transgression, and islands have developed from shallow subtidal shoals off the western coast of Florida (Wilkinson

1975; Davis *et al.* 1979). Barrier islands have been attributed to partial submergence of dune or beach ridges during the Holocene, and drowning of the lower land behind. Hoyt (1967) recognized that some barrier islands are formed by spit detachment, and small, short-lived forms by the upward growth of submarine bars, but he considered that the general absence of shallow neritic sediments, fauna, and open ocean beaches on the landward side of barriers show that these mechanisms have only minor roles. Nevertheless, it has been proposed that beach ridges on the Outer Banks of Virginia and North Carolina are indicative of lengthening, shore parallel spits (Fisher 1968*a*, *b*). The barrier islands of the Mississippi delta plain may also have formed through spit breaching, rather than by the emergence of submarine bars. Abandoned deltas were first transformed into erosional headlands with flanking barriers and breached spits. As submergence continued, transgressive barrier island arcs were formed, as described by Hoyt's detachment mechanism. The final stage involved development of inner shelf shoals, as the islands failed to keep up with the relative rise in sea level (Penland *et al.* 1988).

Colquhoun *et al.* (1968) proposed that there are two types of barrier on the Atlantic Coast of the United States. Primary barriers are formed according to Hoyt's drowned ridge mechanism. They migrate landwards during marine transgressions, and are often reworked and buried by encroaching deltas. Secondary barriers are formed by spit extension from deltas or headlands, and they develop after, and seawards of, primary barriers, during a period of stable or slowly falling sea level. The barrier system of North Carolina originated as a primary form through partial submergence of a pre-existing ridge, which reoccupied the site of an older barrier formed during an earlier stillstand. Extensive secondary barrier spits then developed, and continued migration caused the primary and secondary barriers to coalesce (Pierce and Colquhoun 1970).

Longshore drift and the longitudinal growth of barriers through spit extension may have been too important to accept Hoyt's hypothesis as a general solution to US Atlantic coast barriers (Field and Duane 1977). There are early Holocene barriers on the continental shelf, and backbarrier and possibly barrier sediments are widely preserved on the inner shelf and shoreface. The lagoonal carpet over the mid-Atlantic shelf suggests that most barriers formed on the continental shelf. Although the migrating beachface was nourished by abundant littoral drift, and was able to maintain its equilibrium profile with rising sea level, the unnourished swale behind the beach was flooded (Swift *et al.* 1985; Niedoroda *et al.* 1985). The Core Banks, the southern portion of the Outer Banks of North Carolina, probably originated about 15,000 years ago, as part of an elongating barrier spit, or by mainland beach detachment, far out on the continental shelf. The southernmost part was formed by spit progradation as a result of washover deposition and southerly longshore transport. The Core Banks have migrated landwards up to 6.7 km in the last 6,675 years, although only 1.6 km to 3.2 km in the last 4,000 years (Moslow and Heron 1979).

Instead of continuous shoreface erosion, the barriers off Long Island may have experienced discontinuous retreat by in-place drowning and stepwise retreat of the surf zone. This would have prevented reworking of backbarrier sediments, and would explain their extensive preservation on the inner shelf. There was a chain of barrier islands off southern Long Island about 9,000 years ago, when the sea was 24 m lower than today. About 7,000 years ago the barriers were overstepped by rapidly rising sea level, and the surf zone shifted to a position only about 2 km offshore of the present shoreline. This was the landward margin of the old lagoon—the steep shoreward face of an emerged, late Pleistocene barrier. This barrier then acted as a core for the growth of a new barrier chain, which developed both by detachment and spit progradation from headlands. Overstepping of barriers occurs when there is a rapid rise in sea level and low sand supply, whereas abundant sand supply and slow submergence are conducive to continuous shoreface retreat. Stationary upbuilding or seaward progradation can take place when submergence is slowed or reversed, and/or when there is high sand supply (Rampino and Sanders 1981).

Antecedent topography determines regional slopes, the initial orientation of the coast with respect to winds and waves, provides watersheds and valleys that can be inherited as headlands and embayments, and may control the thickness of Holocene sediments. Modern barrier systems often formed on top of older Pleistocene barriers, or they are located somewhat seaward of them. Most barrier islands on the German North Sea coast, for example, consist of a core of Pleistocene deposits, mantled by Holocene sediments (Streif

1989). In south-eastern Australia, a distinct inner barrier of last interglacial age is separated from an outer Holocene barrier by a lagoon and swamp tract (Thom 1983). Barrier islands on the southern Atlantic coast of the United States migrated landwards and came into contact with relict Pleistocene Silver Bluff bays and lagoons about 4,500 years ago. Holocene beaches were welded directly onto the relict shore in some areas, but where there are major rivers, the barrier islands are separated from the Silver Bluff shore by 3–7 km of marsh and tidal channels—the deltaic deposits of the late Holocene rivers (Oertel 1979). Palaeofluvial drainage networks also provided controlling topographic 'highs' in the antecedent topography. Flooding of a deeply incised drainage system produced baymouth bars in the northern part of the Delmarva Peninsula on the US mid-Atlantic coast, and the position of some inlets was also strongly influenced by incised valleys on the antecedent topography. In the southern part, however, the pre-Holocene surface was less dissected and the barriers are disconnected from the mainland (Demarest and Leatherman 1985; Belknap and Kraft 1985).

Shoreward barrier migration produces transgressive sequences consisting of marine sediments on top of sediments deposited in landward or lagoonal environments. Seaward progradation creates regressive barrier sequences, with landward lagoonal sediments deposited on top of more marine deposits. Transgressions occur when vertical sedimentation rates are less than the rise in sea level, and regressions when the sedimentation rate is greater than the rise in sea level. Although there are regressive barrier sequences on Kiawah Island, South Carolina, and on Galveston and Matagorda Islands, in Texas, most barriers in the world are being driven landwards by the eustatic rise in sea level (Kraft *et al.* 1979). Ideal transgressive systems have a vertical sequence of sedimentary environments identical to the horizontal sequence in the direction of the transgression. This is not always the case, however, particularly where there has been large-scale erosion and sedimentary infilling associated with tidal inlets.

Some barriers and tidal inlets gradually shift alongshore in response to longshore sediment transport. Sand transported along barriers can bypass inlets by either moving downdrift along the seaward portion of ebb-tidal deltas, or by entering the inlets on the flood tide and exiting on the ebb. Longshore migration of tidal inlets can create a series of low, elongated, shore parallel islands in the partially marsh-filled lagoons behind barriers. These islands probably develop during storms, when sand from the flood-tidal shield overtops the adjacent marsh (Cleary *et al.* 1979). Barriers generally migrate landwards through overwash, inlet dynamics, and aeolian processes. Other mechanisms, however, may also be significant in some areas. For example, landward movement of a barrier spit on Tierra Del Fuego is largely due to longshore transport around the end of the spit and back up the bayward side, even though it also has a large number of washover channels (Bujalesky and González-Bonorino 1991).

Overwash transfers beachface sediments to the backbarrier and lagoon. Overwash-dominated barriers tend to be narrow and flat, their foredunes are low and discontinuous, and vegetational recovery on the washover fans is very slow. Where there is a limited supply of sediment, barriers may essentially roll over themselves as they migrate shorewards. Although some islands along the Mississippi Delta Plain rollover, others evolve by in-place fragmentation, involving progressive narrowing, and the formation and widening of tidal inlets. In-place breakup occurs where rollover cannot compensate for rising sea level because of an insufficient supply of sediment, or where islands are too wide to allow storm surges to completely wash over them (McBride *et al.* 1995).

A number of factors determine the importance of sediment transfer through inlets, relative to storm overwash and wind. They include tidal and lagoonal characteristics, which determine inlet dimensions and tidal flow, and waves and currents near inlets, which affect inlet migration and closure. Extreme storms can open or close inlets, they determine the frequency and intensity of overwash events, and they facilitate aeolian transport. Vegetation and moisture conditions are important, and sediment abundance and beach topography partly determine the effect of storm events. Rising relative sea level also facilitates washovers on parts of the Atlantic coast of the United States.

On a barrier on Prince Edward Island, about 52 percent of landward sediment transfer is through tidal inlets, and 48 per cent by overwash and wind, although most of this latter amount occurs at the sites of five former inlets (Armon 1979). The general lack of overwash, other than around inlets, probably reflects the lack of frequent destructive storms on this coast, and the consequently

well-vegetated and continuous dune cover. Tidal inlets are even more important on some of the wider barriers on the open Atlantic coast. It has been estimated, for example, that on Assateague Island and Cape Hatteras, from 72 percent to 82 percent of the sediment moves landwards through tidal inlets, 12–14 percent by overwash, and 6–13 percent by wind. Inlet dynamics are also thought to be mainly responsible for landward sediment transport on Fire Island, New York, and tidal delta sedimentation is about 1.33 times more effective than overwash on the transgressive baymouth barriers of Rhode Island (Fisher and Simpson 1979; Leatherman 1985).

Distinct washover fans develop if overwash is localized at blowouts, former washover channels, or other low points in foredunes. If there is a fairly continuous section of low foredune, overwash produces a coalescing sheetlike deposit, or washover ramp, on the backbarrier. Large amounts of sediment can be deposited during single storms. Hurricanes Ginger and Belle deposited about 30 m^3 and 19.3 m^3 of sediment per metre of breach on Core Banks, North Carolina, and Assateague Island, Maryland, respectively (Dolan and Godfrey 1973; Fisher and Stauble 1977). The overwash surges during another storm carried an average of 20 m^3 of sediment per metre of dune breach on to a subaerial fan and barrier flats on Assateague Island, although the comparable figures for two other storms were only about 4.7 and 5.3 m^3 (Fisher *et al.* 1974; Leatherman *et al.* 1977).

Winds blowing through dunes along overwash corridors may redistribute sand recently deposited in non-vegetated washover fans and flats. If the prevailing winds are offshore, much of the sand can be blown back across the island and returned to the beachface. This also occurs during storms that have a characteristic cyclonic circulation, with a period of strong onshore winds being followed by comparable offshore winds. Overwash may therefore be ineffective in providing sediment to backbarrier marshes and bays in the long term, although sand blown back across islands may help to build up the foredune and backdune areas (Fisher and Stauble 1977; Rosen 1979; Dingler *et al.* 1992).

The type of vegetation growing on the barriers on the Atlantic coast of the United States exerts a strong influence on their development (Godfrey *et al.* 1979). In the north-east *Ammophila breviligulata* is the dominant stabilizer of dunes, whereas

Uniola paniculata, which is better adapted to heat, dominates in the south. *Uniola* does not grow laterally as well as *Ammophila*, however, and does not create as dense a grassland stand. Dune ridges therefore dominate the narrow barriers in the north, while barriers are wide and low in the south, and they have extensive washover flats and low, open, and scattered dunefields. Variations in the distribution of *Spartina patens* and its varieties also play an important role in the stabilization of washover deposits. *S. patens* dominates backbarrier grasslands in the south, and it rapidly colonizes washover deposits, which are also stabilized through aeolian winnowing and the formation of lag deposits of coarse quartz, heavy minerals, and shells, creating characteristically broad backbarrier flats. The tall form of *S. patens* is less frequent in the north-east, and the small, weak-stemmed variant becomes more common. This variety cannot survive substantial burial under sand, and dune-strand rather than salt-marsh vegetation is therefore responsible for the recovery of washover deposits in this area.

The northern islands along the Outer Banks of North Carolina have been developed and stabilized, whereas the southern islands largely remain in a natural state (Dolan 1973; Dolan and Godfrey 1973) (Fig. 5.6). Beach and dune stabilization on the northern islands began in the 1930s, and there are now narrow berms and a high, continuous, and artificial barrier dune. Although the dune protects the highway and towns, it prevents or inhibits overwash, dune movement, the opening of inlets, and natural barrier recession. Furthermore, the artificial dune absorbs the full impact of storm waves and is rapidly eroded. Consequently, instead of saving the islands, stabilization may be threatening their very existence. By comparison, the southern islands have wide berms and a zone of low dunes with numerous overwash passes that help to dissipate storm wave energy, causing sand to be deposited on the inner portion of the islands. The southern islands are therefore able to respond in a natural way to storms and rising sea level by retreating landwards. Nevertheless, Leatherman (1979) considered that barrier dunes do not accelerate beach erosion, but rather serve as a source of sediment for beach nourishment during storms, thereby reducing beach erosion.

Coarse-clastic Barrier Systems

Coarse clastic barriers form a distinctive suite of depositional features on the wave-dominated coasts of the mid-latitudes, especially where waves are reworking glacial and periglacial deposits (Jensen and Stecher 1992; McKay and Terich 1992; Shulmeister and Kirk 1993; Shaw *et al.* 1993; Forbes *et al.* 1995) (Table 5.3). There have been few studies of coarse clastic barriers in Arctic areas, where thermokarst, permafrost, and sea ice are important. Although they may evolve in much the same way as in temperate regions, important differences arise as a result of, for example, short-term increases in sediment supply caused by thermal erosion of ice-rich bluffs, and erosion of the lower

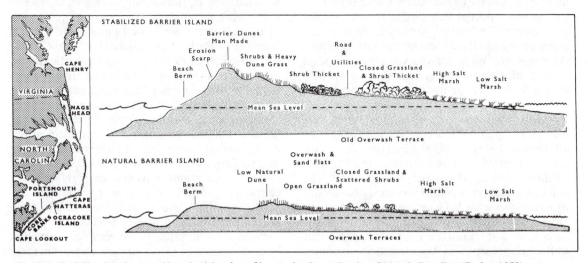

Fig. 5.6. Stabilized and natural barrier island profiles on the Outer Banks of North Carolina (Dolan 1973)

Table 5.3. Comparison of sand and coarse clastic barriers

	Sand	Coarse clastic
Beach and nearshore geometry	Gentle Dissipative Rapid adjustments to process variations	Steep Reflective Slow adjustment to process variations
Barrier Morphology	Extensive dunes Inlets/stream exit common Tidal lagoons common Freshwater lagoons rare Large overwash fans Crestal overtopping deposits rare No backbeach seepage features	Dunes absent or rare Inlets/stream exit uncommon Tidal lagoons rare Freshwater lagoons common Generally small overwash fans, though broad shorewards crestal dislocation by sluicing overwash can occur under extreme events
Sediments	Fine Non-porous Low roughness Low entrainment threshold Normal sediment grading units	Coarse Porous to non-porous due to high internal size variation from stratification High roughness High entrainment threshold Coarsening upward units common Clast form may be as important as clast size in determining sediment mobility and zonation

Source: Carter and Orford 1980.

shoreface and landward transport by ice pile-ups (Chapter 12).

Sediment characteristics are important in barrier development. Most contain varying proportions of clasts and sand, which determine the mobility of the clastic mass and other aspects of beach sedimentation. Coarse clastic beaches with little sand have steep reflective foreshores with a limited morphodynamic range. Sand packed between clasts on the intertidal beachface reduces hydrostatic porosity and permeability, whereas an excess of sand saturates the coarse matrix, and provides a source for barrier top dunes (Forbes and Taylor 1987).

With a headland source of heterogeneous sediment, sorting initially occurs at the proximal or attached end of drift-aligned barriers. The larger clasts are retained as bed armour, and the finer sediment is advected seawards in suspension. The remaining material is graded accorded to its mobility as it moves away from the source (Fig. 5.7). As most clasts move as bedload, the concepts of acceptance and rejection are especially important to their movement and deposition. Sedimentary structures tend to retain clasts that are similar in

size to the material making up the bed, while rejecting those of different size. The rate that clasts move through the grading zone therefore depends not only on their size and shape, but also on their relationship to the background mass. Medium-sized material is trapped by larger material, and it merges with sediment of similar size, whereas low concentrations are able to pass over (overpassing) a frame of finer sediment. Individual clasts that are much bigger than the background mass may sink into it and lose their mobility. Sediment is sorted in the cross-shore as well as in the longshore direction (Chapter 4). With mixed sand and coarse clastic beaches, the sand tends to occupy the lower foreshore, or it forms dunes above the clastic ridge. Sand may also accumulate on the downdrift sector of the grading zone, where it facilitates downdrift overpassing of any clasts that escape from the updrift grading zone. Barriers are therefore often coarse-grained and reflective near headlands, and sandy and dissipative towards the centre (Carter and Orford 1991; Orford *et al*. 1991).

Little recycling take place once sediment has been incorporated into a barrier, and the magnitude of the sediment supply is therefore crucial in

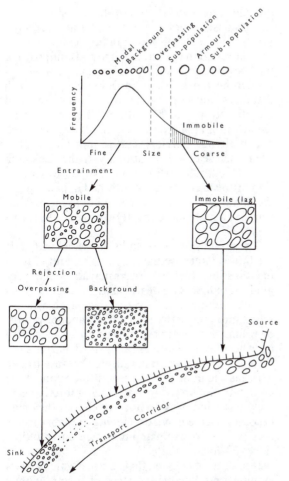

Fig. 5.7. Sorting along a coarse-grained beach. Waves sort and subdivide the initial textural population through clast rejection, producing a graded beach (Orford *et al.* 1991)

determining barrier development. Clast supply changes through time as a result of the diminution of sources, a decline in the rise in sea level, and increasing disruption of littoral transport by developing headlands and, as the drift system matures, by sediment cells. Failure of sediment supplies causes barriers to become massive, solitary features, rather than multi-ridge complexes.

Uprush excursions that just carry over the crest of a barrier deposit sediment on top of the ridge. Overwashing supersedes crest overtopping when enough uprush carries over the crest to create a unidirectional flow on the backslope. Overwashing erodes the upper part of the backbeach, lowers the barrier crest, and deposits sediment on its

landward side. Breaching of high beach crests forms narrow channels that function as conduits into fan- or tongue-shaped sinks. Even greater amounts of water passing over the crest may move it landwards, and deposit sediment across nearly continuous washover fans, which laterally merge into washover flats. Gradual reduction in overwash volume as a ridge increases in height is reflected in the sedimentary sequence in a barrier in southeastern Ireland, where a bottom unit, consisting of sediments deposited by large uprush bores, passes upwards into washover clasts, and then into overtop sediments. Nevertheless, variations in crest height and uprush excursions owing to edge waves, amplification at cusps, wave refraction, and other mechanisms can cause overtop and washover sedimentation to occur at different heights on the same barrier during severe storms.

Coarse clastic barrier beaches enclose brackish or freshwater lagoons. Closure of a lagoonal outlet can have profound effects on barrier dynamics and morphology (Carter *et al.* 1992*a*). Lagoons also discharge seawards through surface channel outlets or by seepage, causing extensive rilling of the beachface in the lower intertidal zone. The general lack of tidal inlets reflects the importance of seepage, and sediment movement is therefore mainly onshore and dominated by overwash processes. Seepage into a lagoon when sea level is higher can form fans along the barrier–lagoon margin, and gradual excavation of seepage hollows helps to undermine the barrier crest.

Most coarse clastic barriers rollover as they migrate landwards. Barrier migration is the result of such factors as high infiltration rates, uprush dominance on steep, rough beachfaces, and high seepage rates, together with the general lack of tidal inlets which would permit sediment to return seawards (Forbes and Taylor 1987). Coarse clastic barriers move rather slowly, however, because of their reflective nature, and the high entrainment thresholds of the sediment. Landward migration has been at least partly in response to rising sea level, although other processes can lower beach crests and increase the frequency of overwash and rollover events. Nevertheless, barrier thickening and seaward progradation can occur where the sediment supply is able to replace onshore or offshore losses, and sustain barrier elongation and stretching. A series of coherent beach ridges is produced on prograding barriers where there is an excessive but intermittent supply of sediment, and a massive, solitary ridge where the supply is more continuous (Carter and Orford 1980; Carter *et al.* 1987).

There are important differences between drift- and swash-aligned barriers. On drift-aligned barriers, strong gradients in such factors as wave height, approach angle, power flux, and longshore currents drive sediment alongshore, from sources to sinks. The shorelines of swash-aligned barriers are parallel to the crests of the incoming waves, and there is no net longshore transport. Drift-aligned barriers only persist when there is an adequate supply of sediment, and they may be transformed into swash-aligned barriers when there is a decline in supply, a change in wave conditions, or a reduction in basement control. Individual barriers may have swash- and drift-aligned segments, however, especially when they are being extended, or are deteriorating (Orford *et al.* 1991).

The continuing supply of sediment to a drift-aligned barrier usually causes extension and/or broadening of its distal end, often through the addition of sets of ridges representing distinct periods of deposition. Longshore transport can also occur in the form of migrating nesses—triangular-shaped sets of ridges created by sediment pulses arising from discrete erosion events, intermittent onshore sediment transport, or spasmodic releases of drift (Orford *et al.* 1991). The probability of a given clast reaching the extremity of a growing drift-aligned barrier decreases as its length increases. The shortfall in sediment supply is compensated by remobilization of sediment, which may cause it to become thinner at its proximal end. Distinct erosion/accretion cells can also develop along a barrier, with redistributed sediment usually accumulating in the downdrift portion of each cell. This can eventually result in breaching and formation of a series of islands, which may reform as a coherent barrier further landwards. As a coarse clastic barrier develops and begins to move into equilibrium, some portions become swash-aligned, and the potential longshore sediment transport reduces to zero. In other areas, sediment grading according to size and shape can balance residual wave gradients, so that net sediment transport is zero even though waves still approach the barrier at an angle (Carter *et al.* 1987).

The form and behaviour of barriers change as they elongate and approach equilibrium. These changes are associated with the transition from forms consisting of unsorted material rapidly deposited under strong wave gradient conditions, to ones adjusted to the wave regime and composed of well-sorted sediment. As the angle of wave approach becomes less oblique, increasing uprush distances cause the barrier to roll over itself and possibly steepen its beachface. Although barriers usually migrate landwards under rising sea level, swash-aligned barriers tend to migrate onshore even when sea level is stationary. If the rise in sea level is too fast for rollover, however, the lower portion of the barrier may be drowned in place, while the remaining upper part continues to move landwards at a more rapid rate (Forbes *et al.* 1991; Ruz *et al.* 1992).

The establishment of links between headlands facilitates lateral sediment transfer, and if sediment is still available, barriers may thicken or prograde. If sediment is not available, however, barriers become thinner and lower as they stretch themselves into arcs between headland anchor points, by rolling over faster near the centre of bays than near headlands. The central parts of barriers are therefore often aligned parallel to the wave fronts, whereas the flanks are adjusted to oblique wave gradients. Solitary ridges advancing into increasingly restricted bays do not stretch, but they may become concatenated, developing as higher forms that can be broken up and reworked into multiple ridges (Orford *et al.* 1991).

It has been suggested that coarse clastic barriers assume a predictable sequence of forms, as they are initiated, established, and finally broken up (Fig. 5.8). Barrier development can be arrested when unfavourable conditions are encountered, however, as when a deep tidal channel prevents extension of a drift-aligned barrier. The first stage is the formation of drift-aligned barriers, which are then either arrested and eventually broken down, or they develop into swash-aligned forms. Swash-aligned barriers either become linked barriers or, in the swash-arrested form, stable barrier spits. Eventual breakdown of barriers can occur in several ways, depending on basement configuration, the presence of fine sediment, and sediment supply. Fringing barriers, flanking cliffs that limit migration, ultimately develop where the barrier is segmented by the basement.

Coarse clastic barriers can develop under rising or falling sea levels (Orford *et al.* 1991; Carter *et al.* 1992*b*). Rising sea level helps to provide adequate

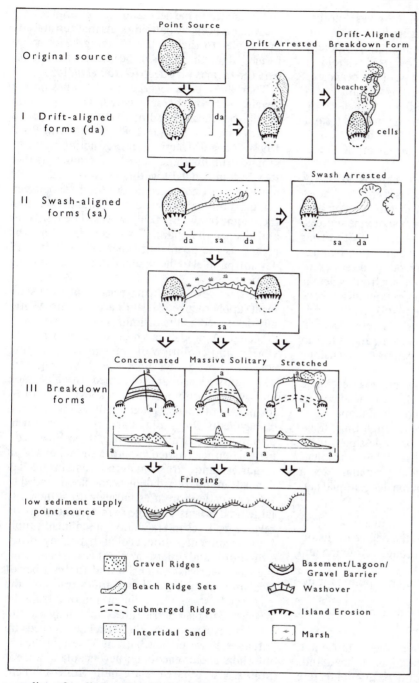

Fig. 5.8. The initiation, establishment, and breakdown of shingle barriers. Drift-aligned, type I barriers may be arrested and eventually breakdown, or they may develop into swash-aligned, type II forms. Swash-aligned barriers can become stable, swash-arrested spits, or linked barriers that can rollover. The configuration of the basement, the presence of fine sediment, and the sediment supply determine the type of breakdown barrier (type III) that develops. The ultimate fringing barrier forms where the basement segments a migrating barrier, although they also develop on rocky coasts with a limited supply of sediment (Orford *et al.* 1991)

supplies of sediment for the formation of drift-aligned barriers, but they also develop with falling sea level, where sediment is supplied from river mouths or deltas. Swash-aligned ridges tend to form when relative sea level is falling, causing shelf materials to be reworked and carried on to the shore. It is more difficult to account for their development when relative sea level is rising. This is because of the apparent lack of a mechanism capable of supplying the fresh sediment needed to construct new ridges along their length. Nevertheless, clasts could be moved along the coast by local, short and poorly refracted, oblique waves. Another possibility is that with a low supply of sediment, the development of equilibrium grading, whereby each clast is always at or below its threshold of movement, arrests the plan shape of the beach at a curvature, before longshore sediment transport is reduced to zero. A third possibility is that coarse sediment, slowly accumulating on intertidal sand around the high tidal level, eventually causes profile adjustment and the formation of a new clastic ridge.

Sea-bed clast sources are exhausted fairly quickly when relative sea level is rising, and most sediment is therefore obtained from erosion and dispersal from point or irregular shoreline sources, including cliffs, platforms, and deltas. With rising or stationary sea level, advance of an erosional front causes sediment to be eroded and reworked. Some moves beyond the advancing front, where it forms barrier spits and tombolos, and fills in bays and estuaries which provide platforms for landward barrier migration. Other sediment is carried seawards behind the erosional front, and is lost to the shelf through in-place drowning or downslope transport.

Much of the recent work on coarse clastic barriers has been on glaciated, paraglacial coasts, which tend to have irregular plan shapes. The degree of irregularity varies through time, however, as headlands become more or less prominent as a result of deposition in adjacent bays and inlets, and differential rates of erosion. Headlands exert a number of important controls on depositional patterns and processes:

a) they are point sources of sediment;
b) their spacing influences the degree of distortion of the wave field through refraction and diffraction;
c) they serve as hinge or anchor points for depositional forms; and
d) they control the rate and pattern of sediment transport along a coast.

Because of wave refraction, the distance and depth of the water between adjacent headlands help to determine the type of depositional feature that develops in an area, although sediment mobility and the strength of the tidal and river flows between headlands are also important. Headlands that are close together and separated by shallow water may be joined together, but as the spacing increases, barrier beaches are replaced by flying

spits attached to the headlands at high angles, or, if the water is deep, by flanking spits that run at fairly low angles to their sides. Fringing beaches are formed along the flanks of headlands if the water between them is very deep (Carter *et al.* 1987).

There have been several recent studies on the development of coarse clastic barrier spits and barrier beaches in eroding drumlin fields (Boyd *et al.* 1987; Carter *et al.* 1987; Nichol and Boyd 1993). Wave diffraction and refraction patterns around drumlin islands largely determine whether they become linked together. Nevertheless, there must also be enough sediment to move between drumlins through water several times deeper than the height of the breaking waves. The narrower and deeper the channel between drumlins, the greater is the probability that lateral spits will develop parallel to the channel, rather than linking barriers.

Barriers develop during periods of high sediment supply, then retreat landwards and are eventually destroyed as the drumlin sediment source is depleted. A model has been developed for the genesis and evolution of the embayed drumlin coasts of Nova Scotia (Boyd *et al.* 1987) (Fig. 5.9). The first two stages are concerned with glaciation and, with the rise in relative sea level, the transformation of former glacial valleys into coastal embayments. Prograding barriers develop downdrift of eroding drumlins in stage three. Finer sediment is either carried offshore or deposited in estuaries, and larger material remains as lag deposits in shoals. Tidal inlets and flood- and ebb-tide deltas develop in estuaries with sufficiently large tidal prisms. In stage four, continued rise in relative sea level and reduction in sediment supply from drumlin depletion, aeolian bypassing, diversion into tidal inlets, and other mechanisms, increase overwash intensity and the number of tidal inlets, resulting in shoreward migration of the barriers. In stage five, sediment removal from the beachface to backbarrier areas ultimately leads to barrier destruction. As the shoreline proceeds up estuaries, however, new barriers are established as intertidal shoals encounter new headland anchor points. These barriers are built from sediment obtained from the destruction of barriers that once existed further seawards within the same embayments, from contemporaneous erosion of drumlin headlands, and from dunes, washover fans, estuarine tidal flats, flood-tidal deltas, and other backbarrier environments. Formation of new barriers

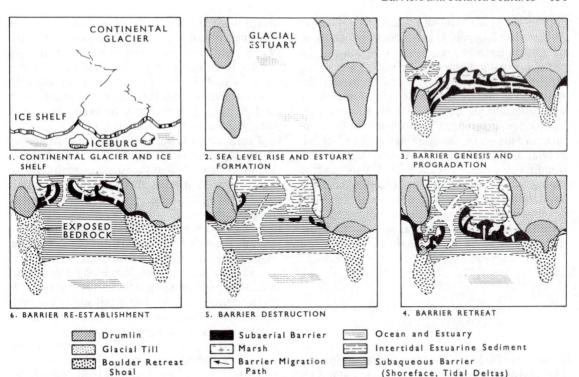

Fig. 5.9. Barrier genesis, destruction, and re-establishment in response to rising relative sea level on the paraglacial coast of Nova Scotia. Stages 1 and 2 represent ice advance and ablation, and subsequent evolution consists of a cyclic repetition of stages 3 to 6 (Boyd *et al.* 1987)

in stage six demonstrates the cyclical nature of barrier formation and destruction on drumlin coasts.

The development of a drift-aligned barrier spit and a swash-aligned barrier beach has been studied on the flanks of Story Head, a drumlin in the Chezzetcook Inlet in Nova Scotia (Carter *et al.* 1990*b*) (Fig. 5.1). The spit initially developed as a flanking form, but with sediment starvation, cannibalization of the proximal portion led to development of a series of beach cells, recurves, and islands. The breaches, however, have healed and reopened several times over the last forty years. There is an abrupt change along the barrier beach on the eastern side of the island, from a reflective, drift-aligned, sharp-crested transport corridor near the drumlin, to a dissipative, swash-aligned, over-wash-dominated beach further away. Rapid transport in the updrift portion of the barrier is therefore replaced by much slower transport at the downdrift end. Reduced littoral transport and stretching of the drift-aligned portion of the barrier may eventually threaten its stability. Shoreward migration of the barrier, which is primarily the result of

the rapid rise in relative sea level, accelerated after 1954, when its lower part became stranded underwater on the shoreface. The barrier remnant was left behind in the surf zone and the upper part moved rapidly onshore, possibly in part because of its reduced volume (Forbes *et al.* 1991). After barrier overstepping, the rollover time—the time needed for a barrier to turn completely over—was sharply reduced, and is now only three to four years.

In southern Ireland, barriers have experienced a fairly slow rise in relative sea level, and the supply of fresh sediment to the shore has therefore been declining. This has resulted in sediment reworking, and the shoreline has become swash-aligned. Movement of finer particles into tidal inlets forms deltas and other subaqueous bedforms, as well as barrier spits and subaerial dunes. Nova Scotia is experiencing rapidly rising relative sea level, and the switching of erosion and deposition zones on this intricate coast is often quite rapid (Carter *et al.* 1992*b*; Nichol and Boyd 1993). Ridge aggradation takes place where there is a local excess of sediment, but barrier morphology must change to counter rising sea level where it is less abundant. This can involve changes in barrier slope, width, or height, or cannibalization or reworking of its sediment to supplement deficits. Rapidly rising relative sea levels therefore lead to barrier destruction, either by remobilization of sediments where breaching is occurring, or less often by overstepping or drowning. The much greater development of barriers on the drumlin coasts of Ireland, in relation to those in Nova Scotia, suggests that rapidly rising relative sea level may also, in some places, inhibit barrier development. On the other hand, when sea level is stationary, major barrier development on drumlin coasts is adversely affected by the limited supply of sediment, and its dispersion in many directions (Carter and Orford 1988).

TOMBOLOS, FORELANDS, AND
CUSPATE BARRIERS

Tombolos (Table 5.1) develop where relative sea level is rising, falling, or stationary, and in a wide variety of tidal environments. They are the result of wave refraction and diffraction on the lee side of islands, which cause longshore currents to converge from each end of the shadow zone, and the shelter afforded to the shadow zone from strong wave action. Tombolos can rapidly develop where there is abundant sediment and strong longshore transport. On Iwo-Jima, where rapid uplift provided large amounts of sediment, a mainland salient advanced towards a group of offshore rocks at an average rate of about 50 m yr^{-1}, producing a 1,700,000 m^2 tombolo over a 33-year period (Shigemura *et al.* 1984).

Simple tombolos can be symmetrical or asymmetrical, depending on the relative strength of the waves and currents approaching from either side. In Boston Harbor, the formation of tombolos connecting drumlin islands may have been influenced by the underlying glacial topography, as well as by wave refraction patterns (Rosen and Leach 1987). Where islands extend for some distance along a coast, tombolos may form at each end of the shadow zone, creating double tombolos. Some of the coarse clastic double tombolos in Maine developed where shallow bays were enclosed by barrier beaches migrating landwards in response to the Holocene rise in sea level (Duffy *et al.* 1989). In

north-western Ireland, one type of tombolo faces the direction of approach of the dominant swell waves, while another, oriented almost at right angles to them, develops where refracted waves bend around islands. The first type forms where the water is shallow and there is abundant sediment, and the second where the water is deeper or sediment is less abundant (King 1972).

Tombolo evolution and direction of growth partly depend upon whether it is constructed from sediment derived from the island or mainland, or from a combination of the two. Single or double tombolos can be formed where sediment, moving tangentially past islands, accumulates to their lee as comet tail spits, which may eventually extend to the mainland. Y-shaped tombolos develop where: cuspate barriers grow out from the mainland to islands; looped barriers, formed by pairs of attached comet tail spits, extend out from islands to the mainland; and comet tail spits or looped barriers on islands coalesce with cuspate barriers on the mainland. On Japanese open coasts, a tombolo grows out of the coast if the ratio of an island's offshore distance to its length is less than or equal to 1.5, a salient develops if the ratio is between 1.5 and 3.5, and the island exerts no influence on the coast if the ratio is greater or equal to 3.5 (Sunamura and Mizuno 1987).

Forelands and cuspate barriers are created as coastlines are reoriented to face the dominant waves. Forelands are salients that extend out from coasts where two dominant swells are in opposition, as, for example, in the sheltered or shadow area behind an island. They also develop, however, where sediment transport is from only one side. The terms 'cuspate barrier' or 'cuspate foreland' have been used to describe structures that enclose lagoons or swampy areas, although they may evolve into forelands if their interiors are filled. Cuspate forelands can develop where longshore transport is mainly from one direction, as when a looped spit curves back onto the land, or, as in the case of double fringing spits, where there are opposing directions of longshore transport. Double fringing spits are formed by the attachment of the distal ends of two spits, or by tombolos that either grew out to islands that later eroded and disappeared, or failed to reach the islands because of strong tidal or other currents (Duffy *et al.* 1989).

Dungeness, on the south-eastern coast of Britain, is one of the most famous and best-studied forelands. This triangular promontory, which contains about 500 coarse clastic ridges, developed under the influence of two dominant swells, one from the south-west, through the entrance to the Channel, and a subsidiary one from the north-east, through the Straits of Dover. Archaeological remains, historical records, beach ridge orientation, and radiocarbon dating have been used to reconstruct the evolution of the foreland, which appears to have been slowly migrating to the east (Long and Hughes 1995).

The cuspate forelands or capes of the eastern United States may represent the larger-scale components of a hierarchical system of crescentic landforms. There has been considerable debate over the origin of these structures, which are generally distributed along the coast at intervals of between about 100 and 200 km. It has been suggested that they developed from deltas formed during the period of low sea level in the last glacial stage, by the action of waves on shoals, and convergence of coastal currents that are possibly associated with secondary rotational cells or eddies along the western margin of the Gulf Stream. Holocene stratigraphy, however, suggests that they have a diverse evolutionary history. The late Holocene depositional history of Cape Henlopen, Delaware, suggests that simple barrier spits, which

form when headlands are being eroded during the early stages of a transgression, evolve into recurved spits, and then, as sand is moved around the recurve onto the mainland, into cuspate forelands (Fig. 5.10). There is no evidence of a deltaic origin for Cape Lookout in North Carolina, which is one of the large cuspate forelands on the mid-Atlantic coast. This feature began to develop between 7,000 and 4,000 years ago, when two barriers came together to form a rapidly retreating headland and shoal system. Slower rates of sea-level rise then permitted initial spit growth and progradation of the Cape, which has been transformed into a prograding cuspate barrier during the last 3,000 years (Fig. 5.11).

Stratigraphic evidence from bore-holes and the distribution of nearshore sediment suggest that two cuspate barriers on the north-western coast of Lake Erie may be erosional remnants of much larger structures that partly developed on top of cross-lake moraines, when lake levels were lower (Coakley 1989). Surficial sediments, beach ridge orientation, and landform elements, however, support the view that Point Pelee, the most westerly of these features, largely consists of sediment carried alongshore from the eroding cliffs to the north (Trenhaile and Dumala 1978).

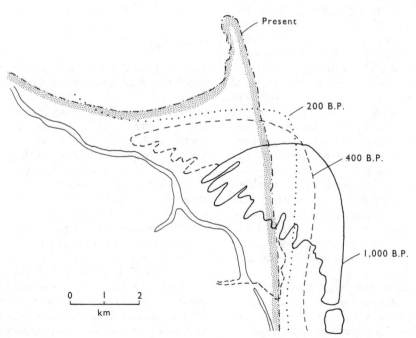

Fig. 5.10. The evolution of Cape Henlopen, Delaware (Kraft *et al.* 1978)

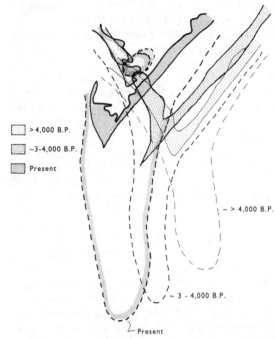

> 4,000 B.P.

~3-4,000 B.P.

Present

~ > 4,000 B.P.

~ 3 - 4,000 B.P.

Present

Fig. 5.11. The development of Cape Lookout, North Carolina (Moslow and Heron 1981)

Cuspate barriers are often found in lagoons where the wave fetch is limited by basin morphology. There are, for example, about thirty in a lagoon on the Chukotskiy Peninsula in the Bering Strait, almost twenty in the open and closed lagoons of St Lawrence Island, Alaska, and six in Nantucket Harbor, Massachusetts (Fig. 5.12). They might develop in a similar way in elongated bodies of water formed by ice floating away from the coast in high latitudes (Rosen 1975). They may be symmetrical or asymmetrical, and paired barriers on either side of lagoons are not uncommon, although they are usually somewhat offset from each other across an intervening channel, rather than being joined at the apices. In the northern Gulf of Mexico, elongated bodies of water have been segmented by the pairing of a variety of forms, including cuspate barriers, a cuspate barrier and some other type of projection, and combinations of unlike structures, including stream, tidal, and washover deltas.

Several explanations have been put forward for their occurrence. Zenkovitch (1959) attributed them to increases in the wave approach angle, and consequently, longshore flow capacity, caused by an initial obstruction to littoral drift. This causes a spit to develop, oriented at the maximum angle for longshore transport. The spit is then transformed into a cuspate barrier by waves approaching from the other side, and another spit is formed in the wave shadow created by the first spit. Repetition of this process produces a series of barriers and, if they grow out from each side of a lagoon and meet in the middle, eventual segregation of a basin. Cuspate barriers have also been related to: deflected lagoonal currents reworking sediment carried over barriers during storms; resonant oscillations associated with standing waves or seiches; and zones of updrift erosion and downdrift deposition associated with rapid increases in wave energy with downwind distance over the first few kilometres in restricted bodies of water.

BEACH RIDGES AND CHENIERS

Mounds or ridges of sand, clasts, shells, volcanic tephra, and other debris are prominent elements of many barriers, deltas, and other accretional landforms. Beach ridges are separated from each other by low-lying swales floored by material of approximately the same grain size as in the ridges, whereas cheniers, which are a type of beach ridge, are separated from each other by extensive areas of fine-grained mudflat sediments.

Beach Ridges

The origin and characteristics of coarse clastic beach ridges are discussed in Chapter 4, and this section will consider only ridges of sand or other fine-grained material. Sandy beach ridges can be tens of metres in width and several kilometres in length, with crests that are up to a few metres higher than the intervening swales. Beach ridge plains contain as many as 200 individual ridges and swales. Subparallel, straight, curved, or recurved ridges may be oriented parallel or obliquely to the modern coast, with older sets erosionally truncated by younger ridges. Ridge sets record changes in wave conditions, storminess, sediment supply, sea level, river mouth and inlet morphology, and other factors (Tanner 1988; Anthony 1991; Mason and Jordan 1993; Nieuwenhuyse and Kroonenberg 1994). Some workers have proposed that successive ridge formation resulted from falling sea level, whereas others consider that it developed while sea level was rising or constant.

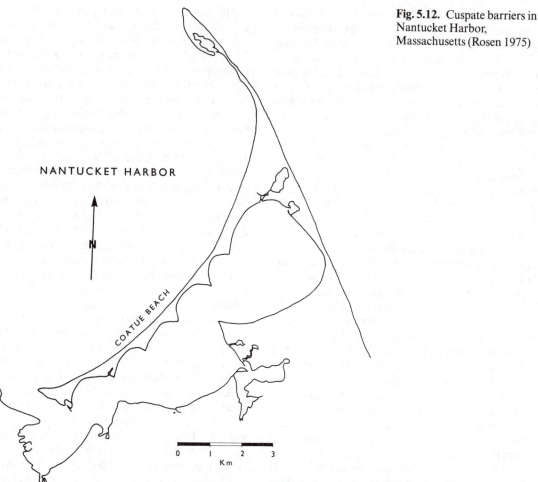

NANTUCKET HARBOR

COATUE BEACH

Each beach ridge represents a previous shoreline, and a series of ridges records the evolutionary stages of an accretionary feature. Ridges containing kitchen middens, shell-rich sands, or fragments of wood can be radiocarbon dated. Shells may be completely leached from old ridges, however, and even young ridges often lack dateable material. Archaeological evidence, historical records, depth of weathering, and other methods have therefore been used to indicate the relative age, or estimate the absolute age, of ridges (Tanner 1988). Based on chronometric records obtained from beach ridges, researchers have postulated shifts in the prevailing wind regime, initiation of the El Niño Southern Oscillation cycle, changes in sea level, rates of isostatic rebound, and the solar cycle periodicity of Arctic storms (Mason 1993). Beach ridge plains have developed in several ways. Some sandy ridges

are marine in origin, others are largely aeolian, and others have been influenced by both marine and aeolian processes.

Although storm waves generally erode and steepen sandy beaches, they may break constructively and form a strong uprush on gentle slopes. On the Mexican Tabasco beach plain, river-borne sand supplied to the coast in summer is built up into ridges by storm uprush in winter (Psuty 1967). Storm washover then moves the berms further inland, until they are isolated from the active beach, and new berms developing at the back of the beach then become the site for the formation of new ridges. Evidence from Australia and Malaysia provides further support for the contention that, under certain conditions, storm waves deposit sand at the back of a beach, while the beach itself is being eroded (Thom 1964; Nossin 1965). Waves also produce beach ridges through the welding of shore-parallel swash bars on to the strongly recurved portion of active spits (Hine 1979; Anthony 1991) (Fig. 5.4). Nevertheless, shelly beach ridges on a barrier spit in northern Australia are neither solely the result of normal beach processes, nor construction of a storm beach. An intertidal gastropod species is continuously incorporated into the beach ridges through normal wave action, but a subtidal bivalve is only introduced to the ridges during storms, possibly when the ridges are submerged and functioning as offshore bars (Shulmeister and Head 1993). A similar explanation may account for shelly ridges formed under low to moderate wave energy conditions in Argentina (Aliotta and Farinati 1990).

Two types of ridge are forming on the Magilligan Foreland in Northern Ireland, as part of the recovery following storm erosion of dune cliffs. One type develops through gradual accretion of swash bars around spring high tidal levels, 30–55 days after a storm. A second type forms when sediment, initially stored in longshore bars, is transported alongshore, elongating the bar and moving it into areas characterized by intermediate and even reflective wave domains. As the bar becomes longer and narrower, sediment starvation causes severance of a downdrift portion, which usually welds on to the beachface and emerges on to the upper beach as a ridge, some 90–150 days after a storm (Carter 1986).

Beach ridges in Venezuela and north-western Florida have also been attributed to accumulation of beach sediment by the action of surf and uprush, rather than by washover, aeolian processes, or other depositional agencies (Tanner 1988). Many ridges were formed with both ends standing in open water, with no attachment at either end to headlands, mainlands, spits, islands, or reefs. Between about five and forty years are needed to build a ridge. A 30–100 year break in construction may allow some aeolian decoration, and if time permits, they can eventually be converted into dune ridges. The ridges probably have a storm wave origin, and their spacing is related to storm frequency and the rate of sediment transport from an offshore source. For a given grain size, ridge height increases with wave height and uprush distance, which are related to the local wave energy (Tanner and Stapor 1972). Other laboratory and field observations have supported this relationship (Sunamura 1989).

The 'Darss' type of beach ridge largely consists of aeolian sand, or wave-deposited material covered by a substantial amount of dune sand. These dunified ridges are common elements of prograding coasts in many parts of the world (Ruz 1989; Martini 1990). Most Australian work on beach ridges has emphasized berm and incipient foredune initiation. Davies (1957) suggested that a berm built by waves during calm weather provides a nucleus on which pioneer vegetation can grow and trap wind-blown sand, forming a beach ridge parallel to the coast. A new berm may eventually be constructed in front of the old berm, allowing the process to be repeated. This hypothesis suggests that frontal dune ridges are the evolutionary descendants of sandy beach ridges. The amount of sand that accumulates on a ridge depends upon the time that elapses before a new ridge forms in front of it, cutting off the supply. The rate of ridge construction increases as storm incidence decreases, although the rate of sand supply and changes in mean sea level and tidal range are also important. Rapidly constructed ridges contain less aeolian material, and they tend to be low, close together, and regular in profile. Slower construction permits greater aeolian deposition, resulting in higher and wider ridges that are generally further apart. Low tidal range and long, low waves are conducive to strong berm development and ridging. Bird (1969) supported the contention that beach ridges originate as berms, although he proposed that the separation of parallel dune ridges results from storm wave erosion and scarping of the developing foredune, which allows a new berm to be built in front (Fig. 5.13).

Hesp (1984) believed that beach ridges are essentially incipient foredunes in south-eastern Australia, although this interpretation has been criticized by some workers (Bird and Jones 1988). He argued that pioneer vegetation does not colonize the crest of berms, as proposed by the berm-initiation hypothesis of Davies and Bird, and he noted that incipient foredunes are forming on dissipative beaches that do not have berms. Foredunes can be initiated by seedling colonization deposited at the height of the spring-tide uprush, where seeds are scattered by winds, and by the growth of rhizomes, stolons, and shoots. Hesp did not accept that the scarping of foredune ridges is responsible for the separation of parallel ridges, which he attributed to biological processes and plant-aerodynamic interactions, with ridge width

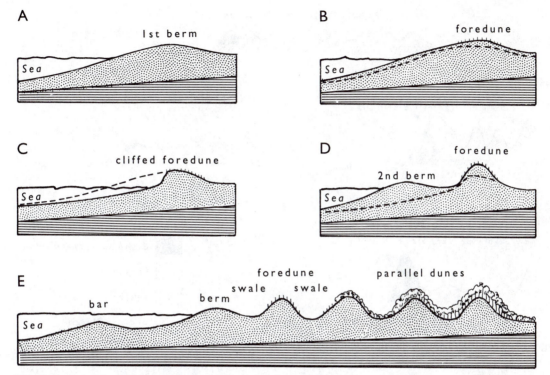

Fig. 5.13. Stages in the formation of parallel coastal dunes (Bird 1969)

decreasing and ridge height increasing with increasing plant density.

Shepherd (1987) proposed that the form of beach ridge plains is a function of changing coastal processes (Fig. 5.14):

a) On coasts with a strongly positive sediment budget, successive foredunes grow rapidly, and each new foredune quickly deprives the low, landward dune of its sand supply.

b) Coasts that prograde less rapidly allow more time for each foredune to grow to a greater size, before it is succeeded by a younger one.

c) When a formerly prograding coast becomes stable or slowly receding, a large foredune will develop and become modified by blowouts. This foredune will be much higher than those further landwards.

d) Rapidly eroding coasts may lack a well developed contemporary foredune.

The internal bedding and morphology of beach ridges provide valuable clues to their origin. Ridges that develop through coalescence of swash bars or berms have a fairly well-defined crest line and sideslopes in the 3° to 10° range. Their internal structure, which is indicative of swash zone processes rather than aeolian or washover effects, is dominated by low angle, seaward dipping, planar beds, oriented approximately perpendicularly to ridge crests. Although some aeolian decoration can be added later, it is not a fundamental aspect of their origin. Ridges formed by the welding of bars are wider and flatter than swash ridges, and they tend to have more of a whale-back shape, with very gentle initial slopes of between 1° and 4°. They have washover fans and deltas on their landward side, and they tend to occur singly or are separated by very wide swales, rather than forming extensive beach ridge plains (Tanner 1988). Their internal structure usually consists of convex-upward beds, with zero dip in the central portions, and a series of steeply landward-dipping foreset units, formed by the spillover and washover processes which add to their height and width. On the Tabasco beach plain, the internal structure of the washover ridges consists of three components: a series of horizontal or gently landward dipping laminae in the upper portion of the ridges; foreset stratification with laminae dipping landwards at up to 28°; and minutely cross bedded, quasi-horizontal laminae

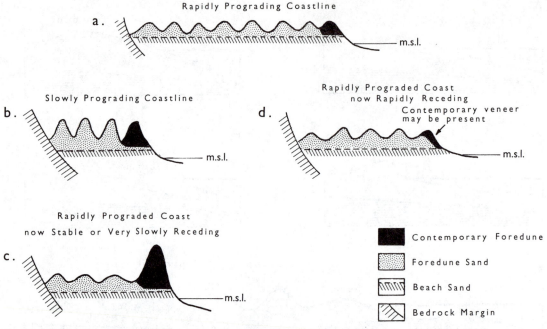

Fig. 5.14. The effect of sediment budgets and prograding rates on beach ridge plains (Shepherd 1987)

on the bottom (Psuty 1967). The complexity of the bedding in most wave-formed beach ridges shows that they are the product of many storms, rather than a single, continuous depositional event.

Cheniers

Chenier ridges consist of sand, coarse clasts, or intertidal or shallow subtidal shells, surrounded and isolated from the shore by fine-grained mudflats or marshes (Fig. 5.15). Some extend down to a firm base of foreshore bottom sand and silty clay deposits below the marsh surface, whereas others rest on top of subtidal or intertidal mud, and are driven landwards over supratidal marsh deposits. Cheniers occur as single ridges, or as clusters of closely spaced ridges that branch out in a feathery, fan-like fashion, diverging and bending landwards as they outline former positions of the shoreline. Cheniers are usually from 2 to 6 m in height, and they range up to 40–60 km in length and 50–600 m in width. The seaward slope is usually greater than the landward slope, although the landward slope is greater on the shelly cheniers of south-eastern England, possibly because of their particularly high mobility (Greensmith and Tucker 1975).

Although most known cheniers are in tropical and subtropical regions, they occur in a wide variety of climatic environments, ranging from the humid tropics to the subarctic. They are also found in tidal environments with ranges up to 10 m in Broad Sound, Queensland and 8.5 m on the Colorado Delta, and down to 0.3–0.5 m in Louisiana, and 0.5–1.5 m on the Guyana coast of South America. Cheniers cannot develop on high wave-energy coasts, however, where fine-grained sediments are carried offshore (Fig. 5.16).

Cheniers were first described from south-western Louisiana and south-eastern Texas, where more than five major sets of ridges occupy a plain that is about 200 km long and 20–30 km in width. The ridges support roads, human settlements, agriculture, and a luxuriant vegation cover. Indeed, the term 'chenier' is a Cajun expression derived from the French word for oak (*chêne*), the dominant type of tree on the crests of the higher ridges.

Chenier plains consist of two or more ridges with intervening muddy, marshy sediments. Otvos and Price (1979) considered that chenier plains develop in two broad types of environment. Gently indented bight coast chenier plains form on open ocean coasts where fluvial sediment is discharged into low to moderate wave-energy environments

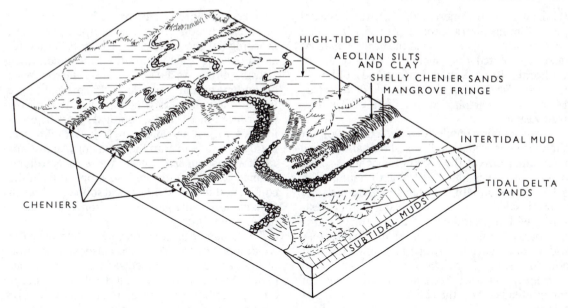

Fig. 5.15. Facies occurrence on a chenier plain (Rhodes 1982)

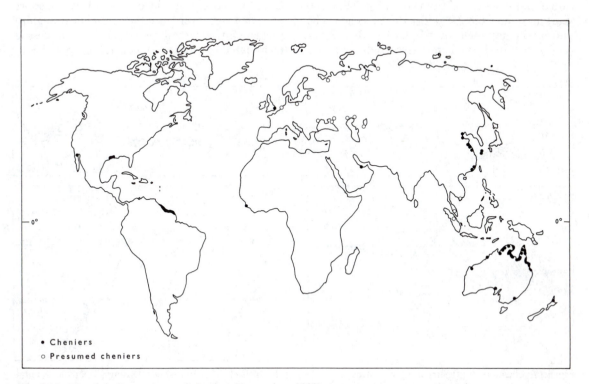

Fig. 5.16. The global occurrence of cheniers (Augustinus 1989)

(Guiana in South America, the Louisiana coastal plain, China, Sierra Leone) (Fig. 5.17). Much smaller chenier plains form at the head of narrow, deeply indented and sheltered bays (Gulf of California, Broad Sound, Queensland), although they can also develop along the side of bays (the Miranda chenier plain in the Firth of Thames, New Zealand).

There are fairly regularly spaced alternations of shoreface-connected mudbank-shoals and erosive interbank areas off the 700 km-long chenier plain of Guyana, Surinam, and French Guiana (Fig. 5.17). The mudbanks, which largely consist of sediments derived from the Amazon, migrate westwards with a periodicity of about thirty years. Cheniers are still forming in the interbank areas today. They develop from the winnowing and concentration of shell material and fine sand near mean low water level on the erosive side of the mudbanks, and the shoreward migration of this material to form emergent longshore bars (Augustinus 1980; Prost 1989). The chenier plain in western Surinam began to develop about 6,000 years ago. Large sets of cheniers were created when progradation was interrupted by periods of erosion, from about 3,850 to 3,200 years ago, and from 2,000 to 1,700 years ago. The fine sand component represents less than 2 per cent of the total Amazon-derived sediment, and

cheniers therefore formed only after long periods of severe erosion. In eastern Surinam, however, more regular chenier formation reflects the abundant supply of local medium-to-coarse sand, which is transported westwards along a narrow zone around mean high water level. In contrast to the fine-grained cheniers which are initiated at low water level, these coarser-grained ridges begin to form at, or just above, the high tidal level, where mangroves help to trap the sediment (Augustinus 1980; Augustinus *et al*. 1989). Coarse- and medium-grained chenier sands are also supplied from the continental shelf and other local sources in French Guiana, where fine-grained Amazon sand appears to be absent (Prost 1989).

There has been considerable debate over the conditions responsible for the episodic deposition of fine- and coarse-grained sediments on chenier plains. Accretion and mudflat development occur during periods of abundant sediment supply, whereas cheniers form during periods of diminished supply, as a result of waves winnowing and concentrating the sand and shell faction from the eroding mud, or from sand carried by longshore transport from nearby sediment sources (Fig. 5.18). Once formed, chenier ridges can be driven onshore by storm washover, until they are high enough to prevent further flooding, or until the process is terminated by renewed coastal progradation.

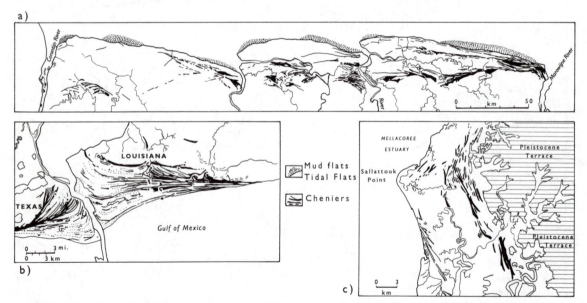

Fig. 5.17. Cheniers in (a) Surinam (Augustinus *et al*. 1989); (b) around the Sabine River Estuary, Texas–Louisiana (Byrne *et al*. 1959); and (c) Sierra Leone (Anthony 1989)

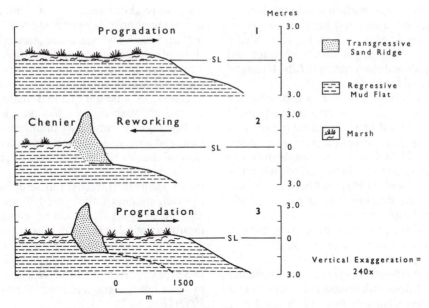

Fig. 5.18. The formation of a chenier plain (Hoyt 1969)

Winnowing can be essentially continuous, or it may occur intermittently, possibly during major storms. The formation of cheniers on the Yangtze (Changjiang) Delta, for example, has been attributed to particularly powerful typhoons, and dated cheniers in northern Australia appear to correspond to major storm events (Lees 1987; Qinshang *et al.* 1989). Nevertheless, chenier plains have far fewer ridges than one might expect on the basis of the number of large storms that they must have experienced during the Holocene.

A variable supply of river-borne sediment may be a necessary condition for chenier development. In Broad Sound, Queensland, for example, cheniers have been attributed to periodic reductions in the amount of sediment supplied from estuaries (Cook and Polach 1973). The supply of fluvial sediment can vary as a result of shifting deltaic distributaries. Mudflat progradation on the Louisiana chenier plain occurred in pulses, corresponding to periods when a major distributary was located in the western portion of the Mississippi deltaic plain. When the course switched away from the chenier plains, reworking of the mudflats by marine processes produced a shore-parallel sand and shell ridge, overlying marsh sediments. Progradation of the chenier plain occurred during the building of the Recent delta plain, however, rather than with the Teche complex, the most westerly of the former

subdeltas of the late Holocene delta plain (Fig. 9.6). The occurrence of individual ridges appears to be largely associated with delta lobe switching within the younger Lafourche complex, and variations in sediment supply from local rivers. A new episode of rapid mudflat progradation is taking place with recent development of the Atchafalaya delta complex—the closest position of an active distributary to the chenier plain since the sea reached its present level (Penland and Suter 1989; Huh *et al.* 1991). Chenier formation on the Colorado Delta may have occurred during periods of reduced sediment supply, and they appear to be forming today because of human-induced sediment starvation (Thompson 1968; Meldahl 1995). On the Victoria Delta in north-western Australia, cheniers developed rapidly from about 2,020 to 1,210 years ago, during periods of reduced mud input to the inner shelf. Although present evidence suggests that this was probably the result of decreased fluvial input owing to climatic changes, delta channel switching by the Ord River may have been a contributory factor (Lees 1992). Chenier development in China has also resulted from river course migration or the switching of distributaries in lower deltaic plains (Xitao 1989). The Huanghe (Yellow) River has discharged at times into the Yellow Sea and at other times into the Bohai Sea. The active delta is presently prograding, while erosion and chenier formation is taking place on the abandoned delta and in other coastal regions that once received sediment from the river. Shelly cheniers up to 10 m in height have recently formed along the abandoned modern delta of the Huanghe River (Wang and Ke 1989).

Changes in climate are also responsible for fluctuations in the supply of fine-grained fluvial sediment. In the Gulf of Carpentaria in northern Australia, prograding mudflats are characteristic of wet periods of high sediment supply, whereas cheniers develop during erosive periods when the supply is low (Rhodes 1982). In Guyana, chenier clusters may have developed during dry climatic periods when large amounts of sand were available, and mudflat progradation during pluvial periods when rivers carried large amounts of fine-grained sediment into the sea (Daniel 1989).

There are other possible explanations for chenier development. Episodic development of the chenier plain in Princess Charlotte Bay, Queensland may be largely attributed to the changing geometry of the Bay (Chappell and Grindrod 1984). As sediment, supplied at a constant rate, fills in a bay, progressive increase in mudflat progradation lowers the frequency of, or even prevents, ridge formation. Chenier formation was also inhibited by falling sea level during the later Holocene, and by a decrease in shell production, which could have resulted from increasing deposition of mud. Mangroves may have played an important role in trapping sediment, and low shelly bars are found today within the mangrove fringe. When the fringe is narrow or extensively damaged, a chenier may pass through it and accrete to an older ridge some distance landwards. Two modes of mud deposition have been identified. Overall progradation rates were fairly low from 2,000 to about 1,200 years ago, when there were about seven chenier building events, averaging almost one per century. This period of 'cut and recovery' was characterized by alternations of faster and slower deposition, and a 100–150 m-wide mangrove fringe that was episodically damaged during ridge-building events. Since the beginning of a period of rapid muddy deposition about 1,250 years ago, mudflat progradation and a wide mangrove fringe have inhibited shell production, and almost completely prevented the formation of chenier ridges. Cheniers in south-eastern England could have resulted from the mass mortality of shellfish (Greensmith and Tucker 1975). Episodic changes in the hydrodynamic regime may also be important, including changes in such factors as wind velocity or direction, and storm frequency, which would have a marked effect on the wave winnowing and transportation processes. In Sierra Leone, cheniers may reflect subtle changes in wave energy and sediment supply (Anthony 1989) (Fig. 5.17).

Some workers have suggested that cheniers and mudflats can develop together, as well as episodically (Rhodes 1982). Woodroffe *et al.* (1983*a*) proposed that erosion, deposition, and migration of cheniers on the Miranda coastal plain in northern New Zealand can occur simultaneously, and are not dependent on variations in sea level or sediment supply. Sand bars, formed on the foreshore sand flat, acquire shell ridges and supratidal storm crests as they migrate landwards and approach the shore. Fine-grained sediments also accumulate in the embayed tidal flats that develop in the sheltered zone in the lee of the ridges, and they may eventually be colonized by mangrove or salt marsh. The chenier ridge may then continue to migrate over these environments, although at a decreasing rate.

6 Sand Dunes

INTRODUCTION

Coastal sand dunes protect the land from extreme waves and tides, and they store sediment used to replenish beaches during and after storms. Large prograding dunefields develop on straight, sandy coasts exposed to prevailing and dominant on-shore winds, while smaller, crescentic fields form at the back of bays enclosed by rocky headlands. Dunes can develop on prograding coasts with a positive net sediment budget (Bird and Jones 1988), and on retrograding coasts with a negative budget (Psuty 1990). They play an enormous role in the coastal sediment budget. About 40 per cent of the sand brought to a beach by longshore drift is blown into a dunefield in South Africa, and between about 20 and 30 per cent in Israel (Illenberger and Rust 1988; Goldsmith *et al.* 1990).

Dune sand can be obtained from a variety of glacial, fluvial, biogenic, and other sources. Quartz is usually the main constituent, but calcium carbonate is common in some areas, particularly in warm tropical and Mediterranean regions. Dune sand is usually medium to fine-grained, and well, to very well, sorted, although it varies considerably within dunes and across dunefields. There is also a tendency for the grains to be more rounded, spherical, and positively skewed than beach sand (McLaren 1981), although there are many exceptions to these characterizations.

The occurrence and characteristics of coastal dunes are determined by the interacting effects of wind, waves, vegetation, and sediment supply (Table 6.1) (Fig. 6.1). Aeolian transport in south-eastern Australia is potentially greatest across wide, gently sloping, dissipative beaches, moderate across intermediate forms, and lowest across steep, narrow, reflective beaches (Short and Hesp 1982) (Chapter 4). There is generally little dune development behind reflective beaches, but dissipative beaches often front large, transgressive dune sheets. Dunes associated with intermediate beach forms range from large parabolic systems in high wave-energy environments, to small blowouts where wave energy is low. High tidal range also exposes wide intertidal zones to the wind during low tide, when the sand can dry out. Although dunes are found in a wide range of environments, most of the largest systems are in the mid-latitudes, behind high to intermediate wave-energy coasts, and facing the prevailing and dominant westerly winds (Fig. 6.2). There is also some good-to-moderate dune development on east-facing swell and Trade Wind coasts, but dunes tend to be less common and smaller in polar and tropical regions (Jennings 1964).

Coarse beaches and the effects of frost and sea ice are partly responsible for poor dune development in high latitudes. Although cold winds may be more effective in moving sediment than warm winds (Pye and Tsoar 1990; McKenna Neuman 1993), annual depositional rates appear to be much less than in temperate regions. Niveo-aeolian deposition of mixtures of sand and snow occurs during the cold season, when winds blow sand onshore, and saltation can be more easily sustained over frozen surfaces. Niveo-aeolian deposits are generally coarser than aeolian sand deposited in summer (Ruz and Allard 1995).

The presence of extensive dunefields in northern Queensland, Sri Lanka, Hawaii, and Ghana suggests that their reportedly poor development in the humid tropics has been greatly exaggerated. The absence of dunes on some tropical coasts appears to be the result of a combination of local factors. A variety of explanations has been proposed, including the effects of dense vegetation at the back of beaches, intense chemical weathering that forms silt and clay rather than sand, coarse beach sand, narrow and steep reflective beaches, almost

Table 6.1. Relationships between dune morphology, shoreline dynamics, wind energy, and sand-trapping vegetation

a) Positive beach sand budget. Low wind energy.	Rapidly prograding beach ridge plain with little dune development.
b) Positive beach sand budget. Wind energy higher than in a) Effective sand-binding vegetation.	Slow beach progradation and development of parallel dune ridges.
c) Positive beach sand budget. Moderate wind energy. Ineffective or discontinuous vegetational cover.	Formation of irregular series of hummock dunes with incipient blowouts and parabolic dunes on moderately prograding coast.
d) Sand supplied to beach is transferred to dunes and bound by vegetation.	Single foredune ridge growing vertically with no change in position of shore.
e) Sand supplied to beach is slightly less than sand supplied to dunes.	Beach is lowered and shore retreats slowly landward. Vegetation damage by salt scalding and wind burn results in blowouts and small transgressive parabolic dunes.
f) Little or no sand supplied to beach. Wind energy is high.	Beach rapidly lowered, increasing exposure to storms and coastal retreat. Vegetation is destroyed and transgressive sand sheets develop.

Source: Pye 1990.

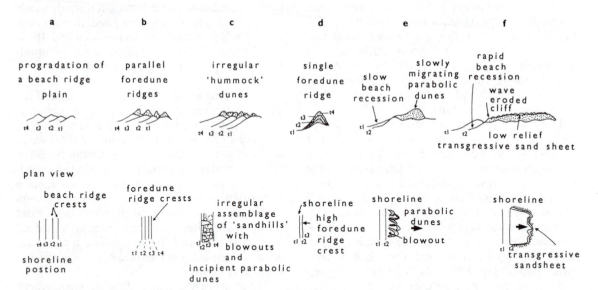

Fig. 6.1. The effect of shoreline dynamics, wind strength, and sand-trapping vegetation on dune morphology. For explanation, see Table 6.1 (Pye 1990)

continuous dampness, and salty, protective evaporation crusts on beaches. Probably the most important single explanation for the lack of dunes along some humid tropical coasts, however, is the fact that tropical winds are weaker and less persistent than those in temperate areas. The characteristic trailing pattern of tropical plants may also be an important factor. In humid temperate regions,

grasses have an irregular distribution and a tussock-like growth form that encourage development of hummocky embryo dunes. The *Ipomoea-Canavalia* community, however, which is an important sand colonizer throughout the tropics, has long creeping prostrate stems (stolons) that cover the sand in a sheet of vegetation, encouraging deposition over a wide area. This produces

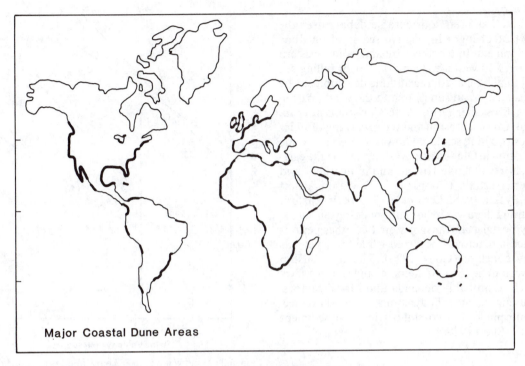

Fig. 6.2. Major coastal dune areas of the world (Carter *et al.* 1990*d*)

low, flattish or berm-like platforms that may make it appear as if dunes are completely absent along a coast (Sauer 1959; Davies 1972; Devall *et al.* 1990).

PERIODS OF DEVELOPMENT

There is widespread geomorphological, pedological, and stratigraphical evidence that periods of dune formation have been interdispersed with periods of dune stabilization and erosion (Bryant *et al.* 1990; Orme 1990; Froidefond and Prud'homme 1991; Lees *et al.* 1993). Soil horizons buried within aeolian sands testify to periods of stability, followed by renewed buildup. Dated layers of buried peat show that there were at least three dune-building periods in western Denmark, for example, between about 400 BC and AD 1750 (Christiansen *et al.* 1990). The increasing greyness and decreasing yellowness of dune sand with age also provides a useful tool in distinguishing between different generations of dunes in an area,

and thermoluminescence techniques now allow dune sands to be dated.

Extensive Pleistocene dunefields, which occur in Australia and the Mediterranean, are generally absent in northern Europe, where most dunes were built at different stages during the Holocene (Bressolier *et al.* 1990; Wilson 1990). Older and Younger Dunes have been distinguished in the coastal stratigraphy of many areas along the southern coast of the North Sea. A podsolized layer separates the two dune sands in some parts of the Netherlands. The low ridges of the Older Dunes are usually attributed to the sub-boreal and early sub-Atlantic periods. The much higher Younger Dunes were formed when rising sea level and increasing storminess destroyed the plant cover, and allowed the wind to erode the Older Dunes. They were built in several phases from about AD 800 to 1850 (Klijn 1990), and there has generally been only minor aeolian deposition since. Aeolian activity may also have varied through time in response to fluctuations in storm frequency. Charcoal contained in buried palaeosoils also suggests that periods of aeolian activity may have been initiated by fire (Filion 1984; Orme 1990; Borówka 1990a, b).

There has been considerable debate over the effect of changes in relative sea level on dune development. In Northern Ireland, south-western France, and western Denmark, low or falling sea level may have encouraged dune development by exposing a broad sandy area to the wind (Wilson 1990; Bressolier *et al.* 1990; Christiansen *et al.* 1990). Coastal erosion and the reworking of older dunes by rising sea level, however, preceded dune formation in Oregon and Washington, on the eastern shores of Lake Huron, and in northern and eastern Australia (Cooper 1958; Davidson-Arnott and Pyskir 1988; Lees *et al.* 1993). In eastern Australia there was major dune building during the early postglacial transgression, but other periods of aeolian activity occurred while sea level was fairly constant (Cook 1986). The relationship between dune formation and changes in sea level therefore probably depends upon local factors, including the morphodynamics of the shoreface, the morphology of coastal barriers, and sediment supply (Short 1988a).

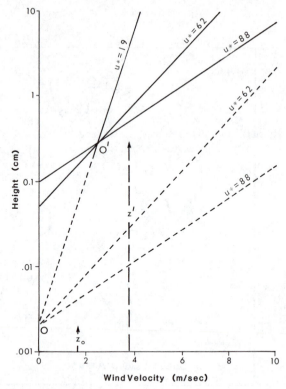

Fig. 6.3. Wind velocity over a bare, fine-grained surface, with (solid lines), and without (dashed lines) sand movement (Bagnold 1941)

SAND TRANSPORT

When the wind blows over a sandy surface, frictional drag creates a near-surface velocity profile that is generally assumed to be logarithmic. The shear velocity of the wind (u_*) is the gradient of the straight line obtained by plotting the wind velocity against the log of the elevation (Fig. 6.3). The horizontal shear stress (τ_0) is determined by:

$$u_* = (\tau_0/\rho_a)^{0.5}$$

where ρ_a is the density of the air. Increases in shear velocity with wind speed and surface roughness therefore result in greater shear stresses or forces being exerted on the surface.

Because of frictional drag, there is no wind movement in a thin layer next to the surface, which is known as the effective surface roughness or equivalent roughness height. For a quiescent surface consisting of closely packed and nearly spherical sand grains, the thickness of this layer (z_0) is estimated to be about 1/30th of the mean grain diameter. Zingg (1953) found that:

$$z_o = 0.081 \log_{10}(D/0.18)$$

where D is the diameter of the sand grain (mm). If pebbles are present, the effective surface roughness also depends upon their spacing, the maximum roughness of about $1/8D$ occurring when the distance between them is about twice their diameter (Greeley and Iversen 1985).

If the wind speed is too low to move grains of sand, its variation in the vertical plane under neutral stability conditions can be expressed by the von-Karman–Prandtl ('law of the wall') equation:

$$u_z = \frac{u_*}{k_n} \log_e \frac{z}{z_0}$$

or:

$$u_z = 5.75\, u_* \log_{10} \frac{z}{z_0}$$

where u_z is the speed of the wind at an elevation z above the bed, and k_n is von-Karman's constant (0.41).

It has been suggested that about three-quarters of the movement of sand in air is by saltation, although the exact proportion varies according to grain size and other factors. Smaller particles, from about 0.1 to 1 mm in diameter, tend to saltate, whereas particles, up to about six times larger normally roll, creep, or hop along the surface. Individual grains, however, can alternate from one transport mode to another. Turbulent fluctuations in wind velocity exert a strong influence on the trajectories of very fine sand and silt (less than about 0.1 mm in diameter), which can be lifted into suspension high above the surface.

The Initiation of Movement

Sand grains are entrained in the air flow when lift, shear stress, and ballistic impact overcome the effects of gravity, friction, and cohesion. Air is a low-viscosity, low-density fluid, and its ability to lift particles off the bed is quite poor. Grains that begin to move through wind shear have diameters equal or smaller than the critical threshold value, yet are large enough to extend up through the z_o layer into the moving wind.

Some workers believe that grains first roll or slide across the surface in response to the direct pressure of the wind, before gaining speed and beginning to saltate (Bagnold 1941; Seppala and Linde 1978; Willetts *et al.* 1991). Others consider that they can also be entrained directly from the ground by lift forces acting vertically upwards (Anderson *et al.* 1991). This could occur through the Bernoulli effect, whereby lift is induced as air accelerates over a grain and lowers the pressure on it, while slowly-moving air beneath the grain generates higher pressures. Grains tend to become airborne after a period of intense vibration, which may be related to burst and sweep sequences in turbulent flow (Chapter 3) (Anderson *et al.* 1991; Willetts *et al.* 1991).

Bagnold (1941) proposed that the fluid threshold or critical shear velocity required to initiate grain movement varies approximately as the square root of its diameter:

$$u_{*cr} = A \sqrt{\frac{gD(\rho_s - \rho_a)}{\rho_a}}$$

where u_{*cr} is the critical shear velocity in cm sec^{-1}; ρ_s is the density of the grain; A is a coefficient—the square root of the Shields parameter—with estimates ranging from about 0.1 to 0.2; and g is the acceleration due to gravity. A range of threshold shear velocities is necessary, however, to initiate

grain movement in sediments with a variety of grain sizes, shapes, sorting, and packing arrangements (Zingg 1953; Nickling 1988).

Rates of grain dislodgement are poorly known at present. Anderson and Haff (1988) proposed that the entrainment rate is proportional to the excess surface shear stress, the amount by which the short period shear stress at the bed exceeds the threshold shear stress. Williams *et al.* (1990) found that the rate of entrainment of homogeneous grains is related to the Shields dimensionless shear stress function (Chapter 3). Aerodynamic entrainment is a stochastic process, however, and even homogeneous sediments have a wide range of threshold shear velocities. Williams *et al.* (1994) emphasized the probable effect of sweep events during fluid bursting cycles in turbulent boundary layers (Chapter 3). Instead of grains suddenly rising vertically from the bed at the moment of aerodynamic dislodgement, they found that they were simply pushed forwards and outwards by the force of sudden, localized gusts, resulting in spatially semi-organized flurries of fifty or more grains. The flurries were then succeeded by quiescent periods at airflow velocities close to the threshold.

Once the first few particles have been aerodynamically entrained, most new entrainment is by ballistic impact, when saltating grains collide with immobile grains on the bed. The impact of descending saltating grains also repositions other grains, and may cause the larger grains to accumulate at the surface (Willetts and Rice 1989; Werner 1990; Sarre and Chancey 1990; McEwan *et al.* 1992). The additional lift provided to a saltating grain by the Magnus Effect, associated with rapid rotation acquired after striking the surface with a glancing blow, is minimal in air (Anderson 1989).

Several 'splash' grains, lying within a few grain diameters of the impact point, are usually thrown into low-energy reptations or trajectories, with ejection speeds that are about 10 per cent of the impact speed. A small fraction of these grains jump up high enough to be accelerated by the wind, and they then start saltating, but the impact speed of the rest is too low to rebound or eject other grains on impact. The reptating population constitutes a considerable proportion of the cloud of sand in transport, and despite their short excursion lengths, their high numbers suggest that grain movement near the ground is more significant than was once assumed (Anderson 1989, 1990; Anderson and Haff 1988; Haff and Anderson 1993; Rice *et al.* 1995). Saltation, reptation, and creep tend to merge into each other, and reptation could be considered a feeble form of saltation, or a vigorous form of creep (Willetts and Rice 1989).

Saltation Clouds

Saltation clouds rarely extend to more than a metre or two above the ground, and most particles travel within a few tens of centimetres of the surface. The shape of saltation trajectories varies according to whether the grains are entrained by fluid lift and drag or by grain impact, or whether they are already saltating and rebounding off the bed. Although some grains may be ejected almost vertically upwards into faster-moving air, mean angles of ascent are between about 34° and 50°. Grains that rebound from the surface during saltation have angles of ascent of between about 21° and 33°, the higher values occurring with coarser grains, whereas grains are ejected from the bed by splashing at angles of between 40° and 60°. Once a grain has reached the highest point in its trajectory it begins to descend, increasing in velocity along a line inclined at between about 9° and 15° to the ground; the angle of impact decreasing with the size of the grain.

The shear velocity, represented by the gradient of the lines, is the same when sand is moving as when it is stationary. The velocity profiles plotted on semi-log paper converge at a new focus (u', z') at a higher level, however, having been shifted upwards by an amount equal to $z' - z_o$ (Fig. 6.3). The height of the new focus is of the same order of magnitude as the height of the surface ripples. It remains to be determined why wind velocity is virtually constant at this level, irrespective of any change in the velocity above, although it has been suggested that it is related to grain diameter, the development of ripples on a mobile surface, or the average maximum height of the saltation trajectories. Bagnold (1941) suggested that kinks in wind velocity profiles, up to about 30 mm above the ground, occurred where the average or characteristic saltation trajectory extracts most of its forward momentum from the wind. Recent work has confirmed the presence of kinks, and their occurrence at the elevation of maximum momentum extraction, although they may reflect the influence of a wide range of trajectories, rather than a single characteristic trajectory (McEwan 1993).

Once grains are moving, the wind speed distribution is expressed by:

$$u_z = 5.75 \, u_* \log_{10} \frac{z}{z'} + u'$$

where (u', z') represent the horizontal (wind velocity) and vertical (height) coordinates of the focus, respectively (Fig. 6.3). Zingg (1953) found that:

$$u' = 8939 \, D$$

and

$$z' = 10 \, D$$

where D is measured in mm and u' is in mm s^{-1}.

The transfer of momentum to saltating grains distorts velocity profiles in dense saltation clouds, causing the shear velocity to fall below the fluid threshold level. Nevertheless, the impact of the descending grains on the beach keeps the saltation cloud moving. This is because the velocity required to continue movement through collision (the impact threshold velocity) is lower than that required to initiate movement by lift or shear (the fluid threshold velocity).

Transport Rates

Numerous empirical and semi-empirical equations have been used to calculate aeolian transport rates (Table 6.2). The equations of Bagnold (1941) and Kawamura (1951) have theoretical foundations, however, and they have been the most widely used. Bagnold's equation, which does not include a threshold term, is based on the loss of momentum from the air to the saltating grains, with a correction for surface creep. Kawamura's equation, which does include a threshold term, is based on the assumption that direct wind shear and the impact of falling grains impart momentum to grains on the bed.

Transport equations are usually of the form (Horikawa *et al.* 1986; Hotta 1988):

$$q = A \, u_*^n + B \, (u_*)$$

where A and n are constants, and B is a function of u_*. Theory, most experimental work, and simulation modelling of grain collision and saltation suggest that n is usually about 3 (Anderson and Haff 1988, 1991; McEwan and Willetts 1991; Sørensen 1991). There is considerable variation in the reported values of the constant A, however, possibly reflecting sand characteristics and differences in the efficiency of the sand traps that

Table 6.2. Some sand transport rate formulae

Bagnold (1941)	$q = B\dfrac{\rho_a}{g}u_*^3\sqrt{\dfrac{D}{D_s}}$	$B = 1.5$ for nearly uniform sand, 1.8 for naturally graded sand, and 2.8 for sand with a wide range of grain sizes.
Chepil (1945)	$q = C\dfrac{\rho_a}{g}u_*^3$	C varies between 1 and 3.1, depending on soil texture.
Kawamura (1951)	$q = K\dfrac{\rho_a}{g}(u_* - u_{*cr})(u_* + u_{*cr})^2$	$K = 2.78$
Lettau and Lettau (1978)	$q = L\dfrac{\rho_a}{g}u_*^2(u_* - u_{*cr})\sqrt{\dfrac{D}{D_s}}$	$L = 4.2$
Zingg (1953)	$q = Z\dfrac{\rho_a}{g}(D/D_s)^{0.75}u_*^{1.5}$	$Z = 0.83$
Hsu (1973)	$q = H(u_*/\sqrt{gD})^3$	$H = \exp(-0.42 + 4.91D) \times 10^{-4}(\text{gf/s. cm})$
Horikawa *et al.* (1986)	$q = H_0\dfrac{\rho_a}{g}U_c^n\sqrt{\dfrac{Da_{50}}{Db_{50}}}(u_* + u_{*cr})^2(u_* - u_{*cr})$	

Note: q is the sand transport rate in grams cm-width^{-1} second^{-1}; ρ_a is the density of air; g is the acceleration due to gravity; u_* is the shear velocity; u_{*cr} is the threshold shear velocity; D is the grain size; D_s is the standard grain size (0.25 mm); and Db_{50} and Da_{50} are the median grain diameters of the underlying bed and the blown sand, respectively; B, C, K, L, Z, H, H_0, and n are coefficients; U_c is the uniformity coefficient of the blown sand (D_{60}/D_{10}, D_x being the grain size at which x per cent of the sand weight is finer), which ranges between 1 and 2 for very uniform sand and 3 and 5 for normal sand.

Sources: Horikawa *et al.* 1986; Sarre 1987; Hotta 1988; Sherman and Hotta 1990.

were used—measured transportation rates generally being higher with horizontal than vertical traps (Hotta 1988). The equations indicate that slight changes in wind speed result in large changes in the amount of sand it can move. Transport equations generally predict sand movement fairly well where there is a steady, uniform flow field with a logarithmic profile, and a flat, extensive, unvegetated surface of dry, well-sorted sand. These ideal conditions are rarely found in nature, however, and there can be marked discrepancies between model predictions and measured rates of movement (Sherman and Hotta 1990; Bauer *et al.* 1990; Anderson *et al.* 1991).

The wind-speed gradient is generally used to determine the surface roughness length and the shear velocity of the wind, which are then used to calculate the shear stress near the bed. As the vertical velocity profile is logarithmic in most laboratory experiments, and over flat surfaces in the field, any two points on the profile can be used to calculate the shear velocity. Portions of shore-normal wind-speed profiles can diverge quite significantly from the logarithmic, however, and one must then decide whether to use a best-fit line or another of the plotted gradients to calculate the shear velocity. Deviations from the true shear velocity are inevitable whatever method is used, and as they are cubed in most transport equations, they cause considerable inaccuracies in calculated sand transport rates (Sherman and Hotta 1990; Bauer *et al.* 1990).

Measurements conducted in wind tunnels and mathematical modelling suggest that saltation systems only require 1–2 seconds to respond to changes in wind velocity (Anderson and Haff 1988; Anderson 1989; McEwan and Willetts 1991; Butterfield 1991). The non-stationary nature of wind velocity therefore presents another problem in the calculation of aeolian transport rates, and suggests that transient rather than steady-state models may be more appropriate for predictive purposes. Aperiodic repetition of strong gusts followed by fairly calm intervals may be particularly significant on vegetated dunes, where gusts of wind can penetrate to the surface of the sand before spreading out (Wolfe and Nickling 1993).

Bagnold (1941) and others derived equations for the sediment transport rate over sloping beds. Although the theoretical models of Allen (1982) and Dyer (1986) (see Chapter 3) were based upon fluid rather than impact thresholds, they also reasonably predict aeolian, fine-grain thresholds over sloping beds. Measured transport rates related to the seventh power of the bedslope, however, suggest that there is a much greater dependency on surface slopes than is predicted by existing theory. This discrepancy may be related to a newly discovered process termed 'impact induced gravity flow' (Hardisty and Whitehouse 1988).

High wind speeds often occur in winter and at other times when the sand is wet from precipitation and spray generated by large, breaking waves. Although transport is much more difficult at these times, moisture does not prevent the movement of sand, especially by strong winds, and it may even be assisted by the splash of raindrops, which throw grains upwards (Borówka 1990a; Jungerius and Dekker 1990). Hotta (1988) and Nickling and Davidson-Arnott (1990) summarized the available information on transport rates on a wet sand surface. The data suggest that:

a) the generation of blown sand on a wet surface is strongly affected by the evaporation rate when the moisture content is intermediate (4–8 percent);

b) the transport rate on a wet surface is only slightly less than on a dry surface if the water content is low (less than 4 percent), but it is dramatically reduced when the water content reaches a critical value; and

c) the wet surface transport rate is low if wind speed is low, but it approaches the dry surface rate with high wind speeds, even with a water content of up to a few percent.

Belly (1964) modified Bagnold's (1941) threshold equation to accommodate the effects of water content:

$$u_{*cr} = A \, (1.8 + 0.6 \log w_c) \sqrt{\frac{gD \, (\rho_s - \rho_a)}{\rho_a}}$$

where w_c is the water content (percent). Hotta *et al.* (1984) suggested that Kawamura's (1951) equation can be used to determine the sand transport rate over wet surfaces, using a modified expression for the threshold shear velocity:

$$u_{*crw} = u_{*cr} + 7.5 \, w_c I_w \quad \text{for} \quad 0.0 < w_c < 8.0\%$$
$$\text{and} \quad 0.2 < D < 0.8 \text{ mm}$$

where u_{*crw} is the threshold shear velocity on a wet surface (cm s^{-1}), and I_w is a dimensionless evaporation factor. I_w is equal to 1.0 if the shear velocity is greater than the threshold shear velocity on a wet sand surface, and from 1.0 to 0, according to the evaporation rate, if the shear velocity is higher than the threshold shear velocity on a dry surface, but lower than on a wet surface.

Evaporation can cause salts to bond grains together, forming a resistant surface crust. This increases the threshold shear stress and significantly reduces the rate of sediment transport. Nickling and Ecclestone (1981) proposed that for NaCl and KCl salts:

$$u_{*crs} = 1.03 \, u_{*cr} \exp \, (0.1027 \, s/c)$$

where u_{*crs} is the threshold shear velocity on a surface containing salts (cm s^{-1}), and s/c is the salt concentration in mg of salt per gram of sand. Later work showed that $MgCl_2$ and $CaCl_2$ salts have a similar effect on threshold values (Nickling 1984). Although salt crusts on marine beaches are usually thin and easily broken, algae and other biological agents can help to bind sand surfaces and dramatically reduce rates of transport (Van Der Ancker *et al.* 1985).

Air temperature affects its density and kinematic viscosity, but the role of thermal stability on the flow of wind over dunes has been neglected. Temperature variations change the thickness of the laminar sub-layer and the shear stress exerted on the bed, and consequently, the rate of sand transport. The thermal gradient usually has a fairly small effect on shear velocity. It may be significant in coastal environments, however, where there can be marked differences in the temperature of the water and sand surfaces, and wind speeds are close to the threshold for grain movement (Sherman and Hotta 1990; Nickling and Davidson-Arnott 1990).

The transport of sand grains is influenced by their size, sorting, shape, and density. Large, dense grains are hard to move, and therefore tend to be left behind as lag deposits. Nevertheless, a surprisingly high proportion of the grains that rise above a saltation cloud are quite coarse, probably because they have more momentum and can therefore bounce higher than smaller grains. The average grain size decreases with height within a saltation cloud, however, although there may be some increase in size within the lower portions of a cloud during high winds. More sand can be carried over a poorly sorted surface consisting of a variety of sizes than over a surface of nearly uniform sand. Because grains can bounce further and higher across hard substrates, saltation across the upper portion of many beaches is enhanced by the presence of pebbles, shells, or other coarse material.

A number of workers have found that particle shape has an important effect on grain movement, but the precise nature of the relationship has not been determined. Experimental results are often contradictory. Some workers have suggested that angular or irregular grains are more difficult to lift from the surface than more spherical ones, while others believe that the opposite is true (Rice 1991).

Most of what we know about aeolian transport was obtained from deserts, where there are large, homogeneous sand surfaces, and in wind tunnels under equilibrium transport conditions. Beaches, however, are often too narrow for winds to become fully saturated with sand (Nickling and Davidson-Arnott 1990; Kroon and Hoekstra 1990). Davidson-Arnott and Law (1990) found that winds just above the threshold velocity require 10 to 15 m to attain maximum sand transport rates, whereas winds greater than 50 km hr^{-1} need more than 40 m. Furthermore, during periods of high winds, beach width is usually reduced by storm surges and high uprush, and more sand may therefore be moved by winds of moderate velocity. Aeolian transport on narrow beaches may also depend on winds blowing obliquely to the coast at low tide, thereby increasing the effective width. The largest foredunes in Hudson and James Bays, for example, occur where the prevalent winds blow at a slight angle to the shore, and aeolian transport effectively ceases on the narrow estuarine beaches of Delaware Bay when the source width is less than 8 m (Martini 1990; Nordstrom and Jackson 1992).

Beach width is important in determining rates of aeolian transport across a beach-dune complex in California (Fig. 6.4). An internal boundary layer developed in the downwind direction when onshore winds were blowing perpendicularly to the coast. Maximum near-surface flow and shear stress occurred at the leading edges of the roughness transitions, and flow competency therefore decreased with distance from the foreshore, before increasing again over the berm crest. Nevertheless, the lack of an upwind source of sediment resulted in much lower transport rates near the ocean's edge than further up the beach. A saltation layer developed up the beach, and fine-grained sediments were carried on to the back-slope of the dune. The transport rate was less than the wind's potential, however, because of the restricted fetch. Coarse grains, prevented from moving further landwards by lower shear stresses towards the back of the beach, accumulated as a semi-mobile lag deposit in front of the foredune, moving intermittently by surface creep in discontinuous ripples (Bauer 1991).

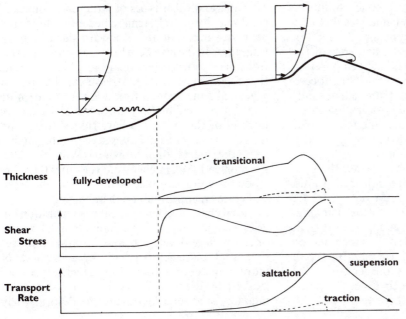

Fig. 6.4. Wind velocity profiles, shear stress, boundary layer thickness, and sand transport over a beach and dune (Bauer 1991)

Bedforms

Asymmetric ripples develop on sandy surfaces when grains are in motion. Ripple crests, which are oriented perpendicularly to the wind, often bifurcate, and they frequently terminate abruptly. Ridges range up to about 50 mm in height, and they are generally spaced from 50 to 150 mm apart. Ripple wavelengths increase with wind speed, although they flatten and disappear when velocities are more than about 12 to 14 m s⁻¹ (Borówka 1990a). Ripples tend to contain grains that are coarser than the underlying sand—the coarsest grains and heavy minerals being concentrated on the crests.

Bagnold (1941) suggested that ballistic bombardment of random projections produces small ripples that are reproduced downwind at intervals equal to the average length of the saltation hop. The relationship between ripple wavelength and the mean length of the saltation trajectories has been questioned (Sharp 1963; Seppala and Linde 1978). Brugmans (1983) developed a hypothesis, based on the Helmholtz theorem, that ripples reflect regular downwind alternations in the velocity or grain size of saltating grains hitting the ground. More realistic models are being formulated as we learn more about the precise nature of grain-bed impact processes. Willetts and Rice (1989) emphasized the role of reptation, and the effect of changing bed relief with ripple development on the nature of grain impact. Saltating grains collide with the ground at angles that vary with the bed slope along a ripple profile, and this affects the speed and angle of emergence of the rebounding grains. Anderson (1990) emphasized the stochastic nature of the saltation process. Ripples develop spontaneously from a flat, sandy bed as reptating grains move in short, low hops, before returning to the surface; the role of true saltation being only to provide the energy needed to mobilize the grains. The initial ripple wavelength, which is about six times greater than the mean reptation length of the splashed grains, increases as the smaller, faster-moving ripples are absorbed by the larger, slower forms. The final mean wavelength, although much greater than in the initial stage, is still much less than the saltation trajectories, and, as noted by Sharp (1963), appears to be determined by the length of the impact-sheltered shadow zone to the lee of ripples, which is proportional to their height.

It has been proposed that most kinds of transverse and longitudinal dune are initiated by wave perturbations and vortices in the atmosphere (Wilson 1972). Surface irregularities could fix the position of transverse, wave-like motions, or they could create them on their lee sides. In either case, shear velocity and sediment movement would increase where the wave descends close to the ground, and decrease where it rises away from the surface. This would produce transverse bedforms and intervening troughs, corresponding to alternating zones of deposition and erosion. Another type of secondary flow may account for bedforms that are longitudinal or parallel to the wind, or perhaps of more relevance to coastal regions, the presence of longitudinal elements superimposed on transverse ridges. There is little proof, however, of a causal relationship between secondary atmospheric circulations and dune formation, which in some cases, appears to be a response to random topographic irregularities, or changes in surface roughness (Pye and Tsoar 1990; Kocurek *et al.* 1992).

Dunes influence the flow of air around them, altering the surface shear velocity and sand transport rates and patterns (Weng *et al.* 1991; Arens *et al.* 1995). Although few detailed investigations have been made of wind movement over sand dunes, the data can be supplemented with information from the modelling of turbulent wind flow over hills and other types of structure. Isolines of wind velocity (isovels) tend to be more compressed near the summit on the windward side of a dune ridge than on the lee side, where the wind rises away from the surface. Numerical modelling suggests that the maximum shear stress near the crest of vegetated dune ridges is from 150 per cent to more than 200 per cent higher than on the plane surface upwind of the ridge. On bare ridges, this could result in sand transport rates several times higher near the crests than upwind (Willetts 1989; Rasmussen 1989). Higher wind velocities therefore erode the windward face of dunes near the crest, and place an upper limit on their height.

Flow separation, which creates a separation envelope containing fairly stagnant air and a reverse or return flow, develops below the crest on the lee side of a transverse ridge, as it becomes higher or more convex, or as the wind velocity increases (Robertson-Rintoul 1990) (Fig. 6.5). As the ridge migrates, sand deposited on the lee side forms a slip

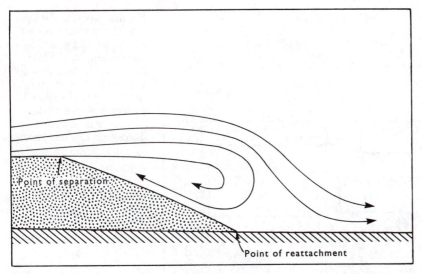

Fig. 6.5. Flow separation over the slip face of a barchan dune (Pye and Tsoar 1990)

face at the angle of repose. The number of saltating grains having the energy to reach any point on the lee side should decrease with the distance downwind from the summit. As a result of the shadowing effect created by the sharp break in slope at the summit, however, the maximum deposition rate on the lee side actually occurs a small distance from the summit (McDonald and Anderson 1995). This is in turn related to the fact that the saltation angle of descent is typically lower than the angle of repose of the sand. The gradient on the stoss side of dunes may be limited by the lack of continuous saltation on ground slopes that approach about 15° against the wind. This gradient is representative of the stoss side of many sand dunes and ripples (Rumpel 1985).

Flow separation does not occur on all dunes. The type of flow on the leeside of a dune depends upon its shape, the angle of approach of the wind relative to the orientation of the dune crest, and atmospheric thermal stability (Sweet and Kocurek 1990) (Fig. 6.6). Strongly separated flows, with low speeds and back eddy flows, develop when winds blow at high angles to the crest of dunes with steep leesides. Attached flows, with much higher leeside speeds, tend to occur when winds blow at oblique angles over dunes with gentle leesides. Atmospheric thermal instability encourages flow separation, and neutral thermal instability favours flow attachment. Attached flows tend to maintain the same direction as the primary flow if the slope of the lee face is less than about 20°, but they tend to be deflected along slopes that are steeper. The

interaction between a dune and the primary wind is still poorly understood, however, and there is no generally accepted model of the air flow over dunes. The situation is made more complex by changes in wind direction. For example, streamlined dunes with a steep lee slope and flow separation beyond the summit, present steep bluff faces to winds that approach from the opposite direction. The resulting stagnation pressure distribution on the upwind face can then generate complex flow fields upwind of the bluff (Greeley and Iversen 1985; Willetts 1989).

Steep leeward slopes are usually absent on vegetated dunes in humid coastal regions. Air flow may help to determine the occurrence and position of the low, intervening slacks, however, which tend to occur where the wind, after rising over the dune crests, comes back into contact with the ground, at a distance downstream equal to about five to ten times the height of the dune. Higher wind velocity in the bottom of the slacks can excavate depressions down to the water table, exposing moist sand or forming ephemeral ponds. Lower plant density or increasing wind velocity causes sand to be transported and deposited over greater distances. Enhanced transportation therefore reduces the height of the dunes while increasing their basal width, so that slacks have less chance of developing, or will be much shallower than where the sand is less mobile (Hesp 1989).

The relationship between wind regime and dune morphology is complex. Low winds may be unable to move sand, while extremely high winds can

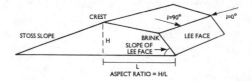

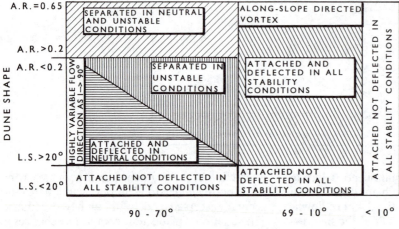

Fig. 6.6. The effect of dune shape (the aspect ratio AR and the lee face slope LS), the wind approach angle relative to the dune crest (i), and atmospheric thermal stability on lee-face flow. L is measured parallel to the primary wind direction (Sweet and Kocurek 1990)

destroy dunes. In some areas, low winds of high frequency are less important in orienting sand dunes than stronger but less frequent storm winds, while the reverse situation may pertain in other places. Generally close correspondence has been found between parabolic dune orientation and vector sand transport resultants in several areas in Britain, Denmark, and northern Queensland, but some modification of the procedure is necessary to account for local circumstances in some areas (Robertson-Rintoul and Ritchie 1990).

INCIPIENT FOREDUNES AND PIONEER VEGETATION

Dunes begin to develop above the spring high-tidal line where the wind is deflected around flotsam, vegetation, and other obstacles. The fixed size of many obstacles limits the size of the resulting dunes, but vegetation is able to grow and facilitate further deposition as the sand accumulates about it. Tapering accumulations of sand, or shadow dunes, form on the lee side of obstacles, particularly where there is an approximately unidirectional wind regime. They can develop as pyramidal ridges

behind rock obstructions or permeable obstacles where the air is slowly moving in the form of swirls and vortices, or behind established, vegetated dunes (Hesp 1981; Clemmensen 1986) (Fig. 6.7).

Incipient (embryo) foredunes develop in the supratidal zone, where sand is trapped within vegetation. These mounds or low terraces of sand, about 1–2 m in height, usually represent the initial stage of dune formation. Vegetation has a crucial effect on saltation clouds. Effective surface roughness can increase to as much as 18 cm in fairly dense vegetation, and saltating grains that cannot rise above this level are deposited. The energy of the falling grains is also absorbed by soft, spongy vegetation, and although descending saltating grains can dislodge other grains, dense vegetation stops them from moving very far.

The form of the wind profile within a stand of vegetation depends upon its height, width, shape, spacing, and arrangement. More sand is deposited over a shorter distance within tall than within short dune vegetation because of reduced flow penetration and turbulent transfer (Hesp 1989). With increasing vegetational density, more of the surface is protected by decelerated flow in the wake region behind each plant, and when the vegetation is close together, the wake of one plant lies in the

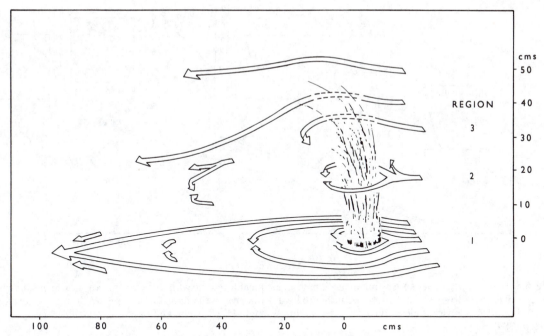

cms
- 50
- 40
REGION
3
- 30
2
- 20
- 10
1
- 0

100   80   60   40   20   0   cms

Fig. 6.7. Wind flow around a *Festuca* or *Ammophila* plant with a free stream velocity of 7 m s⁻¹. Regions 1 and 2 have symmetrically opposed, horizontal reverse flows. There is also a vertical upwind component in region 2, however, and a downwards reversing downwind component within the separation envelope that develops in region 3, where the flow moves through the less dense portion of the plant (Hesp 1981)

wake of neighbouring plants downwind. If the plants are close enough together, stable vortices form between them and the regional flow skims across the top of the canopy. The whole surface is then within the protected wake region, even though there may be large areas of bare sand. Even fairly sparse vegetation may provide substantial protection to the surface, although intrusions into the flow may erode the sand around more widely spaced plants (Willetts 1989; Wolfe and Nickling 1993). A number of workers have derived expressions for the transport of sand in vegetated areas, which confirm that even fairly sparse coastal vegetation can greatly reduce the ability of the wind to erode and transport sand (Niedoroda *et al.* 1991).

The initial and subsequent development and morphology of coastal dunes are strongly influenced by the nature of the pioneer plants, including their mode of colonization (seedlings, shoot production, rhizome development), morphology (height, density, etc.), and growth-and survival-rates, and their response to various environmental factors, including the rate of sand burial, surface

erosion, salt spray, swash inundation, and variations in precipitation, nutrient availibility, and temperature (Bressolier and Thomas 1977; Hesp 1989). In New South Wales, for example, dense, semi-prostrate shrubs trap the sand almost immediately within their windward margins, fostering the growth of asymmetrical, sharp-crested ridges with steep seaward slopes, whereas grasses tend to promote the growth of more symmetrical hummocks (Hesp 1988). *Spinifex hirsutus*, a perennial native grass, forms incipient foredunes that tend to be asymmetrical and wedge-shaped, with long, gentle, seaward slopes. This shape reflects progressively increasing downwind deposition, as the near-surface wind velocity decreases and relative roughness increases across the foredunes (Fig. 6.8).

Seasonal variations in the vegetational cover are important in determining patterns of erosion and deposition. Because of little winter growth and the burial of stems, foredune vegetation in south-western Britain provides the least resistance to the wind in spring, and the greatest resistance in autumn. Therefore, despite similar wind conditions, a significant proportion of the wind-blown

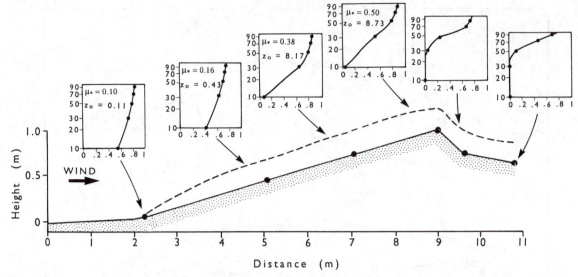

Fig. 6.8. Wind velocity over an Australian, asymmetric, incipient foredune with a 25 cm high cover of *Spinifex hirsutus* (dashed line). Boxes show the ratio of measured wind velocity to the undisturbed upwind flow 4 m above the ground, plotted against the elevation (cm). The roughness length (z_0) is in cm, and the shear velocity (u_*) is in m s^{-1} (Hesp 1983)

sand is carried over the ridge and deposited on the leeward side of the crest in March, whereas most sand is deposited on the windward side in October (Sarre 1989). Maximum sand transport from the beach occurs in autumn in north-eastern Lake Erie, and it is restricted by moisture, ice and snow, and narrow beach width in winter. The vegetational cover is dense enough to prevent much movement within the dunes until winter, however, when the vegetation has been buried or died back (Law and Davidson-Arnott 1990).

Pioneer species tend to have a high tolerance to salt, long, elaborate root systems that can reach down to the freshwater table, and rhizomes that grow parallel to the upper dune surface. Many pioneer species can survive or flourish where sand accumulates at rates of between 0.3 and 0.5 m yr^{-1}, and they become depauperate if burial rates are low. The incipient foredunes of north-western Europe are initially fixed by the perennial grasses *Elymus farctus* (*Agropyron junceiforme* or Sea Couch Grass) and *Leymus arenarius* (Lyme Grass), which have much greater tolerance for salt than Marram Grass. Sea Couch Grass is less common further north, and Lyme Grass is a dominant in boreal regions. The annuals Prickly Saltwort (*Salsola kali*) and Sea Rocket (*Cakile maritima*) are also among the first colonizers in Britain and the

southern North Sea, and Sea Sandwort (*Honkenya peploides*) is also common on the Baltic Coast (Fig. 6.9). Sea Couch and Marram Grass are the major coastal dune species in the North Atlantic. Marram and its American relative, American Beach Grass (*Ammophila breviligulata*), are among the world's most important dune-fixing or building species. They are only really vigorous in accreting environments, and they can survive accumulation rates as high as one metre per year, at least for short periods. They are unable to survive frequent tidal flooding, however, and are normally found further from the beach than more salt-tolerant, but less sand-tolerant, pioneer species (Chapman 1976). Marram and saltwort are usually the first colonizers on Cape Cod and Sable Island on the eastern coast of North America, and Sea Oats (*Uniola paniculata*) and American Beach or Dune Grass on Cape Hatteras.

DUNE MORPHOLOGY

The shape and size of aeolian dunes are determined by at least six factors: sand availability; grain-size distribution; wind energy, velocity distribution, and directional variability; vegetational

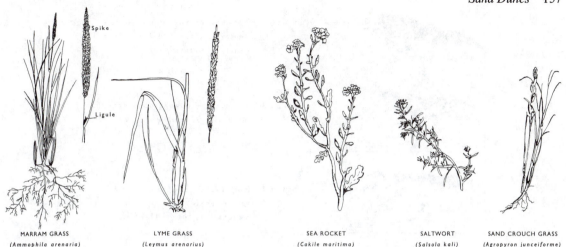

MARRAM GRASS
(Ammophila arenaria)

LYME GRASS
(Leymus arenarius)

SEA ROCKET
(Cakile maritima)

SALTWORT
(Salsola kali)

SAND CROUCH GRASS
(Agropyron junceiforme)

Fig. 6.9. Some dune-binding grasses of northern Europe (Jane and White 1964; Boorman 1977)

cover; the presence or absence of topographic obstacles; and sequential climatic changes (Pye and Tsoar 1990). Vegetation has a pivotable role in the construction of coastal dunes in many regions, but there are also extensive dunefields on arid coasts, including Baja California, Peru, Atlantic Morocco, Namibia, the northern Arabian Sea, the Red Sea, and southern and western Australia. There have been numerous attempts to classify sand dunes, but none is entirely satisfactory (Goldsmith 1985; Olson and Van Der Maarel 1989). Although various criteria have been employed, the basis of many classifications is the distinction between migrating or transgressive dunes that are largely devoid of, or unaffected by, vegetation, and vegetated and essentially stabilized dunes.

Vegetated Dunes

Vegetated dunes are most common in temperate regions with abundant rainfall, but they can also develop on semi-arid coasts if there is an efficient sand-binding flora. Vegetation can fix aeolian deposits as soon as they form, although mobile dunes can persist in exposed areas where there is a large supply of sand, and only partial vegetational cover.

Foredunes and Dune Ridges

Coastal dunefields in temperate regions usually consist of a series of roughly shore-parallel ridges, with intervening troughs or slacks. The continuous, though irregular, ridge crests are often punctuated by low points, including blowouts, through which aeolian transport takes place, and washover sluice channels, which carry flowing water during storms. On some prograding, or formerly prograding coasts, variable thicknesses of fine, wind-blown sand have been deposited over coarser sediment deposited by waves in the form of beach ridges or high tidal berms (Ruz 1989; Martini 1990) (Chapter 5). True foredunes, however, are formed entirely by aeolian deposition. Five types of established foredune have been distinguished in south-eastern Australia, based on such factors as their degree of continuity, blowout development, and vegetational cover (Hesp 1988) (Fig. 6.10). They range from continuous, well vegetated, gently undulating ridges, to remnant knobs of sand separated by large deflation basins, blowouts, and sand sheets.

Foredunes evolve into dune ridges as the coast is extended seawards, or, if there is an incomplete vegetational cover, as the ridges move landwards. Ridges develop on prograding coasts that experience alternating periods of erosion and accretion. This could be associated with the periodic interruption of calm-weather deposition by storm-wave erosion, minor sea-level oscillations, or the intermittent growth of barrier spits and islands (Bird 1990; Borówka 1990b). Ridges only develop when sand accumulates in a well-defined zone within the plant canopy, however, and wide, terrace-like dunes form if there is rapid progradation. Foredune size therefore increases with decreasing rates of beach progradation, and consequently with the time that

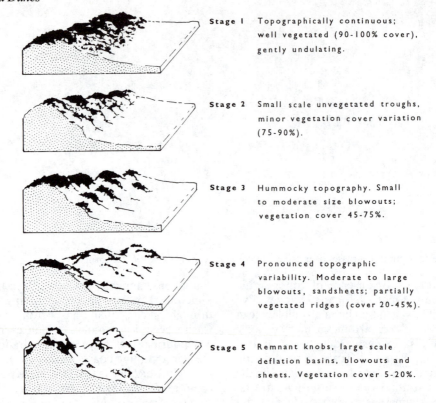

Stage 1   Topographically continuous; well vegetated (90-100% cover), gently undulating.

Stage 2   Small scale unvegetated troughs, minor vegetation cover variation (75-90%).

Stage 3   Hummocky topography. Small to moderate size blowouts; vegetation cover 45-75%.

Stage 4   Pronounced topographic variability. Moderate to large blowouts, sandsheets; partially vegetated ridges (cover 20-45%).

Stage 5   Remnant knobs, large scale deflation basins, blowouts and sheets. Vegetation cover 5-20%.

Fig. 6.10. Foredune morphology and plant cover in south-eastern Australia (Hesp 1988)

the seawardmost ridge remains active (Shepherd 1987; Hesp 1989).

Psuty (1992) modelled foredune development as a function of beach and dune sediment budgets (Fig. 6.11). When the beach budget is positive, continuous progradation sets a limit to the amount of sand that can accumulate on a foredune before it is abandoned. Maximum foredune development therefore occurs where the beach budget is essentially neutral, so that sand can accumulate for long periods of time in approximately the same place. Foredunes can have a positive sediment budget and continue to grow behind beaches that have negative budgets and are being eroded. Slightly negative beach budgets result in intermittent dune scarping and inland migration. Blowouts enlarge and extend inland in foredunes previously constructed behind beaches that have increasingly negative budgets, until they consist only of remnant sand hummocks interdispersed with large washover fans. Sherman and Bauer (1993) also emphasized the interaction of beach and dune sediment budgets, and they modified and expanded Psuty's model to include

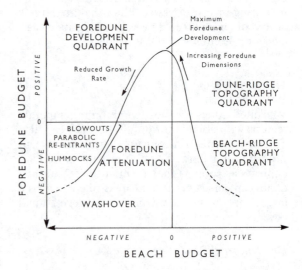

Fig. 6.11. The relationship between beach and dune sediment budgets and foredune morphological evolution (Psuty 1992)

Table 6.3. Conceptual model of beach and dune sediment budgets

Beach budget	Dune budget	Morphology
Positive	Positive	Beach or dune ridges
Positive	Steady state	Indeterminate*
Positive	Negative	Blowouts and deflation hollows
Steady state	Positive	*In situ* dune growth
Steady state	Steady state	Indeterminate*
Steady state	Negative	Blowouts and deflation hollows
Negative	Positive	Dune growth and onshore migration
Negative	Steady state	Indeterminate*
Negative	Negative	Dune erosion and washover

Note: *Indeterminate refers to situations where several types of morphology are possible for the given sediment budgets.

Source: Sherman and Bauer 1993.

steady-state sediment budgets, with little alteration in sediment volume (Table 6.3).

Longshore variations in sediment budgets can also provide an explanation for spatial changes in dune morphology (Psuty 1992). Beach ridges or low foredunes develop near river mouths where there are highly positive beach budgets. The size of the foredune increases with distance from the mouth, until the longshore decline in beach budgets causes less sand to be supplied to the dunes. This results in slower rates of growth, the development of blowouts, and eventually overtopping. Changes in the amount of sediment supplied by rivers, or changes in the position of their outlets, would cause the longshore dune morphology sequence to shift alongshore. Beach budgets along spits often change from negative to positive downdrift (Chapter 5), and the sequence of forms is therefore frequently opposite to that extending from a river source.

Saltation clouds progressively lose energy as they pass over vegetated dune ridges. Only 10–15 percent of the net flux carries beyond the incipient foredune in Northern Ireland, and much of that may occur through gaps in the dune ridge (Carter and Wilson 1990). The first ridge landwards of a foredune, which can be from about 10–30 m in height, has a steep slope on the windward side and a more gentle slope on the landward side. Older ridges become lower landwards, and they lose their straight, shore-parallel form, becoming fragmented by blowouts into small discontinuous sections. Ranwell (1958) estimated that mobile dune ridges in an exposed site in northern Wales need between seventy and eighty years to attain their maximum height and to move far enough landwards, at a rate of 6–7 m per year, to allow formation of new incipient foredunes. Although similar results have been obtained from several other exposed areas, dunes in Picardy and Flanders have migrated by as much as 25 metres in one year, while others, in the more sheltered parts of eastern England and around Lake Michigan, for example, have much lower rates of movement.

Blowouts

Shallow, saucer-shaped depressions or deep, elongated troughs are initiated in dunefields by wave erosion, overwash, lack of aeolian deposition, or deflation of devegetated or poorly vegetated areas (Gares and Nordstrom 1988; Gares 1992). Further development occurs through wind scour and slumping, and avalanching on the side-walls. Smaller blowouts can develop within larger ones, and more complex double or stacked blowouts form where material excavated from a blowout disturbs the vegetation and air flow to its lee, causing a second blowout to develop higher up the slope (Carter *et al*. 1990c). In the Netherlands, because of the rapid colonization and protective role of algae, many small, newly formed blowouts do not survive for more than five years (Jungerius *et al*. 1992). Algae are less effective in preventing erosion of larger blowouts, where longer fetch and consequently higher wind speeds allow effective sand blasting (Van Der Ancker *et al*. 1985; Pluis and Winder 1990).

Blowouts accelerate and modify the wind flowing through them, funnelling the excavated sand inland, where it is deposited in rims or convex, parabolic-shaped lobes along their leeward ends.

The depth and shape of blowouts affect the wind's ability to move sand, and they probably attain aerodynamic equilibrium with the local winds. This would account for the roughly constant width-to-depth ratio of blowouts in the Netherlands, for example, and the cessation of all erosion when they have become about 30 m long (Van Der Meulen and Jungerius 1989*a*). Although it is often assumed that blowouts are extended in a leeward direction, the greatest erosion usually occurs on their windward sides in the Netherlands (Jungerius and Van Der Meulen 1989) (Fig. 6.12). This may be attributed to the wind being most erosive when it first enters a blowout, before it becomes loaded with sand; the turbulence generated by the steep step at the upwind end of many blowouts; and the formation of a protective algal crust at the upwind end (Jungerius *et al.* 1992, Pluis 1992).

Parabolic Dunes

Simple parabolic dunes have a U- or V-plan shape, with two vegetated trailing limbs pointing upwind. There is, however, considerable variation in dune morphology (Fig. 6.13). Parabolic dunes develop from the downwind growth of blowouts in anchored dune ridges (Ritchie 1992). When blowouts form in the lower portion of ridges, their sides, which remain vegetated and stabilized, eventually separate from the bulging, advancing heads. The arms of the blowouts then form ridges that are approximately parallel to the wind, while the separated heads continue to move downwind as parabolic dunes, with their arms facing towards the wind. The process ceases when deflation reaches the water table, or further lowering is prevented by the presence of algae, chemical crusts, exhumed buried soils, or surface lag deposits. Vegetated dunefields are therefore dominated by shore-parallel ridges, and associated slacks (wet dune valleys, swales) oriented either roughly parallel or perpendicular to the ridges, depending, respectively, on whether the primary dune ridges or the trailing, slightly converging arms of the parabolic dunes are more prominent.

Very large parabolic dunes can also be initiated at the coast when sand is deposited too rapidly to be completely stabilized by sparse vegetation. Strong, unidirectional onshore winds drive the dunes hundreds to thousands of metres inland, at rates as high as 13 m yr^{-1}, while patches of vegetation anchor the trailing edges. These very

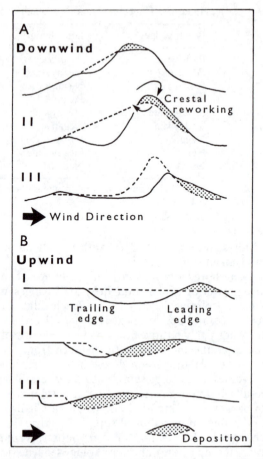

Fig. 6.12. Downwind migration of blowouts when sediment efflux exceeds blowout storage and the form enlarges (A), and upwind migration when downwind infilling is greater than upwind erosion (B) (Carter *et al.* 1990*c*)

elongated parabolic forms are most common in arid and semi-arid regions, although there are generally less-elongated examples in Europe, where the vegetation is denser, the winds more variable, and dune migration is slower (Hesp *et al.* 1989).

Unvegetated Dunes

Mobile, unvegetated dunes are most common in arid coastal regions where there is an abundant supply of sand, strong onshore winds, low rainfall, and sparse vegetation. They can also occur in humid areas where the vegetation has been damaged or removed by, for example, fire, coastal

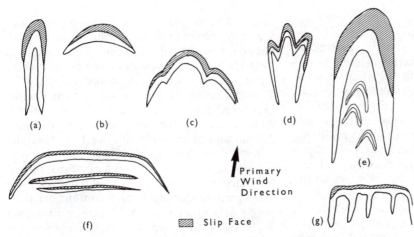

Fig. 6.13. Some types of parabolic dune, (a) hairpin, (b) lunate, (c) hemi-cyclic, (d) digitate, (e) nested, (f) long-walled transgressive ridge with secondary transverse dunes, and (g) rake-like in echelon dunes (Pye and Tsoar 1990)

↑ Primary Wind Direction

▨ Slip Face

erosion, or human activities, or where there is a lack of efficient sand-binding vegetation. In the north-western United States prominent foredunes are developing for the first time with the rapid spread of European Marram Grass, which was introduced to the San Francisco area in 1869 (Cooper 1958).

Transverse Dunes

Ridges with crests roughly perpendicular to the prevailing wind are formed where there is a good supply of sand, a lack of anchoring vegetation, and almost unidirectional wind conditions. They are common in arid coastal deserts (Lancaster 1982; Goudie *et al.* 1987; Tsoar 1990), although they are also found in some humid areas, including the Pacific coast of North America and south-western France (Cooper 1958, 1967; Bressolier *et al.* 1990). Ridge migration rates of between 1 and 9 m yr^{-1} have been reported from southern Australia and South Africa (Short 1988*b*; Hesp *et al.* 1989), and as high as 18 m yr^{-1} in Baja California (Inman *et al.* 1966).

Transverse ridges have crestlines 6–50 m in height and up to a kilometre or more in length, although the Great Dune of Pilat in south-western France is more than 100 m high and 2.5 km long (Froidefond and Prud'homme 1991). Transverse ridges have a fairly gentle windward slope (3° to 15°), and a steep lee slope that usually consists of a convex upper surface above a slip face at, or close to, the angle of repose (about 30°). Some ridges are fairly straight and the crests are regular in height, whereas others, which are sinuous and variable in height, consist of coalescent barchanoid and linguoid elements facing downwind and upwind, respectively.

Barchans

Barchans, which might be considered a type of transverse ridge, are crescentic dunes with their horns or arms oriented downwind. Their crescentic shape develops because aeolian transport is more rapid across and around the side of patches or mounds of sand than across the centres. Low barchans can migrate at rates as high as 30 m yr^{-1}, but rates of 5–10 m yr^{-1} are more typical of larger dunes (Inman *et al.* 1966; Pye and Tsoar 1990). Barchans develop where there is a poor supply of sand, little vegetation, and strongly unidirectional winds. Elongation of one of the horns by secondary winds from another direction may eventually transform them into longitudinal or oblique dune ridges, oriented parallel or at an angle to the prevailing winds, respectively.

Barchans and transverse ridges form quite rapidly on beaches, but they are usually fairly small and quickly disappear when there is a change in the sand supply or the direction of the wind (Bourman 1986; Marqués and Julià 1988). Barchans are characteristic of desert regions, although they are also common on arid and semi-arid coasts, and they developed in south-western France during the more arid Gallo-Roman period (Lancaster 1982; Bressolier *et al.* 1990). They comprise about 22 percent of the dunes on the Łeda Barrier in Poland, and they are found in other humid areas where there is a good supply of sand and an incomplete vegetational cover (Borówka 1990*a*).

Oblique Dunes

If oblique dunes are defined as ridges oriented at an angle to the resultant direction of sand movement, rather than as ridges running at an angle to the coast, then they are not common in coastal regions. In Oregon, however, there are fairly straight-crested, parallel ridges that are transverse to the winter winds, and oblique to the resultant sand transport direction (Hunter *et al*. 1983). Similar types of dune have been identified on northern Padre Island, Texas, and on the Indian coast of Orissa.

Doubts have been expressed on the value of classifying dunes as oblique, longitudinal, or transverse, according to their orientation relative to the resultant sand-transport direction. Apart from the question of whether the gross morphology and orientation of dunes are determined by the transport resultant, there is frequently uncertainty over whether such dunes are in equilibrium with present-day wind conditions. Furthermore, resultant sand-transport directions are often deduced from wind data collected at meteorological stations that are tens or even hundreds of kilometres away, and sometimes in completely different topographic environments (Pye and Tsoar 1990).

The Mendaño

Goldsmith (1985) identified a type of unvegetated coastal dune in the form of a steep, isolated hill up to hundreds of metres in height. The mendaño (Spanish for coastal sand hill) is the result of bimodal or multimodal wind regimes that move sand up to the summit from several directions; there is therefore little net dune migration.

Transgressive Dunefields

Transgressive dunefields or sand sheets are moderate-to-large bodies of generally unvegetated sand that migrate inland or alongshore, often over areas of forest, swamp, marsh, scrub, bedrock, and lagoon. Their initiation and development depend upon the sand supply, wind energy, and the effectiveness of the sand-binding vegetation. Some transgressive dunefields consist of broad, flat-to-undulating sand sheets and domes, whereas others contain transverse and oblique dune ridges, and other mobile, unvegetated-to-partially vegetated dunes. There are transgressive dunefields in semi-arid and humid tropical coastal regions, but they are most common in temperate areas with an abundant supply of sand, particularly behind westerly-facing high-energy beaches exposed to strong onshore winds (Hesp and Thom 1990). They are common, for example, on the Pacific, Gulf of California, and Texan coasts of North America, along the southern coast of South Africa, and in southern and western Australia (Short 1988*b*; Hesp *et al*. 1989).

Large areas of shrub or forest can trap sand and form large vegetated mounds within bush pockets. Transgressive dunefields also contain deflation plains and other erosional features. Erosional aeolian sabkhas develop in arid climates where the wind erodes the surface down to the damp, cohesive sand near the water table, causing evaporitic cements from the saline groundwater to be precipitated at or just below the surface (Fryberger *et al*. 1990). Coppice dunes form where clumps of pioneer vegetation start to grow on deflation surfaces. Shadow dunes may be attached to the downwind ends of the coppice dunes, or more frequently to remnant sandy knobs formed by irregular deflation and erosion. These knobs may also have steep-sided echo-dunes on their upwind side (Hesp and Thom 1990).

The landward side of active transgressive sand sheets often have slip faces that are roughly parallel to the coast. Some are low, convex tongues, whereas others are steep ridges that can attain heights of up to 100 metres or more. Cooper (1958, 1967) used the term 'precipitation ridge' to describe the long, high ridges that form where trangressive sand sheets or dunes are slowly advancing against the edge of coniferous forests in Oregon and California. Precipitation ridges develop where trees or standing water increase sand deposition and vertical accretion at the top of the lee slope, and inhibit rapid deflation from beyond the base. Other precipitation, long-walled, or retention ridges (Fig. 6.13) have been described from near Arcachon in south-western France and in the western Netherlands (Olson and Van Der Maarel 1989), Ontario (Davidson-Arnott and Pyskir 1988), and for about 50 km around the landward boundary of the Alexandria dunefield in South Africa (Illenberger and Rust 1988).

A high water table, periodic flooding, and coarse-grained sand may have prevented dune development on some sand sheets in North America (Kocurek and Nielson 1986). Extensive sand sheets or machairs also developed in the moist, cool conditions of western and north-western Scotland and north-western Ireland (Bassett and Curtis 1985) (Fig. 6.14). Machairs are stable, vegetated sand plains, usually lying behind a narrow fringing foredune. They tend to be fairly level surfaces with gentle slopes, although erosion and re-deposition of the sandy substrate create small hills and scattered blowouts—the prototypes of the bunkers and sand traps of golf-courses. Machairs may have been formed by the infilling of lake basins with sand, or by deflation down to the water table, although slowly changing sea level, strong onshore winds, a failing supply of sediment, and human interference, including pasture and tilling by crofters, probably played an important role in their development (Ritchie 1986; Carter *et al.* 1990*d*).

CLIFF-TOP DUNES

A variety of explanations has been proposed to account for the occurrence of dunes at the top of steep sea cliffs. It has been suggested, for example, that some dunes migrated up more gentle slopes from the landward side, possibly across narrow peninsulas. Others may have formed, essentially in place, when sea level was higher, or climbed up coastal slopes that existed before the cliff was cut. Vortices with strong reverse flows produce single large ridges or echo dunes on the windward side of cliffs or other steep, continuous and impermeable barriers (Tsoar and Blumberg 1991). Although flow separation can prevent sand climbing even some moderately sloping cliffs, gently sloping obstacles, which lack strong vortices, can be progressively buried under featureless sand deposits

known as climbing dunes (Short 1988*a, b*). Sand can be driven on to cliff tops through deep gullies eroded in the cliff face (Jackson and Nevin 1992). On Lakes Erie and Superior, dunes have formed where sand has been eroded from high, inclined bluffs cut in glacial sediments, and deposited in the flow separation zone on top of the cliff (Marsh and Marsh 1987). A similar explanation may account for the occurrence of a large dune on top of a 65 m-high morainic sand and clay cliff in Denmark (Christiansen *et al.* 1990).

SEDIMENTARY STRUCTURES

The internal structure of aeolian sand deposits reflects the nature of the depositional processes, and the configuration of the depositional surface. Dunes with slip faces have well-developed foreset laminae, usually from about 2 to 5 cm in thickness, dipping downwind at the angle of repose (30°–34°). The laminae are grouped together into cross-bed sets (cross-strata) that can be tabular, wedge-shaped, or trough-shaped in cross-section. Tabular crossbed sets are more frequent in the lower portion of dunes, whereas thinner, wedge-shaped sets predominate in the upper portions where cutting and filling takes place in response to changing wind conditions and small migrating bedforms. The sets are separated by horizontal or gently leeward dipping bounding planes which may truncate the more steeply dipping foreset beds. Dunes may also advance over horizontal to gently leeward-dipping bottomset laminae consisting of fine sand blown beyond the dune crest by very strong winds (Greeley and Iversen 1985; Pye and Tsoar 1990).

Hunter (1977*a*) examined the structures within fairly small, unvegetated coastal dunes, largely of

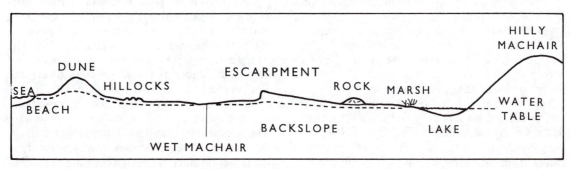

Fig. 6.14. The main features of machair (Bassett and Curtis 1985)

barchan to transverse form. Internal structures vary according to whether the depositional surface was rippled, smooth, or affected by avalanches (Table 6.4). Tractional deposition, which occurs when saltating and creeping grains come to rest, takes place on interdune flats, on the stoss side of dunes, and on lee slopes that are inclined less steeply than the angle of repose. Tractional deposition normally occurs on rippled surfaces. Planebed lamination, which can be produced by tractional deposition over smooth surfaces when wind velocities are too high for ripples to develop, is rarely found in dune deposits. Climbing-ripple structures develop as ripples migrate downslope, upslope, or alongslope, while net deposition is taking place. The angle of ripple climb varies with the rate of ripple migration and the rate of net sedimentation. Net deposition largely occurs on lee slopes, and the highest angles of climb are in the flow separation zones, where saltating grains fall on to the rippled surfaces.

There are two main types of climbing-ripple structure: rippleform lamination composed of wavy layers parallel to successive rippled depositional surfaces; and translatent stratification consisting of even layers parallel to the vector of ripple climb (Hunter 1977a, b) (Fig. 6.15). Translatent stratification is much more common than rippleform lamination in dune deposits. Climbing-ripple structure is dependent on the angle of climb (α) relative to the inclination of the angle between the ripple stoss slope and the general depositional surface (β). The angle of climb is critical if $\alpha = \beta$, subcritical if $\alpha < \beta$, and supercritical if $\alpha > \beta$. Subcritical climbing-ripple translatent strata are by far the most common type in aeolian sediments.

Deposition by other processes generally takes place only leewards of the dune crest. Grainfall deposition generally occurs on smooth surfaces in flow separation zones. The third major type of surface and internal structure is produced by slip face avalanches. Grainflow deposition or avalanching rapidly destroys pre-existing stratification, creating grainflow cross-strata that often alternate with grainfall laminae on small dunes, and on the lower portion of the slip faces of large dunes.

Vegetation influences sedimentation patterns and processes on coastal sand dunes. Shadow dunes, which may develop behind clumps of vegetation, consist of steeply dipping foreset laminae, facing in opposite directions away from the central ridge or crest of the wind shadow (Clemmensen 1986). A chevron-like, cross-bedding pattern may develop, however, when the strength of the wind fluctuates on either side of the dune, causing the ridge to migrate from one side of the dune to the other. The presence of similar structures in coastal foredunes suggests that many developed as shadow dunes. The sedimentary structure of well-vegetated dunes reflects the fact that they tend to grow vertically in place, rather than through slipface deposition and migration. Deposition on vegetated surfaces tends to produce gently dipping crossbeds, with either a unimodal azimuth related to the direction of the prevailing wind, or, as in the case of hummocky dunes, a wide range of dip directions (Goldsmith 1985; Hunter 1977a; Hesp 1989; Pye and Tsoar 1990). More complex structures develop in foredunes that are affected by wave erosion and blowouts, including slump deposits, buried wave-cut scarps, infilled troughs, and the localized development of high-angle crossbeds (Hesp 1988).

Plant types exert an important influence on the internal structure and stratification of coastal dunes. Sets of structures may therefore be characteristic of the depositional processes operating within and to the lee of individual plant communities. A series of plant community facies has been identified on Sable Island in eastern Canada, where dune stratification provides a record of four phases of development associated with ecological succession, and increasing vegetational density (McCann and Byrne 1989; Byrne and McCann 1993).

POST-DEPOSITIONAL CHANGES

Some changes occur to dune sand within a few years of deposition, whereas others take thousands of years. Textural and mineralogical characteristics are gradually altered through physical reworking by surface processes, bioturbation, compaction, weathering, pedogenesis, and cementation. The nature and importance of these changes depend upon such factors as the accumulation rate, the mineralogical content of the sand, and the climate.

Primary sedimentary structures reflect modes of transport and deposition, whereas secondary structures result from post-depositional disturbances. A variety of processes operating on dune slopes can produce contorted bedding, including surface wash, rain splash, soil creep and other mass

Table 6.4. Characteristics of the basic types of aeolian stratification

Depositional process	Character of depositional surface	Type of stratification	Dip angle	Thickness of strata; Sharpness of contacts	Segregation of grain types; Size grading	Packing	Form of strata
Tractional deposition	Rippled	Subcritically climbing translatent stratification	Stratification: low (typically 0–20°, maximum about 30°). Depositional surface: similarly low	Thin (typically 1–10 mm, maximum about 5 cm). Sharp, erosional	Distinct. Inverse	Close	Tabular planar
		Supercritically climbing translatent stratification	Stratification: variable (0–90°). Depositional surface: intermediate (10–25°)	Intermediate (typically 5–15 mm). Gradational	Distinct. Inverse except in contact zones	Close	Tabular commonly curved
		Ripple-foreset cross-lamination	Relative to translatent stratification: intermediate (5–20°)	Individual laminae Thin (typically 1–3 mm). Sharp or gradational, non-erosional		Close	Tabular concave-up or sigmoidal
		Rippleform lamination	Generalized: intermediate (typically 10–25°)		Individual laminae and sets of laminae: indistinct. Normal and inverse, neither greatly predominating	Close	Very tabular, wavy
	Smooth	Planebed lamination	Low (typically 0–15°, max.?)	Sets of laminae: Intermediate (typically 1–10 cm). Sharp or gradational, non-erosional		Close	Very tabular planar
Largely grainfall deposition	Smooth	Grainfall lamination	Intermediate (typically 20–30°, min. 0° max. about 40°)			Intermediate	Very tabular, follows pre-existent topography
Grainflow deposition	Marked by avalanches	Grainflow cross-stratification	High (angle of repose) (typically 28–34°)	Thick (typically 2–5 cm). Sharp, erosional or non-erosional	Distinct to indistinct. Inverse except near toe	Open	Cone-shaped, tongue-shaped, or roughly tabular

Source: Hunter 1977a.

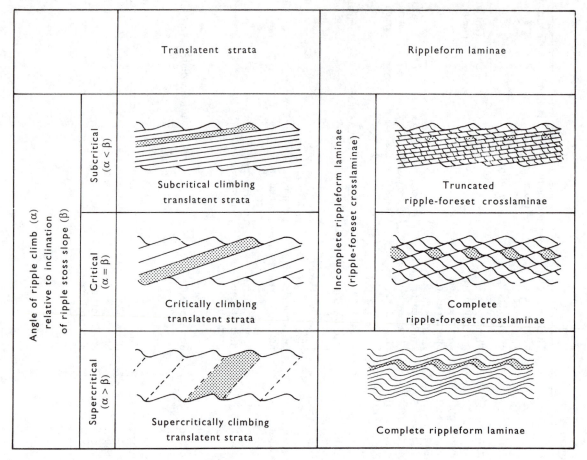

Fig. 6.15. Climbing-ripple structures (Hunter 1977a)

movements, root growth, burrowing, and frost action. Grainflows in dry sand create spoon-shaped depressions and low mounds, but because of rainfall and high water tables, small-scale slumping in weakly cohesive blocks and liquified flows in saturated layers are also quite common in many coastal dunefields (Pye and Tsoar 1990). The shape of coastal dunes varies with the seasons in southern Ontario. Niveo-aeolian sediments accumulate on the lee slopes in winter, when ground frost makes the sand more cohesive and allows surface gradients to increase up to 59°. Rapid melting of the niveo-aeolian deposits can generate mass movement over the frozen dunes in spring, when irregular slopes develop. Ground frost also makes the dune surface impermeable, encouraging fluvial activity and slope wash. In summer, dune profiles are flatter, the windward slopes and crests are more

gently rounded, and slip faces are consistently between 30° and 31° (Law 1990).

The relative importance of wind and water erosion in dunefields depends upon such factors as the strength of the winds, the presence and density of the vegetational cover, the amount of organic matter in the soil, and the degree to which podzolization has cemented it (Jungerius and Dekker 1990). In the Netherlands, gusty winds are less effective in moving grey sand coated by organic matter in the inner dunes, than are the more persistent winds in carrying cohensionless yellow sand, with no organic matter, near the coast. The effect of the wind is therefore greatest near the coast, and also on dune summits, ridges, and other exposed areas, whereas running water becomes more important away from the coast. Indeed, despite traditional concern about wind deflation, up to ten times as much sand is eroded from the inner dunes by running water as by wind (Jungerius 1989). In humid temperate regions, running water washes sand into interdune depressions, causing dune relief to become more subdued with age. Running water also breaches dune ridges, and deeply gulleys dune sands that have been cemented by intense podsolization beneath rainforest and swampland.

Post-depositional modification of dune sands by mineral weathering, leaching, and diagenesis leads to a reduction in mean grain size, poorer sorting, compaction, a decrease in porosity and permeability, and increased inter-particle cohesion. Many of these changes can be attributed to the formation of clay from the weathering of feldspars and heavy minerals, and the infiltration of aeolian dust (Pye 1983). The calcium carbonate content of the sand, which is usually derived from shell fragments and detrital calcite, also decreases as the dunes are leached. The amount of calcium carbonate originally in a dune sand has an important effect on dune development and vegetation. Rain rapidly removes the small amounts of calcium carbonate in the dunes of northern Europe, for example, making them more susceptible to water and wind erosion than those in southern Europe, which have higher proportions of calcium carbonate (Westhoff 1989). Cemented dune sands, or aeolianite, formed by the reprecipitation of dissolved carbonate, are common in arid, seasonally wet mediterranean, subtropical, and tropical climatic regions (Gardner 1983).

Podzolic soils develop on dune sands that are largely composed of quartz. Shell fragments are leached from the upper levels and the yellow iron-oxide stain is removed from the grains. Iron and aluminium compounds and organic matter are washed downwards by percolating rainwater, and deposited at lower levels, creating a slightly cemented, red-brown, yellowish-brown, or black horizon. Where there is very high rainfall, deep podzolic soil profiles can develop on quartz sand dunes within a few thousand years. Their formation has also been encouraged in European coastal dunefields by the large-scale introduction of coniferous trees (Jungerius 1990).

VEGETATIONAL SUCCESSIONS

Vegetated dunes in humid regions usually increase in age with distance from the sea. Pioneer populations occupy the younger dunes that are fairly close to the sea, whereas more mature and complex communities are found on the older dunes further inland. Three 'compartments' have been identified in the dune ecosystems of northern Europe, although the principles are generally applicable to other temperate regions (Van Der Meulen and Jungerius 1989*b*). Geomorphological processes tend to dominate in the seaward, foredune areas where there is shifting sand and active blowouts, whereas the inner dunes are stabilized by grassland and woodland vegetation. Biological processes therefore dominate geomorphological processes on the inner dunes, and there is a well developed A(B)C soil profile. In intermediate areas, geomorphological processes are not active enough to create new landforms, while biological processes are unable to completely stabilize the surface. The soil is organically enriched, but its (A)C profile records phases of erosional truncation and aeolian deposition, which have, in extreme cases, entirely buried or removed soil profiles. In northern Europe, the geomorphologically dominated and intermediate areas correspond to the yellow and grey dunes, respectively.

The type of vegetation on a dune also depends upon climate and many other local factors, but the succession in many parts of the world is from marram to dune grassland to dune scrub and woodland. This transition from pioneer, to intermediate, to mature communities may broadly reflect changes in dune age and environment normal to the coast. Among these changes are those involving the landward depletion of carbonate and available calcium; decreases in salinity, pH, and supply of fresh sand; and increases in the amount of shelter and degree of soil development, including its field moisture capacity and nutrient and organic content (Chapman 1976; Willis 1989). Vegetational successions in dunefields, however, are rarely determined simply by soil maturity and the increasing age of the landscape (De Raeve 1989). Temporal or spatial changes in dune environments can introduce geomorphologically dominated landforms into biologically dominated areas, as, for example, when blowout formation and drifting sand reintroduce younger phases to more mature, herb-covered areas.

Local changes in vegetation, corresponding to the presence of dune ridges and slacks, are superimposed on the broad-scale succession across a field. The water regime produces varied habitats, ranging from permanently flooded or damp, occasionally flooded slacks, to high, arid dune crests. Vegetation is also directly affected by variations in slope temperature according to aspect, and indirectly through its effect on soil moisture. The general landward increase in soil nutrients is particularly marked in the dune slacks, which are enriched by leaching and the flow of water, at the expense of the adjacent higher areas. Although the vegetation in wet dune slacks is therefore quite different from that on the much drier dunes, it is also very variable, depending upon such factors as the depth and period of flooding.

In addition to its geomorphological role in creating wet slack areas, the fresh water table is of great importance to the biological and ecological development of dunefields. Fresh water lies on top of saline groundwater under dunefields in humid regions. Where there is a low lying hinterland, the fresh groundwater table is in the form of a dome, and precipitation is able to maintain underground flow of fresh water to the sea, and to the landward fringes of the dunefield. If the hinterland is high, the water table rises landwards beneath the dunes, and groundwater flows to the field from the uplands. In arid areas, however, the fresh water body is small or absent (Van Djik 1989; Bakker 1990).

DUNE STABILIZATION AND CONSTRUCTION

Dunes can be destabilized by wave undercutting or destruction of the littoral vegetation. Granular avalanching, block falls, or rotational slumping occur on dune cliffs cut by waves, and within a few months the wave-cut scarp may be covered by fallen material and newly blown sand. The scarp-recovery cycle is completed when the rectilinear slope is below the angle of repose, and colonized by vegetation (Carter *et al.* 1990*c*). Although an exposed soil profile may offer significant resistance against erosion, wind deflation and water erosion usually occur where binding vegetation is damaged or destroyed by salt spray, wind, fire, climatic change, and, along the shores of Hudson Bay, by polar bears digging sunning pits (Martini 1990).

With the exception of humans, the animal that has the most effect on European sand dunes is the rabbit. Rabbits dig shallow burrows that cause shallow mass movement, and they deposit the excavated material in small terrace-like mounds on the side of dunes (Rutin 1992). Too many rabbits in

an area reduces the plant cover and increases sand mobility, and species avoided by rabbits may become dominant. A moderate rabbit population stimulates plant diversity through selective grazing, however, creating a dense low sward with shelter sites containing tall grasses. Reduced grazing by domestic and wild animals, and especially the sudden, myxomatosis-induced reduction of rabbit populations which began in 1954, has resulted in dune grasslands becoming overgrown and impoverished, and large areas, especially slacks, being covered in a high, dense scrub of birch, willow, and alder (Westhoff 1989; Boorman 1989). Moderate grazing has therefore become important in maintaining dune-heath vegetation in temperate north-western Europe.

Humans can have a beneficial or deleterious effect on coastal sand dunes (Westhoff 1989; Van Der Meulen *et al.* 1989; Doody 1992). Direct effects include the extraction of sand from dunes for minerals or building purposes, military activities, forestation and deforestation, trampling, off-road vehicles, the construction of sand fences, and the planting of sand-binding species. Dune areas have been destroyed for housing, industry, and roads, and converted into golf-courses. Indirect effects include the introduction of exotic species, air flow disturbance by buildings, increasing amounts of nutrients through acid precipitation, introduction or removal of grazing or burrowing animals, removal of sand from beaches, dredging, the dumping of spoil, construction of groynes, and changes in the water table resulting from forestation or residential and industrial development.

Wet dune slacks once covered about one-third of the entire dune area in the Netherlands. The extraction of groundwater for public water supply, however, caused them to dry out, destroying their rich and diverse flora and fauna. Artificial recharge of prepurified water from the Rhine and Meuse Rivers and from polder canals has restored the original water table, but there has not been a complete return of the indigenous vegetation, possibly because the nutrient-rich river water has encouraged the growth of other, competing plant species (Westhoff 1989; Van Dijk 1989; Van Der Meulen and Jungerius 1989*a*). There has been extensive, and in some cases early, deforestation of previously stable dunes in Europe, Australasia, North America, and South Africa. Deforestation of Polish dunes, which began in the Middle Ages, resulted in the formation of large mobile dunes from the sixteenth to eighteenth centuries, and the destruction of the remaining forest (Piotrowska 1989).

Because of the threat posed by mobile dunes to human settlements, efforts to rebuild and stabilize coastal sand dunes have also had a long history in many parts of the world. The importance of dunes as a natural coastal defence for low-lying land is reflected in laws relating to dune stabilization dating back to the thirteenth century (Van Der Meulen and Van Der Maarel 1989). In Denmark, where drifting sand has buried entire villages, a royal decree was issued in 1539 to prohibit removal of vegetation from dunefields. By the early nineteenth century marram grass and other plants were being used to fix the blowing sand (Jensen 1994). Artificial dune stabilization and construction may be undertaken to strengthen sea defences, eliminate the nuisance of blowing sand, and protect developed areas inland (Van Der Meulen *et al.* 1989; Pye 1990; Pye and Tsoar 1990; Gares 1990). The growth and stabilization of foredunes, however, can prevent sediment reaching dunes further downwind.

Methods of dune construction include the use of brushwood or liquid polymers and other chemicals to prevent the erosion of surface sand, artificial planting or sowing of pioneer grasses, fertilization to increase the mineral status of young dune soils, the use of porous sand fences of perforated plastic or synthetic nylon mesh, wooden slats, wattling, reeds, or branches where it is necessary to accumulate sand rapidly in a particular place, and artificial walkways to provide limited access to beaches while minimizing damage by trampling. Nevertheless, these rehabilitation techniques are capable of providing long-term protection to eroding dunes only if there is an adequate supply of sediment (Piotrowska 1989).

Although sand stabilization must be undertaken for some human activities, including coastal protection and residential development, there has been criticism of the over-enthusiasm of dune managers in seeking to stabilize all the bare sand in dunefields. This reduces morphological variety and species diversity, and their value as aesthetic and nature-watching resources (Nordstrom and Lotstein 1989). There has therefore been some relaxation in the use of stabilization measures in the Netherlands and Denmark in the last few decades (Jensen 1994). The vegetational stabilization of dunes, which represents a static solution to a dynamic problem, may cause further difficulties (Van Bohemen and Meesters 1992). For example, securing the seaward slopes of dunes with vegetation and brushwood fences, or constructing seawalls in front of dunes, impairs or prevents their ability to replenish beaches during storms, which results in accelerated beach erosion.

7 Estuaries and Coastal Lagoons

INTRODUCTION

The term 'estuary' is derived from the Latin *aestu-arium* (*aestus*—tidal flow), which refers to a tidal inlet of the sea. Estuaries have been defined in many ways. One of the most widely used definitions was provided by Pritchard, who described them as partly enclosed bodies of water, freely connected to the open sea, within which sea water is measurably diluted with river water (Pritchard 1967*a*). This definition, which is based on the salinity of the water, extends the use of the term 'estuary' to include a variety of coastal re-entrants, including lagoons formed by a chain of barrier islands.

Many estuaries are young, transitional features that developed from valleys that were incised by fluvial erosion during the last glacial stage, and then drowned by rising sea level in the Holocene. Their initial form is determined by the nature of the coastal topography, but it changes very rapidly as they adjust to present sea level. Estuaries tend to be good sediment traps. Most of the approximately 13.5×10^9 tons of suspended sediment carried by the rivers of the world is deposited close to the continents, and especially in estuaries and other coastal indentations (Milliman and Meade 1983).

ESTUARINE PROCESSES

Tidal waves are propagated in estuaries in response to the tidal rise and fall of the water surface at their mouths. Tides are long waves with a wavelength (L) in shallow water of:

$$L = T(gh)^{1/2}$$

where T is the tidal period of 44,640 seconds (12.4 hours) for a semi-diurnal tide, g is gravitational acceleration, and h is the water depth. The velocity or celerity of a progressive tidal wave (C) travelling up an estuary is given by:

$$C = (gh)^{1/2}.$$

Slowly moving tides in very long, shallow estuaries may fail to reach the landward terminus in the 12.4 hours before the next tide enters at the seaward end. Long estuaries can therefore be occupied simultaneously by several tidal waves. In the Amazon, for example, there can be as many as seven or eight tidal waves at any one time along its 850-km length (Defant 1961).

The energy, amplitude, and velocity of a tidal wave tend to decrease because of friction with the boundaries of the channel. On the other hand, as an estuary becomes narrower and shallower towards its head, the increasing concentration of tidal energy acts to increase the height of the wave. Tidal amplitude (range) can vary along an estuary in three ways, depending on the relative effects of friction and channel convergence (Fig. 7.1):

a) In a hypersynchronous estuary, convergence is more significant than friction, and tidal amplitude therefore increases up the estuary before eventually decreasing. Most estuaries are hypersynchronous, and the strongest tidal currents occur in their middle or landward sections.

b) In a synchronous estuary, the effects of convergence and friction are equal, so that tidal amplitude remains constant up the estuary before finally declining towards its head.

c) In a hyposynchronous estuary, the effect of friction is greater than convergence, and tidal amplitude decreases landwards.

Oceanic tides advance as progressive waves, but they can be transformed into amplified stationary waves by reflection from shoals or from the head of estuaries (Chapter 2). Pure standing waves develop in frictionless, rectangular estuaries when there is a

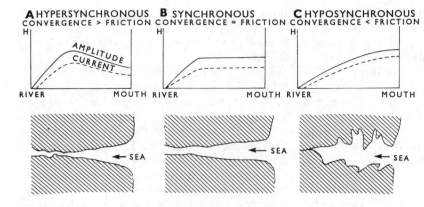

Fig. 7.1. The effect of convergence and friction on tidal range
within estuaries (Salomon and
Allen 1983)

node at their mouth, and their length (l_e) is an odd-numbered multiple (n) of a quarter of the wavelength of the tidal wave within the channels (L):

$$l_e = n \frac{L}{4}.$$

With a perfect standing wave, high water occurs simultaneously along the channel, and the amplitude varies from a maximum to zero every quarter of its wavelength. The highest current velocities occur at midtide, and there are slack periods with zero velocity at high and low water. In a progressive wave, current velocity is in phase with the amplitude, and maximum velocity is at high and low tide, with zero velocity at mid-tide. The tides in most estuaries are a combination of standing and progressive waves, with one dominating over the other according to the geometry of the system. The slack water period therefore occurs an hour or two after high water, and the low water slack period is delayed until after low water. If the fluvial and tidal discharges are comparable, however, the range and timing of the tide in the upper reaches of estuaries can also be affected by variations in river discharge. An increase in river flow reduces tidal range, and it delays the occurrence of high water and advances the occurrence of low water (Godin 1985).

Tidal waves become asymmetrical in estuaries, and the ebb and flood flows are rarely equal. Tidal asymmetry can be expressed in terms of the difference in the peak ebb and flood current velocities; differences in the duration of the ebb and flood flows; and differences in the tidal excursion—the areas under the ebb and flood curves on a time-velocity graph. When the tidal amplitude is a significant proportion of the mean water depth, the crest of the wave travels in considerably deeper

water than the trough. Because of bottom friction, the crest travels faster than the trough and tends to overtake it, producing an asymmetrical wave with a shortened duration of rise (flood) and a lengthened duration of fall (ebb). The associated tidal current velocities become asymmetrical as flood velocities increase and ebb velocities decrease. Truncation of the lower portion of the oceanic tide in very shallow waterways, where the depth is less than the tidal range, also shortens rising tides while increasing the duration of falling tides (Lincoln and FitzGerald 1988). Unequal ebb and flood discharges and velocities can further result from the mixing of river and ocean water, coriolis deflection, and flow separation into distinct channels or zones. If the tidal range is high enough, and the bottom slope around mean sea level is low enough, extreme asymmetry can create a tidal bore, consisting of a fast, upstream-moving wall of water with a steep front.

The asymmetry of estuarine tides is also determined by the extent and morphology of the tidal flats and marshes. It has been suggested that an estuary will have dominant ebb currents, with shorter durations but higher peak velocities, if its area increases with tidal height, while flood currents will dominate if there is little change in area with tidal stage. Estuaries and bays lined with extensive tidal flats and low marshes may therefore be ebb-dominated, whereas estuaries bordered by the nearly vertical walls of high marsh are dominated by flood currents (Aubrey and Speer 1985; Speer and Aubrey 1985). The ebb- or flood-dominance of estuaries and bays may change as the systems evolve. Flood currents could initially dominate bays with fairly steep sides and much open water, but they become ebb-dominated with the development of gently sloping intertidal flats. Flood-dominated storm surges could still bring sediment into bays, however, forming high supratidal marshes and eventually causing the systems to revert to flood-dominated conditions (Boon and Bryne 1981; Lessa and Masselink 1995).

Tidal asymmetry can be represented by the non-linear growth of harmonics and compound constituents of the astronomical tide (Speer and Aubrey 1985). The astronomical tidal spectrum is composed of a large number of constituents, which interact in shallow estuaries (Chapter 2). A rich spectrum of forced motions has been identified (Boon 1975; Aubrey and Speer 1985). The main ones in shallow water involve the principal lunar and solar semi-diurnal constituents, M_2 and S_2, and their harmonic overtides M_4, M_6, M_8 and S_4, S_6, and S_8; where the subscripts represent the number of daily oscillations. The angular speed of the overtides is an exact multiple of one of the astronomical constituents, while the angular speed of a compound tide is equal to the sum or difference of the angular speeds of two or more astronomical constituents. The relative amplitude and phase of the harmonic overtides are responsible for differences in the duration and peak velocity of flood and ebb tides (Fig. 2.18). Friedrichs and Aubrey (1988) found that the growth of the M_4 harmonic is largely responsible for tidal distortion in shallow estuaries with long, narrow channels, and a large tidal range-to-channel depth ratio. In Virginia, for example, differences in the relative duration of the ebb and flood tides can be largely attributed to the addition of the quarter diurnal M_4 harmonic to the M_2 tide (Boon 1975; Boon and Bryne 1981).

Although the velocity and duration of the ebb and flood stages of asymmetrical tides are generally quite different, the flows are rarely directly opposed in sandy estuaries. As the flood tide begins to run into estuaries, new channels are cut to avoid the water still flowing out after low water has passed. This creates a variety of drainage patterns composed of interdigitating, mutually evasive ebb- and flood-dominated channels (Chapter 8).

A variety of residual or non-tidal currents also play an important role in estuarine circulation. Residual currents result from the mixing of fresh and salt water, which generates gravitational circulation consisting of net landward flow of denser sea water in a bottom layer, and net seaward flow of lighter, less saline river water in a surface layer. The strength of this circulation varies considerably,

however, according to differences in estuarine morphology, river discharge, tidal range, atmospheric pressure, and wind and coastal oceanic effects.

Wind-induced currents can affect the entire water column and they may, in some cases, be of equal or even greater importance than those generated by tides and mixing. Winds blowing up an estuary, for example, generate upstream flows that tend to raise the water level. The resulting pressure gradient strengthens seaward flow at an intermediate depth, creating a seaward flowing current between two landward flows. Surface water blown seawards in the Ria de Arosa in north-western Spain, is replaced by sea water flowing landwards along the bottom. This current carries nutrients into the estuary, and helps to support a valuable mussel industry (Chase 1975). Intrusion of saline water into the Choptank Estuary is caused by lateral, wind-induced upwelling of saline water in adjacent Chesapeake Bay (Sanford and Boicourt 1990). In the Potomac Estuary in Chesapeake Bay, winds blowing downstream enhance the seaward flow of water in the surface layer and landwards flow in the bottom layer, but surface flow can be reversed when winds blow upstream. Typical estuarine circulation, involving a seaward flowing surface layer and a landward flowing bottom layer, occurred for only about 43 percent of the study period. Residual currents flowed into the estuary at all depths for 22 percent of the time, while there was reversed circulation, involving inflow in the surface layer and outflow in the bottom layer, for about 21 percent of the time. For the rest of the time, flow was either seaward at all depths, or three layered, with either a landward flow between surface and bottom seaward flows, or a seaward flow between surface and bottom landward flows (Elliott 1978).

Water density varies with temperature as well as salinity. Differences in the density of river and sea water tend to be reduced in winter, when the fresh water is colder, but enhanced in summer when it is warmer than the sea. Although water temperatures in estuaries are usually far less variable than salinity, temperature-related differences in density can be significant in fiords, where a lack of river flow and intense winter cooling cause the denser surface water to sink (Dyer 1973).

An estuary becomes more saline at the landward than at the seaward end if high evaporation causes fresh water to be lost more rapidly than it is replaced by river flow. This creates inverse or negative estuaries with an internal flow opposite to that of normal estuaries. Even in the cool summers of south-eastern England, estuaries with little fresh-water input can become slightly more saline than the sea (Sheldon 1968). Nevertheless, inverse flows are most likely to occur in hot climates. For example, salinity is twice as high in summer in the Laguna Madre of southern Texas than in the Gulf of Mexico (Ward 1980). Inverse flow in San Diego Bay may be intensified by the surface discharge of heated effluent from an electrical generating plant (Hammond and Wallace 1982). The water in Spencer Gulf, South Australia, is very saline during summer, and it becomes much denser than the water on the adjacent shelf when it is cooled in autumn. Gulf water then flows seawards along the bottom and is replaced by a surface inflow of less-saline water (Lennon *et al.* 1987). In tropical Australia evaporation causes a salinity maximum to develop near the mouth of estuaries during the dry season. Downwelling of the denser water then induces classical flow landwards, and inverse flow seawards of the maximum (Wolanski 1986). Evaporation and evapotranspiration from mangals and tidal flats have created seaward tapering salt wedges in the inner portion of estuaries in Senegal. Net upstream discharge has formed barrier spits that are turned upstream, and turbidity maxima that occupy external positions near and beyond the mouth of the estuaries (Barusseau *et al.* 1985).

Other types of current may help to move sediment. For example, helical vortices, with axes parallel to the tidal current, develop near slack water in Southampton Water in southern Britain. The vortices become stronger as tidal flows increase, and they may be responsible for the development of a stable system of furrows on the bottom (Dyer 1982).

The mixing of estuarine water results from bottom friction, internal shear instabilities, and breaking of lee waves, although wind can also play an important role. Internal mixing takes place when wave-like disturbances develop and break across the interface between fresher, less dense water on the surface, and denser saline water on the bottom. Density stratification may be broken down, however, by turbulence generated by tidal flow over the bottom-roughness elements and topographic irregularities. In uniform steady flow, away from boundaries, the relative effect of stabilizing density gradients and destabilizing shear induced turbulence can be determined using the gradient Richardson Number (R_i):

$$R_i = -\frac{g}{\rho}\frac{dp/dz}{(du_z/dz)^2}$$

where ρ is the water density, and u_z the velocity at height z. Theoretically, flow is laminar when R_i is greater than 0.25, and there is little mixing. When R_i is less than 0.25, the flow becomes unstable, and disturbances break at the interface, causing mixing. In the field, however, the transition from laminar to turbulent flow is indistinct and can occur between $R_i = 1$ and $R_i = 0.03$. When mixing is dominated by velocity shear generated at the bed, the bulk or layer Richardson Number (R_{iL}) provides an alternative parameter:

$$R_{iL} = \frac{gh}{\hat{u}^2}\frac{\Delta\rho}{\rho}$$

where h is the water depth, û is the depth mean velocity, and $\Delta\rho$ is the density difference between the surface and the bottom. When $R_{iL} > 20$ there is a stable interface with internal wave activity but little bottom-induced mixing. Modification of the interface by bottom turbulence takes place when R_{iL} is between 20 and 2, and fully three-dimensional turbulent modification of the halocline occurs when $R_{iL} < 2$ (Dyer and New 1986). The interfacial Froude Number, which has the same form as the square root of an inverse Richardson Number, provides another way of considering the influence of density stratification (Dyer 1986) (Chapter 9).

Sediment accumulates in estuaries because of the trapping of silt and clay by the circulation, the effect of settling lags and other associated mechanisms (Chapter 8), and the landward transport of mainly sand and other coarse sediment by waves, and fine-grained sand and silt by asymmetrical tides (Postma 1980). Estuarine sediment undergoes repetitive episodes of circulation, resuspension, and temporary deposition before finally being deposited within the estuary, or escaping out to sea.

Coarse sediment moves along the bottom as bedload, while fine sediment moves as suspended load or, where there is high turbidity, as fluid mud. When settling sediment concentrations become higher than about 10,000 mg 1^{-1}, the hindered settling effect causes the settling velocity of the sediment to decline, because of the upward flow of displaced water (Chapter 3). At very high concentrations the vertical flow is so great that the upward fluid-drag forces on the sediment are equal to the downward force of gravity. A state of dynamic equilibrium, with no vertical movement of suspended sediment, then occurs close to the bed, forming fluid mud. The thickness of the fluid mud layer continues to increase as long as the deposition rate on the upper side of the layer is greater than the consolidation rate on the under side. Fluid mud consists of water with concentrations of suspended sediment ranging from 10,000 up to several hundred thousand milligrams per litre. Layers of fluid mud, ranging from a few centimetres to several metres in thickness, travel considerable distances with the tide. The floor of estuaries can therefore be instantaneously raised or lowered by several metres as these near-bed suspensions suddenly become stagnant or remobilized, according to the fluctuating strength of the tidal currents (Kirby 1988). Fluid mud can be very extensive or locally restricted, and it can be a permanent or temporary feature. It has been identified in many estuaries, including the Thames, Severn, lower Rhine, Ems, and Scheldt in north-western Europe, and the Gironde and Loire in western France, as well as on the inner portion of the Amazon continental shelf (van Leussen and van Velzen 1989; Trowbridge and Kineke 1994).

Tidal asymmetry has an important influence on the movement of suspended sediment. The amount of fine sand, biologically aggregated material, and other relatively coarse sediment that can be carried in suspension is determined by current velocity. Differences in the peak flood and ebb currents are therefore important in the movement of the coarse component of the suspended load. On the other hand, the suspended load is rarely saturated with fine sediment, and it only settles at very low current velocities, and after a significant settling lag (Chapter 8). The movement of fine suspended sediment is therefore most influenced by differences in the duration of the slack water periods that precede the ebb and flood tides (Dronkers 1986).

Erosion is related to bed shear, which, in turn, is related to the square of the current velocity. There is usually a close relationship between current velocity and suspended sediment concentrations in estuaries, although a time lag is generally evident (Nichols 1986; Costa and Mehta 1990). Concentrations are usually higher, however, when estuarine currents are decreasing, than when they are increasing. Flow–sediment hysteresis reflects delays associated with global advective phenomena, and

the response of the sediment to local variations in flow. Dyer and Evans (1989) pointed out that the latter factor produces: a settling lag related to the time it takes a particle to settle to the bed; a diffusion lag owing to the time required for an entrained particle to be diffused to the upper layers in the water column; a threshold lag associated with the erosional resistance of the top layer and the critical threshold shear stress; and a consolidation lag related to the increase in bed shear strength with the time available for bed consolidation. Particles have more time to settle during flood than ebb slack periods, and sediment concentrations for the same current velocity therefore tend to be lower during the slack following the flood tide than the ebb (Nichols and Biggs 1985). Estuarine sediment loads and concentrations can exhibit fluvial and tidal lags or hysteresis. In the Fraser Estuary in southern British Columbia, suspended fluvial sediment loads and concentrations are higher, for a given discharge, when water levels are rising during the freshet or snow melt period, than when they are falling. This reflects the availability of frost-weathered material along upstream channel banks when the snow first begins to melt, then its rapid exhaustion as the discharge declines. On the other hand, because high turbulence causes fine sand to be resuspended, concentrations are greater, for the same current velocity, when the tide is falling and flows are decelerating, than when it is rising and flows are accelerating. There is also hysteresis in the relationship between river discharge and the size of the bedforms. Bedform magnitude increases with the seasonal increase in discharge, but the maximum is attained 2–3 weeks after the peak freshet flows. A similar lag occurs as bedforms decrease in size with the post-freshet decline in river flow (Kostaschuk *et al*. 1989).

TYPES OF ESTUARY

Dalrymple *et al*. (1992) used a three-dimensional prism to represent the evolution of estuaries with changing relative sea level (Fig. 7.2). Progradation, which corresponds to movement to the back of the prism, causes estuaries to fill, and they are converted into deltas if the sediment is supplied directly by the river, or into beach ridge plains or tidal flats if the sediment is delivered by waves or tides, respectively. Transgressions, which correspond to

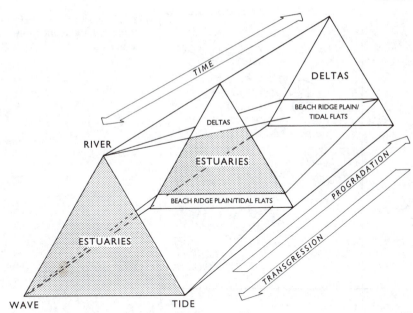

Fig. 7.2. An evolutionary classification of coastal environments. The long axis of the prism represents relative time with reference to changes in relative sea level and sediment supply, and the three edges of the prism correspond to fluvial, wave, and tidally dominated environments (Dalrymple *et al.* 1992)

movement towards the front of the prism, lead to the flooding of river valleys, and their conversion into estuaries. Vertical sections through the prism can be used to classify coastal depositional systems according to the relative importance of fluvial, tidal, and wave processes. The upper section corresponds to the triangular classification of deltas (Fig. 9.8), while the basal portion is conceptually similar to Hayes's (1979) wave–tide classification of barrier coasts (Chapters 5 and 9). The trapezoidal section in the middle provides a framework for the classification of estuaries according to whether they are wave- or tide-dominated. Boyd *et al.* (1992) modified this evolutionary model to produce a ternary process-based classification of clastic depositional environments. Estuaries and other coastal environments are discriminated according to the relative wave, tide, and river power, the degree of coastal embayment, and the rate and source of the sediment supply.

Estuaries have been classified in a number of ways, using a variety of criteria (Fig. 7.3). Most classifications are based on geomorphology and physiography, hydrography, salinity and tidal characteristics, sedimentation, or ecosystem energetics. Pritchard classified estuaries according to geomorphological criteria (Pritchard 1967*b*). Four types were identified:

a) Drowned river valleys (rias) are former river valleys, or parts of river valleys, that were flooded during the postglacial transgression. They retain much of their original fluvial morphology, although they can contain extensive mudflats and salt-marshes and there may be barrier spits at their mouths.

b) Barrier (bar) -built estuaries occur where barriers enclose coastal embayments, forming extensive, shallow lagoons. Sea water enters the embayments through narrow inlets, but their fairly small size, in comparison with the size of the lagoons, tends to reduce tidal action. Partly because lagoons have a large surface area relative to their volume, meteorological forcing, including wave propagation, wind-generated currents, and set-up effects, exerts a strong influence on the mixing of fresh and saline water, and on the pattern of sedimentation (Boothroyd *et al.* 1985; Kjerfve 1989).

c) Fiords were formed by the submergence of glacially overdeepened valleys. They usually have steep sides and almost rectangular cross-sections, with a low width-to-depth ratio. In some fiords the floors consist of a number of basins separated by shallow sills, which may be composed of rock bars with a mantle of morainal or other glacio-marine sediment.

d) A fourth class represents estuaries that cannot be included in the other categories, such as those formed by faulting, folding, or other tectonic mechanisms. Parts of San Francisco Bay and the Tagus Estuary in Portugal, for example, were formed by the slippage of fault blocks.

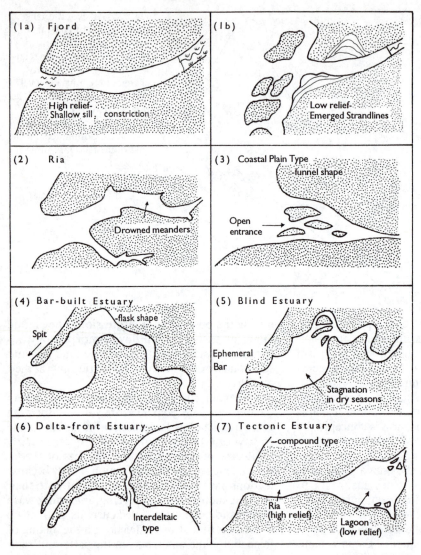

Fig. 7.3. Fairbridge's (1980) physiographic classification of estuaries

A generally more useful classification is based upon the movement and mixing of the water in estuaries (Pritchard 1967*b*; Officer 1976; Bowden 1980) (Fig. 7.4):

a) In highly stratified (salt wedge) estuaries, fresh river water flows to the sea over a wedge of denser sea water. These estuaries generally occur where rivers flow into almost tideless seas. Tidal flow is therefore much less than river flow, and as there is very little turbulence or mixing, the saline water within the wedge is almost motionless. Outside the estuary, the river water may spread out in the sea as a surface plume with a sharp boundary or front.

These fronts are zones of convergence where surface sea and river water sink along the interface between the two water bodies (Simpson and Turrell 1986; McClimans 1988).

If the flow in the upper layer is fast enough, velocity shear at the interface between the upper, landward-flowing and the lower, seaward-flowing layers generate Kelvin–Helmholtz instability waves. These internal waves develop on the halocline, a narrow mixing zone containing an abrupt change in salinity. As these waves break upwards, small amounts of saline water are entrained in the surface layer—this is a one-way process, with no corresponding movement of fresh water

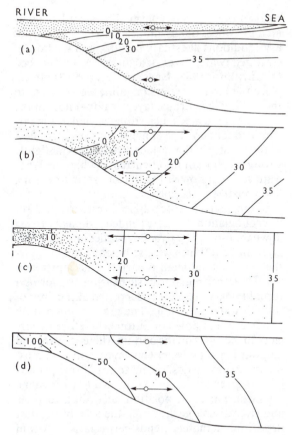

Fig. 7.4. Types of estuary, classified according to the movement and mixing of water: (a) salt wedge, (b) partially mixed, (c) fully mixed, and (d) negative. Numbered lines are salinity levels (in parts per thousand). Dot density represents sediment concentration and the arrows show net sediment movement over tidal ebb and flood (Postma 1980)

downwards. Entrainment-induced increases in the salinity and volume of the upper layer cause it to flow faster, while sea water slowly flows inward in the lower layer to compensate for losses to the surface layer.

Salt wedges change their position according to river flow. When discharge is low in the Southwest Pass of the Mississippi, the salt wedge extends more than 160 km inland, but when it is high, it is only about 1.5 km above the mouth. There is fairly strong tidal flow and high river discharge in the Fraser Estuary in southern British Columbia. The degree of stratification and the position of the salt wedge vary according to river discharge and tides. The estuary is moderately stratified during periods of low discharge and high tidal range, when mixing is enhanced, and highly stratified, with limited mixing, when discharge is high and the tidal range is low. The head of the salt wedge is as much as 30 km upstream when river discharge is low, but it varies between the mouth of the estuary and about 15 km upstream during periods of moderate flow. The head can also propagate as far as 18 km upstream during flood tides, but it is completely flushed out of the estuary during ebb tides (Geyer 1988; Kostaschuk and Atwood 1990).

Deep fiords are highly stratified estuaries, in which a surface layer of fresh water, up to only a few tens of metres in thickness, flows over an extremely thick and almost motionless basin of salt water. Shallow sills at the mouths of fiords prevent the free exchange of water with the ocean, although tides can generate some turbulent mixing. In addition to two-layer estuarine circulation, there can be larger, less dynamic circulation cells at greater depths, and the water may be occasionally replaced by denser coastal or shelf water.

The circulation in fiords is controlled by climatic as well as morphological factors. Estuarine circulation is absent in fiords for most of the year in polar regions, where runoff only occurs for a few months. Winter ice prevents mixing by the wind, and it damps tidal fluctuations, although salt rejection from the freezing mass causes vertical mixing and gravity flows that can eventually affect the middle and lower water layers. Calving is a dominant ablation process in ice-contact fiords, although because of landfast sea ice, sediment rafting may only be of minor importance. Sea ice also carries sediments placed on or in it by the wind, rivers, rock falls, sea floor erosion, bottom freezing, and wave and current washover. As the ice has little mobility upon breakup, however, most of the sediment is not carried very far. Rates of sediment accumulation are generally lower in Arctic fiords than in more temperate latitudes, particularly where there is permanent sea ice, and cold-based glaciers that provide only small amounts of water and sediment (Syvitski *et al.* 1987; Powell and Molnia 1989; Sexton *et al.* 1992; Gilbert *et al.* 1993).

Sedimentation in highly stratified estuaries is dominated by river flow. High discharges, large sediment loads, and low tidal range favour delta formation at the mouth of salt wedge estuaries, where the rivers are freed from the confines of their channels (Chapter 9). Some of the fine suspended sediment carried seawards by the river, however, settles through the halocline into the lower, landward-flowing layer. This sediment is slowly carried back up to the tip of the wedge, where it accumulates in a shoal or bar. The coarser fluvial bedload is deposited a little upstream, where the fresh, seaward-flowing water is forced to rise up over the head of the wedge. Rapid accretion and bar formation therefore occur near the tip of the salt wedge, with the heavier particles deposited just upstream, and the lighter particles just downstream of the salt intrusion limit. An approximately 2 m-thick layer of sediment was deposited over a two week period near the tip of the salt wedge in the Southwest Pass of the Mississippi River, when high river flow prevented dredging. The removal of bars and shoals at the mouth of the Pass has probably intensified the salt wedge, however, and conditions may not be applicable to more natural situations. Nevertheless, although the salt wedge sedimentary model was largely based on conditions near the mouth of the Mississippi River, further support for the model has been provided by investigations conducted in the Fraser River Estuary in British Columbia. They have confirmed the dominant role of riverine processes in controlling suspended sediment concentrations, and the rapid deposition of suspended and coarse sediment near the tip of the wedge (Kostaschuk *et al.* 1992*a*).

b) Most estuaries are moderately stratified or partially mixed. They are usually fairly shallow, and the tides are strong enough to prevent rivers dominating the circulation. Salinity increases seaward in both the upper and lower layers, and there is also a gradual increase with depth. A zone of no mean movement occurs between the surface layer, where fresh water is flowing seawards, and the bottom layer, where sea water is flowing landwards. As downstream flow increases and upstream flow decreases towards the head of the estuary, this zone becomes progressively deeper, usually reaching the bottom in the region of the 2–5‰ salinity isoline: mean flow is downstream at all depths further landwards.

As the water moves back and fore with the tides, friction with the bed generates velocity shear and turbulent eddies. Turbulence transfers fresh water downwards into saline water, as well as moving saline water up into fresh water; it therefore mixes salt and fresh water more effectively than entrainment. Although tides are the most important mixing agent, breaking internal waves also help to weaken the boundary layer. They can be generated by winds, tidally induced changes in boundary conditions at estuary mouths, and the flow of stratified water over irregularities in the bed (Kranenburg 1988; New and Dyer 1988). To replace the large amounts of saline water added to the outflowing surface layer, residual flow in the lower layer must be much stronger than in a highly stratified estuary. The enhanced downstream flow of fresh water in the surface layer results in slightly stronger ebb than flood currents, while the landward residual flow at the bottom strengthens the flood relative to the ebb.

Sediment tends to be trapped in estuaries by two-layer circulation. Fine, suspended sediment carried down the estuary by river flow settles into the lower layer and combines with sediment brought in by the sea. This sediment is carried back upstream, where it either remains in suspension and forms a turbidity maximum, or is deposited at the limit of the salt water intrusion. The greatest deposition under average flow conditions therefore occurs close to the point on the bottom where seaward and landward flowing water converges, although most sediment may be flushed out to sea during periods of high river discharge. As the head of the saline intrusion varies in position according to tides and river discharge, however, the area of greatest deposition actually extends between its upstream and downstream limits.

c) Non-stratified (well-mixed, vertically homogeneous) estuaries occur where tidal currents are strong enough, relative to river flow, to completely remove all vertical differences in salinity. Currents in well-mixed estuaries are directed seawards with ebb tides, and landwards with flood tides. Although perfect, vertically homogeneous estuaries may not exist, this condition is generally most nearly satisfied in shallow estuaries with a high tidal range.

Non-stratified estuaries that are fairly deep and narrow, tend to be laterally as well as vertically homogeneous. In wide, fairly shallow estuaries, however, including the Delaware, Raritan, and Ganges–Brahmaputra–Meghna (Barua 1990), the Coriolis force deflects river flow to the right-hand side in the northern hemisphere, and tidal flow to the left-hand side. Flow is therefore landwards at all depths on the left-hand side, and seawards at all depths on the right hand side. Sedimentation may be greatest near the limit of salt intrusion on the left bank, although it has been questioned whether the Coriolis force has much effect on the transport and distribution of marine and terrestrial sediment (Kjerfve 1989). There is no comparable sediment trap in laterally homogeneous estuaries, and accumulation therefore occurs where the slow, net seaward flow is interrupted by changes in cross-sectional area, islands, tributaries, and other obstacles.

Estuarine circulation varies according to the relative magnitude of river and tidal flows, especially as a result of seasonal changes in river discharge. A moderately mixed estuary may change to highly stratified as freshwater discharge increases, for example, or from moderately to non-stratified as it decreases. The strength of the tidal current is related to the amount of water that is horizontally displaced, and it is always stronger during spring tides, whereas the degree of stratification tends to be greater during neaps. As the width of an estuary increases, more tidal water passes through the larger cross-section, reducing the ratio of river to tidal flow and promoting more complete mixing. Where the depth of a river increases, the same tidal velocities flush a smaller volume of water through the cross-section, and less effective vertical mixing encourages greater stratification. Progression through the sequence from highly stratified, to

moderately stratified, to non-stratified therefore corresponds to increasing tidal flow, relative to river flow, and increasing estuary width and decreasing depth.

Estuarine stratification can be estimated according to the ratio of the volume of fresh water flowing into the estuary during a tidal cycle, to the tidal prism—the volume of water entering the estuary from the sea during the flood tide. The ratio for a highly stratified estuary is about 1, for moderately stratified estuaries about 0.1, and for vertically homogeneous estuaries around 0.01.

The most widely used classification is based on two dimensionless parameters related to current velocity and salinity (Hansen and Rattray 1966) (Fig. 7.5). The circulation parameter is the ratio of the tidally averaged surface velocity, a measure of the river flow, and the saline water mixed in it by turbulence and entrainment, to net river discharge velocity averaged over the cross-section. The stratification parameter is the ratio of the tidally averaged difference in salinity from the surface to the bottom, to the tidally and cross-sectionally averaged salinity. Net flow is seawards at all depths in type 1 estuaries, and upstream salt transfer is affected by diffusion. Type 1a represents non-stratified estuaries, whereas 1b estuaries have appreciable stratification. Net flow reverses at

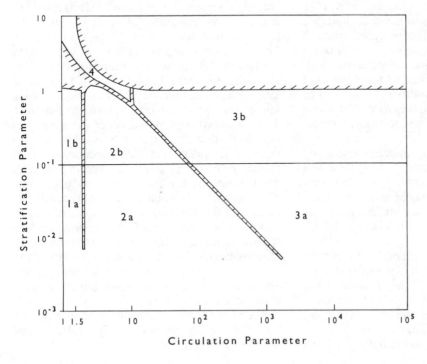

Fig. 7.5. Hansen and Rattray's (1966) estuarine classification, based on current velocity and salinity

depth in type 2, and advection and diffusion make important contributions to the upstream salt flux. Stratification of types 2a and 2b corresponds to type 1a and 1b, respectively. Advection accounts for more than 99 percent of the upstream transfer of salt in type 3 estuaries. The lower layer is so deep in type 3b that, in effect, the salinity gradient and the circulation do not extend to the bottom. Fiord estuaries are usually of type 3b, until mixed to the extent that they assume the low stratification characteristics of type 3a. Type 4 represents highly stratified estuaries.

Pritchard's (1967b) classification was developed for drowned river valleys, but barrier-built estuaries or shallow coastal lagoons respond quite differently to wind stress, tides, river discharge, and surface heating. Lagoon water can range from fresh to hypersaline. Gravitational circulation is generally lacking in lagoons that do not have much river inflow, although vertical mixing by winds and tides is usually better than in deeper drowned river valleys. There may be longitudinal variations in salinity, however, and marked horizontal stratification. Three types of lagoon can be distinguished according to the type of channel(s) connecting them to the ocean (Kjerfve and Magill 1989):

a) Choked lagoons have long, narrow entrance channels, which eliminate or severely dampen the effect of tides. Changes in water level therefore largely reflect hydrologic/riverine cycles, often at the seasonal scale. Choked lagoons tend to occur along medium-to-high wave energy coasts with active longshore sediment transport and low tidal range. Persistent winds can drive sea water in or out of lagoons, and they also help to mix the water. Surface heating can also produce some vertical stratification.

b) Restricted lagoons are connected to the sea by inlets through enclosing barriers. They have well defined tidal circulation and are generally vertically mixed. Circulation in choked and restricted lagoons may reflect longitudinal differences in the density of the water, although the wind also has an important effect.

c) Leaky lagoons are associated with narrow barrier chains. They can extend for more than 100 km along the coast, although they may be only a few kilometres in width. Wide tidal inlets allow free exchange of water with the ocean, resulting in strong tidal currents. Residual bottom currents flow into restricted and leaky lagoons to

compensate for the loss of saline water mixed in with seaward flowing surface layers.

It has generally been assumed that lagoons are depositional sinks that are rapidly filled with sediment. A survey of twenty-two lagoons on the Atlantic and Gulf coasts of the United States has shown that most have attained a state of near-balance between rates of accretion and the rise in relative sea level. Accretion rates in the lagoons, however, are not particularly rapid. The average, short-term rate of 3.1 mm yr^{-1} compares with 5.6 mm yr^{-1} for salt-marshes in the same region, and 2.6 mm yr^{-1} near river estuaries, which are all much less than in mid-latitude fiords and deltas (Nichols 1989).

THE TURBIDITY MAXIMUM

Suspended sediment concentrations in the upper portions of many meso- and macrotidal, mixed and partially mixed estuaries are commonly from 10 to 100 times greater than further landwards or seawards. Indeed, in most estuaries the amount of sediment in the turbidity maximum is greater than the annual supply (Mehta and Dyer 1990). For example, concentrations of up to 10,000 mg 1^{-1} have been reported from the Demerera Estuary in Guyana, more than 10,000 mg 1^{-1} in the Jiaojiang Estuary in China, and more than 50,000 mg 1^{-1} in the Loire (Gallenne 1974; Officer 1981; Li *et al.* 1993). Much weaker maxima can extend for tens of kilometres along estuaries. The turbidity pattern in deep, wide estuaries may be much more complex than in shallow estuaries. In the St Lawrence, for example, several areas of high suspended sediment concentrations are detached from the bottom, downstream of the main turbidity maximum. These secondary maxima appear only once during every semi-diurnal tidal cycle, at the end of the ebb flow or when the currents are usually weakest (D'Anglejan and Ingram 1976). Up to three simultaneous maxima have been observed in the Columbia River Estuary in the north-western United States, and more than one maximum can develop in the Scottish Tay Estuary.

The turbidity maximum can develop in several ways (Nichols and Biggs 1985; Dyer 1986; Jay and Musiak 1994) (Fig. 7.6):

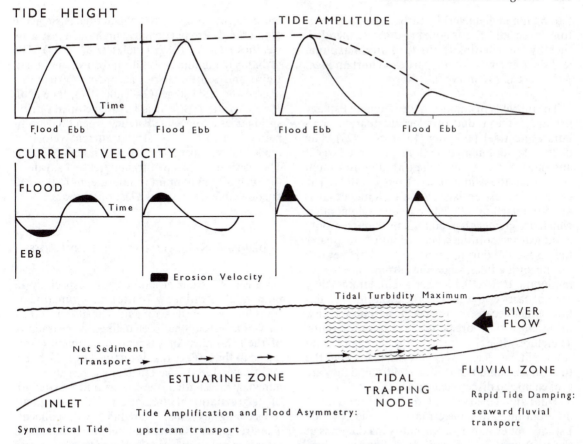

Fig. 7.6. Net tidal transport and suspended sediment trapping in a macrotidal estuary with little or no density circulation. The damping of the tidal wave and currents eliminates flood-velocity predominance in the upper portion of the estuary, and river flow produces net seaward sediment transport. Tidal turbidity maxima develop at the upper tidal limit, usually further upstream than those at the head of the salt intrusion (Allen *et al.* 1980)

a) Residual circulation is thought to be the main process in mesotidal estuaries (British Mersey, Tay, and Thames Estuaries, and Chesapeake and San Francisco Bays), with much of the sediment being trapped in the turbidity maximum near the head of the landward flowing bottom layer.

b) Turbidity maxima can also occur where there is maximum tidal scour and resuspension, or where sediment carried upstream by asymmetrical tidal waves is trapped at the landward limit of the tides (Allen *et al.* 1980: van Leussen and van Velzen 1989). These maxima can develop landwards or seawards of the salt-intrusion limit.

c) Agglomeration and de-agglomeration of fluvial sediment play an important role in maintaining turbidity maxima (van Leussen 1988; Dyer 1989; Gibbs *et al.* 1989) (Chapter 8). Agglomeration of fine, suspended particles can result from the

collision and adhesion of particles in the mixing zone of estuaries. Particle adhesion is caused, or aided by, organic binding, involving polysaccharide mucal films produced by bacteria, algae and other higher plants, and other biological activity. River-borne sediment is already agglomerated when it enters estuaries, and its size initially decreases on contact with salt-water, as organisms consume the organic matter. New agglomerates then develop as the mineral particles are re-glued together with organic matter. New agglomerates then develop as the mineral particles are re-glued together with organic material (Eisma *et al.* 1991). The formation of macro-agglomerates (measured in millimetres) increases the amount of river-borne sediment settling into the landward-flowing bottom layer. These fragile, low-density agglomerates are then broken up into micro-agglomerates (about

5 to 50 μm in diameter) by turbulence in the bottom layer, and the fragments are resuspended, thereby contributing to the turbidity maximum and the trapping of more particles (Burban *et al.* 1989; McCabe *et al.* 1992).

The turbidity maximum in the Jiaojiang Estuary in China is the product of internal estuarine circulation and tidal pumping (Li *et al.* 1993). The dense, clay-dominated agglomerates are largest and most numerous near the turbidity maximum zone, where the salinity ranges from 5 to 10 ‰. The estuary also has very large, loose agglomerates that are dominated by silt. These loose agglomerates, which are greater than 500 μm in diameter, only exist near the bottom, where salinities are greater than 5 ‰, and during neaps, or for short periods during spring tides, when the currents are low to moderate. Unlike the denser agglomerates, they are not carried upwards by the estuarine circulation, either because of their high settling velocity, or because they break up in strong currents. Therefore, while clay particles are aggregated at all tides, silt particles are aggregated only at the turbidity maximum during neap tides, and they are broken up and dispersed at spring tides.

Tidal and density circulation are dominant at different times in some estuaries. In the Gironde Estuary, for example, the turbidity maximum is at the tidal limit during periods of low river flow, about 40 km upstream of the head of the salt water intrusion. Density circulation develops during periods of high river flow, however, and the turbidity maximum moves to the salt-intrusion limit (Allen *et al.* 1980; Allen 1991). Turbidity maxima often develop only during periods of high river flow, but they can be permanent features in estuaries with a high tidal range and a fairly large supply of suspended sediment. The strength and location of turbidity maxima vary with semi-diurnal and spring–neap fluctuations in the strength of the bottom currents (Dyer and Evans 1989). It is fully developed during spring tides and absent during neap tides, for example, in the Korean Keum and Portuguese Tagus estuaries (Lee 1985; Vale and Sundby 1987). Semidiurnal, fortnightly, and seasonal variations in turbidity maximum concentrations are similar, however, in the mesotidal Columbia River Estuary (Gelfenbaum 1983). Large spring–neap variations in suspended sediment concentrations have been reported in the Thames and Tamar estuaries in southern England,

and concentrations in the Scheldt tidal basin in the Netherlands are particularly high during stormy weather (Brinke 1987; Uncles *et al.* 1994). The turbidity maximum moves up and down the Tamar, Loire, and Seine estuaries according to river discharge and tidal stage (Gallenne 1974; Bale *et al.* 1985; Avoine 1986). It is best developed during spring tides in the Seine Estuary, when tidal asymmetry is greatest and strong current velocities erode and resuspend large quantities of sediment. The seaward extension of training jetties has caused the turbidity maximum to migrate more than 40 km down the estuary in the last thirty years.

DEPOSITIONAL PATTERNS AND LANDFORMS

Because of estuarine circulation, especially in moderately stratified estuaries, they function as traps for river-borne sediment. The filtering efficiency of an estuary has been defined as the fraction of the total mass of suspended sediment introduced to the estuary that is retained by it (Schubel and Carter 1984). The efficiency is largely determined by the length of the estuary and the strength of its gravitational circulation. Circulation patterns change as estuaries are filled with sediment. The strength of the river flow, relative to tidal flow, increases as the intertidal volume decreases, and the circulation therefore becomes more stratified or river-dominated. Shortening of an estuary as it is filled from the head also increases the longitudinal salinity or density gradient, which tends to strengthen the gravitational circulation. Depending on the position of the estuary in the sequence of circulation types (Pritchard 1967*b*), increasing river domination could either increase or decrease its filtering efficiency. As an estuary continues to age, the reduction in its length eventually allows salt water to be completely expelled from its basin during periods of high river flow, permitting large amounts of river-borne sediment to reach the sea (Schubel and Carter 1984; Schubel *et al.* 1986).

Dalrymple *et al.* (1992) modelled stages in the evolution of wave- and tide-dominated estuaries (Fig. 7.7). The gradual filling of a wave-dominated estuary is marked by the progressive seaward growth of the bay-head delta, and landward extension of the flood-tidal delta. The central basin eventually disappears, and tidal channels in the flood-tidal delta merge with the river channel (Roy

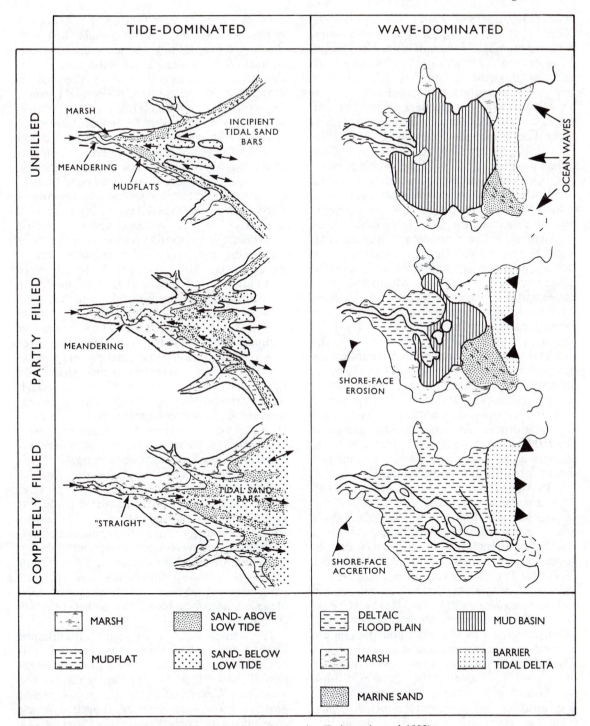

Fig. 7.7. The evolution of tide- and wave-dominated estuaries (Dalrymple *et al.* 1992)

et al. 1980; Roy 1984; Zaitlin and Shultz 1990; Nichol 1991). Bay-head deltas are absent where rivers carry negligible amounts of coarse sediment, however, and their estuaries are infilled almost exclusively from the seaward end (Nichol and Boyd 1993). As tide-dominated, funnel-shaped estuaries are filled, the linear sand bars, or tidal current ridges, become broader, and the meandering zone in the inner estuary moves seawards (Harris 1988; Allen 1991; Allen and Posamentier 1993). Continued progradation beyond the mouth of wave- and tide-dominated estuaries occurs as deltas, beach ridge plains, or open coast tidal flats.

River-dominated estuaries occur where there is a high fluvial sediment supply. Fluvial processes are dominant in the Mgeni and several other estuaries along the wave-dominated, microtidal coast of Natal (Cooper 1993*a*). Severe floods, which occur about every 70–100 years on the Mgeni River, flush out estuarine sediments. Rapid fluvial deposition then restores the river gradient and forms a shallow braided channel within a few months of the flood. Over approximately the next seventy years, vegetation and mud deposition gradually stabilize the banks and bars, transforming the channel into an anastomosing system. This represents a stable, equilibrium morphology that adjusts to the normal range of hydrodynamic conditions, so that the estuary is able to maintain its morphology indefinitely. Nevertheless, some microtidal, river-dominated estuaries are gradually being filled up by progradation of estuarine or bay-head deltas (Sondi *et al.* 1995).

Once an estuary is almost filled, large amounts of sediment can flow out to sea, especially in macrotidal environments. The Huanghe (Huang Ho, Yellow River), for example, which has the largest suspended sediment load in the world, deposits 400 million tons of sediment each year in its lower reaches, while 1,200 million tons are carried out to sea (Wang 1983). The rate at which estuaries are filled largely depends upon the amount of sediment carried by the river, and the amount brought onshore and carried alongshore by waves. Filled estuaries, which have net seaward transport of suspended sediment at their mouths, tend to have high river velocities and high stratification. Salt intrusion, and, if present the turbidity maximum, are restricted to areas near the mouth. Unfilled estuaries are often found in areas with a high tidal range, deep glacial entrenchment, and crustal downwarping, and in semiarid regions

with low sediment input (Fairbridge 1980). They serve as sediment traps, and they tend to have well-developed gravitational circulations, and well-defined turbidity maxima located a considerable distance upstream from the mouth (Officer 1981).

Sediment escapes from the macrotidal Gironde Estuary during periods of high river flow, and during high spring tides, when the sediment deposited in the previous neap tidal period is eroded and resuspended (Castaing and Allen 1981). Variations in the tidal prism have an important effect on hydrological and sedimentological conditions in the Gironde Estuary, which can change from a fairly well-mixed state during spring tides, to a moderately or even well-stratified state during neap tides. During spring tides the water that flows into the estuary on the flood tides cannot be completely excavated on the ebbs. Water therefore accumulates in the estuary as tidal amplitude increases from neap to spring tides, and then flows seaward as the amplitude decreases. This seaward flow probably plays an important role in moving sediment out to sea (Allen *et al.* 1980). High river flow in the Seine Estuary causes the turbidity maximum to migrate into the sea, producing a turbid plume extending up to 30 km offshore. Although the estuary supplies fluvial sediment to the sea, however, it is still a sink for the large amounts of sand carried into it from the sea (Avoine 1986). Wave action at the mouth of this, and other estuaries in exposed environments, probably helps flood currents to move sand landwards. The Taf Estuary in western Wales is being filled with sand brought in by storm waves from the Atlantic, but as it becomes shallower, increasing tidal asymmetry generates stronger tidal currents (Jago 1980). Breaking waves oppose river outflow from estuaries in Papua New Guinea and New South Wales—they modify effluent processes, cause intense mixing of fresh and salt water, and redistribute sediment to form broad, arcuate bars or deltas (Wright *et al.* 1980).

The flushing velocity is a measure of the distance a river can thrust fresh water and sediment into an estuary, and its ability to resist the landward intrusion of salt water. It is defined as the ratio of the mean annual river discharge to the cross-sectional area at the freshwater–saltwater transition, which is assumed to be where the salinity is 1 ‰ at the surface. So much water flows from the Amazon that salt water is unable to intrude into its estuary, even during periods of lower discharge and high spring tides. Water and suspended sediment flow seawards at all depths, although there is vertical stratification and two-layered flow on the continental shelf. The turbidity maximum is also on the shelf, and it extends along the coast to the north-west. Although other large rivers can thrust water and sediment considerable distances, they cannot prevent sea water entering their estuaries (Gibbs 1977).

Rivers in the humid tropics and subtropics have high discharges and suspended sediment loads in the wet season, and low discharges and sediment loads in the dry season (Tucker 1973; Barusseau *et al.* 1985; Cooper 1993*b*) (Table 7.1). When river flow ceases, the Sarada–Varaha Estuary on the monsoonal eastern coast of India changes from a salt wedge/partially mixed to a vertically mixed state, and finally to a negative state from April to June (Sarma *et al.* 1993).

Seasonal changes in ice conditions, waves, sediment dewatering, rain, water temperature, and biological processes exert a strong influence over the transport and deposition of sediment in estuaries in cold environments (Anderson 1983) (Chapter 12). Estuaries are frozen over for almost half the year in subarctic James Bay, and there are enormous seasonal variations in river and sediment discharge. In Rupert Bay, tides and wind stress hinder deposition, and high levels of turbidity are maintained throughout the ice-free period. Although most of the sediment settles to the bottom beneath the ice in winter, it is largely resuspended in the following spring by ice scour, followed by very high river flows (D'Anglejan 1980).

The morphology of sandy estuarine deposits is determined by the interaction of such factors as waves, storms, and tidal currents, and especially by tidal range (Hayes 1975) (Fig. 7.8):

a) Waves, rivers, and winds are the most important elements in microtidal estuaries. Typical features include small river deltas, storm washover fans, and a variety of wave-built elements, including recurved barrier spits, cuspate spits, forelands, and baymouth barrier beaches. Tidal deltas are usually small, although there are some exceptions, including those at Destin in north-western Florida.

b) Tidal currents are of much greater strength and importance in mesotidal estuaries. Large tidal deltas and sand bodies formed by asymmetrical tidal currents are common, and associated barrier islands are short and stubby. Point bar deposits are conspicuous in the meandering tidal channels behind the barriers. Ebb-tidal deltas are formed of sediment deposited by ebb currents seaward of the tidal inlet, while flood-tidal deltas are built on the landward side by flood currents.

c) Tidal currents dominate macrotidal estuaries. These estuaries tend to have a wide mouth and they may extend considerable distances inland. Sand bodies generally develop in the centre of the estuaries, and broad, muddy tidal flats at the margins. Tidal ridges parallel to the currents are the most common type of coarse-grained landform.

Classifications based on the absolute rather than the relative strength of waves and tides are oversimplified. Waves can dominate in macrotidal

Table 7.1. Seasonal variations in the estuaries of Sierra Leone, West Africa

Factor	Wet season	Dry season
Estuary form	Open	May be cut off from the sea (blind)
Wave action	May be intense during storms	Persistent, forming barrier spits
Tidal currents	Ebb-dominated	Flood-dominated
River discharge	High	Low or negligible
Direct climatic effects	Intense rainfall creates channels, and destroys bedforms	Hypersalinity resulting from high evaporation
Salt water	Driven from estuary	Penetrates up estuary
Sediment	Large suspended load carried downstream	Much bed load brought in from offshore and moved upstream by flood tides

Source: Tucker 1973.

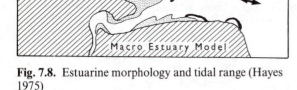

Fig. 7.8. Estuarine morphology and tidal range (Hayes 1975)

the occurrence of tidal deltas or tidal current ridges depends on the tidal prism, the width of the estuary mouth, changes in relative sea level, and the amount and type of sediment available, rather than on tidal range *per se*. One or both of the tidal deltas may be missing where there is a limited supply of sand. Some macrotidal estuaries have tidal deltas, including the Tay in Scotland and the Gironde in France, whereas some microtidal estuaries, including those in Delaware and Chesapeake Bays, have tidal current ridges or bars. Ridges are most common where there is a large tidal prism, and they therefore tend to develop in estuaries with wide mouths, and they can be replaced by tidal deltas in macrotidal estuaries with narrow mouths. Wide estuaries tend to have a complex arrangement of sand bodies and tidal channels. Four to five ebb deltas alternate with three to four flood deltas, for example, across the wide entrance of Chesapeake Bay.

Tidal deltas are broadly similar in a variety of environments (Fig. 7.9). A typical flood-tidal delta consists of: a seaward-facing slope known as the flood ramp; flood channels bifurcating off the ramp; a rim, or ebb shield, running around the delta and protecting portions of it from ebb current modification; ebb spits formed by ebb currents; and spillover lobes formed by unidirectional currents. Ebb-tidal deltas have a main ebb channel flanked on either side by levee-like, channel-margin linear bars. These bars are built by ebb and flood currents interacting with wave-generated currents. The seaward end of the ebb channel passes into the terminal lobe, a crescentic, fairly steep, seaward-sloping body of sand (Hayes 1975, 1980; Boothroyd 1985). Flood currents dominate in marginal channels that separate ebb-tidal deltas from the beaches on either side of inlets. These channels may represent paths of least resistance for flood currents that are confronted by water still flowing quite strongly out of the estuaries, late in the tidal cycle.

The development and morphology of tidal deltas is influenced by stratified flow in estuaries. The flood delta is much larger than the ebb delta at Destin, Florida, for example, because the denser sea water flows along the bottom during the flood tide, where it can move the greatest amount of bed load, while the less dense water in the bay favours the upper layer during ebb flows (Wright and Sonu 1975). Deltas tend to be elongated to one side of inlets where there are strong longshore currents, and

regions and tidal currents in high wave energy environments, and while estuaries can therefore be described as wave-dominated or tide-dominated, no assumptions can be made regarding the absolute value of the required tidal range or the height of the waves (Davis and Hayes 1984). Therefore, although the deposits and landforms of most estuaries generally conform to Hayes's model, there are numerous exceptions. Harris (1988) proposed that

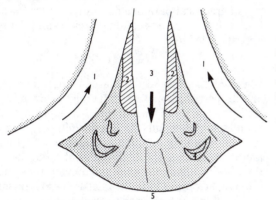

1 FLOOD RAMP
2 FLOOD CHANNEL
3 EBB SHIELD
4 EBB SPIT
5 SPILLOVER LOBE
6 TIDAL FLAT
7 INLET

Fig. 7.9. Ebb- and flood-delta models. The arrows indicate the dominant direction of tidal currents (Hayes 1980)

FLOOD-TIDAL DELTA MODEL

1 MARGINAL FLOOD CHANNEL
2 CHANNEL MARGIN LINEAR BAR
3 MAIN EBB CHANNEL
4 SWASH BARS
5 TERMINAL LOBE

EBB-TIDAL DELTA MODEL

they extend further offshore where tidal inflow and outflow are dominant over longshore transport. The shape of the deltas in the Frisian Islands, off Germany and the Netherlands, reflects the competing influence of wave and tidal currents. Inlets with small tidal prisms have small ebb deltas asymmetrically deflected to the east by waves and longshore currents, whereas inlets with large tidal prisms have large deltas asymmetrically deflected to the west by tidal currents (Sha and Van den Berg 1993). There is a strong correlation between the tidal prism and the amount of sand stored in the ebb shoals of inlets. The ebb-tidal deltas along the Dutch barrier coast, for example, vary in size according to the 18.6 nodal tidal cycle, being greatest during periods of tidal maxima (Oost *et al.* 1993). The volume of sand tends to decrease with increasing wave energy, however, presumably because higher waves are able to drive more sand onshore (Marino and Mehta 1987). On the other hand, ebb-tidal deltas in Florida are larger, relative to the cross-sectional area of the inlets and

the tidal prisms, off wave-dominated than tide-dominated inlets (Gibeaut and Davis 1993).

Tidal currents are largely responsible for sediment movement during summer in the Minas Basin in the Bay of Fundy, eastern Canada. The maximum tidal range in this area is 16.3 m, and the maximum current velocity near the bottom is about 2 m s^{-1} (Knight 1980; Dalrymple *et al.* 1990). As in most tide-dominated estuaries, sand is deposited towards the centre of the bay and mud at the head and in other sheltered places along the margins. Intertidal and subtidal current ridges, consisting of medium to coarse-grained sand, dominate the low tidal flats in the outer portions of the Bay. They are several kilometres in length and are essentially parallel to the tidal currents (Knight 1980). The ridges segregate deep tidal channels into either ebb- or flood-dominated components. Sediment is moved in different directions on either side of the ridges by the ebb and flood tides, causing it to circulate around them; a pattern that has also been reported from other tidal areas.

The width of estuaries in macrotidal or tide-dominated environments often decreases exponentially landwards, creating a characteristic funnel shape. This balances the contrasting effects of convergence and friction on the tidal wave, and helps to maintain its amplitude and reduce the loss of energy. Langbein (1963) proposed that estuaries acquire a funnel shape to minimize the total work accomplished by the water in a tidal cycle, and to equalize the amount of work per unit area. The length of the Ord River Estuary in Western Australia is equal to one-quarter of the tidal wavelength, which therefore has most of the characteristics of a standing wave (Wright *et al.* 1973). Tidal range decreases at a cosine rate from its maximum at the mouth of the estuary, to zero at the upstream limit of tidal influence. The upstream decline in the depth and width of the estuary balances the frictional loss in energy, creating a condition whereby equal work is performed per unit area of the channel bed. Wright *et al.* suggested that macrotidal estuaries have a funnel-shape if the tidal wave exhibits standing wave characteristics, and almost parallel banks if the tidal wave is progressive. This could explain, for example, why convergence is not a significant feature in the macrotidal Bay of Fundy, where the tidal wave is largely progressive.

The unconfined main channel in the central, low-energy portion of tide-dominated systems often has a straight–meandering–straight progression of sinuosities (Ashley and Renwick 1983; Woodroffe *et al.* 1989). The outer straight portion is tide-dominated, and net sediment transport is landwards, whereas the inner straight section is river-dominated and net sediment transport is seawards. Both straight sections contain bars that are attached to the banks and alternate from one side of the channel to the other. The region of tight meanders between the two straight sections, which often has symmetrical point bars, corresponds to the zone of lowest energy and the position of net bedload convergence (Dalrymple *et al.* 1992). Wright *et al.* (1973) attributed the sinuousity of the channel in the landward portion of the Ord River Estuary to the landward increase in tidal asymmetry. The presence of deep channels on the outside of meander bends may be necessary to allow rapid, concentrated ebb flows to carry sediment seawards, thereby preventing the channels becoming clogged by flood-dominant tidal currents. A different explanation has been provided for the meanders in the middle portion of estuaries in mesotidal Chesapeake Bay. As the tidal wave is progressive near the mouth of the estuaries, the maximum flood current is at high tide and the maximum ebb current at low tide. Lateral erosion by these two currents therefore occurs at different elevations. As the tide is increasingly modified upstream, a point is reached where the maximum flood and ebb are at about the mean water level. As lateral erosion by the ebb and flood tides occurs at approximately the same level in the area where the meanders are best developed, it may be a major requirement for their development (Ahnert 1960).

The size of the entrance to estuaries on sandy coasts is related to a variety of hydrological and physiographical parameters. Many workers have found that the cross-sectional area of an inlet is approximated by:

$$A = C\,T_p{}^n$$

where A is the cross-sectional area of the narrowest and deepest section of the tidal inlet below mean sea level, T_p is the diurnal or spring tidal prism, and C and n are constants (O'Brien 1969). Although inlet morphology is also influenced by other factors, including the flood- and ebb-tidal duration, freshwater discharge, and sediment transport (Gao and Collins 1994), the inlet area–tidal prism relationship has been used to predict the response of inlets to dredging, structural modification, and sedimentation. Hume and Herdendorf (1993) found that the A-T_p model is applicable to a wide range of estuarine types in northern New Zealand. Although the exponent (n) is similar for all types of estuary, however, differences in the intercept or C values suggest that the major types of estuary are best represented by individual A-T_p relationships, which can therefore be used to characterize and classify them.

HUMAN IMPACT

Many of the major ports and cities of the world are situated on estuaries. Human interference therefore complicates estuarine dynamics, and erosion and deposition frequently occur in response to anthropogenic activity. This includes an increase in river-borne sediment owing to deforestation, mining and quarrying, and urbanization, a decrease in sediment resulting from dam construction, and the effects of sewage discharge, dredging, dock and marina construction, tidal barrages, and the infilling of tidal flats and marshes (Ustach *et al.* 1986; O'Connor 1987).

Flushing rates, vertical mixing, and density current circulation are sensitive to reductions in the tidal prism caused by marsh reclamation and dock construction, and increases brought about by dredging (Dyer 1972). Dredging in the upper reaches of the Thames Estuary increased the tidal range and the rate of tidal progression, resulting in increased deposition. Removal of the old London Bridge also caused tidal range to increase by about one-quarter. The capacity of the Mersey Estuary has been reduced by training works constructed in Liverpool Bay. This has modified the circulation and contributed to siltation in the upper estuary. Straightening the meanders in the Lune Estuary of north-western England initially increased the depth of the channel. The discharge in the trained channel eventually decreased because of accretion outside the channel, however, causing channel deterioration, reduced flow, and heavy accretion seawards of the trained reach.

The diversion of water from one watershed into another, or other changes involving the amount of fresh water flowing into an estuary, affect the supply of nutrients and sediment, salinity, hydrological regimes, and delta formation. Effective management of fisheries, wildlife, and other estuarine resources requires reliable prediction of the changes that are likely to occur (Funicelli 1984). The 1942 diversion of the Santee River increased the amount of fresh water and sediment brought into Charleston Harbour. This created a salt wedge that helped to trap sediment, resulting in a seven- to ten-fold increase in the rate of deposition, and a substantial increase in dredging costs (Neiheisel and Weaver 1967).

Most estuaries in the United States were in a state of equilibrium, and the ability of the rivers to supply sediment was roughly matched by the ability of the tides to remove it. The dredging of deep navigation channels, however, reduced tidal current velocity, enhanced vertical stratification, and increased salt wedge penetration into the estuaries. Dredging permitted salt water to penetrate more easily along the bottom in the Savannah Estuary, and major shoaling now occurs in Savannah Harbour as a result of fluvial sediment being trapped in the estuary rather than being carried out to sea, and probably because of marine sediment brought in along the bottom by landward flowing currents.

Hydraulic models suggest that channel deepening in Chesapeake Bay and the James River Estuary would increase salinity intrusion and vertical stratification, and it would also reduce salinity variations in the neap–spring tidal cycle in the bottom waters of Chesapeake Bay (Richards and Granat 1986).

Human interference has helped sediment to escape from the Seine Estuary to the sea. After more than 130 years of entrainment, landfilling, jettying, and dredging, the shallow, meandering ebb and flood channels and wide tidal flats have been replaced by a deep, narrow channel. This has concentrated river discharge and ebb flow, and caused the salt intrusion and turbidity maxima to shift seawards (Avoine 1986). As with other subboreal estuaries, the morphology of the Outardes Estuary in Quebec was dominated by the spring freshet. The mouth of the estuary was characterised by a disorganized, braided channel pattern, and high fluvial discharge generated transverse bars, while peripheral channel areas in the upper portions of the estuary functioned during periods of high discharge. Dam construction in the late 1960s drastically reduced the peak fluvial discharge, and the mouth of the estuary was infilled and remodelled by waves and tidal currents (Hart and Long 1990).

Estuaries are traps for pollutants as well as natural sediment. Until fairly recently Chesapeake Bay, which is one of the most biologically productive estuaries in the world, was thought to have an almost unlimited capacity to assimilate human waste. Recent work, however, suggests that there have been dramatic changes during the last century, and especially over the last thirty years. Trapped and recycled pollutants have reduced oyster and fish populations, and submerged aquatic vegetation and over-enriched it in nutrients. High levels of toxic compounds also occur near urbanized areas (Tippie 1984). A tremendous effort will be needed to reduce pollutant loadings to Chesapeake Bay, but this has already been achieved in the Thames Estuary of southern England, which is recovering from centuries of abuse. The trapping and resulting concentration of pollutants brought the estuary to the point of putrescence, but pollution control measures have brought about dramatic improvements in the last few decades, as manifested in the presence of more than one hundred fish species, including salmon (Casapieri 1984).

8 Tide-Dominated Environments

Tidal currents transport large amounts of sediment within the shallow water regions of the world. Ebb and flood currents move in opposing directions, but they often travel along different paths, and because shallow water tides are asymmetrical, there are marked differences in their duration and in the magnitude and occurrence of their peak flows (Chapter 7). Net sediment transport also occurs when tidal currents combine with local, wind-driven currents, flow from rivers, and normal oceanic circulation. In some areas, rotation of the tides adds a longshore component to the simple onshore–offshore movement of sediment.

TIDAL FLATS

Tidal flats are banks of mud or sand that are exposed at low tide. Despite their name, they actually slope seawards between the high and low tidal levels, although the gradient can be very low. Tidal flats terminate at the sea edge or, as in estuaries, on the banks of major tidal channels. The floors of these channels are below the lowest tidal levels, in contrast to much shallower tidal creeks which are roughly graded down to the low tidal level. Tidal creeks on sandy tidal flats tend to have poorly defined banks, fairly symmetrical cross-sections, straight courses, and few tributaries, whereas dendritic drainage patterns, sinuous courses, and point bars are common on muddy flats.

Ebb and flood currents tend to flow along different courses within the main tidal channels. In the North Sea, rapid lateral accretion and sediment reworking by meandering tidal creeks and channels produce gently sloping beds, which are eventually covered by a much thinner capping of intertidal sediment. Sedimentary patterns can be quite different, however, in other parts of the world. In Korea, for example, tidal flats contain less sand, and the drainage channels are more stable than in the North Sea. Vertical sedimentary sequences therefore dominate in this area, and channel-fill sequences play only a fairly minor role (Wells *et al.* 1990; Alexander *et al.* 1991).

Sediment tends to become coarser towards the land in wave-dominated environments, but the opposite is true in tide-dominated areas, where a gradual landward reduction in velocity reduces current capacity and competence. Sandflats and bars in the lower intertidal and subtidal zones usually grade into sandy-mud and mudflats in the higher portions of the intertidal zone, as silts and clays become more prominent (Klein 1985). Tidal flows create a variety of bedforms in the subtidal and lower intertidal zones. Using the classification of Boothroyd and Hubbard (1975), bedforms transverse to the tidal currents include ripples, megaripples, and sand waves, with wavelengths up to about 0.6 m, between 0.6 and 6 m, and greater than 6 m, respectively. Longitudinal forms include: sand ribbons, consisting of strips of fine sand moving over a coarse sand or coarse clast floor; shallow, elongated depressions or longitudinal furrows; and tidal current ridges (sand ridges, sand banks) up to 50 km in length and 6 km in width. Tidal current ridges tend to occur in groups in estuaries, and as solitary near-coastal and banner banks to the lee of headlands or submerged rock shoals (Stride 1982). They are widely distributed in fairly shallow seas wherever there is sufficient sediment, and tidal current velocities of between about 2 and 9 km hr^{-1}.

Fine-grained silts, clays, and organic debris are usually deposited in embayments, inlets, and estuaries, or where barrier spits or other features afford protection from vigorous wave action. Mudflats also front the open sea in several areas, however, including Malaysia, the western coast of Korea, China, south-western India, the north-eastern coast of South America, north-western Florida and Louisiana, the Wash in eastern England, and near Malmo in southern Sweden (Bao-can and Eisma 1988; Wells *et al.* 1990; Alexander *et al.* 1991; Mathew and Baba 1995). Although a variety of other local factors may be significant, the presence of mudflats on many open coasts can be attributed to the occurrence of thick, offshore accumulations of fluid mud on the inner continental shelf and nearshore regions. These muds protect the shoreline by rapidly attenuating incoming wave energy, thereby providing conditions that encourage further deposition.

Mudflats in ice-dominated areas are often strewn with boulders or blocks of ice-rafted peat from the upper inter- and supratidal zones (Chapter 12). They may be covered by mangroves down to mean sea level in tropical areas, but in temperate regions they are only succeeded by salt-marshes at, and above, the high tidal level. Sea grasses and other vegetation growing around, or beneath, the low tidal mark attenuate wave energy, encourage the deposition of suspended particulates, and supress resuspension (Fonseca 1989; Zhuang and Chappell 1991; Asano *et al.* 1992). There are eleven genera of sea grasses, but with the exception of *Zostera*, they are restricted to warm temperate or tropical environments (Davies 1972).

Vegetation may be absent in hot, arid regions. Saline mudflats are common in Australia, and there are abundant evaporites and salt pans on the barren supratidal and upper intertidal zones of the Colorado River Delta (Ridd *et al.* 1988). The sabkhas of North Africa and the southern Persian Gulf are supratidal coastal flats of unconsolidated, salt-encrusted sediments. They generally range between about 8 and 16 km in width, although they are up to 32 km wide in some places. Sabkhas developed as the land gradually prograded seawards, and supratidal sediments accumulated over evaporites, algal mats, and intertidal and subtidal sediments (Evans *et al.* 1969; Oueslati 1992). There are carbonate tidal flats up to 16 km in width on the western, semi-arid coast of Andros Island in the Bahamas. Three distinct zones can be distinguished: a subtidal marine belt; intertidal flats between channels and ponds; and supratidal beach ridges, levees, and marshes, colonized by grasses, mangroves, and dense algal mats (Shinn 1983).

Algal mats trap and bind carbonate sediments on tidal flats in Shark Bay, Western Australia, and evaporites precipitated in the surface sediments also form continuous crusts in the upper intertidal and lower supratidal zones (Logan *et al.* 1974).

Depositional Mechanisms

The deposition of fine-grained suspended sediments has generally been thought to take place during the slack water period at high tide. There is still considerable debate over the precise mechanisms that are involved, however, and increasing evidence that deposition is a continuous process that occurs throughout the tidal cycle. Tidal asymmetry is the main reason for landward transportation and deposition of fine-grained sediment on tidal flats (Chapter 7), although a number of other mechanisms are also involved. In the Dutch Wadden Sea, because of tidal asymmetry and a landward decline in mean current velocity, the period of slack water during high tide is much longer than the corresponding period at low tide. Deposition at the high water stage is also favoured by the much shallower water through which the particles must settle (Postma 1961). Reduced drag in saline flows containing suspended clay also reduces friction factors, and increases the erosion threshold shear velocity, allowing less time for erosion and more time for deposition to take place (Best and Leeder 1993).

Landward transport and deposition of silt and clay on the tidal flats of the Wadden Sea have also been attributed to settling and scour lags (Van Straaten and Kuenen 1957; Postma 1961). Settling lag refers to the time required for fine-grained particles to settle to the bottom through slowly flowing water. Scour lag is a result of the higher velocities needed to resuspend a deposited particle than to keep it in suspension. Scour lag reflects the cohesion provided by organic material, and electrical charges that bind fine sediments together. Sediment is more difficult to move once it has become consolidated. This occurs through the expulsion of water under pressure, particularly in the seven-to-ten days around the neap tidal period when weaker currents are unable to remove all the sediment brought in by the previous spring tides (Sanford and Halka 1993).

Sediment sinks to the bottom when the incoming tidal current becomes too weak to carry it in suspension, but it continues to move landwards during the settling lag. Because of scour lag and the landward shift of sediment into areas where the tidal currents are weaker, the water mass that deposited the sediment is unable to resuspend it on the ebb flow. Some sediment is deposited at the end of the flood tide where the ebb currents are too slow to resuspend it, but the rest is picked up by water masses that occupied more landward positions than the water in which it was originally suspended. Near the end of the ebb flow, this sediment begins to settle landwards of its original position, although much of it remains in suspension (Fig. 8.1).

The explanation for the deposition of silts and clays in the Wadden Sea is applicable to other situations where the average current velocity and water depth decrease landward. These characteristics are common in tidal environments, although in estuaries and other places where there is significant fresh water runoff, fine sediment may be introduced by an incoming current flowing along the bottom (Chapter 7). Even allowing for tidal asymmetry and lag effects, fine-grained material tends to flow in and out with the tide, and few particles settle to

the bottom as discrete grains during the slack water period. Suspended particles can be agglomerated into larger units, however, which have higher settling velocities than the individual grains.

Tidal flats are occupied by shrimps, polychaetes, mussels, oysters, copepods, clams, scallops, tunicates, barnacles, and other planktonic and benthic organisms that obtain food by filtering suspended matter through particle-retentive organs. Suspension feeders trap large amounts of suspended organic and inorganic sediment, and then excrete the residual inorganic matter as mucus-bound faecal or psuedo-faecal pellets, that typically settle at rates of between 1 and 20 mm s^{-1}, according to their density and size. Suspension feeders can deposit huge amounts of suspended sediment. Three bivalve species account for 39 per cent of the annual deposition in the intertidal zone on a tidal flat in Maine (Anderson *et al.* 1981). Biodepositional rates are about 6 mm yr^{-1} in a commercial oyster bed in Chesapeake Bay, and a tunicate, *Molgula manhattensis*, can deposit sediment at rates up to three times higher. Two common marine decapods, *Callianassa major* and *Onuphis microcephala*,

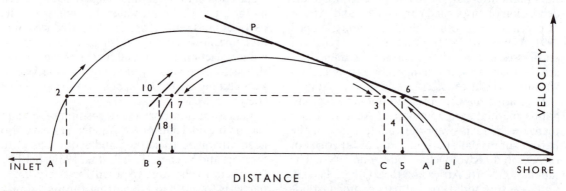

Fig. 8.1. The graph shows the path and velocity of a water mass as it moves in and out of a tidal inlet. Each water mass moves along one such curve. Stages are numbered to illustrate the sequence of particle suspension and deposition during a tidal cycle. The distance–velocity curves are asymmetrical, even though the tide at fixed points is assumed to be symmetrical. The maximum current velocity at each point along the inlet is represented by the tangent (P), which meets each curve at the point attained by the mass of water at half tide. The horizontal dashed line is the velocity needed to initiate and sustain particle suspension.

The settling lag effect. Water mass A increases in velocity on the flood and attains the critical velocity to suspend a particle (2) at point 1. The particle starts to settle at point C, although it is still carried further landwards by the current and actually reaches the bottom at point 5.

The scour lag effect. The sediment particle cannot be eroded by the same water mass (AA′) after the tide has turned, because it only attains the required velocity after it has moved further down the inlet, beyond the particle. The particle is therefore eroded by a more landward water mass (B′) and is carried towards the inlet to point B. It starts to settle at point 7 and reaches the bottom at point 9. The particle is therefore transported landward from point 1 to 9 during one tidal cycle. Because of the landward decrease in the average velocity of the tidal current, particles may eventually reach a point, after a number of these landward transport cycles, where they cannot be entrained by the ebb flow (Nichols and Biggs 1985) deposit layers of faecal pellet mud at rates of up to 4.5 mm yr^{-1}, producing up to 12 metric tons of mud per square kilometre per year (Pryor 1975).

Mineral particles can also be strongly glued together by organic material, especially by muco-polysaccharides produced by bacteria, algae, and other higher vegetation (Van Leussen 1988). Organic attachment creates micro-agglomerates consisting of mineral grains and organic material, and looser macro-agglomerates composed of micro-agglomerates and single mineral particles (Chapter 7). Mucus and other sticky material secreted by microbes and larger organisms also influence the chemical and physical properties of bottom sediments, which can affect their erodibility (Montague 1986).

Electrolytic flocculation may occur in estuaries or deltas when the repulsive negative charges on river-borne clays are overcome by the addition of positively charged ions in salt water. Suspended sediment would then become attached as a result of intermolecular or Van der Waals forces. The effect can only be significant, however, where high concentrations of clay or other electrochemically suitable sediment ensure frequent collisions, and the water is calm enough to allow agglomerates to settle. Collisions can occur as a result of Brownian motion of thermally agitated water molecules, turbulent mixing, or differential settling velocities (Lick *et al.* 1993).

Some workers believe that the role of physio-chemical flocculation has been exaggerated, particularly as the electrostatic stability of suspended particles is greatly influenced by the almost ubiquitous presence of metallic and organic coatings. Nevertheless, other workers consider that flocculation can occur where there is mixing of waters with only slight differences in salinity. In the Gironde Estuary in western France, for example, stream sediments coagulate in the upper estuary when they encounter water with a salinity of only 0.2‰ (Gibbs *et al.* 1989). In the Wadden Sea, the settling and scour lag mechanisms are effective only when concentrations are high enough to permit flocculation and settling during the high water slack period. Therefore, net suspended sediment transport is landwards only when wind-generated waves stir up the bottom and produce high sediment concentrations (Pejrup 1988).

Although the transport and deposition of fine-grained sediments are usually attributed to tidal currents, the contribution of waves should not be overlooked. Enormous amounts of Amazon-derived muds are transported in suspension and deposited along the Guiana coast of South America by waves and currents. Short, wind-generated waves also resuspend fine-grained sediments on tidal flats (Anderson 1983). As resuspension is most effective in shallow water, wave action in the slack water period preceding the ebb plays an important role in counteracting the landward movement of fine-grained sediment at high water. Waves can roll blocks of mud into armoured balls where there are seasonal or other cycles of accretion and marine erosion. The core of these balls can be derived from marsh erosion, slumping of muddy cliffs, or the widening of desiccation cracks by waves (Kale and Awasthi 1993).

SALT MARSHES

Sediment supply, tidal regime, wind–wave climate, and changes in relative sea level largely determine the location, character, and dynamic behaviour of salt marshes, although marsh vegetation also plays an important role (Allen and Pye 1992). Salt marshes begin to develop when tidal flats attain a level, relative to the tides, that allows colonization by salt-tolerant, subaerial vegetation. Salt marshes extend, according to the degree of exposure, from about the mean high water neap tidal level up to a point between the mean and extreme high water spring tidal levels. They terminate seawards in bare intertidal flats, where the duration and frequency of tidal inundation limit plant growth, and landwards where salt-tolerant plants cannot compete with terrestrial species. Salt marsh sediment usually consists of heavy or sandy clay, silty sand, or silty peat. Marshes in New England, south-western Ireland, southern England, and around the Baltic contain peat that accumulated in freshwater or brackish bogs, on gradually subsiding coasts, with a limited supply of silt.

Salt marshes are found throughout temperate regions, but they are also more extensive in the tropics than has generally been recognized (Fig. 8.2). They commonly occur in subtropical and warm temperate regions on the landward side of mangrove forests, where they are only occasionally flooded by the tides, as in southern Japan, New Zealand, southern Australia, Florida, and the US Gulf Coast. In the tropical Americas, salt marsh

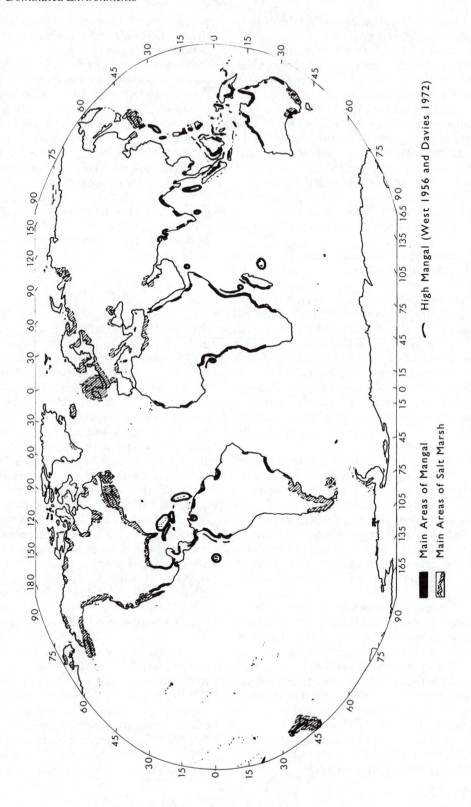

Fig. 8.2. Mangal and salt marsh regions of the world (Chapman 1977a)

Main Areas of Mangal

Main Areas of Salt Marsh

— High Mangal (West 1956 and Davies 1972)

plants colonize mudflats around the fringes of mangrove woodlands, saline soils within or along their landward edges, and where the mangrove has been cut or degraded. They are absent in areas of optimal mangrove development, however, possibly because the forest canopy prevents much light reaching the ground.

Six types of salt marsh have been distinguished in Europe: estuarine, including deltaic; Wadden, which lie in front of alluvial coastal plains fronted by barriers, shallow bays, or tidal flats; lagoonal; beach plains on the lee side of barrier spits or barrier islands; bog; and polderland. Their distribution is strongly related to tidal range. Deltaic and more especially lagoonal forms are common in the microtidal Mediterranean, for example, whereas beach plain, Wadden, and estuarine forms are characteristic of the meso- and macrotidal coasts of the eastern North Sea, the British Isles, France, and the Iberian Peninsula (Dijkema 1987).

Tides bring sediments and seeds into sheltered coastal areas and disperse them over marsh surfaces. Tidal range largely determines the thickness of the sediment that can accumulate on stable coasts, although the turbidity of the water is also important. In some areas a distinction can be made between the higher and lower portions of salt marshes, with the demarcation occurring at about the mean high water level. Low marsh occupies the upper middle intertidal zone. It is submerged more than 360 times a year, and is never continuously exposed for more than nine days. The older high marsh in the uppermost portion of the intertidal zone is flooded less frequently, and can be exposed for periods of ten or more days.

Some marshes end at the steep banks of drainage channels, whereas others are fronted by mudflats. Intertidal mudflat–salt marsh geomorphology varies according to the wave energy. Salt marshes act as energy buffers and they supply sediment to mudflats during storms, in a similar way that eroding dunes provide sand to beaches. The flattening of beaches during storms, which allows maximum energy dissipation, is matched, in tide-dominated environments, by lower intertidal gradients resulting from vertical erosion and retreat of the edge of salt marshes, and accretion on the fronting mudflats (Pethick 1992).

Marshes often terminate abruptly seawards in a scarp or cliff, although the edge may be dissected by deep rills and intricate branching networks that dissipate moderate wave energy on open coasts.

Marsh cliffs could develop on marshes that are unable to advance further seawards, yet are able to continue building upwards. The resulting break of slope would eventually become a cliff if it became a locus for scour. They could also reflect differences in the trapping efficiency of the vegetation, although it might also be sharpened by erosion (Davies 1972). Other kinds of cliff are primarily the product of erosion and marsh retreat by storm waves and currents.

In western Britain and western France marshes often form a series of terraces, each terminated by a small scarp or grassy ramp, descending in steps to the sea. The scarps represent former cliffs that were cut at the marsh edge by erosion, and were then gradually buried as mudflats and salt marshes developed in front of them. Cliff formation and abandonment was the result of alternating erosional and depositional episodes, possibly reflecting changes in relative sea level, wave climate, or the course of tidal channels. In western Britain, cliffs are steep and up to 5 to 10 m in height in the cohesive, muddy marshes of the Severn Estuary, and gentle, grassy slopes in the sandy marshes of Solway Firth and Morecambe Bay (Allen 1989) (Fig. 8.3). Root density rather than clay content, however, is the main factor in the Netherlands (Van Eerdt 1987).

Drainage Systems

Most marshes are drained by creeks that form a strikingly dendritic pattern. The system is generally empty at low tide, but full or overflowing during high spring tides, when water floods out on to the marsh surface. The initial drainage pattern is partly inherited from the ancestral tidal flat, and probably reflects minor topographic irregularities during the early stages of development. Creeks can be converted into subterranean channels or pipes by vegetation growing over them and becoming entangled, thereby helping to trap sediment and floating debris.

Mudflat creeks are formed by direct fluid shear stresses exerted against the substrate, small waves washing against the banks, and rotational slides. They are fairly indistinct during the early period of marsh development, but once vegetation becomes established, deposition increases and creeks are deepened through scour and bank buildup, and widened by runoff erosion. Stream piracy occurs

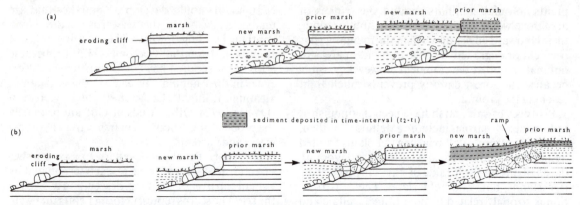

Fig. 8.3. The development of abandoned salt marsh cliffs in western Britain: (a) the muddy Severn Estuary; and (b) the sandy Morecambe Bay–Solway Firth systems. Sandy marsh cliffs, however, can sometimes develop in the same way as muddy marshes (Allen 1989)

when creeks intercept, and levees can develop along the banks where coarser sediment is deposited and trapped by vegetation during overbank flooding (Pestrong 1972; Steers 1977; Collins *et al.* 1987).

On clay or silty-clay marshes, large numbers of minor creeks form a tree-like drainage system with long, often roughly parallel, branches. Creeks tend to be simpler and less numerous on sandy and peaty marshes (Chapman 1976), and lateral undercutting and collapse of creek walls occur very readily, forming narrow creeks running through secondary marsh created from fallen blocks within wider channels. Salt marsh creeks tend to be more sinuous and meanders are better defined than those cut in tidal flats. Drainage creeks are particularly tortuous on *Spartina*-dominated marshes in southern England and in the eastern United States, possibly because it grows in isolated clumps, on land without any distinctive slope. Tidal range also affects drainage systems. Marshes have a complex network of shallow, winding and branching creeks in areas with a small tidal range, but in macrotidal areas they tend to be deep, parallel systems, running perpendicularly to the shore (Chapman 1976).

Water usually moves very slowly over the surface of large marshes because of vegetation, the lack of channelization, and gentle gradients. Sheet erosion is therefore less important on salt marshes than on mudflats. Most marsh erosion probably occurs on the outer banks of meandering creeks, whereas deposition takes place on point bars on the inner banks. In addition to lateral movement of

meandering creeks, there may be headward growth at the end of the stream bed. Headward growth is most active on sandy substrates, however, and it may not occur at all on many finer-grained marshes. Headward erosion is generally most rapid during the early stages of marsh development, and it ultimately reaches a state of dynamic equilibrium with rates of channel closure by bank collapse and vegetational infilling (French and Stoddart 1992).

There are several similarities between terrestrial streams and intricate salt marsh and other tidal creeks, including their mode of growth, the presence of meanders and dendritic drainage, and geometric relationships between creek or stream order and their corresponding frequency (number) and length (Knighton *et al.* 1992). There are also important differences in the hydraulic geometry of streams and marsh creeks, however, which can be partly attributed to fundamental differences in their flow regimes. Marsh creeks experience flow in two directions, and they can be dry for hours or days on end. Furthermore, creeks can be filled much more frequently than stream channels, although in contrast to streams, creek flow at the bankfull stage may be very slow. Nevertheless, although the mechanics are somewhat different, creek morphology in mature marshes can be just as finely adjusted to tidal dynamics as terrestrial streams are to flow discharge and other factors (Pethick 1980; Ashley and Zeff 1988).

Tidal flow in salt marsh creeks is strongly asymmetrical. In the eastern United States and in San Francisco Bay, for example, maximum velocity and discharge occur within about one to two hours before and after high water, rather than midway between the high and low slacks. The asymmetry of creek flow is caused by asymmetrical tidal waves entering the system (Chapter 7), and the morphology and storage characteristics of the drainage basin (Boon 1975). It could also reflect differences in the total amount of water carried by the flood and ebb tides, caused by such factors as rainfall, evaporation, soil uptake, storage in marsh depressions, and creek drainage of water that flowed on to the marsh over the seaward edge (Dankers *et al.* 1984).

Although marsh creeks are occupied by opposing ebb and flood currents, the dominance of one over the other may create an essentially unidirectional pattern of erosion and deposition. Flood flows are generally stronger than ebb flows in New England (Lincoln and FitzGerald 1988; Stevenson *et al.* 1988), but the peak discharges and velocities are from 25 percent to 100 percent greater on the ebb than the flood in the south-eastern United States (Boon 1975; Letzsch and Frey 1980; Zarillo 1985). The ebb flows of over-marsh tides also attain higher maximum velocities than flood flows in San Francisco Bay. Two explanations have been proposed to account for ebb-aligned drainage systems. The first is that because of attenuation of flow velocities towards headwaters during floods,

only the lower portion of channel systems experiences effective bidirectional flows. As sedimentation continues on the vegetated interfluves, more and more of the water flows into embryo creeks, which are extended headwards during ebbs. The second model is based on the assumption that ebb flows dominate creek systems, with stronger flood flows occurring only during storms. Observation of the ebb-aligned creeks in eastern England suggest that the two hypotheses are complementary, in that channel-forming processes are both spatially varied and temporally intermittent (French and Stoddart 1992).

The nature of flow asymmetry in salt marsh creeks depends upon a number of factors, including the amplitude of the tidal cycle in relation to the elevation of the marsh surface (Reed 1987; Stoddart *et al.* 1989; French and Stoddart 1992). The marsh surface operates as a topographic threshold, separating two distinct flow regimes. The fairly steady flood and ebb flows of below-marsh tides, that fail to reach the bankfull level, have essentially symmetrical velocity distributions, although the flood velocities may be slightly higher than the ebb. Over-marsh tides, that flood out on to the marsh surface, have well-defined velocity maxima just above bankfull on the flood, and just below bankfull on the ebb (Fig. 8.4). High velocity and discharge are associated with water suddenly

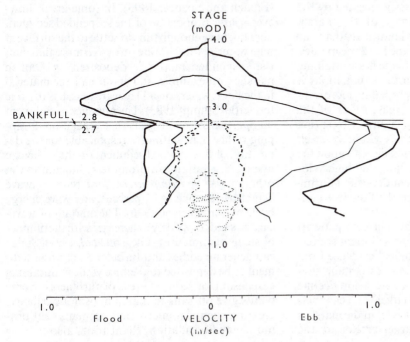

Fig. 8.4. Velocity-stage curves in a salt marsh creek in eastern England. Maximum current velocity for tides that rise above the marsh surface occurs just above the bankfull stage on the flood, and just below bankfull on the ebb (Stoddart *et al.* 1989)

flowing out on to the marsh surface on the flood, and draining back into creeks on the ebb. Although the highest overbank tides usually attain maximum velocity and discharge on the ebb in eastern England and the Netherlands (Dankers *et al.* 1984; French and Stoddart 1992), maximum flows were recorded on the flood during a storm in eastern England (Bayliss-Smith *et al.* 1979).

Sediment Transport and Deposition

Flow asymmetry and velocity pulses in marsh creeks can entrain and transport large quantities of sediment, especially during storms. As the marsh surface is raised, however, fewer and fewer tides can generate the high stresses needed to entrain sediment in the channels. In the Netherlands, all sediment was imported to a salt marsh when flow was restricted to the creek, but the situation was less clear for the movement of coarse (> 50 μm) material by higher storm tides. Although ebb currents are too weak to resuspend sediment over the surface of the marsh, the creeks are maintained by the resuspension of bottom sediments by strong ebb flows (Dankers *et al.* 1984).

Suspended sediment movement through a marsh creek in south-eastern England is very variable, both in magnitude and direction (Reed 1988). Large spring tides, with pulses generated by the alternate flooding and drainage of the marsh, moved much more sediment landwards than the net ebb transport of many of the lower neap tides. This may be partly attributed to the flood attaining higher velocity maxima than the ebb, and, as it occurs at a higher water level, propelling more sediment and water through the creek. Much of the sediment deposited on the marsh surface by spring tides is derived from storage areas within the creek itself. Local patterns of accretion depend upon the frequency of over-marsh flooding, and high velocity pulses to mobilize sediment deposited during the previous neap tides, when the flow was contained within the creeks.

There is a strong relationship between patterns of accretion and proximity to a sediment source. Mathematical modelling, and field evidence from the Severn Estuary in south-western Britain, suggest that grain size and surface elevation decline outwards, from the main and larger tidal channels to the bedrock margin (Allen 1994). In the marshes of Norfolk, England, the larger creeks are the

sources, or the means of delivery, of the sediment. Smaller creeks are much less effective in distributing sediment over the marsh surface (Stoddart *et al.* 1989). Over-marsh tides in eastern England have higher sediment concentrations than below-marsh tides, particularly on the flood. A sample of sediment concentrations over ten tidal cycles of varying magnitude showed that all of them imported sediment to the marsh, averaging 43 percent of the total flood transport for those inundating the marsh surface, and 10 percent for those that did not leave the creek system. The vegetated marsh surface therefore constitutes a very efficient sediment sink (French and Stoddart 1992). Nevertheless, although marshes have functioned as sediment sinks for thousands of years, some are now losing sediment. While tidal asymmetry in New England generally favours the landward movement of sediment, for example, dominant ebb flows tend to move sediment seawards in the south-eastern United States (Stevenson *et al.* 1988).

Deposition from over-marsh tides probably involves gravitational settling and flocculation during slack water, downward diffusion, interception by vegetational surfaces, and bio-deposition by filter feeding organisms (Pye 1992). Direct settling, at median rates of 3 to 8×10^{-4} m s^{-1}, removes much of the over-marsh suspended sediment within a short distance of a creek in eastern England (French and Spencer 1993). In some areas, however, a large proportion of the suspended sediment does not have enough time to settle to the surface at high water, before the tide turns. No more than half the suspended matter is deposited on Danish marshes, for example, and about half the material in the flood waters on a Delaware marsh is too fine too settle (Stumpf 1983). The presence of sediment on *Spartina* stems suggests that vegetational trapping rather than settling is responsible for the deposition of fine-grained sediment on the Delaware marsh. Vegetation can promote sedimentation in other ways. *Spartina alterniflora* reduces wave height by as much as 71 percent, and wave energy by 92 percent (Wayne 1976). The amount of wave-carried sand trapped by it increases with the number of stems per unit area (Gleason *et al.* 1979). Stalks can generate eddies and turbulence, causing sediment to be deposited to their lee while maintaining suspension, or scour in areas of turbulence. Some plants give off salts to maintain osmotic balance, creating chemical micro-environments that promote clay flocculation. Plant roots also help to bind sediment, thereby promoting the accumulation of clayey, cohesive sediments.

Marsh accretion rates have usually been determined using historical records, anthropogenic pollutants, introduced marker horizons of brick dust, coloured sand, or some other distinctive material, vertical stakes, dated horizons, and precise surveys. Recent investigations have used radioisotopic dating techniques, especially those based on ^{137}Cs and ^{210}Pb. Accretion rates depend upon such factors as the tidal or wave energy, tidal range, the elevation of the marsh surface, vegetational characteristics, and the amount and type of available sediment. True accretion is the amount of buildup minus the amount of settling or compaction. Settling is most significant in the higher parts of marshes, where there is less accretion, more organic matter, and more frequent drying out.

Marshes are flooded less frequently, for shorter periods, and they are covered by shallower water as they are built up. Accretion rates therefore decrease with increasing marsh elevation, and with increasing distance from creeks. On Sapelo Island, Georgia, vertical accretion is about 9 mm yr^{-1} on the low marsh, compared with 4.5 mm in the transitional zone, and 1 mm on the high marsh (Letzsch and Frey 1980). In Norfolk, England, vertical accretion rates vary from less than 8 mm yr^{-1} near large channels, to less than 1 mm yr^{-1} on higher surfaces far from the creek network (French and Spencer 1993). Complex spatial variations over fairly short distances may therefore introduce considerable error in estimating mean marsh accretion rates from a limited number of point samples. Most temperate marshes in Europe, North America, and New Zealand accrete at rates of between about 0.6 and 25 mm per year, although they may be somewhat higher along channel levees (Reed 1988; Craft *et al.* 1993; French *et al.* 1994). Much higher deposition rates have been reported in some areas, however, ranging up to 25 mm in a single tide on the marshes of the Bay of Fundy.

The depth and duration of tidal inundation are probably the primary controls on accretion rates by marsh elevation. Calculations based on the relationship between marsh elevation and age suggest that a ten-year old marsh in Norfolk is accreting at a rate of 17 mm yr^{-1}, compared with 1.2 mm yr^{-1} on marshes 200 years old, and less than 0.02 mm yr^{-1} on marshes more than 500 years old (Pethick 1981). In the sixteenth century vertical accretion on the marshes of the Ems–Dollard Estuary in the Netherlands is estimated to have been about 17 mm yr^{-1}, and between 13 and 14 mm yr^{-1} at the end of the nineteenth century. Between 1950 and 1970 the annual accretion rate was negligible on marsh surfaces more than half-a-metre above the mean tidal level, and about 10 mm yr^{-1} between the mean tidal level to one-and-a-half metres below (Dankers *et al.* 1984).

Sedimentation rates have been used to calculate the age of marshes. Chapman (1974) estimated that a New England marsh needed 600 to 700 years to build up well into the *Juncus gerardii* stage, whereas 70 years was sufficient in north-eastern Scotland and 100 to 125 years on Baltrum in the Frisian Islands. Minimum estimates of the time needed for salt marshes to reach maturity in the British Isles, Denmark, and New England, as defined as the point at which accretion is balanced by compaction or settling, range from 10 to 330 years (Ranwell 1972); partly enclosed marshes usually developing faster than those on more open coasts. These estimates do not, however, include the time required for formation of the preceding tidal flat.

Many studies have shown that vertical accretion is promoted by a slow rise in sea level, with some of the material being derived from erosion and trimming back of the seaward edge. The low marsh on Rhode Island is accreting at rates ranging from about 1.5 to 1.7 times the rate of sea level rise (about 2.6 mm yr^{-1}), while accretion rates on the high marsh is about the same as the rise in sea level (Bricker-Urso *et al.* 1989). Marshes in South Carolina, which are accreting at about 1.3 to 4.5 mm yr^{-1}, also appear to be keeping pace with a rise in sea level of about 3 mm yr^{-1} (Sharma *et al.* 1987).

Coastal marshes are at risk from the predicted rise in sea level resulting from global warming (Chapter 1). Most marshes in the north-eastern United States and in eastern Britain are accreting at rates equal or greater than the local relative rise in sea level, whereas many marshes in the south-eastern United States and Gulf of Mexico have rates that are less than the rise in sea level (Stevenson *et al.* 1986; Craft *et al.* 1993). Subsidence, changes in the locus of deltaic sedimentation, and human modification of the fluvial system, including the building of dams that trap suspended sediment, are causing severe erosional problems in Louisiana (Ramsey *et al.* 1991). Although deteriorating marshes in the Mississippi delta plain are accreting faster than the stable marshes in the same region, they are failing to keep up with subsidence and the rise in relative sea level. In south-western Louisiana, for example, where marsh accretion is about 8 mm yr^{-1} and coastal submergence 12 mm yr^{-1}, waterlogging is killing marsh plants, leading to the formation of open bodies of water. Rising sea level is thought to be the primary reason for marsh erosion in Chesapeake Bay, with most losses occurring as the channels migrate and widen, and the drainage network increases in density as it adjusts to the larger tidal prism. Some interior ponds form in seemingly random locations, possibly as a result of heavy grazing by musk-rats, but others develop where small creeks are undergoing headward extension (Kearney *et al.* 1988).

Pans

Many marshes have numerous shallow, unvegetated depressions, or pans, that are filled with water by the high spring tides. Eddies of flooding tides, water seepage, and small waves generated within the pans tend to produce well-rounded shapes, that can be elongated, crescentic, elliptical, branched, winding, or roughly circular. Primary pans were formed during initial marsh development, when there was irregular plant colonization and numerous bare areas. As the level of the vegetated marsh gradually increased, plant growth in the barren areas was discouraged by rushing water, waterlogging, or the development of very saline conditions following summer evaporation. Secondary pans develop on secondary marshes, which form at the foot of erosional cliffs at the edge of retreating marshes. They develop in the same way as primary pans, although they may be less regular in shape, owing to the random distribution of the eroded clumps of vegetation (Chapman 1974, 1976). Sinuous, elongated channel pans are also a type of secondary pan formed by slumps of growing vegetation blocking small creeks, or by the exhumation of underground pipes. Many creeks become redundant and are abandoned as marsh surfaces increase in elevation, allowing a smaller number of deeper channels to accommodate tidal flows (Pethick 1969).

Increasing pan frequency with marsh height in England suggests that they develop on mature as well as embryonic marshes and tidal flats. This could result from subdivision of existing pans, especially through the growth of vegetation, although the fact that they are most common at the seaward margin of marshes suggests that they may develop from bare patches created by the deposition of vegetational debris (Pethick 1974). Warming first suggested, in the early part of this century, that pans are produced under putrefying masses of debris and algae deposited on the marsh surface by spring tides. In the Atlantic marshes of the United States bare areas, or barrens, may result from the destruction of marsh vegetation under trash left by melting snow in spring. In the north-eastern United States large, irregularly shaped pans develop from the decay of surface turf caused by inadequate drainage, standing water, and possibly salt concentration by evaporation. Initial depressions could be produced by reduced sedimentation in inaccessible areas, compaction and settling of the underlying peat, and enclosure by levees around drainage channels. Pans can also be formed in many other ways, including breakup of the winter ice cover in high latitudes (Chapter 12) (Fig. 12.10); tidal eddies around driftwood logs, boulders, and other obstacles (Reed 1988); and vegetationally induced differences in peat accumulation (Collins *et al.* 1987).

Marsh Vegetation

As marsh elevation increases, one type of vegetation may be replaced by another, producing strong shore-parallel banding. The vertical range of individual species often overlap, however, and unless marsh gradient is quite high, or there is a large tidal range, species competition and variations in salinity, drainage, and other factors usually create a complex vegetational distribution, with plant communities merging to form a vegetational mosaic, rather than a series of well-defined zones (Ranwell 1972; Gray 1992). Differences in vegetation may be the result of local topographic irregularities, or because of the different ability of species to trap sediment, the reason for the irregularities. It has been shown, for example, that patterns of erosion and accretion on a marsh in eastern England are influenced by seasonal changes to the *Puccinellia, Spartina,* and *Salicornia* cover (Hartnall 1984).

Although vegetation on individual salt marshes can be very diversified, it is often dominated by grasses belonging to one or two species. A number of topographic, sedimentary, and climatic factors determine the type of vegetation that grows on a marsh, but tolerance to tidal immersion is particularly important. Many *Spartina* species, for example, can survive tidal submergence for more than three hours, whereas *Sueda* and *Puccinellia* are far less tolerant of prolonged inundation (Ranwell 1972).

Nine major regional salt marsh groups have been identified: arctic; northern European; Mediterranean; eastern North American; western North American; Chinese, Japanese, and Pacific Siberian; South American; Australian and New Zealand; and tropical (Chapman 1974). Each region is characterized by a number of species that have a fairly wide range. Several dominant genera, including *Juncus, Spartina, Salicornia, Arthrocnemum,* and *Plantago*, are widely distributed on salt marshes around the world, although the distribution of individual species is more restricted. *Spartina* is most common in the Atlantic from north-western Europe to north-western Africa, and from eastern Canada to Brazil, and it has been introduced to other areas by humans. *Salicornia* may be the most widespread salt marsh genus, but it becomes increasingly dominant in the lower latitudes (Davies 1972). Unlike most halophytic or salt-tolerant marsh plants, *Salicornia* can only grow in saline environments.

Monocultures of *Spartina alterniflora* dominate low marshes on the eastern coast of North America, except in the far north, but a greater variety of succulents, rushes, and grasses occupy high marshes. *Juncus gerardii* and *J. balticus* are abundant in the higher marshes of New England and southern Canada, and the former is common as far south as Chesapeake Bay, while *J. roemerianus* is dominant further south. *Puccinellia americana* is dominant in the Bay of Fundy, whereas *Spartina patens* is common on the higher marsh areas along the entire eastern coast (Chapman 1976; Reimold 1977). More mixed communities on the western coast of North America include *Salicornia virginia, Puccinellia pumila,* and *Carex lyngbyei* in the Pacific north-west, and *Spartina folioso* and *Salicornia virginia* further south.

Marsh vegetation in northern Europe is fairly varied, with *Salicornia* species passing landwards into low marsh plants that include *Sueda maritima, Salicornia perennis, Halimiones* species, and *Puccinellia maritima*. Species of *Festuca, Juncus,* and *Carex* grow in the high marshes (Fig. 8.5). Sandy marshes are usually dominated in Europe by *Puccinellia*, but many mudflats have been invaded or planted with *Spartina townsendii—S. anglica*, an exceptionally vigorous hybrid complex that was first discovered in 1870, at Hythe on Southampton Water, Hampshire (Chapman 1974) (Fig. 8.5). It is thought to have developed as a hybrid between *Spartina maritima*, which is now fairly rare, and the American *Spartina alterniflora. S. townsendii* is a sturdy perennial grass, sometimes growing over a metre in height, with roots extending up to a metre below the surface. There are two forms of the

Fig. 8.5. Some salt marsh plants of eastern England (Jane and White 1964)

Spartina townsendii

Salicornia herbacea

Festuca rubra

plant. The one originally termed *S. townsendii* is a sterile F_1 hybrid, which can spread only through its rhizomes and dispersal of plant fragments. There is also a fertile species which is able to spread by seeding. This species has been named *S. anglica*, but as the two species are difficult to distinguish, the term *S. townsendii* (*sensu lato*) is often used. Initial clumps spread rapidly, forming a continuous sward drained by tortuous and steep-sided creeks (Tubbs 1977).

Arctic and subarctic marshes generally consist of short, sparse vegetational stands with low species diversity and simple zonation. The grass *Puccinellia phryganodes* is the primary marsh community, with *Carex* species becoming prominent at higher levels. *Carex subspathacea*, for example, dominates the higher levels of marshes in northern Europe, and in northern Canada as far south as the St Lawrence.

Human Impact

Salt marshes have been used for waste disposal, and large areas have been reclaimed for housing, industry, and airports. Marsh grass was once cut for hay, and its peat has been used for lawns. Reclaimed mudflats and salt marshes have been used for grazing in Europe, but because of their high organic content and fertility, they are increasingly being converted into arable land. Coastal marshes can be reclaimed using squares of brushwood and earthen groynes to trap clay and silt in settling fields, and embankment or poldering can also be used to isolate them from the tides. One of the largest reclamational projects took place in the Zuyder Zee in the Netherlands, but large estuarine barrages have been proposed for many other areas. Great concern has been expressed over the effect of the loss of marshland on birds, fish, oysters, and other marine organisms, however, and despite intense economic pressures in some areas, it is now recognized that coastal marshes are important and productive ecosystems, and there is new interest in their preservation.

MANGALS

The term 'mangal' refers to a community of mangroves—trees and shrubs, together with a few associated lianes, palms, and ferns, that colonize tidal flats in the tropics. Mangals occur in river-, tide-, and wave-dominated environments (Woodroffe 1990). They flourish on tidal shorelines with low relief and fairly low wave energy, and are especially common in the brackish waters of estuaries and deltas, although they are also found in depressions behind cheniers, on the broad mudflats of shallow-water carbonate banks, and in the shelter of barrier spits, barrier islands, and coral reefs. Although mangroves are halophytic, most species are facultative to a wide range of salinities. Most grow quite well in fresh water, although they may be stunted, and they are often distributed along river banks extending well inland. Mangroves may therefore be excluded from terrestrial environments not because of an inability to survive, but because they cannot compete with species that are not burdened with the additional features necessary to grow in saline, waterlogged environments (Tomlinson 1986).

Some mangrove species are able to tolerate much more frequent inundation than salt marsh vegetation. Mangals can therefore extend from about the high spring tidal level to a little above mean sea level (Macnae 1968; Bunt *et al.* 1985). They lack the pans of some salt marshes, but they can contain lagoons and pools. Creek systems may be just as prominent in mangals as in salt marshes, although the banks are sometimes formed of roots rather than sediments, and they are often quite gentle.

Mangroves can grow on allochthonous sediments deposited by rivers or tides, or on organic autochthonous material deposited *in situ* beneath the mangal. Small, crooked forms can grow on stable sand and rock, and attenuated mangroves grow on coral cays in the Caribbean and on the Low Wooded Islands of the Australian Great Barrier Reef (Chapter 10) (Fig. 10.13). Mangroves have also spread over reef flats on many Pacific islands, but they are best developed on emergent flats that are at, or above, mean sea level (Woodroffe 1987).

Mangroves range from less than 1 metre up to more than 30 metres in height, depending upon soil, drainage, and climate. The tallest mangroves in an area are often found along the banks of creeks, where there are well-drained soils and nutrient-rich overbank deposits (Thom 1967; West 1977). The distinction has been made between low mangal, where the trees are rarely more than 10 metres in height, and high mangal where some species, particularly of *Bruguiera* and *Rhizophora*, attain heights of 40 metres or more. High mangal borders tropical rain forest, and it is only found where precipitation is well distributed throughout the year (West 1956; Davies 1972; Chapman 1975) (Fig. 8.2). As mangroves can extract fresh water directly from the sea, the occurrence of high mangal in the perennially wet tropics may therefore reflect the large amounts of fine-grained sediment carried by rivers in these areas. Alternatively, because the highest and floristically most diverse mangals are found where salinities are somewhat lower than sea water, their distribution could reflect nutrient supply by freshwater runoff, or reductions in salinity by stream flow and rainfall (Bunt *et al.* 1982, 1985).

Although mangals are essentially tropical, there are outliers in subtropical regions washed by warm ocean currents, including South Africa, Victoria in southern Australia, southern Japan, and southern Florida (Tomlinson 1986). They generally replace salt marshes within 28°N and 25°S of the equator in the New World, and within 28°N and 38°S in the Old World, and they commonly coexist with salt marshes in subtropical regions (Fig. 8.2). Mangals flourish where the temperature of the coldest month does not fall below 20°C, although *Avicennia* species tolerate minima of 12.7°, 10°, and 15.5°C in Florida, Brazil, and the northern Red Sea, respectively. Frost damage occurs at the northern limit of mangals in Florida, and at their southern limit in New Zealand. Therefore frost probably largely controls mangrove distribution, although *Avicennia marina* var. *resinifera* survives in the Auckland area, where temperatures occasionally fall to –2°C (Lange and Lange 1994).

Mangroves are absent in Baja California, parts of the Pacific coast of South America, on the southwestern coasts of Africa and Australia, and in some other arid areas. This could reflect the lack of river-borne silt or the inhibition of seedling growth by upwelling cold water. Recent changes to the hydrology in parts of southern Florida have led to the salinization of soil and water, and landward expansion of mangroves into areas that once supported freshwater marsh (Ball 1980). Mangal growth is inhibited, however, by hypersaline conditions. In humid areas, the soil is leached between flood tides and the most saline soils are therefore on the seaward edge of mangals. In arid or tropical regions with a dry season, evapotranspiration causes a strong landward increase in salinity, and barren, hypersaline flats are often found on the innermost portion of mangals (Wolanski *et al.* 1980; Semeniuk 1981; Guilcher 1989). Some saline Australian flats were created in the past when there was a fall in relative sea level. Others developed when cyclones destroyed the vegetational canopy and exposed the ground to sunlight, increasing the evaporation rate and precipitation of salts that prevented recolonization (Spenceley 1976). On the other hand, strong winds and wave action can clear detritus from blocked drainage channels, allowing sea and fresh water to flush hypersaline areas.

Vegetational Distributions and Characteristics

There are about sixty–seventy species of mangrove (Chapman 1976; Teas 1983; Tomlinson 1986; Duke 1992) (Table 8.1). Most are in south-eastern Asia and Micronesia, and only about one-fifth occur around the Atlantic. The number of species also tends to decrease with distance from the equator. The genera *Rhizophora*, *Avicennia*, *Acrostichum*, and to a lesser extent *Bruguiera*, are widespread, although only the first two are common to both the Old and New Worlds. Six broad geographical areas, each characterized by significant mangal uniformity, have been distinguished: the New World, western Africa, eastern Africa, Indo-Malaysia, Australasia, and Oceania (western Pacific) (Chapman 1977*a*).

Four genera, *Rhizophora*, *Avicennia*, *Laguncularia*, and *Conocarpus*, dominate in the western hemisphere and western Africa, although they are not exclusive, nor do they occur in all areas. The discovery of forty-five species of mangrove in north-eastern Australia demonstrated that floral diversity is much greater in this region than had been previously recognized (Bunt *et al.* 1982). *Avicennia marina* is the only species in temperate Australia, however, and only *A. marina* var. *resinifera* grows in northern New Zealand. *Heritiera littoralis* is characteristic of the Oceania group (Chapman 1977*a*), and *Kandelia candel* is the only species in southern Japan. Cold currents and low water temperatures prevent mangrove growth along portions of the western American coast. The Indo-Malaysian group, extending from Pakistan to Papua–New Guinea, is the richest and most diverse floristic group. There are about fifty species on the coasts of Malaysia, Indonesia, and New Guinea, and the trees are especially tall and luxuriant (Chapman 1975, 1977*b*). Some workers believe

Table 8.1. Dominant mangrove species forming communities

Dominant community	Occidental Zone						Oriental Zone											
	GC	Fl	PE	NB	EB	WA	RS	EA	WI	EI	SL	BB	Ir	IM	PN	Ph	RI	Au
Avicennia germinans	—	—	—	—	—													
Rhizophora mangle	—	—	—	—	—													
Conocarpus erectus	—	—	—	—	—													
Laguncularia racemosa	—	—	—	—	—													
Rhizophora racemosa		—	—	—	—													
Rhizophora harrisonii	—				—													
Avicennia germinans/ Avicennia schaueriana		—																
Pelliciera rhizophorae	—																	
Avicennia africana					—													
Laguncularia racemosa/ Conocarpus erectus					—													
Avicennia marina							—	—	—	—	—	—	—	—	—	—		—
Avicennia marina/ Rhizophora mucronata							—											
Rhizophora mucronata							—	—	—	—	—	—	—	—	—	—		
Rhizophora mucronata/ Bruguiera spp.							—											
Sonneratia alba							—							—	—			—
Ceriops tagal							—							—	—			
Sonneratia apetala								—		—								
Xylocarpus obovatus/ Heritiera littoralis							—											
Bruguiera gymnorrhiza														—	—	—		
Mixed mangrove							—	—	—	—	—	—	—	—	—	—		—
Avicennia alba								—		—	—							
Rhizophora mucronata/ Rhizophora apiculat								—		—	—							
Avicennia spp./ Excoecaria agalloch								—	—	...		...	...	...	...	—		
Lumnitzera racemosa							—	...		—		...	...	...		—		
Aegiceras corniculatum/ Ceriops tagal							—											
Avicennia officinalis																—		
Excoecaria agallocha								—	...	—					—			
Nypa fruticosa (palm)/ Acrostichum aurem							—	—					...	...	...			—
Bruguiera cylindrica														—				
Bruguiera parviflora														—				
Sonneratia caseolaris/ Nypa fruticans (palm)														—	—			
Rhizophora stylosa/ Rhizophora apiculat																		—
Rhizophora stylosa														—				
Excoecaria agallocha/ Xylocarpus obovatus																		—
Excoecaria agallocha/ Heritiera littoralis																		—
Heritiera fomes										—	—							
Lumnitzera racemosa/ Ceriops tagal							—				—	—				...	...	—
Aegiceras/ Aegilitis spp.															—			—

Table 8.1. (*cont.*)

	Occidental Zone						Oriental Zone											
Dominant community	GC	F1	PE	NB	EB	WA	RS	EA	WI	EI	SL	BB	Ir	IM	PN	Ph	RI	Au
Aegiceras corniculatum/ Camptostemon schultz															—	—		
Aegiceras corniculatum																—		
Bruguiera gymnorrhizal Heritiera littoralis																		
Kandelia candel																	—	
Rhizophora apiculata							…	…	…	…	…	…	…	…	…	…	…	—

Notes: GC = the Gulf of Mexico and the Caribbean; F1 = Florida; PE = western Panama, Mexico, Colombia and Equador; NB = northern Brazil; EB = eastern Brazil; WA = west Africa; RS = the Red Sea; EA = east Africa; WI = western India; EI = eastern India; SL = Sri Lanka; BB = the Bay of Bengal; Ir = the Irrawaddy; IM = Indo-Malaysia; PN = Papua–New Guinea; Ph = Philippines; RI = Ryukyu Islands; Au = Australia. Dotted lines show where species are also present.

Source: Chapman 1977*a*.

that the diversity of the Indo-Malaysian region indicates that it was the centre of origin and dispersal for modern mangroves (Macnae 1968; Chapman 1975). Others consider that mangal distribution is the result of the modification of a formerly widely distributed biota by tectonic and climatic events, speciation, and extinction. A similar debate concerns the bio-geographical distribution of corals and seagrasses, the other main shallow water sessile assemblages in the tropics (McCoy and Heck 1976; Woodroffe 1987).

Mangroves are uniquely adapted for survival and reproduction in a hostile environment. The large, spear-shaped seedlings of *Rhizophora* and *Bruguiera*, for example, allow them to drop down and become embedded in the mud below the parent tree, while the bean-like seedlings of *Avicennia* float and are carried off by coastal currents. Many mangrove species are viviparous—the seeds germinating within fruits that are still attached to the mother tree. This protects them from the toxic effects of chlorides and other marine salts, and allows them to colonize areas as soon as they drop or are carried by currents on to suitable substrates. Mangroves employ a variety of mechanisms to combat salinity and the stresses induced by waterlogging. Salt glands in the leaves of *Avicennia*, *Aegiceras*, and *Aegialitis*, for example, allow them to secrete excess salt through their sap, so that it can be washed away by rainfall. On the other hand, *Rhizophora*, *Laguncularia*, *Sonneratia*, and *Conocarpus* prevent salt entering their sap by generating abnormally high osmotic pressures in their leaf cells. Dilution resulting from increased water content or succulence may also be used to prevent high

ion concentrations building up in the leaves (Walsh 1974).

Most mangroves have shallow but extensive root systems, and few persistent tap roots. To aerate waterlogged soils, *Avicennia*, *Sonneratia*, *Xylocarpus*, and *Laguncularia* species employ shallow, horizontal roots with pneumatophores—numerous periscope-like root extensions that stick up above the surface. They are usually thin and less than 30 cm in height in *Avicennia*, but they become quite thick in *Sonneratia*, and in exceptional cases attain heights of up to 3 metres (Tomlinson 1986). The horizontal roots of *Bruguiera*, *Ceriops*, and *Lumnitzera* loop above the surface in knee-like bends. A tangle of arching prop roots extend out from the trunks of *Rhizophora* and *Acanthus*, and to a limited extent in *Bruguiera*, *Ceriops*, and *Avicennia alba* and *A. officinalis* (Fig. 8.6). These subaerial roots contain airholes which allow gaseous exchange with the atmosphere, as well as providing mechanical rigidity. Aerial stilt roots descend to the ground from the upper branches, and become entangled in the prop roots. *Aegialitis*, *Kandelia*, and some other mangroves which lack subaerial root systems employ other aerating structures, but *Conocarpus* and other genera lacking 'breathing' roots tend to grow furthest inland.

Mangals often exhibit a zonal arrangement, with species distributed parallel to the seaward edge or to the tidal channels. Although zonal distributions have developed along the Guiana coast of South America, they tend to be less pronounced in the New World than in the Old, possibly because the former has fewer species. In the New World *R. mangal* along the outer edge and along tidal

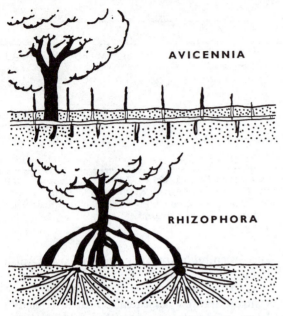

Fig. 8.6. *Avicennia* and *Rhizophora* root systems (Chapman 1975)

channels is followed successively inland by *A. germinans*, *L. racemosa*, and in places where it is not reached by tides, by *C. erectus* (West 1977). *Avicennia* is the pioneer along accreting coasts in northern South America (Vann 1980). The sequence in the Indo-West-Pacific often consists of *Avicennia* or *Sonneratia* at the seaward edge, followed successively inland by *Rhizophora*, *Bruguiera*, *Ceriops* thickets or *Ceriops* in *Bruguiera* or *Rhizophora* forests, and finally by a variably forested, landward fringe of Nypa Palm or Barringtonia associations, which may be absent in arid areas (Macnae 1968). There are numerous exceptions to the simple zonal models, however, and mangals often consist of interdigitating associations rather than simple zonal patterns (Thom 1967; Bunt *et al*. 1982).

The local distribution of mangrove species reflects variations in salinity, tidal exposure, microtopography, substrate characteristics, and depositional and erosional processes, as well as shade, chemical excretions that inhibit the growth of competitors, and other biotic factors involving species interactions (Macnae 1968; Chapman 1977*a*; Woodroffe 1987; Day *et al*. 1989; Semeniuk 1994). Some *Avicennia* species are more tolerant of very saline water than *Rhizophora* (West 1956; Chapman 1975; Steers 1977), and possibly because of

shallow, rapidly developing root systems, they may also be more tolerant of rapid sedimentation (Giglioli and Thornton 1965; Vann 1980). In Colombia, *Avicennia* usually colonizes mudflats before *Rhizophora*, which only thrives in sheltered places, in quiet brackish and salt water, or where barriers provide protection from wave action (West 1956). *Avicennia* species are usually the pioneers in the Indo-West-Pacific, although this role is assumed by *Sonneratia* in some of the warmer parts of the region, and by *Rhizophora racemosa* in West Africa (Giglioli and Thornton 1965; Macnae 1968; Coleman *et al*. 1970). *Avicennia* is often found on heavily oxidized clay soils in the Americas, and it grows on coarse sand in Colombia, although it tends to be replaced by *Laguncularia* in the West Indies (West 1956, 1977; Chapman 1976, 1977*a*).

Mangroves change the characteristics of their local area, eventually providing conditions that are more suitable for other species, which then replace them. Some workers therefore believe that the zonal arrangement of mangrove species is symptomatic of a temporal succession—gradual replacement of communities with the build up of the land, culminating in a climax terrestrial forest. Others consider that mangroves are opportunistic colonizers of available habitats, with distributions that reflect variations in energy and substrate conditions, rather than temporal changes induced by the plants themselves. Geomorphological factors, for example, play important roles in determining mangrove species distributions on accreting micro- and macrotidal deltaic coasts in south-eastern Mexico and north-western Australia, lagoonal environments in El Salvador, and on the inner reefs of northern Queensland, and on an eroding coast in north-western Australia (Thom 1967; West 1977; Thom *et al*. 1975; Stoddart 1980; Semeniuk 1980).

Geological Development

Mangrove shorelines become transgressive units when sea level is rising, and regressive units when they colonize and possibly help to prograde marine substrates. Rising sea level caused mangals to migrate landwards over the western margins of the Florida Everglades and Big Cypress swamps, and mangrove peat was deposited over freshwater calcitic marls and peats. The coast stabilized or prograded as sea level rose less rapidly between about 3,200 and 3,500 years ago, and a regressive sedimentary sequence, consisting of biogenic sediment, which provides the foundation for numerous low mangrove islands, or thick mangrove peats, was deposited over the transgressive sequence (Parkinson 1989). In Florida Bay regressive sedimentation and mangrove colonization of mud banks may have helped to transform them into islands, although some developed around supratidal precursors consisting of eroded remnants of tidal flats or topographic highs that remained above sea level throughout the Holocene (Quinn and Merriam 1988). A transgressive–regressive sequence has also been identified in the Florida Reef Tract, elsewhere in the Caribbean (Turmel and Swanson 1976; Woodroffe 1983), and in eastern Australia (Grindrod and Rhodes 1984). In a wetland in western Jamaica the area covered by fresh water vegetation has dramatically increased within the last 5,000 years, at the expense of man-

groves. Although this partly reflects slowly rising sea level, the effect may have been exacerbated by climatic change (Hendry and Digerfeldt 1989).

Mangals form as narrow fringes along the banks of estuarine channels in northern and north-western Australia. Many macrotidal estuaries provide evidence of widespread mangrove forests during the 'Big Swamp Phase', about 6,000 years ago, although extensive mangals appear to have been absent in mesotidal estuaries (Knighton *et al.* 1992; Woodroffe 1993; Woodroffe and Chappell 1993; Chappell 1993) (Fig. 8.7). Mangrove swamps in the South Alligator and other estuaries in southern Van Diemen's Gulf, were replaced by grass-and-sedge covered floodplains about 5,500–5,300 years ago, when sea level had stabilized and the surface was high enough to prevent further tidal flooding (Woodroffe 1990; Woodroffe *et al.* 1993) (Fig. 8.8). In the macrotidal Cambridge Gulf region of

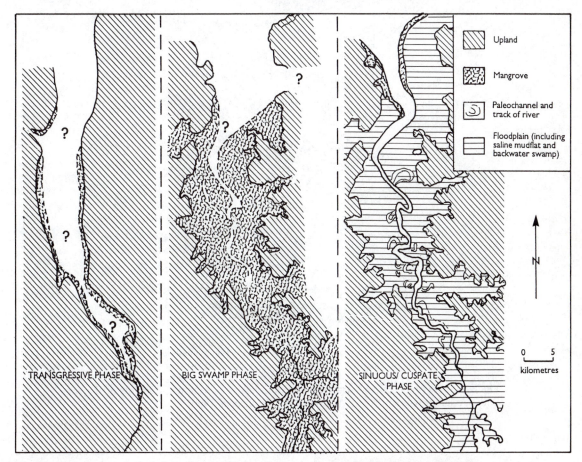

Fig. 8.7. Holocene evolution of the South Alligator River, northern Australia (Woodroffe 1990)

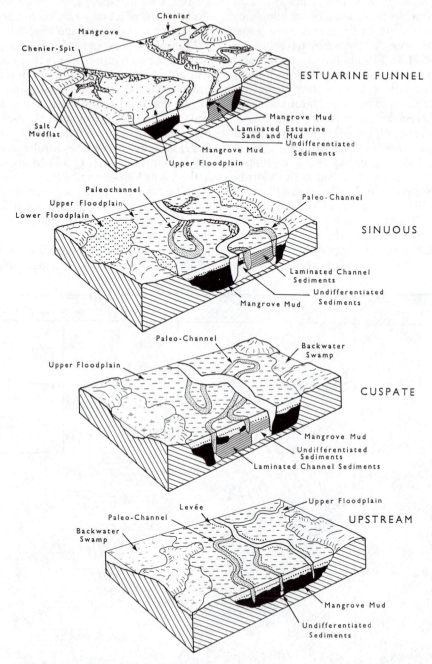

Fig. 8.8. Morphological and stratigraphic relationships in different segments of the South Alligator River, northern Australia (Woodroffe *et al.* 1990)

north-western Australia tidal flats were built up as sea level rose to near its present level, between about 5,000 and 6,000 years ago. The flats were extensively colonized by mangroves, but as they rose to the high spring tidal level, hypersaline conditions developed as a result of decreasing inundation and increasing desiccation. Mangroves on the landward side of the flats began to die, and as hypersalinity became more severe with continuing accretion, they were increasingly restricted to stream banks and the fringes of the coast, with bare mudflats landwards (Thom *et al.* 1975). Despite quite different conditions, the pollen record shows that a similar transition from mangrove to bare tidal flats occurred as hypersaline conditions developed on a prograding chenier plain in Princess Charlotte Bay in north-eastern Australia (Grindrod 1985). Stumps up to a metre in diameter imply that mangroves grew much more vigorously than today in the macrotidal Fitzroy Estuary between about 7,400 and 6,000 years ago. A reduction in the length and intensity of the rainfall season may therefore provide a better explanation for the reduced extent of mangals and the disappearance of high mangrove forest in this area, than changes in sedimentation and sea level (Jennings 1975).

Whereas mangroves in northern Australia were replaced by saline mudflats in semiarid areas, and by freshwater herbaceous floodplains in monsoonal areas, they were succeeded by freshwater peat swamp forest in the wet equatorial climate of Malaysia (Kamaludin 1993). Mangroves also appear to have persisted longer in south-eastern Asia than in Australia, remaining in many areas until after 4,500 years ago. This could reflect the fact that sea level in south-eastern Asia rose to several metres above its present level in the mid-Holocene, whereas it may have remained at its present level during the last 6,000 years over much of northern Australia (Woodroffe 1993).

Geomorphological Role

There has been much debate on the geomorphic role of mangals, but little actual research. Some workers believe that the main function of mangroves is to occupy sites where silting has already taken place (West 1956; Scholl 1968; Steers 1977), whereas others believe that they prograde into shallow water, hasten deposition, and facilitate coastal progradation (Chapman 1975, 1976; Wells and Coleman 1981).

There is growing evidence that mangroves help sediment to accumulate, thereby promoting vertical accretion. Once they are established, they also become an important source of organic material, help to bind deposited sediment, and provide an effective baffle to waves and tidal currents. Sediment accumulates preferentially around the roots of *Rhizophora mangle* in Bimini Lagoon in the Bahamas (Scoffin 1970). Average rates of accretion, corrected for consolidation, are 1.6 mm yr^{-1} in fairly infrequently inundated *R. mangle* and *A. germinans* swamps in south-western Florida, and 2.4 mm yr^{-1} in frequently inundated lagoonal and freshwater-dominated riverine mangals of *R. mangle*, *L. racemosa*, and *A. germinans* in south-eastern Mexico (Lynch *et al.* 1989).

The spacing of prop roots and pneumatophores, in relation to the strength of the tidal current, must play an important role in determining whether scour or deposition takes place. The roots and rhizomes in an *Avicennia* community in New Zealand, for example, are too far apart to retain much of the mud washed between them, and accretion is less than 0.2 mm yr^{-1} (Chapman and Ronaldson 1958). Rods of variable spacing have been used to simulate mangrove pneumatophores in northern Queensland (Spenceley 1977). The results suggest that small rootlets help to bind and stabilize sediment, rather than providing a quiet environment for its initial deposition. In Victoria, however, there has been sustained accretion under an *Avicennia* mangrove fringe, where pneumatophore densities can range from 300 to more than 400 per square metre around the trunks (Fig. 8.9). In contrast, unvegetated mudflats seawards experience alternating periods of erosion and accretion. The use of pegs suggests that pneumatophores strongly influence depositional patterns by providing a calm, filtering environment, a conclusion that supports earlier work conducted in north-eastern Australia (Bird 1986).

Mangals may help to protect the coast from storm waves. A tropical cyclone caused little damage to coastal areas protected by mangroves in Queensland, for example, whereas up to 14 metres of recession occurred in adjacent areas where they were absent (Hopley 1974). Mangroves failed to protect the coast from storms in other areas, however, even though they did not suffer serious damage themselves. This situation occurred on Mauritius, where fringes of *Rhizophora mangle* along the sheltered portions of the coast were

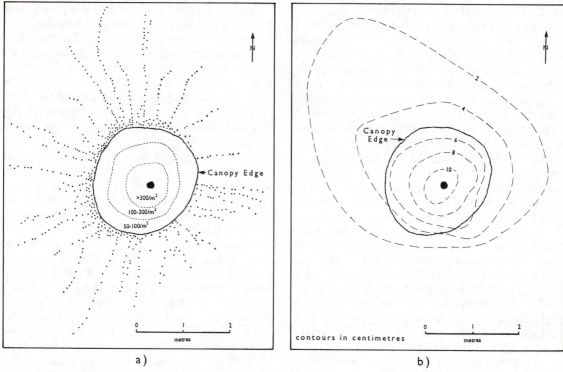

Fig. 8.9. (a) Pneumatophores and canopy, and (b) substrate contours of an *Avicennia marina* tree in Victoria, Australia. The leeward accretional spur at this site reflects the local prevalence of south-easterly wave action (Bird 1986)

almost unscathed by a cyclone, although the vegetation behind it was battered and washed out (Sauer 1962). In Western Australia, wave action is strong enough, in and behind mangals, to allow coarse clastic ridges to pass through them (Jennings and Coventry 1973).

The hydrodynamics of mangals are less well understood than in temperate marshes and estuaries. Peak ebb flows may be 20–50 percent greater than peak flood flows in estuaries with large fringing mangals. Strong ebb flows flush out sediment from creeks and help to maintain deep tidal channels, and groundwater flow, facilitated by decaying roots and biological burrows, helps to prevent excessive salt accumulation from evapotranspiration. Friction associated with high vegetational density in swamps retards flows, and encourages sediment trapping. Outflow velocity in a mangal in northern Queensland, for example, was at least 2,000 mm s[-1] at the mouth of a creek, following the flooding of the marsh surface by spring tides, while the unchannelized flow through the mangroves was less than 100 mm s[-1]. Mangrove density therefore exerts an important influence on the geometry of drainage channels. A mathematical model suggested that a decrease in vegetational density increases the export of sediment and results in erosion, creek widening, and accelerated meandering and braiding, whereas an increase in density causes siltation and narrowing of the creek (Wolanski *et al.* 1980, 1992).

Human Impact

Human activities are causing irreversible damage to mangals over large areas, although sparse populations and enlightened conservational practices have preserved essentially virgin stands in some places. The management of mangals has been hindered by the lack of reliable data on their value to humans, and they have usually been neglected and relegated to the category of wastelands. Agriculture, brackish water aquaculture ponds, and in some areas solar salt production compete with mangroves for the upper intertidal zone, and they have been reclaimed for housing, industrial estates, and garbage dumps near urban centres. Although mangals have been converted into rice paddies, the texture of the substrate is generally unsuitable for most agricultural equipment, and oxygenation of the previously aerobic soils makes them highly acidic (Christensen 1983; Tomlinson 1986; Adegbehin and Nwaigbo 1990).

Mangal detritus contributes to the marine food web, and many commercially important organisms spend crucial portions of their life-cycles in them. Crabs, fish, and shrimp are caught far away, however, and their importance may therefore not be fully appreciated. Although honey and wax are collected in some areas, their direct role as a food source is less important, and they are generally considered suitable only to prevent starvation. Although some fruits, seedlings, and other parts of the plants are edible, they are designed, through their leathery texture and high tannin content, to resist herbivores.

Mangrove wood is used for fuel, charcoal, poles, thatching, pit props, and railway sleepers. *Rhizophora* is particularly valued for firewood because its heavy wood burns with even heat, while producing little smoke. *Rhizophora* also produces high-quality charcoal, which is made on a commercial scale in south-eastern Asia. It is not good material for sawn-wood timber, however, and whereas *Heritiera fomes*, *Xylocarpus*, and other members of the back-mangrove community are more suitable, they usually grow in small, inaccessible areas that preclude large-scale commercial operations. Mangroves are commonly used to make poles for scaffolding and for the construction of rural houses, because of their durability and resistance to termites.

The low floristic diversity of mangals, combined with fairly cheap water-borne extraction, encourage their use for industrial purposes. In some areas, mangroves have been used to make matches, newsprint, and wood chips for the production of rayon and pulp. Sugar and industrial ethanol have been extracted from Nypa Palm, and its leaves are used for thatching and cigarette paper, and they may have potential uses in the production of high-strength paper. Oils extracted from the trees also have some limited medical applications. The development of synthetic and other alternatives, however, has generally ended the commercial use of mangroves for tannins and dyes, although they continue to be exploited for local purposes.

9 Deltas

INTRODUCTION

The term 'delta' was first recorded almost 2,500 years ago, when the historian Herodotus noted its use by the Greeks of that period to refer to the fertile, triangular-shaped, alluvial deposits at the mouth of the River Nile. Lyell introduced the term to the geological literature in 1832, in his *Principles of Geology*. Deltas played an important role in the development of classical civilizations. They provided food, water for irrigation, and because of economical access to the sea and the land they became focal points for commerce. Present geological interest in deltas partly reflects the fact that ancient deltaic sediments provided reservoirs for a large portion of the world's oil and gas reserves. Although large populations on deltas continue to testify to their agricultural and commercial value, they often suffer catastrophic natural disasters, from river flooding and marine inundation during storm surges.

Marine deltas are the product of fluvial and marine processes. They develop where the velocity of streams decreases as they flow into the sea, reducing their ability to carry sediment. They occur in a variety of environments, ranging from low-energy coasts with low tidal range and weak waves, to high-energy coasts with macrotidal ranges and strong waves (Fig. 9.1). Although some deltas are river-dominated and others wave- or tide-dominated, different mechanisms can operate on different portions of the same deltas.

Marine deltas develop where rivers supply sediment to the coast more rapidly than it can be removed by the sea. Tectonically active coasts and those close to drainage divides generally lack large drainage basins, and consequently large deltas. In a survey of fifty-eight major rivers with large catchment areas, almost 57 percent were along the trailing coasts of continents, and 34.5 percent along coasts fronting marginal seas and protected from the open ocean by island arcs. Only 8.6 percent were along tectonically active collision coasts, which usually have drainage divides close to the sea (Inman and Nordstrom 1971).

Deltaic sediments range from gravel to clay, depending upon the characteristics of the drainage basin and the sediment yield. Grain size influences stream gradient and channel pattern on delta plains; sediment mixing behaviour as it leaves river mouths; whether shorelines are dissipative or reflective; and deformation and resedimentation processes on subaqueous delta fronts (Orton and Reading 1993).

Although the annual sediment load in most large river systems increases with annual precipitation and discharge, its distribution may be more important than the total amount. Braided rivers, in which the generally fairly low annual discharge is concentrated into a short flood period, move more coarse material than meandering rivers, which usually have higher, more equally distributed discharges (Coleman and Wright 1975). This could explain why the braided, more-variable flow of the Mekong River carries more sand across its delta than the meandering Mississippi, although it could also be related to the much smaller size and higher average elevation of the Mekong drainage basin (Kolb and Dornbusch 1975). The surfaces of most large deltas are built up by annual flooding, but as the rainy season usually occurs in a delta before a river floods, sediment-laden flood waters may be denied access to areas that are already flooded with rainwater (Santema 1966). Inundation of the Irrawaddy Delta by heavy rainfall, before and during the start of the monsoon season, for example, prevents the silt-laden river overflowing and spreading out over the delta when it attains its bankfull stage. Most of the huge sediment load is therefore discharged into the sea (Rodolfo 1975). In tropical regions deltas

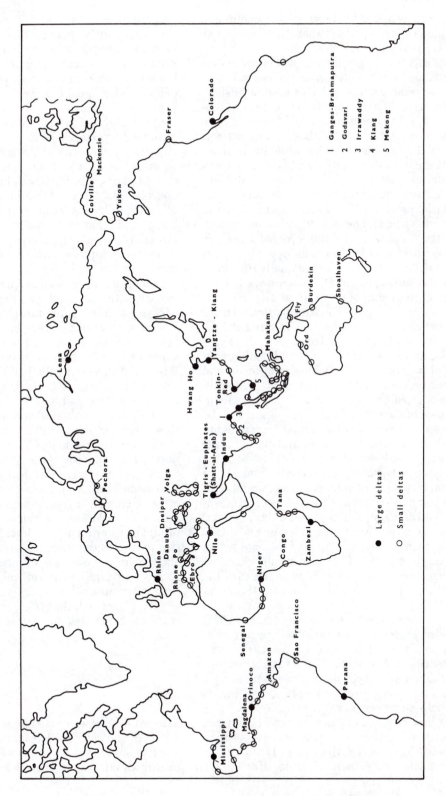

Fig. 9.1. Major world deltas

are generally flooded to much greater depths and for much longer periods than their temperate counterparts (Fosberg 1966).

The subaqueous portion of a delta provides the foundation for progradation of its subaerial form. Fine, suspended sediment builds shallow offshore platforms that favour rapid progradation (Galloway 1975). The finest material is deposited from suspension on the prodelta, the seawardmost portion of the subaqueous delta, while the coarsest sands are laid down at the mouths of active distributaries. Deltas prograde as long as sediments are deposited faster than subsidence and removal by marine processes. Coarser sand is laid down over finer prodeltaic and delta front sediments, producing the classical coarsening-upward sequence. Subaqueous levees develop on the uppermost parts of the delta front, and they eventually rise above the water surface as accretion continues, thereby extending the banks of the distributaries seawards. The subaerial portion of a delta is often a veneer over much thicker prodeltaic clays and deltaic front sands, silts, and clays. It can generally be subdivided into a lower plain that extends to the upper limit of tidal action and is influenced by fluvial and marine processes, and an older, fluvially dominated upper plain which is essentially a seaward continuation of the river valley.

Delta plains consist of accreting areas containing active distributary channels, and inactive areas that were abandoned when the lower course of the river adopted a shorter, more efficient route. Wave action, the effects of subsidence, and rapid sediment compaction often result in erosion of abandoned areas. Compaction of thick clay and mud accumulations between stream channels is usually greater than in the sandier sediments deposited in, or near, the channels. Old channels may therefore continue to be raised above the general level of the delta surface long after abandonment (Santema 1966). Deltaic coastal recession usually creates a coastal barrier, beach, or dune complex, but where subsidence rates are high, rapid marine encroachment results in deposition of shallow marine sediments across the abandoned delta surface, without significant reworking of the regressive sequence.

Natural levees, overbank splays, and minor channel systems occur next to the channels. Most of the area of active deltas, however, lies between the distributary channels. There are shallow open or closed bays on the Mississippi Delta and in other areas where there is rapid distributary progradation and subsidence, and low nearshore wave energy. Bays may be filled in with marsh or mangrove deposits, however, where there are slow rates of subsidence. There are numerous freshwater lakes between distributaries in many Arctic deltas, and evaporite or barren salt flats in the Shatt-al-Arab and other arid areas with a high tidal range. Tidally influenced deltas in wetter areas, including the Ganges–Brahmaputra, tend to have an intricate network of tidal creeks with densely vegetated mudflats. There are widely spaced, chenier-like beach ridges with broad, intervening areas of tidal flat or marsh on the Mekong and other deltas that experience periodically high wave energy and rapid progradation, whereas continuous sandy plains with closely spaced beach ridges or dunes develop where wave action is stronger and more persistent.

Delta plains are eroded and drowned by a rapid rise in relative sea level, and their lower alluvial valleys are filled with transgressive sediments. Coasts stabilize or prograde during stillstands or periods of only slowly rising sea level, and rivers begin to fill the lower portion of their valleys. Bayhead deltas eventually build out into shallow coastal bays, forming lacustrine delta complexes that evolve into shelf-phase delta plains on the continental shelf. Geological studies on the Mississippi River Delta and continental shelf show that there were several hiatuses in its development during the Holocene. At least three delta plains formed on the continental shelf during Holocene stillstands. These sediments are separated by transgressive erosion surfaces that formed when the depositional phase was terminated during periods of rapidly rising relative sea level. The data suggest that the cycle stops when sea level rises faster than about 2 cm yr^{-1} over a few centuries, resulting in submergence, the loss of wetland, and coastal erosion. Alternatively, when relative sea level rises at less than 2 cm yr^{-1}, the cycle constructs new wetlands, estuarine bays, and barrier islands (Penland *et al.* 1991).

PROCESSES

Seaward-flowing water spreads out into the marine waters of the receiving basin at the river mouth. The river mouth system consists of the outlet and, further seawards, a distributary mouth bar.

Effluent Dynamics

The outlet and distributary mouth bar influence, and are in turn influenced by, effluent dynamics. In river-dominated situations, effluent diffusion, sediment dispersion, and bar geometry depend upon the relative importance of (Coleman and Wright 1975; Wright 1985; Coleman 1982):

a) turbulent diffusion associated with the inertia of the issuing river water;
b) turbulent friction between the effluent and the bed immediately seawards of the mouth; and
c) outflow buoyancy resulting from differences in the density of the issuing and ambient water.

The relative importance of the three forces varies according to outflow discharge and velocity, outlet geometry, amount and grain size of the sediment load, and the degree of density stratification.

The term 'homopycnal' is used to refer to outflows of essentially the same density as the water in the receiving basin, and 'hypopycnal' for outflows of significantly lower density than basin water (Bates 1953). Hyperpycnal outflows of very saline issuing water, or water with very heavy suspended sediment loads, are denser than basin water and therefore sink beneath it. This can occur in hot climates where, because of high evaporation rates, the outflow from estuaries and lagoons is sometimes more saline and denser than sea water (Chapter 7). There are, however, little data on the depositional processes and patterns of hyperpycnal environments.

The degree to which effluent behaves as a turbulent or buoyant jet is largely determined by the densimetric (interfacial) Froude number F':

$$F' = \frac{u}{(ygh')^{0.5}}$$

where u is the mean outflow velocity of the upper layer (in the case of stratified flows); h' is the depth of the density interface; and y is the density ratio, as given by:

$$y = 1 - \frac{\rho_f}{\rho}$$

where ρ_f and ρ are the density of fresh and sea water, respectively.

Low F' values suggest that buoyant forces are dominant, whereas high values are indicative of inertia-dominance and turbulent diffusion. Turbulence and inertia are suppressed when F' is less than or close to 1, and fully turbulent effluent diffusion takes place when F' is equal to, or greater than, 16.1.

The simplest river-mouth model is probably fully turbulent, homopycnal outflow that experiences no interference with the bottom. Inertial forces are dominant when outflow velocity is high, density contrasts are negligible, and there is fairly deep water immediately seawards of the mouth. The effluent spreads and diffuses as a fully turbulent jet, and turbulent eddies cause the issuing and ambient water to exchange fluid and momentum. Turbulent jets of inertia-dominated effluent have a lateral spreading angle of only about 12°, and lateral sediment dispersion is therefore limited to a narrow zone. As the outflow disperses and decelerates, the coarsest sediment is deposited around the lateral margins and ends of the core of constant velocity. This produces a narrow, lunate-shaped distributary mouth bar (Bates 1953) (Fig. 9.2). Under ideal Gilbert-type conditions, deposition of suspended sediment forms horizontal bottomset beds. As the river mouth progrades, basin-dipping foreset beds develop as the coarser bed load is deposited just beyond the point at which the effluent begins to spread out. Advancing channel deposits eventually cover the foreset beds with horizontal topset beds. Gilbert-type deltas are common in marine areas where coarse bed load is deposited in sheltered and fairly deep water. Steep foreset slopes are usually absent, however, where suspended sediments accumulate in shallow basins and form gently sloping deltas.

High outflow velocity and flow inertia, especially in the wet season, are partly responsible for the narrow, single-finger delta of the monsoonal Solo River in north-eastern Java (Hoekstra 1993). Inertial affects are usually matched or exceeded, however, by the effects of turbulent bed friction or buoyancy. Deposition gradually reduces water depths at the mouths of nearly all rivers, introducing the effects of bottom friction and restricting turbulent diffusion and expansion of the effluent to the horizontal plane. Friction acting with lateral or plane jet diffusion increases the deceleration rate of the effluent, and the lateral spreading angle increases to about 16° to 17°.

Rapid expansion of the effluent in friction-dominated, homopycnal river mouths initially forms a broad, arcuate radial bar. With continuing sedimentation, subaqueous levees develop beneath

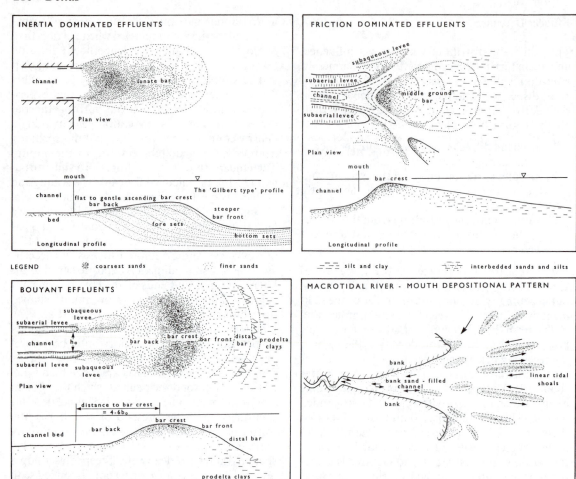

Fig. 9.2. River mouth forces and bar types (Wright 1977)

the lateral boundaries of the spreading effluent. The levees inhibit further expansion of the effluent, and channelization occurs as the central portion of the bar grows upwards. This takes place along zones of maximum turbulence, which tend to follow the levees, and a bifurcating channel develops with a triangular, middle ground shoal or bar between the diverging channels (Fig. 9.2). The middle ground bar, channels, and associated features of friction-dominated river-mouth deposition are similar to those of flood-tide deltas that form on the lagoonward side of tidal inlets (Chapter 7).

Where there is strong vertical density stratification in and near the outlet, river water spreads out as a thin buoyant layer over denser sea water, and is isolated from the effects of bottom friction (Chapter 7). This situation occurs, for example, at

the mouth of several distributaries on the Mississippi, Danube, and Po deltas. Other river effluents that are generally inertia- or friction-dominated may experience buoyant conditions at certain periods of the year. Buoyant effluents have pronounced frontal boundaries with the ambient water, and strong radial spreading and reflection of internal waves off this boundary can create a number of weaker, interior concentric fronts or rings of smaller radii (Garvine 1984).

Jet flows lose momentum and spread out faster where there is high bottom friction (Wang 1984). The lateral expansion rate of buoyant effluents is therefore greater than for fully turbulent jets, although less than for friction-dominated effluents. As the freshwater effluent decelerates and spreads out from the river mouth, deposition of coarse sands close to the outlet and finer-grained sediments further seawards creates a bar, which spreads out radially from the river mouth. The bar crest with buoyant effluents is situated further seaward than for other types of effluent, because of less rapid flow deceleration (Fig. 9.2). It provides an effective barrier to salt wedge intrusion in the Po River Delta, and it is usually able only to invade the distributaries at high tide and during periods of very low river flow (Nelson 1970). Internal waves cause intense salt water entrainment and vertical mixing in buoyant effluents. This results in rapid deceleration and deposition of coarser sediments over the bar crest. Super-elevation of lighter effluent over denser salt water drives internal circulation in the form of dual helical cells. Weak flow convergence near the bottom of the effluent and at the base of the helical cells produces narrow, finger-like bars. Secondary flows probably also inhibit divergent flaring of the subaqueous levees, encouraging formation of deep, narrow distributary channels (Wright 1985).

Grain size influences mixing behaviour at river mouths (Orton and Reading 1993). Fine-grained suspended load encourages salt wedge intrusion and development of buoyancy-dominated mixing patterns. Water depths are generally quite shallow in mixed load and bed load fluvial channels, and the sediment grade and large range of available grain sizes causes turbulent bed friction to become pronounced. In coarse-grained, bed load-dominated alluvial systems, the density of the sediment-charged outflow is often similar to the basinal water and, if the water is deep enough, avalanching of the rapidly deposited bed load results in Gilbert-type foreset beds. If flash floods and sediment gravity flows figure prominently, however, the outflow will be denser than the basinal water and initial mixing with ambient waters will be more limited.

Waves, Tides, and Currents

Waves can have a strong influence on effluent behaviour. Seaward-flowing effluent oversteepens incoming waves and frequently causes premature breaking. The outflow also helps to reduce wave phase speed, and may, according to the bottom topography, enhance wave refraction around it. Moderate to high waves breaking in the effluent region encourage intense vertical mixing, which reduces stratification and buoyancy effects. They increase the rate of effluent deceleration, and can temporarily trap and impound it, creating locally abnormal set-ups near stream outlets.

Waves erode most deltas, sorting and redistributing river-borne sediment, and remoulding it into beaches and barriers. Although deep-water wave climate partly determines the role of waves, slope gradient off deltas is of greater significance. High rates of deposition and rapid progradation can produce very low slopes in front of river-dominated deltas, but slopes tend to be much greater, and frictional attenuation of waves much less, off wave-dominated deltas.

Delta configuration and landforms depend upon the degree to which river-borne sediment is able to overwhelm the ability of waves to rework and redistribute it. Waves are most able to redistribute fluvial sediment where the quantities are fairly low. This could partly explain why the Rhone Delta, which has a similar wave environment to the Mississippi Delta but only about one-twentieth of its sediment discharge, lacks the latter's pronounced birdfoot appearance (Russell 1942). The ability of waves to redistribute fluvial sediment can be represented by the discharge-effectiveness index, the ratio of average discharge per unit channel width, to average nearshore wave power per unit wave crest (Wright and Coleman 1973) (Table 9.1; Fig. 9.3). Index values are high for river-dominated deltas. They have extended, digital distributaries and long, linear bodies of sand oriented at high angles to the coast. Areas between the distributaries are occupied by broad marshes, open bays, or tidal flats. With increasing nearshore wave power and decreasing river discharge, deltas become gently arcuate protrusions, with regular shorelines and often beach ridges. Wave-dominated deltas have low index values. They have the straightest shorelines and the best-developed interdistributary beaches, dunes, barriers, and beach ridges, oriented approximately parallel to the coast.

Temporal variations in the relationship between river discharge and wave power must also be considered (Wright and Coleman 1973; Coleman and Wright 1975). If, as on the Ebro and Danube deltas, maximum river discharge occurs when wave power is greatest, sediment is immediately worked on to the shoreline and deposited in beaches close to the river mouths. This should result in fairly regular and smooth progradation. If, as on the Niger and Nile deltas, maximum wave power occurs some time after maximum river discharge, sediment can

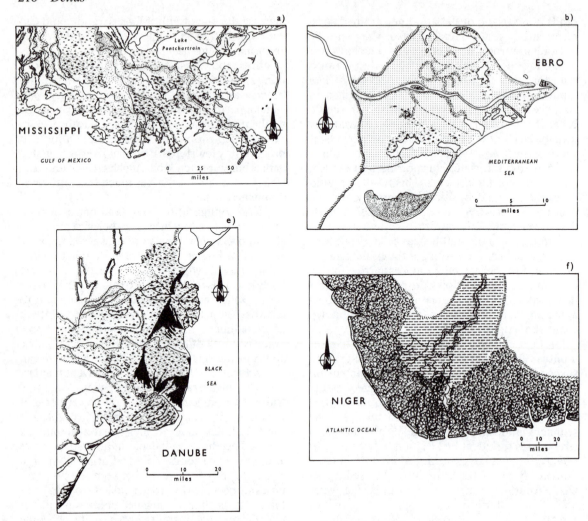

Fig. 9.3. The geomorphology of seven major deltas (Wright and Coleman 1973)

be carried along the coast by tides and currents, before being worked back onshore by waves. Deltaic coasts flanking river mouths are then characterized by barriers and beach ridges.

There have been several attempts to simulate delta development (Muto and Steel 1992). Komar (1973b) found that in wave-dominated environments changes in shoreline curvature maintain a condition whereby waves are just able to transport river-borne sand along the coast—lower angles being required with increasing distance from the river mouth because of the smaller amounts of sediment reaching these areas. A theoretical and experimental study suggested that when waves are

normally incident, deltas prograde symmetrically around river mouths, but asymmetrical growth occurs with obliquely incident waves, with faster growth occurring on the downdrift side. Deltas are eroded, particularly on the updrift side, when longshore currents remove sediment faster than it is supplied by rivers, and they may eventually be replaced by straight beaches, inclined so that waves tend to approach them normally (Refaat and Tsuchiya 1992).

Strong tidal currents modify depositional processes and patterns in macrotidal environments. Tide-dominated deltas have at least three characteristics (Wright 1985):

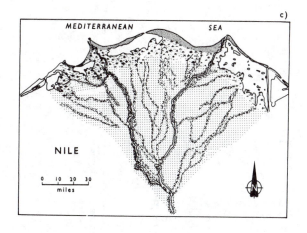

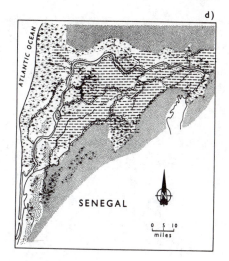

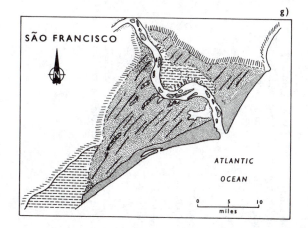

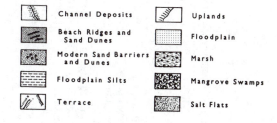

a) strong tidal mixing, which eliminates vertical density stratification and buoyant effects;
b) rapid bidirectional currents over the delta front and in river mouths; and
c) considerable vertical and horizontal variation in the position of the land–sea interface and the zone of marine–fluvial interaction.

Tidal currents usually transport bed load in ebb- and flood-dominated courses, separated by linear tidal ridges. Tidal ridges may replace distributary mouth bars as the main form of sand accumulation at the delta front (Fig. 9.2). Ridge crests are often exposed at low tide, and they may become predominantly subaerial as progradation continues. Although ridges normally develop approximately parallel to river outflow, they can be parallel to the coast at the mouth of rivers that flow into narrow straits.

Flood currents are generally stronger than ebb currents because of tidal wave asymmetry (Chapter 7). Bed load migrates upstream in tide-dominated areas, with ebb-dominated transport limited to deeper channels, where the flow is concentrated during falling tides. Sand brought in by the tide accumulates in river channels, and fills abandoned channels, whereas clay normally fills channels where the tidal range is lower. The rotational nature

Table 9.1. Delta morphology and the mean annual discharge–wave power relationship

River	Coastal and river mouth configuration	Delta shoreline landforms	Delta plain landforms	Nearshore wave power (erg/sec)*	Discharge ×1000 m³/sec	Discharge effectiveness index
Mississippi	Very indented coast with multiple extended digitate distributaries ('birdfoot' delta).	Indented marsh coast. Few, poorly developed sand beaches.	Marsh, open and closed bays.	1.34×10^4	17.69	1.00
Danube	Slightly indented coast with protruding river mouths.	Marsh, with sandy beaches next to river mouths.	Marsh, lakes, and abandoned beach ridges.	1.40×10^4	6.29	2.14×10^{-1}
Ebro	Smooth shoreline with single protruding river channel.	Low sand beaches and large spits with some dunes.	Salt marsh and a few beach ridges.	5.09×10^4	0.55	4.87×10^{-2}
Niger	Smooth, arcuate shore with multiple, slightly protruding river mouths.	Nearly continuous sand beaches along coast.	Marsh, mangals, and beach ridges.	6.59×10^5	10.90	8.03×10^{-4}
Nile	Gently arcuate, smooth coast with two slightly protruding distributary mouths.	Wide, high sand beaches and barriers with dunes. Beach ridges at distributary mouths.	Floodplain with abandoned channels and a few beach ridges. Hyper-saline flats and barrier lagoons near coast.	3.21×10^6	1.47	5.86×10^{-4}
São Francisco	Straight, sandy shore with one, slightly constricted, river mouth.	Wide, high sand beaches and large dunes.	Stranded beach ridges and dunes.	9.97×10^6	3.12	2.37×10^{-4}
Senegal	Straight coastline with extensive barrier deflecting river mouth. No protrusion.	Wide, high sand beaches and large dunes.	Large, linear beach ridges and swales, with dunes.	3.77×10^7	0.77	4.75×10^{-5}

Notes: * Wave power values are given per centimetre of wave crest. The discharge effectiveness index is normalized to maximum discharge.

Source: Wright and Coleman 1973.

of reversing tidal currents may favour deposition on one side of an outlet over the other. In front of the Colorado Delta, all the mud transported seawards by the ebb tide is moved to the western side of the Gulf of California by clockwise rotation of the tide, during the change to flood conditions (Thompson 1968; Meckel 1975).

Delta growth rates are highest where rivers carry large amounts of sediment into low-energy basins (Allen 1970). Although the Ganges discharges about 450 million tons of sediment per year, strong tides at the head of the Bay of Bengal prevent rapid growth of its delta. On the other hand, the deltas of the Godavari and other rivers on the eastern coast of India, where tidal currents are weaker, have been rapidly prograding (Nagaraja 1966). Ocean currents can also carry river sediments great distances along the coast. This occurs where currents are as strong as river outflow, so that there is little loss of sediment as the river enters the sea. Although annual sediment discharge of the Amazon is more than a billion tons, seaward protrusion of its 'delta' has been inhibited by the strong Guiana Current, which carries sediment north-westwards for hundreds of kilometres. Although most deltas have a subaerial and subaqueous component, the Amazon lacks subaerial expression, and the subaqueous delta, which extends for hundreds of kilometres offshore, may therefore represent the general case where large rivers discharge their loads in energetic environments that prevent accretion to sea level (Nittrouer *et al.* 1986). At a much smaller scale, ephemeral subaqueous deltas are formed by occasional river flooding along the wave-dominated coast of central California (Hicks and Inman 1987).

Wind

Along high wave-energy coasts with onshore winds, the entire subaerial delta may consist of wind blown sand moving over the delta plain as a broad transgressive sheet. Nevertheless, winds are generally most important in generating coastal currents that can move significant amounts of sediment (Coleman and Wright 1975). When strong onshore winds generate set-ups at the coast, surface currents moving shorewards are countered by bottom-hugging currents moving offshore, or subparallel to the coast. Set-down conditions, involving lowering of the water level at the coast

by persistent offshore winds, produce offshore surface currents that are countered by bottom-hugging currents moving onshore, or subparallel to the coast. Persistent alongshore winds tend to drive currents that move sediment away from river mouths.

Subaqueous Mass Movement

Creep, landslides, and other mass movements can play an important role in the development of the subaqueous portion of deltas. Bulk sediment movement creates features that are quite different from those formed by the individual movement of particles. In the latter case, slopes at the angle of repose are fixed, and transport rates adjust to maintain the slope. Slopes formed by bulk sediment movement are variable, however, and they adjust to provide the transport rate necessary to accommodate a given supply of sediment (Kenyon and Turcotte 1985).

A variety of subaqueous landforms testifies to the importance of mass movement on the Mississippi Delta. Nearly circular collapse depressions and more elongated slides have occurred on the floors of gently sloping, and interdistributary bays. On the upper parts of the delta front large, rotational slumps feed elongated mudflow gullies, with overlapping mudflow lobes further downslope. Larger gulley systems in the lower portions of the delta front feed mudflows that travel over the edge of the shelf on to the upper continental slope (Prior and Coleman 1979; Roberts *et al.* 1980).

Subaqueous mass movements are important on the steep, subaqueous slopes of coarse-grained deltas (Nemec 1990*b*; Kostaschuk *et al.* 1992*b*; Hart *et al.* 1992). Boulder avalanching, coarse-grained inertia flows, turbidity currents, slumps, and translational slides are dominant in the fiords of British Columbia, and they transport large amounts of bottom sediment for considerable distances. Subaerial debris torrents generate subaqueous debris avalanches if they reach the shoreline with enough momentum to continue below the water level. Delta morphology and sediment distribution are greatly affected by episodic, high-energy events, including earthquakes and seasonal flooding owing to rainfall, snow and ice melting, and occasional *jokulhlaups*, involving the sudden drainage of ice-dammed lakes (Weirich 1989; Prior and Bornhold 1990).

BARS AND CHANNEL PATTERNS

Five types of river mouth bar have been identified (Coleman and Wright 1975; Coleman 1982):

a) Radial bars are common at friction-dominated outlets with rapid effluent spreading.

b) Lunate bars have a deep scour pool immediately seaward of the river mouth, and a shoal or bar further seaward. They tend to develop in streams with high gradients, where inertial forces are dominant.

c) Middle ground or bifurcating bars with prominent subaqueous levees are usually found where friction and buoyancy are significant.

d) Subaqueous, jettied bars, with sands restricted to the immediate vicinity of the outlet, develop where inertial forces are dominant and buoyancy plays a minor role. The mouths of the rivers are usually deep and often intruded by sea water.

e) Tidal current ridges and swale bars, usually oriented parallel to the river channel, develop where there are strong, bidirectional currents (Chapters 7 and 8).

Deltas may contain more than one type of bar if different factors dominate at the mouths of their distributaries. The Bella Coola Delta in British Columbia, for example, has a radial bar at the high tidal distributary outlet, where frictional forces dominate, and a lunate bar at the low tidal outlet, where there are significant inertial, frictional, and buoyant forces (Kostaschuk and McCann 1983).

Depositional features at wave-dominated river mouths are similar to ebb-tide deltas (Fig. 7.9). A broad, sandy bar develops slightly seawards of a river mouth because of rapid decreases in flow velocity and sediment transporting ability, and it assumes an arcuate form as it is smoothed by waves. Rapid deposition on the flanks of the effluent produce broad, shallow subaqueous levees, akin to the ramp-margin shoals of tidal inlets. Dissipative surf over the shallow platform moves sediment shorewards in the form of swash bars that weld on to the shore adjacent to the mouth, and cause it to become constricted. The situation is similar for oblique waves, except that longshore currents and littoral drift skew the forms downdrift.

The contribution of distributary mouth bars to delta progradation therefore depends upon such factors as the number, density, and stability of the channels and their outlets. Three types of deltaic

channel pattern have been distinguished (Coleman and Wright 1975):

a) Systems that bifurcate successively downstream tend to occur where there is rapid subsidence, weak waves, low offshore slopes, small tidal range, and fine-grained sediment loads. Deltas have a large number of river mouths, and they are often so close together that their sand deposits merge to form a continuous sheet. Distributary middle ground and radial bars are most common. Examples include the Mississippi, Orinoco, and Volga rivers.

b) Deltas with channels that alternately bifurcate and rejoin tend to occur where there are erratic discharges, intermediate wave energy, high tidal range, and fairly steep offshore slopes. They have fewer active river mouths than in type (a), and distributary mouth sands rarely merge. These systems favour formation of subaqueous jettied bars, although some radial forms may also develop. Examples include the Irrawaddy and Nile rivers.

c) High wave action, high tidal range, and steep offshore slopes favour the development of systems consisting of a single or a few channels originating from a common point at the head of the delta. River mouths usually have bell-shaped and lunate bars or long, linear tidal ridges (Coleman 1982). Examples include the Mekong and São Francisco rivers.

RECEIVING BASINS, SWITCHING PATTERNS, AND DELTA MORPHOLOGY

The gross form of a delta is influenced by the geometry, tectonic stability, and rate of subsidence of the receiving basin (Coleman and Wright 1975). Its tectonic stability partly determines the growth rate and thickness of deltas. Thick accumulations can develop where subsidence rates are high, whereas deltas prograde more rapidly, but as thin sheets, over flat, tectonically stable platforms. Subsidence and compaction in the Mississippi and Ganges–Brahmaputra deltas, for example, have resulted in sediment accumulations of up to 120 m and 150 m in thickness, respectively. On the other hand, sediment accumulations in the Mekong Delta, which is underlain at fairly shallow depths by stable crystalline basement rocks, may be less than 60 m thick (Morgan 1970).

The topography of the continental shelf is also important (Coleman and Wright 1975). There are submarine canyons near the mouths of the Ganges–Brahmaputra, Indus, Congo, São Francisco, and other modern deltas. Large amounts of sediment are trapped and funnelled offshore by these canyons, and deposited in deep water as enormous submarine fans. The fan at the base of the canyon off the Ganges–Brahmaputra Delta is several times larger than the subaerial delta, and another off the Rhone Delta contains unconsolidated sediment more than 1,000 m in thickness.

There is an apparent, though not yet quantified, relationship between delta shape, channel migration, switching patterns, and the geometry of the receiving basin. Three types of migration pattern have been identified (Coleman and Wright 1975) (Fig. 9.4):

a) Some rivers, including the Huanghe (Yellow, Huang Ho) and Ganges–Brahmaputra, have changed their courses well upstream, and occupied new alluvial valleys. Channel switching generally occurs where there are intermediate shelf slopes, high, persistent wave energy, and high tidal range. Associated deltaic channel patterns are usually of the complex braided or single types.

b) Two or more courses may alternate in relative importance, so that one subdelta is eroding while the other is prograding. Switching occurs when the prograded distributary loses its gradient advantage through over-extension, so that river flow switches to another distributary. This pattern occurs where there are steep offshore slopes, low subsidence rates, and variable wave power and tidal range. Examples include the Danube and Red River deltas.

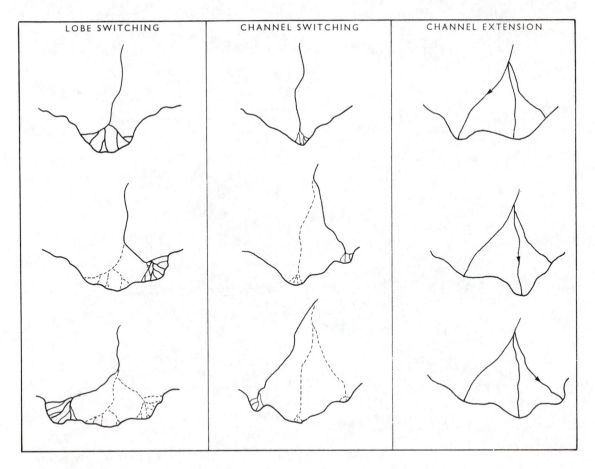

Fig. 9.4. Delta switching patterns (Coleman and Wright 1975)

c) As an outlet advances seawards, the river's path to the sea lengthens, and its gradient and capacity to transport sediment decrease. This causes the bed of the river to be built up, resulting in more frequent flooding, and increased sedimentation on the interdistributary surfaces (Santema 1966). The lower course of the river eventually shifts to a steeper course when channels become overextended and inefficient. Lobe switching produces delta plains consisting of a series of overlapping and interfingering lobes. It is always associated with bifurcating channel patterns, and tends to occur where offshore slopes and wave power are low, and tidal range is generally less than 2 metres.

Many deltas consist of a series of active and abandoned depositional lobes. Four have been identified on the Nile Delta (Fig. 9.5), on the delta plain of the Spanish Ebro River, and on the Texan Guadalupe Delta, which is forming in a quiet, shallow bay behind a barrier island (Donaldson *et al.* 1970; Maldonado 1975; Coutellier and Stanley 1987). At least seven major lobes have formed on the Mississippi Delta within the last 5,000 years (Fig. 9.6). Each lobe experienced a cycle of growth and abandonment, followed by subsidence and coastal erosion (Kolb and Van Lopik 1966; Penland *et al.* 1988). Lobe formation and abandonment were accompanied by corresponding changes in

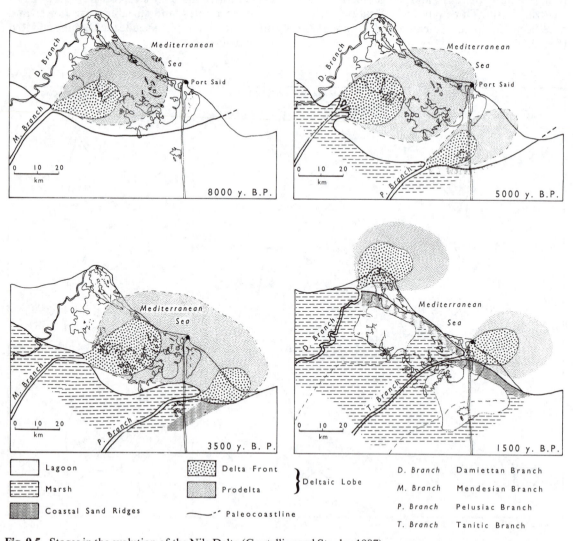

Fig. 9.5. Stages in the evolution of the Nile Delta (Coutellier and Stanley 1987)

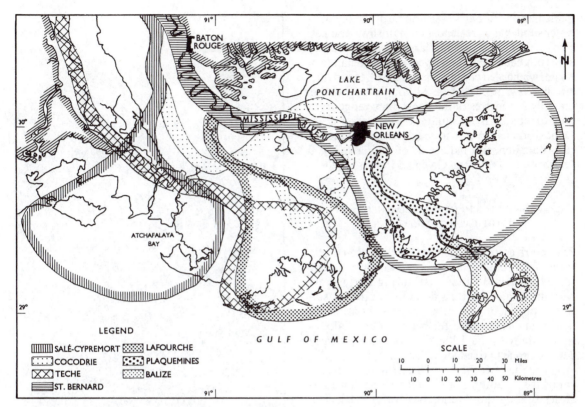

Fig. 9.6. Depositional lobes of the Mississippi Delta, listed in order of occurrence with the Sale–Cypremort and Balize the oldest and youngest, respectively (Kolb and Van Lopik 1966)

tidal prisms, sediment supply, and tidal inlet morphology (Levin 1993). The modern birdfoot delta, which has developed in the last 600–800 years, has almost completed the depositional portion of its cycle. The Atchafalaya River is now the main distributary of the Mississippi, and it carries about 30 per cent of its flow. It began building a marine delta in 1952, but it remained largely subaqueous during the first twenty years of its development. The floods of 1973–5 supplied large quantities of sediment, however, and encouraged formation of subaqueous and subaerial levees, resulting in the rapid development of a subaerial expression. Although the delta is still growing today, control locks prevent the Atchafalaya from capturing most if not all of the flow of the lower Mississippi, and they are therefore limiting the growth of the deltaic lobe into the bay (Shlemon 1975; Van Heerden and Roberts 1988).

There have been more than twenty large-scale changes in the position of the Huanghe River channel over the last 2,000 years, and its mouth has moved five times in this period (Xue 1993; van Gelder *et al.* 1994). Ten large superlobes, which have developed from changes in the lower channel, consist of smaller lobes that have resulted from shifts in the distributaries. Whereas each superlobe was active for a few hundred years, the constituent lobes on the modern delta complex were created and abandoned about every twelve years. Because of the very high sediment load, and the shallowness, microtidal range, and sheltered nature of the receiving basin, active lobes can prograde by as much as 1,500 m yr^{-1}.

Smaller lobes, on the order of 160 km^2 in area, infill the interdistributary bays on the Mississippi Delta. These shallow-water subdeltas or crevasse splays result from overflow and breaching of levees. The cycle is completed by waning deposition, sediment compaction, inundation by marine water, and the formation of open bays. They are therefore similar to the cycles experienced by the major deltaic lobes, although they generally require only between about 100 and 150 years for completion

(Coleman and Roberts 1989). Some of the shallow-water subdeltas were deposited in freshwater lakes. They have helped to fill in large interdistributary areas bounded by natural levees in the Atchafalaya Basin, which also contain meanderbelt, lacustrine, and backswamp deposits. Although these deltas are only 3–5 metres in thickness, they range from tens to hundreds of square kilometres in area. Deposition is rapid, with each deltaic sequence recording depositional events of only about 100 years duration (Tye and Coleman 1989*a*, *b*).

DELTAS IN GLACIATED AND HIGH-LATITUDE REGIONS

Sea and river ice assume important roles in arctic areas. Wide, braided streams carry coarse, poorly sorted material to the coast during fairly short periods of thaw. If rivers flood before the coastal ice has broken up, fluvial sediment is deposited on fast ice and later rafted further offshore or along-shore. Many Arctic deltas therefore lack a clearly defined prodeltaic sediment distribution.

The deltas of the Colville, Kuparuk, Saga-vanirktok, Canning, and other rivers have co-alesced along the Arctic coast of Alaska. The tundra-covered delta plain of the Colville River is dominated by shallow, oriented lakes, thaw ponds, pingos, thufur, and patterned ground. More than 40 percent of the annual discharge, and three-quarters of the total inorganic suspended sediment load, is discharged during a three-week period around the spring breakup. Much of the sediment therefore overflows over and on to offshore ice, al-though part of the sediment-laden water drains down through the ice (Naidu and Mowatt 1975). In the Mackenzie Delta in north-western Canada, shallow, interconnected lakes contain water trapped by higher ground around the intricate net-work of channels, although some may be thermo-karst features created by ground ice melting (Fig. 9.7). The delta can be flooded during breakup as a result of storm surges or ice jams (MacKay 1970). Ice jams are not, however, restricted to the Arctic. They form almost every year in the delta of the Danube River, for example, and are particu-larly hazardous in the vicinity of the river-mouth bars, where shallow depths hinder ice breakup (Vagin 1970).

Deltas often develop very rapidly in fiords as a result of intense erosion and sediment production

Fig. 9.7. Thermokarst and other lakes on the Mackenzie Delta, north-western Canada

(Bogen 1983; Powell and Molnia 1989). Meltwater discharging from en- or subglacial tunnels that terminate below sea level forms subaqueous out-wash fans, which may develop into ice-contact deltas. Deltas and associated sandurs at the head of Arctic fiords are strongly influenced by the lack of stabilizing vegetation, strong winds, glaciers, inter-mittent discharge patterns, and periglacial pro-cesses and landforms. Frequent channel switching tends to distribute sediment fairly uniformly across the front of sandur or braid deltas.

Deltas at the head of fiords in temperate regions have more of the characteristics of their open-ocean counterparts. Dense vegetation helps to stabilize river banks, and the channels tend to be fewer, but deeper and narrower, than in Arctic regions. Vegetational cover and wet climates also inhibit aeolian action. Temperate areas have a greater ability to store precipitation than Arctic regions, and their discharge events are longer and more complex. Chemical weathering is also more important in temperate than Arctic regions, and temperate rivers therefore carry much more clay-sized material. The head of temperate fiords are zones of high sedimentation and rapid delta formation and progradation. Their delta plains generally extend from one side of a fiord to the other, but their length varies according to the tidal range and the thalweg slope of the river. Broad radial mouth distributary bars form as a result of rapid bed load deposition at the lift-off point at the head of the salt wedge (Syvitski *et al.* 1987) (Chapter 7).

MODELS OF DELTAIC DEPOSITION

Deltas can be classified in a number of ways, according to the purpose of the classification and

the criteria that are used (Nemec 1990*a*). Genetic geomorphological classifications attempt to relate geomorphic features to natural factors (Volker 1966). Galloway (1975) identified three extreme forms representing fluvial-, wave-, and tide-dominated deltas, and the characteristics of the intermediate forms. This classification has been extended into a fourth dimension, to include the effect of the dominant grain size of the sediment delivered to the delta front (Orton and Reading 1993). Coleman and Wright identified six major types of delta on the basis of their gross sand body geometry, although most are probably intermediate between the extreme forms (Coleman and Wright 1975; Coleman 1982; Wright 1985) (Fig. 9.8).

a) River-dominated, type I deltas consist of elongated distributary mouth bar sands, oriented approximately perpendicularly to the overall trend of the coast. These protrusions have been termed bar-finger sands in the modern birdfoot delta of

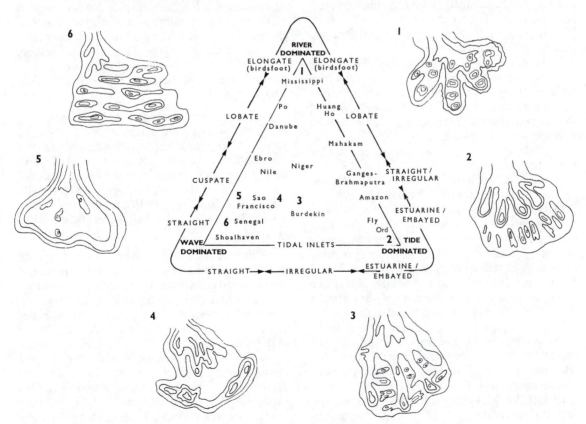

Fig. 9.8. Delta morphology and classification according to the relative importance of river, wave, and tidal processes (Wright 1985)

the Mississippi, where seaward progradation of the principal distributaries has created thick, elongated bodies of sand, up to 24–32 km in length, and from about 6 to 8 km in width. The sands are enclosed by huge wedges of clay and silt that occupy the intervening shallow basins. Differential compaction of the soft, water-rich, prodeltaic clays beneath the bar sands has produced a series of sand pods that are more than 75 metres in thickness. Bar sands at the mouths of active distributaries have also displaced prodeltaic and older shelf clays. Vertical displacements, sometimes of more than 100 m, have intruded the clays into the sand above, forming 'mud lumps' that may reach the surface and create small, low islands around the mouth of distributaries, sometimes almost overnight. Although most islands are soon destroyed by waves, a few have been incorporated into prograding marshlands (Morgan *et al.* 1968; Gould 1970). Type I deltas develop in areas with a low tidal range, very low wave energy, low offshore slopes, low littoral drift, and a high, fine-grained, suspended load. Examples include the Mississippi, Parana (Brazil), Dneiper (USSR), and Orinoco (Venezuela) river deltas.

b) Tide-dominated type II deltas have broad, seaward flaring and finger-like channel sand protrusions. They are fronted by sandy tidal ridges that are produced by tidal deposition and reworked fluvial sediments at distributary mouths. These deltas occur in areas with a high tidal range and strong tidal currents, low wave energy, and low littoral drift. Examples include the Ord (Western Australia), Indus (Pakistan), Colorado (USA), and Ganges–Brahmaputra (Bangladesh) river deltas.

c) Type III deltas are influenced by both waves and currents. Tidal currents form sand-filled river channels and tidal creeks oriented approximately normal to the coast, but waves redistribute riverine sand to form beach-dune ridge complexes and barriers parallel to the coast. Although one of the most common types of delta, they vary greatly according to the relative strength of the waves and tides. Type III deltas are found in areas with intermediate wave energy, moderate to high tides, and low littoral drift. Examples include the Irrawaddy (Burma), Mekong (Vietnam), and Danube (Romania) river deltas.

d) Type IV deltas prograde into lagoons, bays, or estuaries sheltered by offshore or baymouth barriers. They consist of finger-like bodies of sand deposited as distributary mouth bars, or they can merge to form an extensive sand sheet. Intermediate wave energy and low offshore slopes, sediment yield, and tidal range favour the development of type IV deltas. Examples include the Brazos (Texas), Appalachicola (Florida), and Horton (Canada) river deltas.

e) In type V deltas, river-borne sands are redistributed by waves to form extensive sand sheets. These deltas are characterized by high lateral continuity of sands, and delta plains largely consisting of beach ridges and dunefields. Distributary bar deposits are restricted to the immediate vicinity of the river mouth, and they are quickly resorted and remoulded by waves into the shoreface or beach. These deltas develop where there is persistent, moderate to high wave energy, low littoral drift, and moderate to steep offshore slopes. Examples include the São Francisco (Brazil), Grijalva (Mexico), and Godavari (India) river deltas.

f) The coasts of wave-dominated type VI deltas are completely straightened by waves. They consist of numerous sandy barrier spits trending parallel to the coast, alternating with fine-grained, abandoned channel fills. They occur in environments with strong waves, strong, unidirectional longshore transport, and steep offshore slopes. Examples include the Senegal (Senegal), Shoalhaven (NSW), and Tumpat (Malaysia) river deltas.

Traditional delta classifications have been mainly concerned with shallow-water deltas built by large, mixed, or suspended load rivers with low gradient distributaries. There has been a great deal of recent interest, however, in the form and development of coarse-grained deltas (Nemec and Steel 1988; Orton and Reading 1993; Chough and Orton 1995). Two major types have been distinguished (Fig. 9.9):

a) Fan-deltas develop where fans prograde directly into standing water. Fans, which represent the subaerial component of fan-deltas, are conical, lobate, or arcuate in shape, with a steep subaerial slope of between 1° and 5°. The sediment is deposited by channelized and unchannelized water and debris flows, mudflows, landslides, and other gravity-induced movements, and it is largely coarse-grained, angular to subrounded, and poorly sorted. Fan-deltas are found in a variety of tectonic and climatic environments, but they are most common along collision coasts, where there are young, high-relief mountains, narrow coastal plains, and steep offshore profiles or deep shelves.

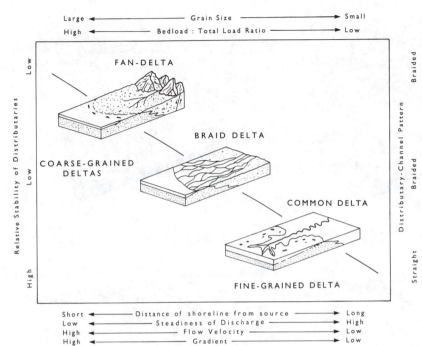

Fig. 9.9. Morphological, hydrological, and sedimentological differences between coarse- and fine-grained deltas (McPherson *et al.* 1987)

b) Braid deltas (Nemec 1990*a*) are produced by progradation of braided streams into standing bodies of water. They are not as steep as fan-deltas, their sediment is generally more rounded and better sorted, and boulders and cobbles are much less common. They are much larger than fan-deltas, ranging up to hundreds of km² in area, compared with tens of km² or less for fan-deltas. Braid deltas are also more common, particularly in middle and high latitudes where there is high precipitation, a ready supply of coarse sediment, and sparse vegetation. Braid deltas in these environments are often associated with glacio-fluvial sandurs.

Fan-deltas and braid deltas can develop from subaqueous conical deltas, created where fans or bed load-dominated rivers terminate in fiords or other deep-water environments. Conical, deep-water deltas essentially lack subaerial distributary plains, and virtually all the sediment is deposited below water. They are very steeply sloping, with gradients of as much as 35° in gravels. Huge amounts of sediment can be deposited, and they can prograde considerable distances without any subaerial expression, although cone growth and basin shallowing may eventually facilitate development of other types of coarse-grained delta (Nemec 1990*b*) (Fig. 9.10).

Postma (1990*a*, *b*) incorporated coarse-grained forms into a general classification of deltas prograding into low-energy basins. Four types of feeder system were distinguished, ranging from streams that largely carry gravel, to those with a large suspension load (Table 9.2). Each of the feeder systems can discharge into deep or shallow water.

Coarse-grained deltas often have Gilbertian foreset, topset, and bottomset bedding (Nemec 1990*a*, *b*) (Fig. 9.10). The most suitable conditions for the formation of steeply inclined foreset beds, where slipface processes dominate, include bed load transport to the river mouth, fairly large depths of water immediately seaward of the mouth, and inertia-dominated spreading of the effluent as an axial turbulent jet. If the water is too shallow, turbulent diffusion is restricted to the horizontal plane, and effluent diffusion is friction-dominated. This encourages formation of a delta or distributary mouth bar, with a gently sloping profile of less than a few degrees.

Deposition of coarse bed load on the front of deep water deltas causes the upper portions of the delta slope to advance faster than the lower portions. Deep-water, Gilbert-type deltas with type A and B feeder systems (Table 9.2) are therefore modified by gravity flows operating on their

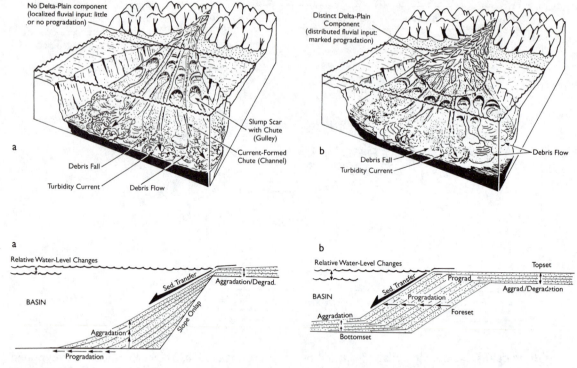

Fig. 9.10. Two types of steep-face, coarse-grained delta, (a) conical, subaqueous form without a subaerial distributary plain, and (b) Gilbert-type with a distinct deltaic plain (Nemec 1990*b*)

over-steepened fronts, and they acquire a tangential shape (Kenyon and Turcotte 1985). On the other hand, convex profiles may develop on fine-grained deltas if progradation of the delta front lags behind aggradation of the delta slope by settling suspended sediment. As these slopes tend to be unstable, however, the profile of the delta may also gradually become tangential downslope (Postma 1990*a*).

HUMAN IMPACT

Human activities have caused major changes in delta dynamics and morphology. Many deltas are receiving less water and sediment as rivers are dammed for irrigation, flood control, and power generation (Meckel 1975; Poulos *et al.* 1994). Because of damming and irrigation, the Ebro River is now unable to supply enough sediment to its delta, and coastal recession is taking place (Jiménez and Sánchez-Arcilla 1993). Until about 1905 the Colorado River supplied more than 150 million

tons of mud and sand each year to the Gulf of California. River diversions and especially sediment trapping by dams, however, have prevented much sediment reaching the Colorado Delta since that time.

Although much human activity is deleterious to deltas, it has sometimes been responsible for their formation or growth. The Ombrone River Delta in north-western Italy was initiated when the population growth in the Imperial Roman period led to deforestation, agricultural intensification on the adjacent uplands, and extensive soil erosion. The delta prograded by about 2 km during the Middle Ages, when the land was being severely exploited, although the Black Death, which halved the population of Tuscany, may have been responsible for an erosional phase during this period. The delta prograded by a further 2 km between the sixteenth and eighteenth centuries, when developing industry and rapid population growth resulted in further deforestation. The delta began to erode during the latter half of the nineteenth century, however, because of human activities including river-bed quarrying, land reclamation, reservoir

Table 9.2. Delta processes in low-energy basins

Feeder/ Distributary system	A	B	C	D
Physiographic setting/alluvial system	Faultblock, mountain fronts, fiord margins, very steep slopes and ephemeral discharges. Generally gravel-dominated.	Glacio-fluvial sandurs, steep braidplains and braided rivers. Often gravelly.	Moderate gradient, often vegetated braidplains and braided rivers. Gravel and sand.	Low gradient, commonly vegetated coastal plains and meandering rivers with well developed levees. High suspension/ bedload ratio.
Delta plain processes	Landslides, mass flows, unconfined stream flow.	Poorly confined and unconfined stream flow with bulk sediment transport during floods.	Mainly confined flow in fairly stable channels. Channel splitting at delta front through small mouth bars.	Confined stream flow in stable channels. Large mouth bar systems
Delta front processes	Land derived mass flows.	Inertia/friction-dominated outlet, hyperpycnal flow during floods.	Inertia/friction/ buoyancy-dominated outlet with mouth bars.	(Inertia) friction/ buoyancy-dominated outlet with mouth bars.
Delta slope processes	Land and delta front derived, confined and unconfined, hyper-pycnal and mass flows (debris flows).	Land and delta front derived, confined and uncon-fined, hyperpycnal and mass flows (turbidity currents).	Land and delta front derived, confined hyper-pycnal and sediment flows (turbidity currents).	Land and delta front derived, confined, hyperpycnal and sediment flows (laminar and turbulent).
Prodelta processes	Land and slope derived mass flow, density currents, hemipelagic sediment action.	Density currents, hemipelagic sedi-mentation. Common slope derived mass flow.	Hemipelagic sedi-mentation, turbid-ity and density (muddy) currents.	Hemipelagic sedimenta-tion. Density currents.

Source: Postma 1990*b*.

construction, decline in upland agriculture, and reafforestation (Innocenti and Pranzini 1993). Some deltas developed following river diversion. The Soto River in north-eastern Java was diverted to a new course in about 1890. The river, which has a large, fine-grained load, has subsequently built a single-finger delta that is already 12 km long, having increased in length at an average rate of 70 m yr^{-1} (Hoekstra 1993).

The Mississippi Delta, with its associated wetlands and estuaries, is a resource of national importance. It is under threat, however, as a result of uncoordinated decision-making and human interference. The natural, self-maintaining nature of the delta once guaranteed continuity of these resources. This is expressed, for example, in the way that new deltaic lobes replaced older lobes, with progradation and aggradation in the new lobes coexisting with deterioration in the old. Human activities have seriously undermined this natural balance.

The loss of wetland has accelerated during the last fifty years, and it now averages about 0.8 per cent per year. Losses occur through the development of shallow ponds by flooding, and erosion of the marsh and the expansion of bays between distributaries or deltaic lobes (Reed 1991). There are different opinions on the reasons for these losses, however, and the degree to which they can be attributed to human influences. Erosion is certainly a response, in part, to natural processes (Williams *et al.* 1991). The area is in a sedimentary basin that is naturally subsiding as a result of thermal cooling and downwarping of the crust, excessive sediment loading, compaction of unconsolidated sediments, and the withdrawal or solution of salt in deep, thick beds. Erosion can also be attributed to natural deltaic cycles involving channel switching and the formation, abandonment, and erosion of deltaic lobes.

The Cubits Gap subdelta or bayfill developed when a ditch, excavated in 1862 by the daughters of Cubit, an oyster fisherman, created a crevasse break. Nevertheless, most human interference has led to erosion and environmental degradation in the Mississippi Delta. These activities include deforestation of the enormous drainage basin, dam construction, and the building of levees and other attempts to confine the Mississippi River for navigation and flood control. These measures reduced the suspended sediment load by a little over 40 percent from 1963 to 1989. The sediment is now funnelled through three major passes into the deep waters of the Gulf of Mexico, and with the elimination of overbank flooding it can no longer build new land in the lower delta, or help to compensate for the effects of subsidence and erosion. The building of highways and the dredging and construction of canals for navigation and access to petroleum-exploration sites have also affected the distribution of flood waters and resulting aggradation of the marsh. Natural drainage courses have been blocked, and canals now allow salt water to penetrate into freshwater marshes. The extensive canal network in the coastal zone has caused major changes in salinity, runoff, and tidal exchange patterns. Further degradation and wetland loss have resulted from land reclamation and urban encroachment, and from increasing subsidence caused by the withdrawal of groundwater, and probably hydrocarbons from shallow oil and gas fields (Gagliano and Van Beek 1975; Coleman and Roberts 1989).

About 90 percent of the Egyptian population lives on the Nile Delta, and it contains almost all the country's cultivated land. Human activity has been modifying the delta since pre-dynastic time, but its greatest effect has been experienced during this century. The first Aswan Dam was built in upper Egypt in 1902, and was increased in height in 1920, and the Aswan High Dam was constructed in 1964. There has also been intensified irrigation, channelization, and land reclamation in the northern portion of the delta during the last two decades. The Nile once carried between about 120 and 140 million tons of sediment each year, approximately half of which was deposited along the Nile Valley and on the delta, and half was carried into the sea. Almost no fluvial sediment now reaches the delta and adjacent continental shelf. This has resulted in accelerated coastal erosion, and marine encroachment in areas that were rapidly prograding as recently as the late nineteenth and early twentieth centuries.

It has been estimated that the Aswan High Dam resulted in a three- to fivefold increase in average erosion rates on the delta. Erosion is most rapid on promontories formed at the mouths of the two active distributaries. The Damietta promontory is eroding at a rate of about 10.4 m yr^{-1}, and the Rosetta promontory by 106 m yr^{-1}. Nevertheless, the dam has only exacerbated a problem that existed long before it was built. Sediment deposition on the Nile Delta began to decline in the late nineteenth and early twentieth centuries because of a natural decrease in river discharge, and the effects of the first Aswan Dam and other artificial river controls. Delta erosion affects coastal cities, ports, and resorts, and it destroys valuable arable land. It is responsible for contamination of coastal freshwater aquifers, and it threatens barriers that separate coastal lakes from the sea. The possible rise in sea level associated with global warming poses a particularly severe threat to the Nile and other densely populated deltas that are already experiencing the effects of natural subsidence (Khafagy *et al.* 1992; Warne and Stanley 1993; Fanos 1995).

10   Coral Reefs

Coral reefs extend over about 2×10^6 km^2 of the tropical oceans, making them the largest biologically constructed formations on Earth. Of the areas covered by coral reefs and associated communities living down to depths of 30 m, about 30 percent are between northern Australia, Indonesia, the Philippines, and continental Asia; 25 percent in the Pacific Ocean; 24 percent in the Indian Ocean; 6 percent in the Red Sea and Persian (Arabian) Gulf; 14 percent in the Caribbean Sea and the North Atlantic; and 1 percent in the South Atlantic. More than half the world's coral reefs are therefore in south-eastern Asia and the Indian Ocean (Smith 1978).

Extensive reef development is largely restricted to the western sides of the tropical oceans. There are only a few small reef areas in the eastern Pacific—along the coast of Central America—and with the exception of the Houtman Abrolhos Islands off western Australia, they are much less abundant and well developed in the eastern than in the western Indian Ocean. They are also generally poorly developed in the Atlantic Ocean outside the Caribbean, although there are a few small structures on the coast of western Africa (Laborel 1974). The reefs along the coast of Brazil are largely composed of encrusted beachrock, coralline algae, and vermetid gastropods, although shallow bank reefs have formed through the coalescence of isolated, mushroom-shaped columns of coral.

Reefs consist of a living framework with internal cavities partially filled with sediment, and a surrounding area of reef-derived skeletal grains. The living portion of a reef, however, consists of only a thin organic veneer. There are therefore two basic processes in reef development, the building of the framework, mainly by colonies of stony corals and coralline algae, and production of unconsolidated or cemented carbonate sediment. Coral colonies consist of millions of living, closely packed polyps, ranging from about 1 mm up to 20 cm in diameter, according to the species. Once the framework has been formed, physical and bioerosional processes begin to break it down. The encrusting, binding, and cementation of the organic framework and its internal sediment increases reef relief and rigidity.

The terms 'hermatypic' and 'ahermatypic' (from the Greek *herma* meaning a rib or reef) have been used to distinguish: reef-building from non-reef building corals; shallow-water from deep-water corals; tropical from high-latitude corals; and those that contain dinoflagellate algae within their tissue, from those that do not. There has been some confusion over the fact that some non-zooxanthellate corals build reef-like structures, whereas some zooxanthellate corals do not. The suggestion has therefore been made that hermatypic should only be used for corals and other organisms that contribute significantly to the framework of reefs—as defined as biogenic structures rising to the surface of the sea (Schuhmacher and Zibrowius 1985).

Small, slowly growing deep sea corals without zooxanthellae survive at depths down to 6,200 m, where there is complete darkness and temperatures as low as $-1.1°C$. They occur at the edge of the continental shelf off Ireland, Britain, and France, and off the coast of Norway, where they have constructed coral banks up to 61 metres in thickness and almost 4 km^2 in area. They also grow in many other areas, however, extending from tropical to polar regions.

ECOLOGY

Corals are carnivorous suspension feeders that use their tentacles to trap living zooplankton in the water travelling over them. Although the main reef-building organisms are anthozoan corals of the class Scleractinia, minor contributions are made by some other corals, including the stoloniferan *Tubipora*, the coenothecalian *Heliopora*, and the hydrozoan *Millepora*. Scleractinian corals secrete an external skeleton of aragonitic calcium carbonate (Fig. 10.1). The column wall, oral and basal discs, and tentacles have three distinct layers of tissue: the ectoderm, mosogloea, and endoderm. In most reef-building corals the cells of the endoderm contain vast numbers of zooxanthellae (up to 30,000 cells per mm^3)—unicellular, yellow-brown, dinoflagellate algae of the family Dinophyceae (Wells 1956; Goreau *et al.* 1979; Muscatine 1990; Preobrazhensky 1993). The zooxanthellae, which live symbiotically with the coral, have several important roles:

a) they produce organic carbon through photosynthesis, which corals use as a source of energy;

b) they produce much more oxygen than they consume, but as corals do not appear to experience oxygen deficiency, this factor is probably of less importance than carbon fixation;

c) they promote calcification and growth of coral skeletons;

d) they provide a source of food for the corals; and

e) they help to remove metabolic coral waste.

Scleractinian corals employ a wide variety of reproductive and dispersal mechanisms. Sexual reproduction is usually seasonal, with breeding occurring during brief periods of the year. Most species spawn eggs and sperm for external fertilization and development, but planula larvae are brooded within the polyp in some species, and in a few of these cases the larvae are produced asexually. Other asexual processes create new colonies, including fragmentation of established colonies, budding and longitudinal or transverse fission, detachment of groups of polyps as drifting balls, and the dissociation and dispersion of single polyps from adult colonies.

Coral larvae have limited swimming ability, and their dispersion and settling patterns are largely determined by current strengths and directions, their position in the water column, and the duration of the planktonic phase. Brooded larvae are usually released at an advanced developmental stage, and they tend to settle much more quickly than externally developed larvae, which require at least four to six days before they are able to settle.

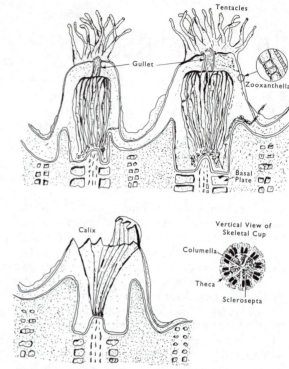

Fig. 10.1. Coral polyp anatomy and skeletal structure (Goreau *et al.* 1979; Hopley 1982)

Delayed settlement greatly increases the potential for externally developed larvae to disperse beyond their natal reefs. Wide geographical distribution of most scleractinian species is often attributed to larvae dispersal by ocean currents. Some workers consider that corals are poor dispersers, however, and they have proposed other biogeographical and evolutionary explanations, involving, for example, the effect of tectonism, glacial sea levels, and rafting by floating objects (Harrison and Wallace 1990).

When larvae find a suitable substratum, they attach themselves to it and metamorphose into a juvenile polyp. It has been generally assumed that corals only colonize firm substrates, including bedrock and other corals. Recent work, however, has shown that bank-barriers in the eastern Caribbean, fringing reefs in northern Queensland, and reefs in the Java Sea and the Bay of Batavia developed on sediments (Hopley and Barnes 1985; Johnson and Risk 1987; Macintyre 1988).

Corals are attacked by some types of mollusc, crustacean, fish, and echinoderm. The large, multi-armed Crown of Thorn starfish, *Acanthaster planci*, feeds on coral, and it has caused severe damage to reefs in the Indo-Pacific, although it has not been identified in the Caribbean Sea or the Atlantic Ocean. Periodic infestations, which have been occurring since the mid-1950s, and possibly since the eighteenth century, have been well documented on the Australian Great Barrier Reef. There has been widespread concern that recent outbreaks could have been triggered by human activities, possibly including reductions in predator populations by overfishing, and over-collection of the shell of the giant Triton *Charonia tritonis*, which is one of *Acanthaster's* main predators (Endean and Cameron 1990; Bradbury 1990). Today, however, most workers believe that outbreaks are a normal, recurrent phenomenon, reflecting particularly suitable climatic or oceanographic conditions for breeding and diffusion of larvae (Grigg 1992).

Recovery of devasted reefs largely depends upon the continued growth of survivors, and recruitment of planula larvae. The recovery period varies with the number of survivors, the availability of recruits, their success in establishing themselves, and their growth rates. Although recovery is often very slow, coral communities are more resilient than was once believed. On Guam, for example, coral communities devasted by *A. planci* in 1968–9 had largely recovered twelve years later (Colgan 1987). Rapid recovery of devasted reefs, however, has been disputed. There has been heavy mortality of massive corals on the Great Barrier Reef, including colonies up to several hundred years old. These large corals cannot be quickly replaced, and

complete recovery of a reef will therefore take as long as the age of the oldest members of the population (Endean *et al.* 1988).

Corals are able to assume a variety of forms, in response to the influence of a large number of environmental and genetic factors (Stearn 1982; Done 1983). The terms 'massive', 'ramose', 'foliaceous', and 'encrusting' refer to colonies consisting of thick masses or heads, branches, thin overlapping sheets with small basement attachments, and thin sheets on the substrate, respectively. The terms 'bulbous' and 'vertically platy' are also used to describe colonies with convex upper surfaces tapering down to a narrow base, and in the form of vertical plates, respectively.

Each zone on a reef contains a variety of coral shapes, but the growth form of single colonies tends to reflect the exposure and depth of the site. Nevertheless, growth forms cannot be simply attributed to one, or even a few, environmental factors (Chappell 1980*b*) (Fig. 10.2), and although distribution patterns can be identified on individual reefs, no universal pattern has been recognized. Reduction in light intensity with depth may be responsible for changes in the form of the Caribbean coral *Montastrea annularis*, which is generally massive and hemispherical in water less than 10 metres deep, and broad and foliaceous at greater depths (Fagerstrom 1987). Variations in water turbulence, however, may be largely responsible for depth-dependent changes in the growth form of branching corals. Hydrodynamic theory

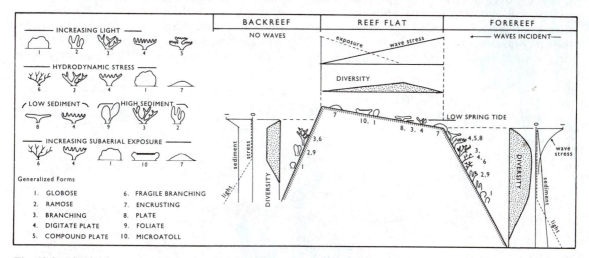

Fig. 10.2. Idealized representation of variations in coral community forms and diversity across a reef in response to changing environmental stress (Chappell 1980*b*)

suggests that corals can adopt one of two adaptive strategies to resist strong waves and currents. As wave- and current-induced stresses increase, branches can either be progressively eliminated until the colony has assumed a more compact, massive, and finally encrusting form, or they can reorient their branches to face the waves or currents. Corals exposed to strong wave action may therefore have one set of branches facing the forward and the other the backward directions of wave thrust (Graus *et al.* 1977) (Fig. 10.3). Corals also increase their resistance to high mechanical stresses by reducing the porosity of their skeletons. Branched colonies may develop strong, low porosity skeletons to compensate for their fragile geometry, although increases in effective porosity by even moderate organic boring can reduce coral strength by half (Chamberlain 1978).

Other organisms contribute calcium carbonate to coral reefs. Most of the main carbonate encrusting algae, of which *Neogoniolithon*, *Porolithon*, and *Lithophyllum* are the dominant genera, belong to the rhodophytic Corallinaceae family (Adey and Macintyre 1973; Bosence 1983). They are particularly successful on the high, turbulent crests of reefs, where grazing organisms and other competing fleshy algae cannot survive. The Chlorophytic (green algae) codiaceae family, of which *Halimeda* is the best known genus, is prominent in sheltered lagoonal environments in the tropics, especially in the Caribbean. The calcium carbonate contributed by green algae, as well as by molluscs, foraminifera, sponges, and echinoderms, largely accumulates on reefs as sediment. Bryozoans are not major constructors of the reef framework, but they are one of the first organisms to become attached to coral structures, which they strengthen by filling in cav-

ities and encrusting the underside of ledges and coral heads.

ENVIRONMENT

Large-scale physical factors, including tectonics, sea-level history, and temperature, determine whether and where reefs develop, and together with antecedant topography, they initially control the gross morphology of the reefs. At intermediate scales other variables, including light, wave energy, nutrients, sediment level, and predation, are largely responsible for the surficial character of reefs and the depositional fabric of their interiors.

Reefs can be dominated by one of four competing groups of sessile photosynthetic organisms. Corals tend to be dominant where there is moderate wave action, intensive grazing, and a very low nutrient supply. Microalgae tend to dominate reefs where there is low physical disturbance and low nutrient levels, whereas larger leafy or fleshy algae dominate where nutrient levels are high. Coralline algae, which are usually unaffected by moderate-to-high nutrient levels, tend to dominate environments with moderate-to-heavy grazing or wave action (Littler and Littler 1985).

Coral and other faunal species-richness decreases towards higher latitudes. There are, for example, about 330 coral species in the northern and central Great Barrier Reef, about 252 at the southern end of the outer reef (22°S), 237 at the isolated Bunker–Capricorn Group at the southernmost limit of the reef province (23°30'S), 111 in the southern Coral Sea (29°50'S), and only 60 on

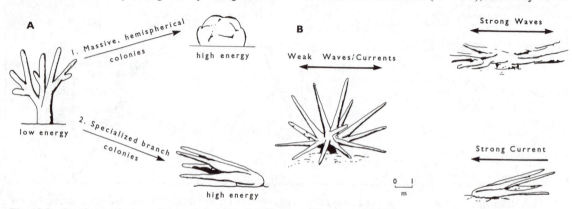

Fig. 10.3. Coral growth strategies to minimize hydraulic stress (Graus *et al.* 1977)

Lord Howe Island, the most southerly reefs in the southern hemisphere (31°33′S). There is a similar southerly decline on the western coast of Australia, although rapid decreases only occur near the thresholds of reef growth. The coral cover also decreases towards higher latitudes, and calcareous and fleshy macroalgae become more abundant with declining herbivore populations and grazing rates (Crossland 1988; Hopley 1989).

Sea temperature is the main factor determining the global distribution of coral reefs. The usual upper and lower limits for growth for most species are about 33°–34°C and 17–18°C. Water temperature lower than 18°C for at least part of the year is the main reason for the lack of reefs along the eastern coasts of oceans. Trade winds combine with the Coriolis force to generate currents that flow to the north-west in the southern hemisphere and to the south-west in the northern hemisphere. The warmer water carried away from the western coasts of continents is replaced by cooler water rising up from depths of 100 to 150 m. Upwelling water inhibits reef development along the eastern sides of oceans, while warmer water flowing across the oceans allows them to develop in anomalously high latitudes on their western sides. For example, reef development on Bermuda, the eastern coast of Africa, on the Ryukyu Islands off southern Japan, and on Lord Howe Island in the Tasman Sea can be partly attributed to warm water carried by the Gulf Stream, and the Agulhas, Kuro Shio, and East Australian currents, respectively.

Reef construction is inhibited where zooxanthellae do not receive enough sunlight for photosynthesis. Major reef building is therefore generally limited to the zone extending between the mean spring and neap low tidal levels down to depths of 25 to 40 m, although maximum growth for most corals occurs at depths of less than 10 m. In areas of fairly high tidal range, organisms on reef flats must be able to tolerate fairly regular exposure during periods of low spring tides. Widespread coral mortality can occur when very low tides coincide with heatwave conditions or heavy rainfall. Negative surges during tropical storms also lower water levels and expose reefs to heavy rainfall, and more prolonged exposure results when breaching of coarse clastic ramparts drains the ponded reef flats behind (Hopley 1982).

The expulsion of zooxanthellae from coral tissue causes a loss in colour, and in some cases mass mortality (Brown and Howard 1985; Williams and Bunkley-Williams 1988; Bunkley-Williams and Williams 1990). Bleaching can result from stresses induced by high or low temperatures, low water levels, turbidity, sedimentation, parasites, and changes in salinity, or human factors such as thermal discharges, dredging, and various forms of pollution. There are often considerable differences, however, in the occurrence and degree of bleaching between and within reefs, and among species (Gleason 1993).

The extent and increasing severity of recent bleaching cycles suggest that they may reflect global warming during the 1980s. There have been three major periods of global bleaching, in 1979–80, 1982–3, and 1986–8. The last, which affected corals in the Pacific, Indian, and Atlantic Oceans, and in the Caribbean Sea, was the most severe and widespread ever recorded. There was also widespread and prolonged bleaching and mortality in the Caribbean, and in the central, western, and eastern Pacific in 1982–3. This circumtropical event has been attributed to the meteorological and oceanographic consequences of a very intense El Niño–Southern Oscillation event, which resulted in five to six months of particularly high sea-surface temperatures. Resulting mortality levels in the eastern Pacific were greatest in areas that normally experience lower and more variable temperatures, rather than in areas with persistently high temperatures (Glynn *et al.* 1988, 1993).

Corals can survive where salinity (S) is from 30 to 38 (‰), and even up to S = 41. Growth is restricted, however, in the more saline water of very hot areas, and where there is low salinity near the mouths of large rivers. Water turbidity, which affects its clarity and reduces the depth to which corals can grow, is generally more important than salinity. It is especially significant in south-eastern Asia, where the Ganges–Brahmaputra, Yangtse–Chan Jiang, and other rivers supply about 70 per cent of the world's sediment supply to the oceans. Some corals are equipped to cope with high rates of accretion, whereas others are not. Coral mortality caused by a volcanic eruption in the Marianas, for example, varied according to differences in the degree of openness of the colonies to settling ash (Eldredge and Kropp 1985). Reefs affected by high turbidity in Puerto Rico have reduced coral cover, low species diversity restricted to effective sediment removers, and an upward shift in zonation with a shallow lower limit to coral growth (Acevedo and Morelock 1988).

Most reefs in the southern hemisphere, outside the equatorial zone, are affected by the south-east Trades, although there is seasonal variation in the Indian Ocean, with north-east Trades in winter and the south-west monsoon in summer. There are light, variable winds throughout the year in the equatorial belt off western Africa and central America, and from eastern Africa eastwards to the International Dateline. A variety of wave environments is created by superimposition of energetic swell waves and more local, Trade Wind-generated waves. Swell and Trade Wind waves generally approach from the south in the southern hemisphere, and there are marked differences in the leeward and windward side of reefs. In the northern Pacific, however, southerly swell and waves generated by the north-east Trade Winds provide roughly equal amounts of wave energy on the opposite side of reefs. Waves generated by the north-east Trades dominate in the western Atlantic, and they are probably enhanced by swell from the northern storm wave belt. North-easterly waves also dominate in winter in the northern Indian Ocean, but the south-westerly monsoon reinforces the southerly swells in summer. Southerly swell is probably dominant in equatorial regions. Major tropical storms are rather infrequent in equatorial regions, but they are important in many areas more than about 10° north and south of the equator (Stoddart 1971a).

The direction of wave approach strongly influences reef morphology. Reefs usually develop best on the windward side of islands, especially in areas that experience unidirectional Trade Winds. The edge of reefs facing the dominant wave direction tend to be straighter and more continuous than on the leeward side, which may consist of a series of isolated patches. Intermediate- to high-energy sectors around Grand Cayman Island in the Caribbean, for example, have reefs with steep fronts and well-developed spur and groove systems, whereas low-energy sectors have less steeply sloping profiles with patch reefs and extensive accumulations of sediment. Several biological explanations have been proposed for the better growth of coral and coralline algae on windward coasts (Wells 1956, 1957), but the main reason may be the inability of weaker waves to remove fine sediment in leeward areas (Adey and Burke 1977).

Waves reflect, refract, and break at the steep seaward front of coral reefs, and they are strongly attenuated as they propagate as bores through the shallow water over rough reef flats. Water depths of 2 to 3 metres occur over reefs at high tide in the Great Barrier Reef, but in the Caribbean, where the tidal range is rarely more than 0.5 metre, more than 80 percent of a wave's energy is lost on breaking at low tide, and about 70 percent at high tide. Nevertheless, wave breaking creates strong surges that are capable of carrying coarse sand across reefs under normal Trade Wind conditions. Infragravity waves can also modulate sea-surface slopes between the fore-reef and lagoon, and they may be important agents of mass transport across reefs (Roberts et al. 1988; Young 1989; Gourlay 1994; Symonds et al. 1995). Despite the effect of submerged reefs in dissipating storm wave energy, incoming wave groups can induce resonant, long-period oscillations with the natural frequency of the reef basin, producing considerable increases in wave height and damaging storm surges (Nakaza et al. 1990).

Although large areas do not experience hurricanes, including most of the equatorial zone, they are important elements of tropical regions, especially on the western side of oceans. Hurricanes and their equivalents can generate winds of more than 120 km hr^{-1}, with gusts greater than 300 km hr^{-1}; extreme wave conditions; storm surges that raise water levels by up to 5 m; and intense rainfall. Major tropical storms cause considerable damage to coral reefs. The effects can vary enormously, however, depending upon storm characteristics, proximity of reefs to storm centres, the height, size, and composition of coral islands, the occurrence and extent of beachrock and other cemented material, the amount of bioerosion, and whether areas behind the reefs have been cleared of natural vegetation (Stoddart 1971a; Woodley et al. 1981; Boss and Neumann 1993; Bythell et al. 1993). Storm damage is most severe on the outer edges of reefs that experience the full force of the waves. Reef islands (cays) can be completely removed, especially where they are low, small, and sandy, but other islands are affected by retreating shorelines, spit destruction, sand stripping, and tree uprooting (Stoddart 1974). In French Polynesia massive, shallow-water colonies are well adapted to strong wave conditions, and storm damage is therefore greater at depths ranging from about 12 to 20 m. Damage also occurs on reefs at much greater depths, when fragile colonies are shattered by large amounts of coral and other material rolling down steep slopes. In 1982–3 cyclone-generated avalanches destroyed 50 to 100 per cent of the living coral down to 40 m on the exposed sides of Tikehau and Takapoto atolls in the Tuamotus (Laboute 1985).

There should be fundamental differences between the structure, diversity, and history of reefs that are repeatedly devastated by tropical storms, and those that are not (Stoddart 1985). For example, one might expect the distribution of fast-growing ramose and slow-growing massive corals to be partly determined by the intensity and frequency of storm events. Damage is usually greatest to unattached corals and fragile branching forms, and least to massive corals growing on gently sloping surfaces; foliaceous corals are also suprisingly resilient.

Coral mortality can be delayed or sustained for several months after a major storm, and secondary mortality can be much greater than the initial damage. The rate of reef recovery varies according to the site, the degree of destruction, and the life-history strategies of individual corals (Laboute 1985; Brown and Howard 1985). Surviving polyps on dispersed fragments of skeleton can repopulate the substrate asexually, but if no live polyps remain, recovery is dependent on slow resettlement by planulae (Fagerstrom 1987).

Hurricanes throw up coral rubble that is later transported landwards by weaker storms, and eventually used for shoreline replenishment and island construction. Storms deposit boulders, coral rubble, and carpets of sand and gravel on reef surfaces. Huge blocks, generally consisting of single coral colonies, are dislodged from the reef-front below the wave-resistant upper few metres and thrown on to reef flats. Rubble zones and single and multiple ramparts are formed during tropical storms. In 1972 Cyclone Bebe generated waves that constructed a huge rampart of coral rubble and clasts, 18–19 km in length, 30–40 m in width, and up to 4 m in height, on Funafuti Atoll in Tuvalu. Low, discontinous rubble areas were formed initially in the more sheltered parts of the island, and other rubble zones developed later as the rampart migrated landwards and became lower. In 1958 the small but intense Typhoon Ophelia also created a number of temporary and more-permanent features on Jaluit Atoll in the Marshall Islands, including rubble and gravel ridges on the oceanward side of the reef flat; thick, often tongue-shaped, sheets of gravel across the islets; and erosional channels, often terminating in outwash features,

on the lagoon-side of the islets (Scoffin 1993).

Some reefs are in very active seismic areas, especially in south-eastern Asia. Damage occurs through ground- and water-movement, and in some cases from tsunamis. Although massive corals appear to survive hurricanes better than more fragile branching forms, the faster growth and more rapid regeneration of branching corals may favour them in areas of frequent earthquakes.

SURFACE FEATURES AND SEDIMENTS

The reef crest or algal ridge divides reefs into two main units: the outer or seaward portion of the reef, or fore-reef, which is dominated by waves down to about –15 m, and by currents at greater depths; and the inner or landward section, or back-reef, which is dominated by tidal currents and local turbulence. The outer slopes of coral reefs are generally very steep. They may be only a few metres in length where reefs rise from a sandy sea floor, but they descend for thousands of metres to the sea floor on oceanic reefs. There are terraces, up to several kilometres in width, on the upper few tens of metres of the slope in some areas, and there may also be caverns and overhangs in the upper few metres (Fig. 10.4).

The outer edges of reefs often consist of a series of narrow, sand-floored grooves and intervening, elongated spurs or buttresses, usually oriented at high angles to the reef crest, extending from the surface to depths of 10 to 20 m. On Bikini (now Pikinni) Atoll, for example, grooves are about 15 to 90 m long, 1 to 3 m wide, and from 2 to 7.5 m deep, and the spurs are 7.5 to 15 m or more in width (Tracey *et al.* 1948; Emery *et al.* 1954). The grooves are generally straight, but they can be forked, usually in a downslope direction, and a secondary system of grooves sometimes runs at right angles to the primary forms. Grooves and spurs are characteristic of fore-reefs exposed to high wave energy, and they act as natural breakwaters that help to disperse incoming energy. They are usually best developed on the windward side of reefs, but they are sometimes as well defined on leeward coasts, and can even occur on the shores of lagoons.

Some spurs, especially in the Indo-Pacific, are built of, or have a veneer of, coralline algae, whereas others, particularly in the Caribbean, consist of coral (Tracey *et al.* 1948; Shinn *et al.* 1981). In the

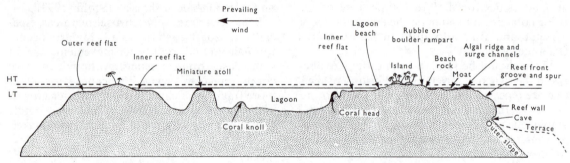

Fig. 10.4. Morphology of an Indo-Pacific atoll rim (modified from Stoddart 1969, 1971a)

Persian Gulf spurs developed from coral growth on tails of sediment formed in the lee of patch reefs (Kendall and Skipwith 1969). In other areas, however, spurs are residual features resulting from groove erosion, and are unrelated to the growth of corals or algae (Newell *et al.* 1951). Some spur and groove systems combine elements of erosion and construction. This could result from groove erosion by sand scour, while oxygen and nutrients in the turbulent water of the developing channels encourage organic accretion on the intervening spurs. Spurs and grooves in Belize and on the Great Barrier Reef may be growth forms inherited from rills and runnels formed by freshwater solution on the antecedent platform (Purdy 1974; Davies 1977). To account for the much greater size of grooves, compared with subaerial rillenkarren, however, it would have been necessary for some to have continued to grow while others were progressively eliminated. Spur and groove systems can therefore develop in a number of ways, but once an initial form has been produced, its relief is accentuated by coral and algae growth on the spurs, and, through the movement or deposition of sediment, erosion or prevention of growth in the grooves.

Some grooves continue landwards into or through the algal ridge, extending onto the reef flat up to 100 m from the edge of the reef. The growth of encrusting coralline algae over these surge channels facilitates the development of blowholes. Crater-like mounds of living algae, up to 9 m in diameter and 1 m in height, have developed around their mouths on Bikini and Eniwetok atolls (Tracey *et al.* 1948).

Coralline algae build low ridges on the windward side of mid-oceanic atolls and other reefs exposed to vigorous wave action (Adey and Macintyre 1973) (Fig. 10.4). They are particularly prominent in the central and western Pacific, but are less common in the Indian and Atlantic oceans. Algal ridges normally rise to less than 1 m above the low tidal level. They are up to about 25 m in width, and they have steep seaward and gentle landward slopes. Ridge height determines the degree of protection afforded to the coastline from wave attack, and it may therefore have an important effect on shoreline morphology. On Tongatapu Island, Tonga, low wave-energy coasts have low algal ridges, wide reef flats with numerous living corals, and many beaches. Surge channels are few and uncovered, and blowholes are poorly developed. High-energy coasts have high algal ridges, few beaches, and erosive shore platforms (reef flats) that are above high tidal level and lack living corals. Almost all the surge channels are covered with algae, forming spectacular blowholes (Nunn 1993).

Ridges of growing coral or algae may coexist with, or be replaced by, rubble zones, low-angled ramps of coral detritus with long, gentle seaward slopes and steep or abrupt landward slopes. Ramparts, up to several metres in height, develop where the rubble is better sorted, particularly on the windward side of reefs. Typical ramparts have a landward slope of between 40° and 60°, and a seaward slope of about 10°, and they often impound water on the reef flat at low tide. Coarse clastic tongues, up to hundreds of metres in length, may extend across the reef from their arcuate inner edges. Ramparts are modified by shoreward migration, breaching, and eventual stabilization by sea grasses, algae, and cementation. Sawtooth profiles or basset edges are formed by erosion of lithified ramparts, which exposes the arcuate foreset beds of the cemented clasts beneath.

Reef flats vary from less than 100 m up to several kilometres in width. Their inner, landward surfaces tend to be much looser, less cemented, and much more porous than the outer portions. They may be covered by thin, mobile sheets of sand, gravel, or rubble, and consolidated reef flat may alternate with long stripes of sand and gravel. Narrow zones of deep water (moats) occur in the outer portion of some flats. Indo-Pacific flats are usually regularly exposed during low-water spring tides, whereas exposure in the Caribbean is generally infrequent. Differences in reef flat exposure may reflect the generally smaller tidal range in the Caribbean, but mushroom-shaped rock remnants also suggest that Indo-Pacific reef flats are planed rock surfaces eroded from older, higher reefs. Coral growth is more important on Caribbean reef flats, but although there are also large amounts of *Halimeda* and other green algae, encrustation of loose debris by coralline algae and other organisms is more important in the Indo-Pacific.

Upward growth of massive subtidal corals stops when they reach the level at which they are exposed for prolonged periods during low spring tides. This produces microatolls, subcircular coral colonies up to several metres in width, with living sides and dead upper surfaces (Stoddart and Scoffin 1979). Most consist of massive, rounded corals, such as *Porites*, although branching corals create less compact forms. Microatolls are natural tidal gauges. New growth takes place over dead surfaces when relative sea level rises, and lateral growth produces small terraces on their sides when it falls. Changes in growth rates and forms can be determined from annual bands that develop within corals with variations in the density of their skeletons. This information has been used to date changes in Holocene sea level and moated water levels, and the occurrence of seismic events on tectonically active coasts (Woodroffe *et al.* 1990; Woodroffe and McLean 1990).

The sediment on coral reefs ranges from large blocks of reef-rock to sands and finer material derived from organic matter and mechanical disintegration (Milliman 1974; Boss and Liddell 1987). A distinction can be made between immature sediments, whose size, shape, and sorting largely reflect the skeletal characteristics of the responsible organisms; mature sediments that are roughly in equilibrium with dynamic conditions; and lag deposits that are incapable of being moved (Hopley 1982). The character of reef sediments strongly reflects the nature of the contributing organisms, and especially the micro-architecture of their skeletons. Coral clasts, for example, gener-

ally consist of sticks of branching corals, although more rounded clasts are derived from small, hemispherical colonies. Sand can be formed from foraminifera, the breakdown of coral sticks, and flakes of *Halimeda* and other calcareous algae, which tend to produce grains in the shape of flattened plates.

Beachrock, an indurated, sometimes casehardened, sedimentary formation, helps to stabilize beaches and islands on coral reefs (Scoffin and Stoddart 1983; Dalongeville 1984). Beachrock is formed by lithication of sediment in the intertidal and low supratidal zone, incorporation of bottles, cans, and World War II debris testifying to its rapid development in some areas. The main component is usually calcareous sandy material—with large amounts of *Halimeda* in the Caribbean and coralline algae and foraminifera in the Indo-Pacific— cemented by magnesium calcite and aragonite (Milliman 1974). Cementation occurs beneath the beach surface, and extensive outcrops therefore result from prolonged periods of beach migration. Beachrock often retains the typical 5° to 30° dip of internal beach laminations, sloping seawards as a series of flagstones up to a few metres in width. The rock is very thin in microtidal environments, but it can be up to several metres thick in macrotidal seas. The terms 'beach or reef conglomerate' are used for beachrock containing coarse clasts, including volcanic rocks on oceanic islands, as well as fragments of coral and older beachrock.

Differences in morphology, structure, and cement suggest that beachrock can be formed in a variety of ways. Cementation has been attributed to organic activity involving microbial action, decay of organic matter, or algae, and inorganic precipitation from groundwater or sea water. Most recent work supports the contention that beachrock is generally produced by precipitation of an aragonite cement from sea water. Lithification takes place at depth within the beach, following surface heating and evaporation, or sea water percolation. Cay sandstones, however, develop above the high water level, probably at the water table, with a calcite cement derived from ground water. Erosion and exposure of the horizontal layers of lithified sediment often form ledges on retreating coasts (Tayama 1952; Stoddart and Steers 1977).

Emerged reefs and lithified sediments may have been formed during periods of higher sea level, especially in the Holocene and during the last interglacial stage. Although there is growing evidence of Holocene sea levels 1–2 m higher than today in the Indo-Pacific Province (Chapter 2), however, it is difficult to distinguish ancient reefs a little above present low tidal level from contemporary conglomerate platforms formed from coarse clasts elevated during storms.

CALCIUM CARBONATE PRODUCTIVITY

It has been estimated that reef production precipitates, at least temporarily, about half the calcium delivered annually to the oceans. This is partly offset, however, by losses incurred through dissolution, bioerosion, and movement of sediment off reefs. Recent work on the Great Barrier Reef has shown that rates of bioerosion vary according to stages of growth. Dead coral on immature reefs experiences rapid rates of erosion by grazers, and a large proportion of the carbonate precipitated in the framework is reduced to sediment. Dead coral is protected on mature reefs by reduced grazing and high encrustation, resulting in increased rates of substrate excavation by boring. Boring alters rather than destroys reef frameworks, however, by creating internal sites for carbonate precipitation. With prolonged substrate survival, the growth history of reefs is characterized by complex cycles of boring, encrustation, sedimentation, cementation, and new framework construction (Kiene 1988).

It is important to know whether:

a) reef productivity (calcification) rates and vertical growth were fast enough to allow them to keep up with rising sea level during the Holocene;

b) whether productivity rates will allow reefs to survive predicted rates of rising sea level in the future;

c) whether productivity rates have changed since the sea reached its present level; and

d) whether variations in productivity across reef surfaces are responsible for variations in the thickness and relief of the Holocene veneer (Hopley 1982).

Estimates of the rate of calcium carbonate accumulation have been based upon the population and growth rates of reef organisms; the rate of calcium carbonate deposition in coral skeletons; changes in the reef topography; vertical accumulation rates from dated bore-hole samples; and

decreases in the alkalinity of sea water passing over reefs as calcium carbonate is extracted from it.

The diameter of individual massive corals can increase by 1 cm each year, and the length of the branches of branching corals can increase by 10 cm each year (Stoddart 1969). Smith and Kinsey (1976) estimated typical calcification rates of about 4 kg m^{-2} yr^{-1} for reef flats, and about 0.8 kg m^{-2} yr^{-1} for lagoons or complete reef systems. The latter figure was considered to be representative of the calcification rate of more than 90 percent of the world's coral reef area. Later work suggested that areas of very high cover can have calcification rates of about 10 kg m^{-2} yr^{-1}, whereas sand- and rubble-dominated areas have rates of about 0.5 kg m^{-2} yr^{-1} (Kinsey 1981). More-recent work has provided general support for these estimates (Kinsey 1985; Kinsey and Hopley 1991). There are few data on growth rates and productivity of individual species with increasing depth, although it is usually assumed that they rapidly decrease (Bosscher and Schlager 1992). In Jamaica the gross production of the massive coral *Montastrea annularis* is 1.025 kg m^{-2} yr^{-1} at a depth of 1 m, 0.510 at 10 m, 0.318 at 30 m, and 0.293 at 50 m (Porter 1985).

Although calcification rates become markedly seasonal towards the limits of reef growth, it remains to be determined whether annual rates decrease with increasing latitude. Present evidence is contradictory. On the Houtman's Abrolhos Islands off the coast of Western Australia, the average production rate is about 12 kg m^{-2} yr^{-1}, a figure that compares very favourably with rates in lower latitudes. Reefs may therefore disappear beyond certain latitudes because of increased biological competition, especially with macroalgae, rather than because of low productivity. On the other hand, calcium carbonate production declines with increasing latitude in the Hawaiian Archipelago, which is also near the latitudinal limits of reef development (Smith 1981; Johannes *et al.* 1983).

Assuming that calcium carbonate density is 2.9 g cm^{-3}, and the average porosity of reef sediments is 50 percent, calcification rates of 10, 4, and 0.5 kg m^{-2} yr^{-1} correspond to potential upward growth rates of 7, 3, and 0.35 mm yr^{-1}, respectively. These values are similar to recorded rates of growth, which usually fall between about 1 and 8 mm yr^{-1}. Higher maximum accumulation rates have been recorded, however, including 12 mm yr^{-1} on the Alacran Reef off the northern Yucatan Peninsula,

15 mm yr^{-1} on St Croix, and 14.3 and 20.8 mm yr^{-1} on a fringing reef on the Pacific coast of Panama (Adey 1975; Macintyre *et al.* 1977; Glynn and Macintyre 1977). Rates of growth are increased further by development of more porous reef frameworks under rapid growth conditions (Kinsey 1985). Although vertical growth dominates while reefs are submerged, lateral growth becomes more important when reef surfaces reach the low tidal level. Lateral growth has therefore been fastest during the last 5,000 years, with rates generally ranging from 2.5 to less than 0.2 mm yr^{-1} (Fagerstrom 1987).

The development of a reef is determined not only by growth rates, but also by the intensity of the destructive processes, including wave action and bioerosion, and sediment reworking and transportation. On a fringing reef on the western coast of Barbados, for example, annual production of about 206 tons of calcium carbonate is offset by a loss of 123 tons (Scoffin *et al.* 1980). On St Croix in the Virgin Islands average carbonate production on a fringing reef is 1.21 kg m^2 yr^{-1}. The reef fabric retains 0.91 kg m^2 yr^{-1}, of which 45 percent is in the form of detrital sediment. Of the 0.65 kg m^2 yr^{-1} of sediment produced by bioerosion, about one-third is probably washed away during storms (Hubbard *et al.* 1990). Unless most sediment produced by erosion stays on the reef, calcification measurements are not directly comparable to rates of vertical and lateral accretion recorded in bore-holes. In the Pacific, where the sediment carried landwards over reef flats generally stays on the reefs, the relationship between calcification and accretion rates is usually quite good, but it is more variable in the Caribbean, where sediment tends to be carried seawards down the fore-reef slopes into deeper water (Hubbard 1985).

Reef Turn-on and Turn-off

The relationship between calcium carbonate productivity and changing environmental conditions provides the basis for the concept of reef turn-on and turn-off, corresponding respectively to the initiation and termination of reef growth. Turn-on implies attainment of calcification rates that allow vertical and/or lateral reef accretion. It requires suitable environmental conditions for long-term reef growth, and an adequate breeding population of the appropriate organisms. The deviation of a single factor from its appropriate range can result in the loss of a suitable environment or the necessary biotic community, causing a conclusive and usually prompt turn-off.

The turn-on/turn-off concept operates at a variety of spatial scales. At the local scale, corresponding to individual reefs, growth could be initiated or terminated by changes in the location of a river mouth, or by movement of a tectonic block during a major seismic event. At the regional level, extending for hundreds of kilometres, turn-on could occur where long-term volcanic activity, associated with an oceanic hot-spot, creates a suitable foundation for reef growth, whereas turn-offs could be induced at this scale by changes in climate. The provincial scale covers tens of degrees of latitude or longitude. Changes in climate and sea level between glacial and interglacial stages could induce reef turn-on and turn-off at the provincial level over a few thousand years, whereas changes over millions of years could reflect the movement of suitable substrates into and out of areas that are conducive to coral growth. The effect of various factors may not be consistent throughout a scale unit. For example, whereas a rise in sea level may gradually turn-on a reef whose growth has been limited by sea level and the absence of a suitable substratum, it could gradually turn-off a reef that was near the lower depth limit of growth.

The turn-on/turn-off concept is also applicable at a variety of temporal scales. Short-term turn-offs can be generated by events of low duration, such as storm devastation or *Acanthaster* infestation. Although short-term events may have essentially temporary effects, they can combine with other factors to cause more permanent turn-off. Turn-offs can also result from events of longer duration, including fluctuations in weather patterns and anthropogenic activity. The effects of sea level changes, plate tectonics, and the evolution of scleractinian corals are felt over increasingly longer periods of time. Short-term events are largely expressed at the local, or at most, regional scales, whereas the impact of events of longer duration occur over much larger areas (Buddemeier and Hopley 1988).

Coral growth rates should have enabled reefs to keep up with rising relative sea level caused by sea floor spreading and other long-term geological mechanisms. Reefs could have drowned, however, during brief periods of rapidly rising sea level, or when environmental changes reduced rates of reef upgrowth. Higher nutrient levels in surface waters increase bioerosion, encourage the growth of fleshy algae, and by stimulating plankton growth, reduce available light. Increasing nutrient supply could therefore transform reef budgets from net deposition to net erosion (Hallock 1988). On the Great Barrier Reef long-term phosphate enrichment of a patch reef reduced calcification rates by more than half (Kinsey and Davies 1979). Potential calcification rates should also have allowed reefs to keep pace with rising Holocene sea level, and drowning may therefore be attributed to changes in the environment (Hallock and Schlager 1986). The drowning of barrier reefs off Florida and the Virgin Islands may have been related to increases in turbidity and nutrient levels caused by soil erosion on the flooded platforms (Macintyre 1988). At a much larger scale, more intense oceanic circulation and upwelling in the early Holocene may have caused world-wide increases in nutrient levels, and consequent supression of reef growth (Kinsey and Davies 1979).

THE ATLANTIC AND INDO-PACIFIC PROVINCES

Reef-building coral thrive in two main areas—the Atlantic and Indo-Pacific Provinces. The dominant reef-building corals and crustose algae are generally the same in the two Provinces, and they have the same relationship to morphological and energy zones, and roughly equivalent growth rates (Adey and Burke 1977). Although the number of identified species in the Caribbean continues to grow, however, the total is still much less than in the Indo-Pacific. The Indo-Pacific probably has at least 500 coral species from about 80 genera, compared with perhaps as many as 70 scleractinian species in the western Atlantic. Whereas two of the most important coral genera, *Acropora* and *Porites*, are each represented by three species in the Atlantic Province, there are about 150 of the former and 30 of the latter in the Indo-Pacific (Fagerstrom 1987). The greatest number of genera are around the Philippines, Indonesia, New Guinea, and the Great Barrier Reef—which alone has 350 species from about 60 genera (Hopley 1982; Veron 1985) (Fig. 10.5). Climatic modelling suggests that lower species diversity in the Caribbean, compared with the Indo-Pacific, did not result from greater cooling during the Pleistocene (Climap Project Members 1976). Part of the explanation may therefore be that fewer species simply reflect the much smaller area of shallow coastal water in the Caribbean (Adey and Burke 1977).

Reef development in the southern Caribbean is much better than in the cooler northern Caribbean. Seven genera tend to dominate Caribbean reefs: the scleractinians *Acropora*, *Montastrea*, *Porites*, *Diploria*, *Siderastrea*, and *Agaricia*, and the hydrozoan *Millepora*. The massive coral *Montastrea annularis* is the dominant species on many reefs, although the elk- or moosehorn coral *Acropora palmata* may be dominant in the northern Caribbean (Milliman 1973). Because of thicker branches, *A. palmata* can grow in shallower water and in areas exposed to higher wave stresses than the more fragile *A. cervicornis*. Because of fewer species in the Caribbean, the complex associations of the Indo-Pacific tend to be replaced by zones dominated by a small number of species (Geister 1977). Morphological zonation is also less well developed on Caribbean reefs, probably owing to moderate wave energy, low tidal range, and the fairly recent attainment of present sea level (Hopley 1982).

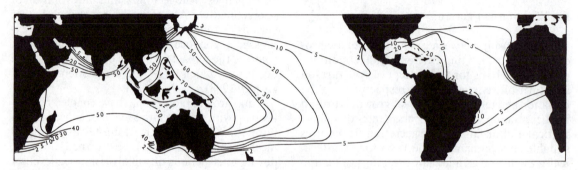

Fig. 10.5. Distribution of hermatypic coral genera (Veron 1985)

There are only about ten true atolls in the Caribbean (Milliman 1973). They have shallow lagoons (depths 10–15 m), and leeward sides that often consist of coral heads that rarely break the surface, rather than solid, continuous reef flats. The lack of Caribbean atolls may be attributed to the apparent lack of hot-spots, which produce atoll chains on migrating plates in the Pacific and possibly Indian Oceans, and the occurrence of late Tertiary volcanism and uplift, rather than subsidence, in the eastern Caribbean. Similar, non-subsiding areas near plate boundaries also lack atolls and barriers in parts of the Indo-Pacific. Karstic processes, during periods of lower sea level, may therefore have been at least partly responsible for Caribbean atolls that did not develop on subsiding volcanic basements.

It is now realized that Caribbean reef development and growth are comparable, proportional to its size, to the Indo-Pacific. Algal ridges, once thought to be absent in the Caribbean, are quite common where Trade Winds are strong and constant, especially between Martinique and Guadeloupe, where the reefs are open to North Atlantic swell (Adey and Burke 1976; Bosence 1984). The belief that the best reef development occurs on Pacific atolls has also been challenged by evidence that vertical accretion during the Holocene was two to three times faster on many Caribbean reefs than in the Pacific. This may reflect the generally greater depths of antecedent substrates in the Caribbean, and much higher wave energy levels in the Pacific (Adey 1978). Kinsey (1981) proposed that elevated antecedent platforms and the early stabilization of relative sea level allowed Pacific reefs to grow up to sea level before their Caribbean counterparts, resulting in slower rates of accretion. Differences in accretion rates are therefore probably the result of differences in sea-level history, tectonics, and wave environment, rather than reflecting any basic difference in the ability to precipitate carbonates in the two regions.

THE ORIGIN OF CORAL REEFS

Until fairly recently, discussions on the origin of coral reefs were dominated by the debate between advocates of Darwin's subsidence theory and Daly's theory of glacial sea level control. It is generally accepted today, however, that to account for reef development and morphology, one must consider the combined effects of subsidence and changes in sea level, during and preceding the Pleistocene epoch (Steers and Stoddart 1977).

The Subsidence Theory

In a series of articles published between 1835 and 1842 Charles Darwin proposed that as high volcanic islands slowly subside, fringing reefs are transformed into barrier reefs, and then, as the volcanic summits sink below sea level, into atolls (Fig. 10.6). The main contribution of this theory was to account for atolls rising from great depths on the ocean floor, even though corals only flourish near the surface.

The Glacial Control Theory

Daly was concerned with the effect of changes in relative sea level on reef development. In a series of

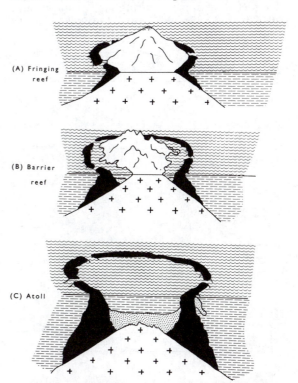

Fig. 10.6. Darwin's subsidence theory, involving the transformation of fringing reefs into barrier reefs and then into atolls on a subsiding volcanic foundation (Scoffin and Dixon 1983)

publications that appeared between 1910 and 1948 he proposed that the foundations of modern coral reefs were formed during periods of low Pleistocene sea level. The theory was based on the belief that the floors of lagoons on atolls and barrier reefs are fairly level and approximately accordant, at depths of between –50 and –90 m. Formation of these surfaces was initially attributed to marine abrasion during glacial stages, and later to deep, preglacial weathering. Whatever the origin of the erosional surface, however, it was envisaged that once sea level began to rise, coral recolonization and vigorous growth on the outer rim would produce a barrier reef or atoll, depending upon whether part or all the land mass had been planated (Fig. 10.7).

Although Daly drew attention to the effect of changing sea level on reef development, his interpretation of the responsible processes are no longer accepted. There are a number of basic problems with the glacial control theory, including the lack of cliffed headlands behind barriers, sea temperatures that were not low enough to prevent protective reef development during glacial stages, and glacial sea levels considerably lower than –50 to

–90 m in some areas. In any case, lagoon depths are much more variable than Daly realized (Tayama 1952), and they often have concave profiles that are inconsistent with marine erosion and cliff retreat (Steers and Stoddart 1977). Coastal and subaerial limestones probably erode too slowly to planate large emerged reefs during glacial stages. This is supported by the occurrence of an inverse relationship between the depth of drowned banks and their summit area in Hawaii. Because of marine and subaerial planation, small emerged banks were submerged below the minimum depth for coral growth during periods of high sea level. Larger, unplanated banks remained above the critical depth, however, and continued to provide foundations for coral growth (Grigg and Epp 1989).

The Antecedent (pre-existing) Platform Theory

Hoffmeister and Ladd (1944) proposed that atoll and barrier-reef formation can be attributed to coral growth on submerged platforms, especially

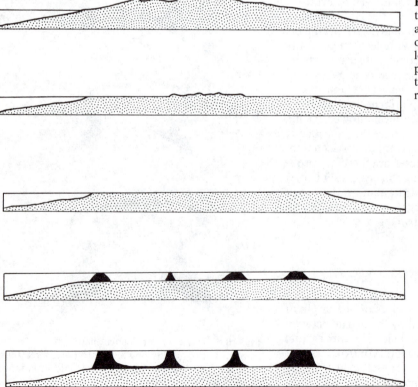

Fig. 10.7. Daly's theory for the development of atolls as a result of marine planation of islands during periods of low glacial sea level, and postglacial coral recolonization on the outer rim with rising sea level

around the favoured margins. The antecedent platform theory is based upon the premiss that reefs can utilize any surface as a foundation, as long as it is at a suitable depth. If climatic and ecological conditions are appropriate, reefs can then grow up to the surface without any change in sea level. Reefs grow on platforms with erosional, depositional, volcanic, and tectonic origins, and especially on top of older reefs that were subaerially eroded during periods of low glacial sea level.

The Karstic Theory

Proponents of the karstic theory consider modern reefs to be thin accretions over older reefs that were modified by karstic processes, while they were emerged. One can assume that all reefs, at or near sea level today, are growing on substrates that were emergent on many occasions during the Pleistocene. The karstic theory can therefore be classified as a variant of the glacial control and antecedent platform theories.

Purdy (1974) extended and modified the karstic theory. He proposed that the shape of many, if not most, reefs is karst-induced, although enhanced by rapid rates of carbonate deposition on the residual prominences. Rainfall solution would be most rapid towards the interior of the surface of exposed, steep-sided limestone platforms, and inhibited by rapid runoff around the edges. These karst-eroded surfaces, with raised rims and central solutional depressions, are transformed into atolls when drowned by rising sea level and colonized by coral. Barrier reefs may also be karst-induced. This results from lateral solution undercutting emerged reefs along their seaward-sloping contact with underlying, non-carbonate rocks. Seaward retreat of the limestone cliffs leaves behind flat, karst-marginal plains. Lagoons are formed when the marginal plains are flooded by rising sea level. Drowned karst towers within the lagoons provide foundations for the development of patch reefs, whereas cone karst and other intricate relief on the outer platforms form complex reef systems, and their solution rims create the outer barriers (Fig. 10.8).

Reef complexes consist of a variety of materials. Slower solution of the permeable rubble and open framework of reef fronts helps to create rims, whereas higher groundwater residence times and more rapid solution of the less permeable, lagoonal sediments produces karst plains (Strecker *et al.*

1986). The shape of postglacial reefs may therefore reflect differential solution of the various facies, which emphasizes constructional differences in relief between reefs and the floors of their lagoons. As subaerial weathering generally lowers limestone surfaces by less than 0.5 mm yr^{-1}, however, there may not have been enough time for karst landscapes of necessary relief to have developed during the last glacial stage. This is particularly true of fairly dry areas such as northern Barbados, where sinkholes occupy only about 1 percent of the area of the raised reefs (Day 1983). On the other hand, karst relief could have been acquired progressively during periods of low glacial sea level through the Pleistocene (Hopley 1982; Spencer 1985).

The rims of many islands and emerged coral structures in the Pacific appear to have resulted from solution of unrimmed carbonate banks. For example, lagoons on Mayotte Island in the western Indian Ocean, Truk in the Caroline Islands, Gambier Island in eastern Polynesia, and Clipperton Atoll off the Pacific coast of Mexico contain deep, steep-sided karstic pits and upstanding coral knolls. Further evidence of drowned karst is provided by the submerged sinkholes or blue holes off Belize, in the Bahamas and Florida Straits, and in a few places on the Great Barrier Reef. Mangaia in the Cook Islands is an eroded volcanic island surrounded by an elevated rim of limestone, 0.7–2 km in width. Volcanic rock is separated from the limestone by a series of swampy depressions, lying at the foot of limestone cliffs up to 60 m in height. The gross features are clearly erosional, and the junction between the rim, swamp, and volcano is characteristic of a karst-marginal plain. The barrier-like rim and associated depressions around the island therefore probably developed in the manner proposed by Purdy (Stoddart *et al.* 1985).

Modern reefs appear to mantle a highly pitted karst surface in the Sudanese Red Sea, and parts of the Great Barrier Reef may also be underlain by an ancient, dissected karst surface. Most reefs developed on topographic highs that had been subjected to karstic processes, but apart from a few blue holes, and possibly some inter-reef channels, there is no evidence of large karstic features, such as marginal plains and towers, on the Great Barrier Reef. Seismic and bore-hole evidence also suggest that the large barrier reef off Belize does not have an antecedent karstic origin. Reef foundations in the southern part of the region are diverse, and

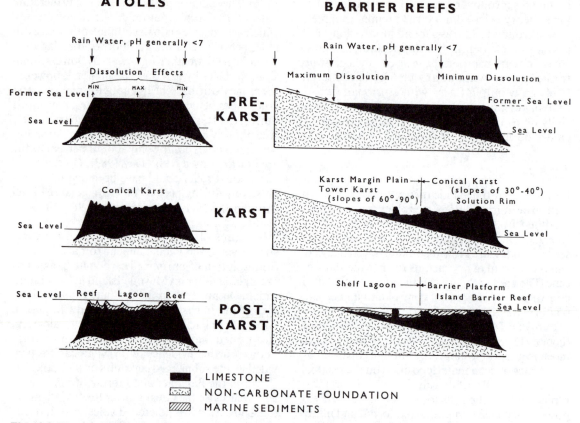

Fig. 10.8. Purdy's (1974) antecedent karst theory for the origin of atolls and barrier islands

include fluvial and deltaic sediments, as well as an eroded, possibly karstic basement. Holocene reefs grew on topographic highs consisting of late Pleistocene reefs and carbonate mounds, as well as on deltaic lobes in the inshore lagoon area, and possibly marine mud or sand banks in coastal areas (Choi 1981).

Bore-hole and Related Data

Deep drilling and geophysical surveys have made it possible directly to evaluate the subsidence theory, which implies that shallow-water reef limestones extend to considerable depths beneath atolls and, to a lesser extent, barrier reefs (Fig. 10.9).

Oceanic Reefs

The Royal Society (London) commissioned drilling on Funafuti Atoll in Tuvalu between 1896 and 1898. Two holes penetrated up to 343 m of limestones and dolomites, but although they did not reach the volcanic basement, later seismic work suggested that it lies at depths of between 550 and 770 m on Funafuti and nearby Nukufetau. Two holes drilled in 1947 on Bikini Atoll in the Marshall Islands also failed to reach volcanic foundations, despite penetrating to a maximum depth of 780 m. Geophysical data have indicated that the average depth of the volcanic basement on Bikini is about 1,300 m.

In 1951 Eniwetok in the Marshall Islands became the first atoll to be successfully drilled to its volcanic foundation. The island consists of more than 1,250 m of shallow-water limestones on top of a basaltic volcano that rises 3.2 km off the ocean floor. Deep drilling has subsequently identified great thicknesses of shallow-water calcareous formations on Mururoa, Fangataufa, Midway, and other atolls around the Pacific (Ladd *et al.* 1967; Guillou *et al.* 1993). On Mururoa Atoll, the site of

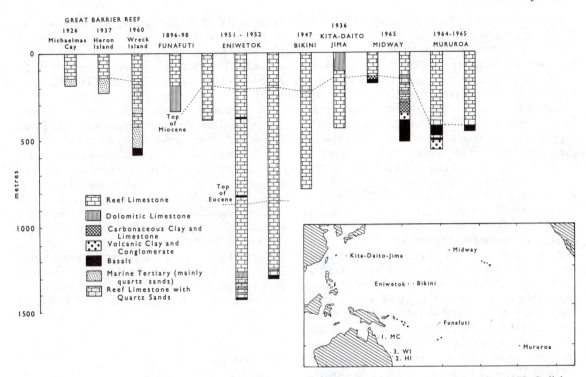

Fig. 10.9. Cores from Pacific atolls and the Great Barrier Reef (Ladd *et al.* 1967; Steers and Stoddart 1977; Guilcher 1988)

French nuclear tests, thirteen holes have been drilled to depths of more than 175 m, and about 100 have penetrated to between 15 and 30 m. Geophysical profiles have also been made, and probably more is now known about its development than any other atoll. The data obtained from this atoll provide strong support for Darwin's theory, with only slight discrepancies involving longer fringing and shorter barrier stages than are predicted by his model (Buigues 1985). Changes in sea level may be sufficient to account for the formation of the 80–100 m thick barrier reef of New Caledonia, and possibly the barrier on Moorea in the Society Islands, without recourse to subsidence (Chevalier 1973, 1977). Less data have been obtained from the Indian Ocean, but continual subsidence of an Eocene volcanic foundation has allowed more than 2,133 m of largely shallow-water carbonates to accumulate in the Maldives (Purdy 1981).

Great thicknesses of shallow-water limestones on oceanic atolls prove the essential validity of the subsistence theory. It does not necessarily follow, however, that atolls develop from barrier and fring-

ing reefs, or that their rims are produced by differential growth rather than karstification or other geomorphic processes. Local tectonic and environmental factors must have played an important role in determining reef structure on subsiding foundations. The elevated Pleistocene reefs on New Guinea and Atauro Island suggest that gently sloping or shelving reefs, lacking a reef flat, develop where relative sea level rises much faster than the growth rate, whereas a potential growth rate much faster than the rise in relative sea level produces an ever-widening fringing reef. Barrier reefs develop where there is less of a difference between the two factors (Chappell 1983).

Shelf Reefs

There has also been a great deal of subsidence on continental shelves. The carbonates underlying the Bahama Bank are much thicker than on any atoll which has been investigated. A bore-hole on Cay Sal went through 5,700 m of carbonates, and another, on Andros Island, went through 4,446 m of shallow-water, possibly entirely lagoonal, carbonates and evaporites (Dietz *et al.* 1970; Lynts 1970). Two bore-holes have been drilled through 959 m and 1,219 m of carbonate deposits beneath reefs off the Belize coast (Steers and Stoddart 1977). Several moderately deep holes have also been drilled in the Great Barrier Reef. The presence of between 115 and 150 m of shallow-water corals in the southern and central barrier reef province suggest that subsidence may have occurred in this area, although they are approximately within the range of Pleistocene changes in sea level. Nevertheless, the reefs largely developed on the topographic highs of the pre-Holocene surface, which probably resulted from the growth of reefs during former periods of high sea level (Hopley 1982).

The subsidence theory provides a far less satisfactory explanation for reef development on continental shelves than in deep oceanic regions. This can be partly attributed to the great size of continental shelves and slower rates of subsidence. Movements of the Earth's crust have been more important than eustatic changes in sea level in the long-term development of oceanic reefs, but eustatic effects have had a greater influence on the formation of thinner, younger shelf reefs. During periods of low sea level, coral migration down the steep sides of oceanic reefs may have allowed growth to continue at lower levels, although reductions in area and site variety would have caused severe environmental stress, and local species extinction (Stoddart 1976). The effects of low sea level were more severe on shallow continental shelves, however, where there was deep stream incision, karst development, and the death of almost all the coral (Fagerstrom 1987).

Discontinuities

There are submarine terraces on the seaward slopes of reefs in many areas, and in the lagoons of some atolls. Simulation modelling suggests that they were not cut by waves, but are the product of the interaction of growth, subaerial erosion, and subsidence during glacio-eustatic sea level cycles (Paulay and McEdward 1990). Although this suggests that their formation did not occur during periods of stable sea level, reef terraces can often be correlated with leached unconformities and calcite cements of freshwater origin identified in boreholes. These horizons were formed during periods of emergence, when solutional processes operated on exposed limestone surfaces. On Etiwetok, for example, a discontinuity at –12 m separates coral that is only 6,000 years old from coral about 120,000 years old. This break in deposition corresponds to the period of low sea level and reef emergence during the last glacial stage (Thurber *et al.* 1965).

The presence of near-surface discontinuities shows that Holocene limestones form only a thin veneer over older reef-rock. Holocene accumulations are from 7.4 to 30 m in thickness on the central and northern Great Barrier Reef, and between about 7 and 8 m on New Caledonia (Thom *et al.* 1978; Davies *et al.* 1985; Cabioch *et al.* 1989; Johnson and Risk 1987). The Holocene section is at least 22.7 m thick, and probably up to 33.5 m, in the Alacran Reef off northern Yucatan. The rapid accumulation of branching corals has also produced a Holocene coral cap more than 26 m in thickness on the Houtman Abrolhos Islands, off western Australia (Macintyre *et al.* 1977; Collins *et al.* 1993*a*, *b*).

Plate Tectonics

Darwin's subsidence theory can now be placed within the context of plate tectonics (Scott and Rotondo 1983*a*, *b*). Volcanic basements progressively subside as they move from mid-oceanic ridges or intraplate hot spots towards subduction zones (Fig. 10.10). Although the subsidence theory emphasizes vertical movements of the basement, horizontal migration has occurred at rates two to three orders of magnitude greater. Average migration rates for Pacific archipelagos, for example, have been between about 80 and 150 mm yr^{-1}, compared with rates of 0.02 to 0.03 mm yr^{-1} for Darwinian subsidence (Stoddart 1976).

Pacific islands are grouped into south-east–north-west oriented archipelagos, subsiding and increasing in age as the deep sea floor drifts to the north-west (Chubb 1957). A good example of the succession is provided by the Society Islands, which are moving west-north-westerly at about 125 mm yr^{-1}, from a hot-spot represented by at least four active volcanoes in the Tahiti–Mehetia region (Fig. 10.11). Mehetia, which last erupted in historical times, has a narrow, discontinuous fringing reef. It is succeeded along the axis of movement, from Tahiti to Maupiti, by more dissected and extinct volcanoes with fringing reefs, then by fringing and barrier reefs, and finally by the three small

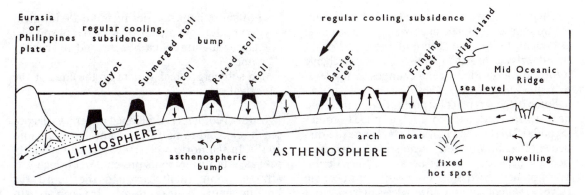

Fig. 10.10. Model of reef evolution on the Pacific Plate. Fringing reefs are progressively transformed into barrier reefs, atolls, submerged atolls, and guyots on the subsiding plate as it moves towards the subduction zone (Scott and Retondo 1983a, b; Guilcher 1988)

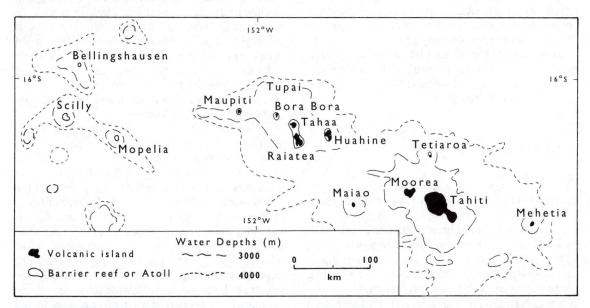

Fig. 10.11. The transformation of reef types in the Society Islands, as they move to the west-north-west (Brousse 1985)

atolls of Mopelia, Scilly, and Bellingshausen (Brousse 1985; Pirazzoli 1985). Similar successions occur along the Hawaiian chain and in the Caroline Islands, and although some other chains have more complex arrangements of islands and reefs, most coral structures on the Pacific plate are consistent with the general model.

Isostatic adjustment to the temporary load of a volcano on the lithosphere may produce a peripheral depression or moat, and a rise or arch further away: this effect could elevate and deform atolls (McNutt and Menard 1978; Scoffin and Dixon 1983; Spencer *et al.* 1987). Other anomalies arise where moving plates override bulges in the asthenosphere, which are associated with thermal rejuvenation at hot spots (Pirazzoli and Montaggioni 1985; Pirazzoli *et al.* 1988).

There are stairways of elevated coral terraces on New Guinea, Atauro near Timor, the Ryukyu Islands, Barbados, around the Red Sea, and in other regions of tectonic upheaval near plate boundaries. The spectacular series on the Huon Peninsula of Papua New Guinea, for example, consists of more than twenty fringing or barrier reef terraces, some with associated deltas, extending up to more than 600 m above present sea level (Chappell 1974). The dating of these and other raised reefs has provided some of our most reliable information on Pleistocene changes in sea level (Chapter 2).

Some biogeographical aspects of coral reefs can be structured within the concepts of plate tectonics (Stoddart 1976). The main obstacles to the dispersal of shallow marine organisms are land barriers and deep ocean basins. Because of their depth, areas of new crust spreading out from oceanic ridges represent zones of low species diversity and impoverished biotas. Continental mountain ranges close to subduction zones are also effective barriers to species dispersal. On the other hand, island arcs over subduction zones are major dispersal routes, and they tend to have very diverse biotas. The species diversity of coral reefs on the broad, tropical shelves and islands of south-eastern Asia may therefore be partly attributed to convergence of the Indo-Australian, Asian, and Pacific plates (Valentine 1971).

Changes in Relative Sea Level

All reefs probably emerged when sea level fell in the Tertiary, with formation of the Antarctic ice sheet. Tertiary reefs in subsiding areas eventually obtained a new veneer of coral as they sank below sea level. Karst may have developed on exposed reefs where subsidence was slow, and thick limestones accumulated where upward growth compensated, during interglacial high sea level stages, for subsidence during glacial stages. In stable and uplifted areas, new fringing reefs developed against the seaward slopes of the original platforms. These younger reefs were later exposed by further drops in sea level during glacial stages. New reefs developed at that time on the edges of the shelf, and grew up to interstadial sea levels, about 25 m below today's (Adey and Burke 1977). The windward side of many islands in the eastern Caribbean therefore have (Adey 1978; Macintyre 1988):

a) emergent reefs of Pleistocene and Tertiary age;
b) shallow Holocene bench or fringing reefs consisting of less than 6–10 m of coral and algae built on the margins of shallow, wave-cut Pleistocene terraces;

c) offshore bank-barrier reefs, 10–20 m thick, composed of coral or algae on bar or spit-like accumulations of carbonate sand and rubble; and
d) a submerged reef system on the edge of the shelf.

Modern coral reefs developed under a variety of sea level conditions (Davies and Montaggioni 1985). In the north-western Caribbean sea level rose asymptotically to its present level, which was attained about 2,000 years ago: the record was broadly similar in the south-western Caribbean. In the South Pacific there was a rapid rise in sea level to about 1 m above today's level, which was attained about 6,000 years ago, and a subsequent decline to its present position. The rise in sea level was even faster in eastern Australia, until about 7,500 years ago, when a more gradual rise brought it to the present level about 6,500 years ago. There is a lack of agreement over whether the sea subsequently maintained its present level throughout this period, or whether it rose about a metre higher and then fell to today's level.

Three types of reef have been distinguished in the Caribbean and western Atlantic, according to their response to the Holocene rise in sea level (Neumann and Macintyre 1985). Keep-up reefs grew upwards fast enough to keep pace with the rise in sea level. Most of these reefs therefore consist of shallow-water communities throughout their depth. Catch-up reefs gradually grew up to sea level. Those that began to grow in shallow water consist of an upward deepening sequence deposited when upward growth was less than the rise in sea level, and a shallowing sequence corresponding to a later period when upward growth had become faster. Catch-up reefs that originated in deeper water have only an upwards shallowing sequence. Give-up reefs fell behind rising sea level, and were eventually drowned. They therefore consist of an upward deepening sequence capped by deep water facies. Similar terminology has been used for the Great Barrier Reef (keep-up, katch-up, and screw-up), but in addition to the growth signatures, start-up refers to fairly recent colonization of substrates under significant depths of water (Davies and Montaggioni 1985; Davies *et al*. 1985).

Holocene start-up growth has occurred under more than 20 m of water on the Great Barrier Reef, but it has not been identified in the Caribbean, where a 10 m water column appears able to prevent colonization of the bottom. No screw-up strategies have been positively identified on the Great Barrier Reef, and although a few reefs were of the keep-up type, most growth was characterized by katch-up. Three types of katch-up reef have been distinguished: katch-up 1, which grew up to sea level before it stabilized; katch-up 2, which caught up with sea level only after it had stopped rising; and katch-up 3, which only began growing after sea level had stabilized. The atolls of the Tuamotu and Society Islands in French Polynesia developed in a similar way, with katch-up 1 and 2 strategies predominating. Tarawa Atoll in Micronesia and Eniwetok and Bikini Atolls in the Marshall Islands are also of the katch-up type, which appears to have been dominant in the Pacific, whereas keep-up, catch-up, and give-up types are common in the Caribbean.

TYPES OF REEF

Darwin's division of reefs into atolls and fringing and barrier forms is still the basis for many classifications, although some kinds of reef cannot be accommodated within this simple scheme. The system is least suitable for reefs on continental shelves, which are much more variable than oceanic reefs in shape, and biologic and geomorphic structure. Oceanic reefs have volcanic foundations, whereas shelf reefs usually have foundations of continental crust, which vary greatly in size, shape, and location. Furthermore, as shelf reefs are closer to major land masses, they experience greater variation in physical and chemical conditions (Fagerstrom 1987).

Fringing Reefs

Fringing reefs extend out from the coast. They tend to be narrow where the submarine slope is steep, and wide where it is gentle. They are usually distributed asymmetrically around islands where there is considerable variation in the strength of the waves, and they can be replaced by sandy tails on the more sheltered sides.

Fringing reefs contain numerous mini-lagoons in some areas. They have been attributed to enclosure by rapidly growing digitate corals, the closure of passes through the reef front, and karstic processes. Fringing reefs in the Red Sea are breached by sharum (singular sharm) or marsas, long, narrow, and deep embayments cut by rivers during periods of low sea level. They usually continue seawards as drowned valleys to depths of 40 m or more, and landwards as ephemeral streams or wadies.

Classical concepts suggest that fringing reefs simply grow outwards and upwards while attached to steeply dipping land. Drilling has shown that some originated offshore, however, eventually becoming attached to the land through infilling of the lagoon, or progradation of the shoreline behind (Easton and Olsen 1976; Braithwaite 1982b). This could occur where coral growth is inhibited at the shore by muddy fluvial sediments or sediment movement by storm waves. On the Ryukyu Islands, for example, growth of a reef crest had formed a barrier separated from the mainland by a shallow moat, about 5,000 years ago. There was then rapid growth on the back of the crest and expansion of the reef flat into the moat (Takahashi et al. 1988). Shoreward accretion over back-reef sediments has also been documented in the Caribbean (Shinn 1980, Macintyre et al. 1983, Macintyre 1988). Shoreward accretion explains why the interior of reefs often largely consists of accumulated detritus and reworked material, rather than in situ coral and algal skeletons. The degree to which fringing reefs are dominated by detritus or framework material may be partly determined by local wave energy. Fringing reefs in protected areas tend to be dominated by detritus, but they are largely composed of framework in high-energy environments (Macintyre et al. 1981; Davies et al. 1985; Montaggioni 1988; Hubbard et al. 1990).

Drilling and evidence from raised reefs show that most fringing reefs are veneers of coral over platforms with non-reef origins. The thickness of some fringing reefs, however, suggests that they have a much longer and more complex history. A few fringing reefs have been studied in some detail. A fringing reef in Hanauma Bay, a breached volcanic crater in Hawaii, began to develop about 7,000 years ago, during the Holocene transgression, but most upward accretion took place between 5,800 and 3,500 years ago, at an average rate of 3.3 mm yr^{-1}. Seaward growth has occurred at a rate of 22.2 mm yr^{-1} during the last 3,000 years (Easton and Olsen 1976). A fringing reef at Galeta Point in Panama is at least 14 m in thickness, and it has formed over a Miocene siltstone foundation during the last 7,000 years. Average accumulation rates of 3–4 m per thousand years for the main reef framework allowed it to keep-up with rising sea level, although there was a marked decline in deposition rates during the later stages of development (Macintyre and Glynn 1976).

Some reefs, which have been called almost-barriers, are intermediate forms between fringing and barrier reefs. Boat channels are fairly shallow depressions between a reef flat and the mainland (Tayama 1952; Bird and Guilcher 1982). They develop where the deposition of sediment from the land inhibits coral growth near the coast, although they occur in Saudi Arabia, where only a small amount of sediment is discharged from the land (Guilcher 1985a). Other explanations may therefore be appropriate in some areas. For example, the 25 m-deep boat channel off the Sudanese coast is a horst and graben structure defined by major faults. Karst activity during a period of low sea level may account for closed, steep-sided depressions, 10–28 m in depth, in a 1,000 m-wide fringing reef in the Red Sea (Braithwaite 1982a; Schroeder and Nasr 1983).

Barrier Reefs

Barrier reefs are fairly narrow, elongated structures, separated from the land by lagoons that are generally less than 30 m in depth. Some are fairly continuous, whereas others are fragmented into smaller units. There are notable examples in southwestern Madagascar, around 150 km off northern and southern Mayotte Island in the western Indian Ocean (Guilcher 1971; Coudray et al. 1985), around 1,500 km off northern, western, and southern New Caledonia, on Bora Bora, and along the northern side of the two main Fijian islands. The barrier on Guadeloupe Island in the Lesser Antilles is 30 km long, and the large barrier off Belize in the western Caribbean is more than 200 km in length (Stoddart 1965a; Bouchon and Laborel 1988). The bank-barriers of the eastern Caribbean, which are intermediate forms between fringing and barrier reefs, are separated from the shore by a lagoon up to several kilometres in width and 10 m in depth (Adey and Burke 1977; Adey 1978).

Some workers use the term 'almost-atoll' to describe barrier reefs surrounding islands that are almost completely submerged. Truk in the eastern Caroline Islands, which has about twenty high volcanic islands or islets protruding above the surface of the lagoon, is probably the best example in the world. Ladd (1977) has objected to the use of the genetic term 'almost-atoll', however, as well as to similar terms including 'almost-barrier reef' and 'near-atoll', which he considered to be undesirable in the absence of proof that reefs evolve through the three classical Darwinian stages.

There are submerged or partially submerged barriers in the Indian and Pacific Oceans between India and Sri Lanka, off the southern coast of easternmost New Guinea and north-western Madagascar, and off Tahiti, Fiji, Vangunu in the Solomon Islands, and Tutuila in Samoa. There are also submerged barriers in the Caribbean. One off south-eastern Florida is at least 95 km in length, and its crest is at depths of 15 to 30 m (Lighty et al. 1978). Another is at depths of 11–15 m off St Croix in the Virgin Islands (Adey et al. 1977; Macintyre 1988). Give-up reefs of the eastern Caribbean may have failed to develop into modern barriers because of high turbidity and nutrient levels caused by flooding and erosion of the soil and lagoonal sediments on the shelf. By the time normal conditions had returned, the water may then have been too deep to allow recolonization by *Acropora palmata*, the dominant shallow-water species in this vigorous wave environment. The more fragile *Acropora cervicornis* flourishes in the lower wave-energy environment of the western Caribbean, however, and the ability of this species to accumulate rapidly down to depths of 20 m allowed reefs to 'catch-up' to rising sea level. Rapid accumulation therefore resulted in formation of thick, late Holocene accumulations in this region, as on the barrier reef off Belize and on the Alacran Reef off the Yucatan coast (Macintyre et al. 1977; Macintyre 1988).

Multiple barrier reefs occur in some areas, including Mayotte Island in the Indian Ocean, and in the Pacific Ocean on Vanua Levu and Viti Levu in Fiji, New Caledonia, Truk in the Caroline Islands, Wallis Island in Polynesia, off the Queensland coast, and around New Georgia in the Solomon Islands, where a third barrier exists in places (Guilcher 1976; Pichon 1977). Inner barriers, which are usually much shorter than outer barriers, may have developed from fringing reefs formed on the subsiding mainland after formation of the outer barrier. Multiple barriers may develop when subsidence occurs in stages, separated by periods of faulting. The inner barrier on Mayotte grew on the landward side of a fault scarp that bordered the inner, subsiding area, and a similar explanation has been proposed for the multiple barrier reefs of New Caledonia (Guilcher 1976, 1988; Coudray *et al.* 1985). Alternatively, the inner barrier of Mayotte may be an expression of tower karst, resulting from the formation of a karst-marginal plain (Purdy 1974).

Conditions are usually much less favourable for coral growth in lagoons than on the wave-washed coasts of outer barriers. Therefore if inner barriers result from differential growth rather than solution, lagoonal reef development must have been facilitated by a variety of local factors. These could include: high tidal range and wide passages through the outer barriers to allow swell to pass over and through the barriers at high spring tides; strong currents moving through the passages; and lagoons that stretch in the direction of the dominant wind, therefore providing suitable conditions for wave generation and water circulation (Guilcher 1976).

Atolls

Atolls consist of fairly continuous reefs that surround one, or a number, of central lagoons. A total of 425 atolls have been identified in the world, not including some shelf or bank atolls. Most are in the Pacific, including eighty in the Tuamotus Archipelago, which has more than any other chain in the world (Chevalier 1973), and Kwajalein in the Marshall Islands, the world's largest, with dimensions of 120 km by 32 km. The greatest concentration in the Indian Ocean extends from the Chagos Bank northwards through the Maldives and the Laccadives (Stoddart 1973). There are only about ten atolls in the Caribbean Sea (Milliman 1973).

There are also many known examples of drowned atolls. Most are within a few tens of metres of the surface, although the Darwin Guyot, between Hawaii and the Marshall Islands, is at a depth of 1,253 m (Ladd *et al.* 1974). There are also thought to be several elevated atolls in the Pacific, ranging from tens of metres up to 200 m above sea level. It is still a matter of debate, however, whether their outer rims and central depressions are the result of karstification, or Darwinian subsidence and differential growth.

There are a few possible examples of double or nested atolls. Coral pinnacles create an inner ring in the lagoons of Nukuoro and Ant Islands in the eastern Caroline Islands, and on Taongi in the Marshall Islands, and they form a second inner ring on Greenwich in the eastern Caroline Islands, which therefore appears to be a triple atoll (Tayama 1952). Similar explanations, involving tectonic movements or currents entering lagoons through passages, have been proposed for nested atolls as for multiple barrier reefs.

The size and shape of atolls are controlled by the morphology of their volcanic foundations, which can consist of single peaks or clusters of cones (Wiens 1962). The basements of giant atolls, including Kwajalein, could consist of two volcanoes rather than one, and the same explanation could account for the figure-eight shape of Woleai in the western Carolines. Atoll shapes are commonly elliptical or oval, pear-shaped, or essentially rectangular with bulging corners, and they can therefore depart very markedly from a simple circular or annular form. Despite these variations, however, most atolls are elongated, with a mean length-to-width ratio between 1.5 and 3 (Stoddart 1965*b*).

The shape of atolls can be modified by landslide scarring, wave action, and other mechanisms. There are usually significant differences in the windward and leeward sides of atolls in the Trade Wind belts. Spurs and grooves provide sawtooth edges to reefs on the windward side of atolls, for example, where the surf is strong and persistent. Waves throw up sediment and form low isles on the windward side of atoll rims in the Society and Tuamotu Islands in the southern Pacific, but they are almost continuous around Diego Garcia in the Indian Ocean, where the winds are more seasonally variable (Stoddart and Taylor 1971).

Reticulated atoll and barrier-reef lagoons consist of a honeycomb pattern of basins and anastomosing ridges. The lagoon on Mataiva Atoll, at the western end of the Tuamotus, has about seventy basins averaging about 8 m in depth and ranging between 100 m to more than 2 km in diameter, and separated by ridges that are under 0.1–0.8 m of water. The basin–ridge pattern has been attributed to karstification of the reef while it was emerged (Delesalle 1985). The lagoon of Alacran Atoll off the Yucatan Peninsula is also reticulated (Purdy 1974), and there are other examples on the Great Barrier Reef (Maxwell 1968), in the Society Islands, and on Christmas Atoll in the Line Islands, where there are about 500 shallow, hypersaline lakes (Valencia 1977).

The lagoons of atolls and barrier reefs may be connected to the sea through deep passes that are navigable by ocean-going vessels. These passes, which were probably cut by streams draining lagoons during periods of low sea level, are generally restricted to the leeward sides of islands, in part because those in more exposed locations were filled in with sediment. The number of channels per atoll averages 1.5 in the Marshalls, 1.1 in the Carolines, and only 0.46 in the Tuamotus, where they are known as avas (Wiens 1962). In the Tuamotus, avas are up to several metres in depth, whereas hoas are shallower but much more numerous. Hoas are channels cut through atoll rims, possibly along joints formed by reef compaction. Some function as overflow or overwash channels, only connecting lagoons with the ocean during stormy weather, whereas others are graded to the level of the reef flat. Hoas run between elevated, linear islets or motus. In the Marshall Islands, hoas are less common and avas are more numerous than in the Tuamotus.

The distinction has been made between oceanic atolls built on volcanic foundations rising from the deep sea floor, and atoll-like rings of narrow, elongated reefs in the shallower water of continental shelves. Shelf- or bank-atolls are not as common as oceanic atolls, but they do occur in many parts of the world, including the Great Barrier Reef, north-western Australia, Indonesia, the western Caribbean, and the Gulf of Mexico (Ladd 1977). Although morphologically similar to oceanic atolls, shelf-atolls are genetically quite different. They tend to be crescentic in shape in the Caribbean, with open lagoons and very poorly developed reefs on their leeward sides, and, irrespective of their size, the depth of the lagoons only ranges from 5 to 15 m. The atoll-like shape of several reefs off the coast of Belize may have been derived from karstic foundations (Purdy 1974). As reefs grew up to the surface from antecedent platforms on the Great Barrier Reef, accretion occurred most rapidly on the windward sides, forming crescentic reefs. Atoll-like reefs then developed as shallow lagoons were gradually enclosed by sediment moved by winds, waves, and currents to the lee side of the reefs (Hopley 1982, 1983).

Bank Reefs

A confusing array of terms is used for platform-like reefs that lack prominent lagoons. Most workers classify them according to their size, but there is a lack of consistency in the use of the terms. Guilcher (1988) used 'bank reefs' as the collective term for these features. Some are very large. The Great Chagos Bank in the Indian Ocean, for example, is 150 km long and 100 km wide (Stoddart 1971b). The term 'patch reefs' can be used for small bank reefs. They are found in the lagoons of atolls and barrier reefs, although they are, perhaps, most numerous on continental shelves. Patch reefs are usually submerged, although their summits can extend up to the low tidal level. James (1983) reserved the term 'patch reef' for structures 5–50 m wide and 3–6 m high; 'table reef' for those 50–500 m wide and 5–20 m in height; and 'pinnacle reefs' for reefs 5–50 m wide and 6–20 m high.

The lagoons of some atolls and barrier reefs contain large numbers of pinnacles and knolls, usually with living coral on their summits (Steers and Stoddart 1977). Pinnacles are spire-shaped with vertical or overhanging sides, whereas knolls are mound-shaped structures that taper upwards, with side-slope gradients that are usually lower than 45°. Some knolls are less than 3–15 m in diameter, and 1–6 m in height, whereas others are more than a kilometre across at their base, and they rise to the surface from the deepest parts of lagoons. Some knolls and pinnacles are probably coral-veneered remnants of karstification, but others could be the result of differential coral growth.

Algal cup reefs or boilers have raised rims running around central depressions or micro-lagoons that are a few metres in depth. These reefs, which are awash at low tide, are up to 12 m in height and a few tens of metres in diameter. They occur along the seaward margins of pronounced breaks of slope in Bermuda, Yucatan, and Brazil. In Bermuda, the reefs are built entirely of crustose coralline algae, vermetid gastropods, and the encrusting coral *Millepora*, but in the Mediterranean similar forms are merely veneers of encrusting vermetids and algae over eroded blocks of Pleistocene aeolianite (Ginsburg and Schroeder 1973; Safriel 1974; James 1983).

Faros

The edges or rims of large atolls in the Maldives, off the western coast of India, consist of a series of small ring-shaped reefs, or faros (fareos, atollon). There are similar reefs on the rims and in the lagoons of atolls and barriers on south-eastern

New Guinea, Mayotte Island, New Caledonia, Fiji, and Indonesia, and one might, on morphological grounds, include some of the crescentic reefs and shelf-atolls of the Great Barrier Reef (Ladd 1977). Some are 10 km or more in length, but most are smaller, and their lagoons are usually between 4 m and almost 40 m in depth. Small faros tend to be circular, but the larger ones are generally elongated parallel to the edge of the atolls or barriers. Their origin is controversial, although some are the result of sediments recurved around the leeward side of reefs by wave refraction, thereby enclosing small lagoons (Guilcher 1971).

Ridge Reefs

Guilcher (1988) proposed that long, narrow coral ridges, parallel to the axis of the Red Sea, are a distinct type of reef. There are shallow and emerged ridge reefs (shabs) on the eastern, western, and northern sides of the Red Sea, huge ridge reefs in deeper water off Saudi Arabia and Sudan, and an elongated ridge atoll built on a horst off Sudan. The swarms of ridge reefs on the Farsan Bank of western Saudi Arabia, which are 600–700 m in height, could also be coral-mantled horsts, formed during slow subsidence caused by divergence of the African and Arabian plates and the Sinai platelet. Ridge reefs therefore appear to be a product of the particular tectonic environment of the Red Sea, and no counterparts have been identified elsewhere.

THE GREAT BARRIER REEF

The Great Barrier Reef runs for about 2,300 km off the coast of Queensland, Australia, on a shelf ranging in width from about 300 km in the south to less than 30 km in the north. Reefs occupy the whole of the shelf in the north, but they only occur in the outer to middle portions in the south. The Great Barrier Reef is the largest, most continuous, most complex, and one of the richest reef environments in the world. It is not a barrier reef in the classical Darwinian sense, but rather a rich assemblage of more than 2,500 individual reefs, ranging in size from small pinnacles to massive structures up to 25 km in length and 125 km² in area (Hopley 1982; Parnell 1988).

Because of gradual northern migration of the Indo-Australian plate into tropical and sub-tropical waters, the Great Barrier Reef is a discontinuous, northward-thickening carbonate wedge, that is laterally continuous with terrigenous fluvial and deltaic sediments. Reefs in the northern portion of the Barrier Reef are between 1,000 and 1,500 m in thickness, and they have been growing for about 15 million years. The reef complex in the central part of the Barrier Reef began to develop about 4 million years ago, on top of an extremely thick fluvio-deltaic sequence. The reefs are only 250–300 m thick, and they contain a series of karstic unconformities that developed during periods of low sea level. Reefs began to develop on quartz sands about 2–3 million years ago in the southern part of the Barrier Reef, and they are less than 150 m in thickness today. Pleistocene reefs that formed during periods of high sea level were subaerially eroded and dissected by fluvial channels during glacial stages. Although there was probably not enough time for the formation of most major karstic features, some steep-sided, gently meandering channels could have originated as karst gorges (Hopley 1982; Davies and Montaggioni 1985; Davies 1988).

Multiple shallow bore-holes have been used to determine the thickness, internal structure, and accumulation rates of fringing and other types of reef, and to investigate the nature of their three-dimensional growth. Most modern reefs developed during the Holocene transgression on fairly shallow, pre-Holocene reefal foundations, whose morphology could be karstic or constructional in origin. Holocene reef limestones are usually fairly thin. Thicknesses range from 7.4 to 14.3 m beneath four reefs in the southern Great Barrier Reef, although more widespread geophysical surveys suggest that they range between 7 and 23 m. Most reefs began to grow between 9,500 and 5,500 years ago, depending on the elevation of their foundations. Vertical growth rates varied between 0.7 and 10.7 mm yr^{-1} on their windward sides, and 0.6 to 8.3 mm yr^{-1} on their leeward sides. Upward growth declined, however, as the reefs caught up with sea level between about 6,000 and 4,000 years ago, and lateral growth then became significant (Hopley *et al.* 1978; Marshall and Davies 1984; Hopley and Barnes 1985; Johnson and Risk 1987).

Hopley (1982, 1983) classified reefs on continental shelves with particular reference to the Great Barrier Reef. The classification is based upon the degree to which reefs have retained the inherited morphology of their antecedents. Coral growth on an antecedent platform enhances its relief, until it reaches sea level. Once sea level stabilizes, carbonate productivity is no longer required for vertical accretion, and there is an increase in the amount carried from the productive windward margins to the leeward side of the reefs. This material causes lateral accretion, which first creates and then fills up the lagoon, obliterating the inherited relief. Maxwell (1968) had previously postulated, as did a number of Japanese workers (Burke 1952), that lagoons were formed by central degeneration of the reefs.

The stage of reef development depends upon a number of factors. Reefs growing on platforms at fairly shallow depths probably have the greatest lateral, as opposed to vertical, development, and they most effectively mask the underlying morphology. High carbonate productivity also helps to mask the underlying relief. Variations in Holocene sea-level curves are also of great importance. Significant lateral accretion and masking of underlying topography are only possible where sea level has been stable for thousands of years. Vertical accretion on topographic highs tends to dominate reef development on subsiding foundations, helping to retain or emphasize the morphology of the antecedent platform. The ratio of lagoon size to the area of the productive reef margins increases with the size of the reef. Large reefs may therefore require more time than small reefs to fill in their lagoons and mask the underlying topography (Davies and Kinsey 1977; Davies *et al.* 1977).

Most medium-sized reefs (diameters between about 1.75 and 3.25 km) grew on antecedent platforms that had only one major central depression, although there may also have been many smaller karstic depressions. Large reefs (diameters > 3.25 km) developed on large platforms that had several major depressions, and they therefore tend to have more than one lagoon. Small reefs (diameters < 1.75 km) grew on platforms that probably lacked a major central depression, and lagoons were generally absent or quickly infilled.

Hopley's classification distinguishes three phases in the Holocene development of medium-sized reefs (Fig. 10.12):

a) The juvenile phase is characterized by colonization of topographic highs on antecedent platforms. Submerged shoal reefs develop as sea level rises faster than the upward rate of accretion, enhancing the inherited relief. Irregular patch reefs

form in the late juvenile phase, particularly on the windward sides, as parts of reefs attain present sea level.

b) The general reef surface reaches modern sea level in the mature phase. Reef flats develop over topographic highs on the antecedent surfaces, and especially around the margins of the reefs. Crescentic reefs are produced by patch reefs coalescing on the windward sides. Lateral sediment movement gradually encloses central areas to the lee of the crescentic reefs, forming atoll-like reefs with lagoons. Lagoon infilling, which begins in the later stages of maturity, masks the original relief and widens reef flats.

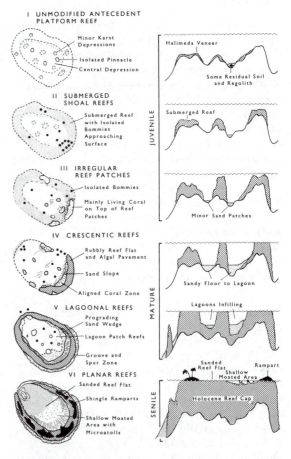

Fig. 10.12. Classification and evolution of atoll-like shelf reefs on medium-sized antecedent platforms, with particular reference to the Great Barrier Reef. More than one lagoonal cell can develop on larger reefs, but the lagoonal stage may be absent on smaller reefs (Hopley 1982)

c) In the senile phase, planar reefs develop as a result of lagoon infilling and coalescence of lagoonal patch reefs. Lateral sediment movement from the windward side causes reefs to grow downwind. Extensive sand cays and ramparts can develop on the reefs in this advanced stage of development.

There are long, narrow ribbon or linear reefs along the outer perimeter of the northern Great Barrier Reef. They are commonly between 6 and 8 km or more in length, but only 300–800 m in width. Ribbon reefs are often recurved inwards at each end by wave refraction, and separated from each other by narrow passes. Despite their distinctive shape, they probably developed in much the same way as other types of reef, although they lack lagoons. Ribbon reefs developed on narrow foundations which could have originated as fringing reefs or other shoreline features formed when sea level was lower than today, solution rims on karst-eroded platforms, or, particularly where the reefs are transverse to the edge of the shelf, as structural features.

There is a spatial pattern in the distribution of the different types of reef (Hopley 1982). Mature and senile forms, for example, dominate in areas where the sea attained its present level as early as 6,000 years ago. Nevertheless, although differences in accretion rates, depth of antecedent platforms, and other factors influenced the development of modern reefs, abrupt changes in reef morphology across the region may be indicative of structural dislocations, suggesting that Quaternary neotectonic movements have played a very important role on the shelf.

CORAL CAYS

Low coral islands, or cays, are composed of biogenic sediment or young sedimentary rocks on reef platforms. All islands on continental shelves tend to be low, but they can also occur in oceanic regions close to high volcanic islands. Low islands temporarily store sediment for reef systems. The amount of sediment in the cays increases or decreases according to vegetational growth and cementation, changing reef morphology, and changes in climate and sea level. Reef size, shape, and orientation are important in determining whether a cay will develop. Cay formation requires an ample supply of material, and most importantly, wave and current patterns that concentrate sediment on the reef. On some of the larger of the Great Barrier Reefs, for example, wave refraction doesn't provide the required centripetal action, and sand is spread out fairly evenly over the surface (Hopley 1982; Gourlay 1988). Cays may also develop more easily where there is a low tidal range (Stoddart and Steers 1977).

There are few islands on the outer reefs of the Great Barrier system, and on the barrier reefs of Mayotte, New Guinea, and New Caledonia. They are common, however, on the barrier reef off Belize. Some atolls have no islands whereas others, such as Diego Garcia, have an almost continuously emerged rim. Dissection of emerged rims has created between 100 and 300 islands on many of the Tuamotu atolls. Most atolls in the Marshalls, Gilberts, and Tuvalu have more than twenty islands, whereas those in the Carolines usually have less than ten. Of 125 atolls surveyed in the Pacific, six have a virtually complete land rim, twenty-two have from half to two-thirds of the rim occupied by land, thirty-four have one-third to one-half, and fifty-five have less than one-third (Wiens 1962). On about 72 percent of Pacific atolls, less than half of their rims are therefore occupied by land. The Trade Winds are important in building low islands in the northern Pacific, where they are twice as common on the windward as on the leeward side of atolls. Island distributions are more complex in the Indian Ocean, however, because of seasonal reversals in wind direction.

Much of a wave's energy is lost through breaking and reflection at the front of a reef. Coarse sediment is therefore deposited in ramparts and coarse clastic cays near the reef margins. Finer material, carried across the reefs by slowly attenuating, transmitted waves, can form sand cays in the centre or towards the lee of reefs, where refracted waves converge (Hopley 1982). Coarse clastic cays on the windward side of reefs are probably more stable than sand cays. When vegetation becomes established on a cay, roots and surface creepers help to bind the material together, and it traps wind-blown sand and particles carried by overtopping storm waves. This allows the island to grow above the level of the highest waves. Increasing stability of the cay may also permit beachrock and cay sandstone to form, further enhancing its stability.

There have been several attempts to classify coral cays, particularly on the Great Barrier Reef.

Stoddart and Steers (1977) recognized six basic types:

a) sand cays, vegetated or unvegetated;
b) coarse clastic (shingle) cays, vegetated or unvegetated;
c) sand cays with clastic (shingle) ridges (motu), usually vegetated;
d) mangrove cays, with or without low, dry areas;
e) mangrove cays with windward sand ridges; and
f) low wooded islands.

Sand cays are low, mobile features that tend to form on the lee side of host reefs, although their size and position vary according to the wave-refraction pattern (Fig. 10.13). Strong centripetal sediment transport produces crescentic or oval cays, but some have more complex shapes, often including spits extending out from their lee side. Unvegetated cays can be up to a few hundred metres in length, and they are often found on barrier reefs and lagoonal patch reefs. Unvegetated linear sand cays, usually less than 1 metre in height and a few metres in width, occur on the large- to medium-sized reefs of the Great Barrier Reef. Linear cays are mobile, and many are little more than ephemeral sand banks that are frequently overtopped by uprush. They develop either where refracted wave trains meet along the length of long, narrow reefs, or where there are seasonal reversals in wind direction (Hopley 1982). The summits of most vegetated sand cays are between 1 and 3 m above the high spring tidal level. The larger islands may be occupied by tall broadleaf woodlands, whereas smaller islands usually have a variety of grasses, herbs, vines, and low shrubs.

Coarse clastic (shingle) cays usually develop on either the windward margins of large reefs, or in the central portions of small, exposed reefs (Fig. 10.13). Some linear cays are parallel to the reef front and are a type of rampart, whereas others are normal to the reef front and are essentially ana-

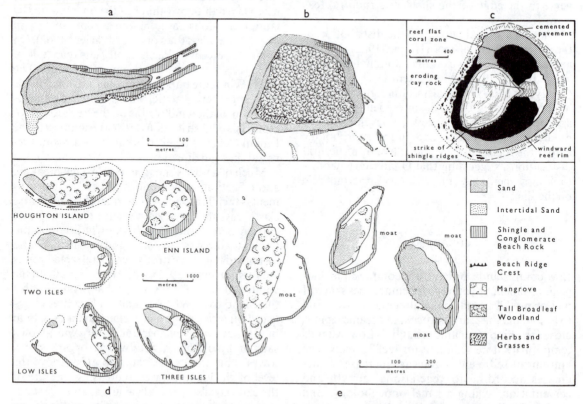

Fig. 10.13. Coral cays on the Great Barrier Reef, (a) Waterwitch, a sand cay (Stoddart *et al.* 1978); (b) East Hope, a vegetated sand cay (ibid.); (c) Lady Elliot Island, a vegetated shingle cay (Hopley 1982); and (d) low wooded islands (Stoddart 1965*a*). Figure (e) depicts moat islands of the southern barrier reef lagoon off Belize (ibid.).

logous to tongues extending out from the lee side of ramparts. They range from low, mobile mounds up to larger, more stable and complex structures consisting of recurved clastic ridges, and cemented beachrock and conglomerate (Hopley 1982). Vegetation only develops in the more stable areas.

Coarse clastic islands provide shelter for accumulation of finer sediments on coral reefs, and they are usually transformed into motus (the Polynesian term) or sand cays with clastic ridges (Stoddart and Steers 1977). Motus are most often situated on the inner half of the reef flat, although some occur closer to the seaward edge of the reef. Motus are the most common type of island on atolls in the Indian and Pacific Oceans, and they are the largest and most stable type of reef island. They normally consist of a seaward ridge of coarse clastic material, several metres in height, with spits of finer material extending from their ends towards the lagoons. A lower area of sand can also form a small ridge to the lee of the clastic ridge, and there may be standing water in the enclosed depressions between them.

Mangrove or mud cays are absent over much of the coral seas, although they are common in the Caribbean, East Indies, Melanesia, and parts of Micronesia, and there are a few examples on the Great Barrier Reef. These islands are not protected by clastic ramparts, and they can only develop on top of high reefs in low energy environments, where mangroves colonize shoals and promote further deposition. Mud cays only extend slightly above high tidal level, and the ground is often waterlogged, although it may be drier in the central portions, or on the windward or leeward sides.

Mangrove cays with a windward sand ridge have been described from Belize, Jamaica, the Marquesas in Florida, and the Bahamas. Moving from the windward to the leeward side, they consist of: a sandy shore with strand vegetation, often with beachrock and sometimes dunes; a sand ridge area several metres high with woodland and coconuts; a mangrove transitional zone with invading woodland; and a leeward mangrove zone. The islands are usually situated hundreds of metres from the edge of shelf reefs.

Low wooded islands occupy a high proportion of the surface of small patch reefs. The term was first used by Captain Cook to describe islands on the Great Barrier Reef that consist of clastic ridges close to the windward edge of the reef; a simple sand cay on the leeward side; and a sand flat or open water, sometimes colonized by mangroves, between the rampart and the cay (Fig. 10.13). There are similar islands in Jamaica, Belize, and in Jakarta Bay and the Java Sea (Stoddart 1965*b*; Stoddart *et al.* 1978; Hopley 1982). The moat-islands off Belize, which consist of windward coarse clastic ridges, shallow, open water and mangrove, and leeward cays, are analogous forms (Fig. 10.13). On the Great Barrier Reef, the most common type of low wooded island has separate, well-defined sand and clastic units and a fairly restricted area of mangrove, but in some cases mangrove covers the sand cay and is connected to the coarse clastic cay. Other islands have almost all the characteristics of low wooded islands, but they lack the central reef flat. The lee side of these islands largely consists of clastic ridges, and mangrove is limited to the intervening depressions (Stoddart *et al.* 1978).

The term 'cay' (or 'key') has also been applied to some slightly emerged reef islands, especially in Florida, the Bahamas, and Bermuda (Stoddart and Steers 1977). The Florida Keys, for example, are composed of reef limestones raised up to 1 m above the high tidal level. Many of the cays in the Bahamas and off the coast of Belize are about 2 m in height, and are composed of Pleistocene reef-rock of possible Sangamon age (Milliman 1973). Others, extending up to as much as 30 m above sea level, consist of lithified or partly lithified Pleistocene sand dunes.

On the Great Barrier Reef, high reefs near to the coast have low wooded islands, lower reefs further seawards have sand and coarse clastic cays, and the lowest, most seaward reefs have no islands at all. There have been several attempts to account for this distribution. It has been proposed that it reflects systematic variations in the height of reef surfaces, resulting from tilting normal to the mainland coast, or differences in wave exposure and possibly changes in sea level. Hopley (1982) attributed the distribution of island types to differences in the height of the reefs across the shelf, tidal range, cyclone frequency, and exposure to oceanic swell. Other areas provide support for the exposure hypothesis, although they do not have the same variety of islands (Stoddart and Steers 1977). Islands on the barrier and lagoonal reefs of Belize, for example, have a similar distribution to those off the Queensland coast. Sand is carried across reef flats exposed to high wave energy or extreme tidal range, preventing the formation of islands. Coarse clastic and sand-clastic cays develop on slightly less exposed reefs, sand cays in more protected environments, and mangrove and mangrove-sand cays in areas with low wave energy. Moat-islands are formed in deeper lagoons in areas of moderate wave energy, but they may develop wherever there are constant, unidirectional winds, and rapid changes in wave energy within short distances (Stoddart 1965*a*).

HUMAN IMPACT

Two main schools of thought on the stability of coral reefs have important implications for the assessment of human impact. Some workers regard the reef ecosystem as being stable, because of its ancient origin and complex evolution in an environment where conditions have been uniform for long periods of time. Despite the complexity of the system, which provides some buffering potential within certain limits, they believe that it is very fragile when pushed beyond those limits. Human activities are therefore thought to produce catastrophic and probably irreversible changes to coral reefs. Other workers believe that the reef ecosystem is unstable over time and heterogeneous in space. They consider that reefs are constantly modified by natural perturbations, and are in different stages of recovery from various sources of disturbance. According to this view, reef ecosystems are not as fragile as was once thought, and anthropogenic disturbances are generally of secondary importance to those resulting from natural causes (Brown and Howard 1985; Dahl and Salvat 1988; Grigg and Dollar 1990).

Coral communities and reefs are threatened by a variety of human activities, including dredging, mining, land clearance, effluents from desalination, nuclear-weapon testing, oil, chemical and sewerage pollution, thermal pollution from electrical generating stations, careless anchoring and boat grounding, and the collection of precious corals and other marine organisms (Dahl and Salvat 1988; Rogers *et al.* 1988; Weber 1994). Considerable damage has been caused to coral reefs in many areas, although the response to human interference is complex, and surprisingly little damage has occurred in some cases (Brown and Howard 1985). Nevertheless, most accessible reefs in the South Pacific are deteriorating, and there has been significant deterioration of Caribbean reefs since 1975 (Dahl 1985; Rogers 1985).

Explosives left by American forces after World War II have been used for fishing on many islands in the South Pacific. This severely damages coral reefs and lagoons, and recovery may be limited to an annual 1–2 percent increase in the living cover (Alino *et al.* 1985). Dynamite is also used for fishing on reefs in the Indian Ocean and Indonesia. Chlorine bleach and other poisons used to catch fish are toxic to corals (Chesher 1985). Damage can also occur as a result of more-efficient fishing techniques. The disappearance or replacement of extensive coral reefs by algae on the Chinese island of Hainan, for example, have been attributed to dredging, siltation, and the removal of almost all fish by the use of fine-meshed nets (Wu 1985). On the other hand, Pacific reefs have proved to be fairly resilient to wartime damage. In the lagoon of Truk in the Carolines sunken Japanese ships are covered by a growth of marine organisms, while quays built on fringing reefs in the inner parts of the lagoon of Bora Bora are being covered by nearshore vegetation. In a few cases human influence has encouraged reef development. In Hawaii the number of newly settled corals increases exponentially with proximity to the effluent outlet of an electrical generating station, presumably in response to water temperatures up to 8°C higher than normal (Coles 1985). On Guam, however, thermal effluent from power plants, 4–6°C above ambient temperatures, is deleterious to coral growth (Neudecker 1981).

The type of vegetation growing on coral islands is important in determining the effect of storms. Dense natural vegetation protects islands from damage by high waves, and it also traps sediment during storms. There has been extensive replacement of natural vegetation with coconuts, however, which have an open structure that is easily penetrated by sea water, and a dense but shallow root system that is undermined by sand sapping. The lack of ground vegetation in many plantations exposes the surface to sand stripping and channelling, and islands therefore tend to be eroded and destroyed by storms (Stoddart 1971*a*). Removal of natural vegetation to build tourist amenities and prevent harbouring of insects, rats, and other undesirable fauna, is likely to play an increasingly important role in the future.

Coral, sand, and gravel are mined and dredged for road construction and urban development in some areas. Apart from the direct impact and complete destruction of reefs by dredging, the effects of high turbidity and sedimentation caused by the constant resuspension of fine, residual particles persists for many years after dredging has ceased. Decreased light, consumption of energy for sediment cleansing, and possibly a reduction in planktonic food cause coral mortality, low coral diversity, and the vigorous growth of algae (Gabrie *et al.* 1985; Brown and Howard 1985).

Reefs are damaged by soil erosion, mining, and the use of pesticides on the land behind. Siltation, refinery effluent pollution, high heavy-metal concentrations, and other problems associated with mining and processing of low-grade lateritic nickel near to coral reefs are likely to increase in the future. In New Caledonia open-cast mining of nickel on hilltops has polluted lagoons with fluvial sediment (Guilcher 1985*b*). On Mayotte Island in the western Indian Ocean population growth and increasing cultivation have led to heavy soil erosion on the steep slopes, and an increase in the amount of terrigenous sediment reaching the lagoon. This has resulted in a marked decrease in its calcium carbonate content in the last thirty years, and massive death and burial of corals under mud.

High nutrient concentrations encourage competitive growth of fleshy and filamentous algae, and inhibit coral development. Eutrophic conditions associated with sewage discharge are therefore detrimental to reef development. Increases in turbidity and nutrient supply from polluted groundwater, and runoff from outhouses, septic tanks, and industrial waste killed reefs in Venezuela (Weiss and Goddard 1977). In 1978 pressure by the public and scientific communities forced local government and the military to discontinue the discharge of large amounts of sewage into Kaneohe Bay, in north-eastern Oahu, Hawaii. Response of the depressed coral populations was slow at first, but there had been a dramatic recovery in abundance, distribution, and variety by 1983. The previously explosive growth of green algae, which had overgrown and killed many coral communities, has experienced a correspondingly marked decline (Maragos *et al.* 1985).

There has been a big increase in the international trade in coral and shells since 1970. As reefs in Japan and Taiwan became exhausted, coral-fishing ships began operating in other waters, off Haiti, Australia, and the Philippines. Commercial coral collecting in Australia is strictly controlled, and it poses little threat to reefs at present levels of exploitation. It has proved more difficult to enforce

a total ban on coral collection in the Philippines, where the industry once employed over 1,000 people and was responsible for millions of dollars in exports (Oliver and McGinnity 1985). The destruction of coral colonies on the Great Barrier Reef and in Mauritius by the starfish *Acanthaster planci* may partly reflect removal or reduction in numbers of its main predators, particularly through over-collection of the shell of the giant triton *Charonia tritonis* (Endean and Cameron 1990). Nevertheless, other workers have argued that pre-collection population levels of the shellfish were probably too low to control starfish numbers, and there is increasing evidence that outbreaks may be linked to the removal of fish predators as a result of increasing intensification of reef fisheries. Large triggerfish and pufferfish, for example, appear to be responsible for the dispersion of starfish aggregations in the Sudanese Red Sea, where there have been no large starfish outbreaks (Bradbury 1990). On St Croix in the Caribbean, reduction in predator species through overfishing and shell collection may have caused an increase in the population of the sea urchin *Diadema antillarum*, which resulted in coral overgrazing (Ogden *et al.* 1973). Mass mortality of this echinoid throughout the Caribbean from 1983 to 1984, apparently for natural reasons, led to increased algal growth on some reefs (Lessios *et al.* 1984).

Coral destruction is often the result of a combination of causes. On Martinique in the Caribbean coral is buried under sediment from sewage generated by urban and industrial development, hillslope deforestation, and coastal road works. In Thailand, mining, dynamite fishing, bottom trawling, coral collecting, and industrial and domestic pollution contribute to coral deterioration. Land reclamation, dumping of garbage, and the building of roads, buildings, and other facilities for the rapidly expanding tourist industry have also caused heavy siltation and reef degradation (Sudara and Nateekarnchanalap 1988). Coral reefs on the Japanese Ryukyu Islands have rapidly deteriorated, and most are now dead or dying. There has been little recovery from continuing predation by *Acanthaster planci*, and the problem is now compounded by economic development, with further degradation caused by land reclamation, deforestation, levelling of land for dams and pineapple and sugar cane fields, and construction for industry and tourism. Soil erosion has greatly increased with removal of natural vegetation and the paving of stream courses, and terrestrial sediment is now accumulating on beaches, lagoons, and reef slopes (Muzik 1985).

At one time subsistence economies conserved reef systems for the general benefit of the community. Once the customs and traditional conservation practices had been swept away by industrialization and the introduction of cash economies, however, there was subsequent degradation and pollution of reefs, and their resources were exploited until depletion. It has therefore become necessary to set up marine parks and reserves (Salvat 1981). Measures taken to arrest coral decline on the Great Barrier Reef represent the best example of the comprehensive management of reefs. These efforts have culminated in the creation of the Great Barrier Reef Marine Park, the largest in the world with an area of 348,700 km^2 (Dinesen 1988; Craik *et al.* 1990). The park region is the focus of a variety of often competing interests, including tourism, scientific research, commercial and sport fishing, shipping, oil exploration, and conservation. Some areas of the reef are to be preserved in their natural state, to be used only for scientific research. Controls are implemented through zoning and the use of permits to control the type and location of activities, and periodic closure of particular reefs and waters. Management of reef users is implemented on a daily basis through education, surveillance, and enforcement.

Economic pressures on reefs are more severe in developing areas, and conservation policies are therefore more difficult to enforce (Salvat and Soegiarto 1981). Although Sri Lanki already has legislation to prevent reef degradation, it is rendered ineffective by lack of enforcement, political restraints, and the priorities of a developing nation (De Silva 1985). Officials and ordinary people have to be made aware of the importance of coral reefs: this is much more difficult to accomplish in developing countries. A local or village-level approach to coastal management may be more realistic in the Philippines than one based solely on national regulations (McManus *et al.* 1988). A successful programme of public education has also been carried out in a wide variety of communities in the Sudan (Schroeder 1981; Nasr 1985). Financial assistance is needed to set up marine parks in developing

countries. Conservation is difficult where indigenous people increasingly depend upon reefs for their livelihood, and where development pressures encourage unregulated use for quick financial gain. It is unrealistic in such regions to expect reefs to be temporarily protected or set aside as preserves until proper research can be conducted. Reef-management policies must therefore be formulated to recognize these economic, social, and political realities (Holthus 1985). Politicians need to be convinced of the economic benefits of marine parks, although there is presently little relevant documentation. The profitability of marine parks depends upon full exploitation of their tourist potential. Benefit-to-cost ratios of 11 : 1 and 9½ : 1 were determined for parks in the US Virgin Islands and in Costa Rica, respectively, but although these studies suggest that some marine parks are economically profitable, many are not (Van 'T Hof 1985).

The indirect effects of human activity possibly pose an even greater threat to coral reefs and islands than the more direct and immediate effects of human interference. It is predicted that sea level will rise in the next fifty to 100 years as a result of global warming (Chapter 1). In the last 100 years reefs have been adjusting to an average rise in sea level of between about 1 and 2 mm yr^{-1}, but this could, according to most current estimates, increase to between 2.6 and 8.9 mm yr^{-1} up to the year 2080. The higher estimates, which are similar to the rise in sea level during the postglacial transgression, would result in the drowning of reefs. Although they could maintain themselves if the lower estimates prove to be more accurate, higher temperatures could cause widespread coral mortality. Global warming and the rise in sea-surface temperatures may also increase the frequency and intensity of hurricanes, and alter their distribution. Rising sea levels threaten the quality and existence of freshwater lenses, which in turn affect the vegetation and habitability of reef islands (Stoddart 1990). Predicting the possible effects of greenhouse gases, however, is very controversial, and it is possible that there could be a substantial fall rather than a rise in sea level, and that increased storm intensity could build up the land and increase the supply of fresh water through precipitation (Emanuel 1987; Freitas 1991).

11 Rock and Cohesive Clay Coasts

Rocks and cohesive clays form a high proportion of the world's coasts, and even many sand and coarse clastic beaches are underlain by shore platforms and backed by marine cliffs. These coasts have been neglected in the contemporary process-oriented coastal literature, where the emphasis has been on beaches and other coastal features that quickly respond to changing environmental conditions.

ROCK COASTS

Despite the increasing application of modern analytical techniques, geochronometric dating, and physical and mathematical modelling, and careful measurement of processes and erosion rates, we can still only speculate on the mode and rate of development of rock coasts (Trenhaile 1987; Sunamura 1992).

Processes

The marine and subaerial processes that operate on rock coasts have varied through time, with changes in relative sea level, climate, and other factors. Although our ability to identify and measure processes has improved, we are still largely ignorant of their precise nature and relative importance. Increasing rates of economic development provide a practical need to understand the dynamics of these coasts. The acquisition of quantitative data, however, has been hindered by very slow rates of change, the importance of high intensity–low frequency events, exposed and often dangerous environments for wave measurement and subaqueous exploration, and the lack of access to precipitous or heavily vegetated cliffs. Furthermore, even a complete understanding of contemporary erosive processes would not be enough to account for the development of coasts that frequently retain vestiges of former environmental conditions, that were often quite different from today.

Wave Action

Mechanical wave erosion usually dominates rock coasts in storm and vigorous swell wave environments, and although waves are generally much weaker in polar and tropical regions, they still play an important role in removing the products of weathering. The term 'mechanical wave erosion' encompasses a number of processes (Fig. 11.1). Their relative importance is usually determined from ambiguous morphological evidence. Fresh rock scars and coarse, angular debris consisting of joint blocks and other rock fragments suggest that quarrying is usually the dominant form of erosion in storm wave environments. Rocks are dislodged by water hammer, high shock pressures generated against structures by breaking waves, and, probably most importantly, by air compression in joints and other rock crevices. These processes require the alternate presence of air and water, and are therefore most effective in a narrow zone extending from the wave crest to just below the still-water level.

Abrasion occurs where rock fragments or sand are swept over rock surfaces. Gently sloping abrasional surfaces are generally much smoother than those subjected to wave quarrying, although grooves develop if abrasion is concentrated along joint planes. Although abrasion is not as closely associated with the water level, its efficacy rapidly decreases below the water surface. Only large waves can move particles beneath accumulations that are more than about 10 cm in thickness, but if they are too large particles are taken into suspension, and there is less contact with the bedrock (Robinson

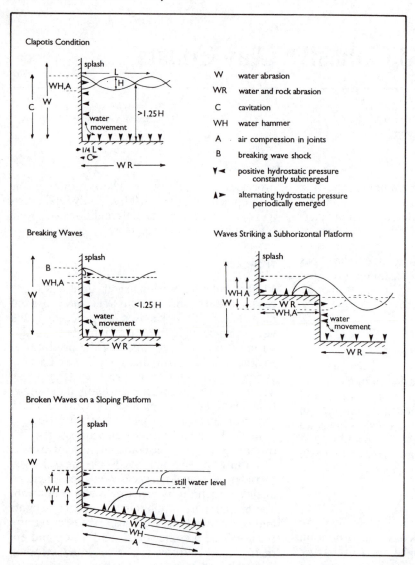

Fig. 11.1. Processes of mechanical wave erosion and their distribution on rock coasts (Sanders 1968)

1977*a*, *b*). Potholes are approximately cylindrical depressions that develop where large clasts are rotated by swirling water in the surf or breaker zones. They tend to develop in the upper intertidal zone where abrasives are trapped at the foot of scarps, and in structural or erosional depressions. Widespread pothole development also occurs in calcareous rocks, through the inheritance of corrosional hollows.

The magnitude and distribution of the forces exerted by waves on coastal structures depend upon deep-water wave characteristics, tidal elevation, and submarine topography. Waves breaking directly against cliffs or other steep, natural structures may be more effective erosional agents than broken waves, but they occur much less frequently. Most mathematical models suggest that standing, broken, and breaking waves exert the greatest pressures on vertical structures at, or slightly above, the water surface—where the most important mechanical wave erosional processes also operate. As tides control the elevation of the water surface, they also determine where erosion occurs. The level of greatest wave erosion on a rock coast must therefore be closely associated with the elevation most frequently occupied by the water surface. This is at, or close to, the neap high and low tidal levels. Wave action is increasingly concentrated at, and between, the neap tidal levels as the tidal range decreases (Trenhaile 1987) (Fig. 11.2).

Although the vertical distribution of wave energy and mechanical wave erosion is related to the tidal distribution of the water surface, it may be skewed towards the upper portions of the tidal range. This is because wave action is most vigorous during high tidal periods, when there is deeper water in the nearshore zone. The largest, most erosive waves also occur during storms, when the water surface is meteorologically raised above the tidal level. In microtidal environments, the level of maximum erosion may be elevated to the spring high tidal level or above. Only vigorous storm waves, which operate at higher elevations than weak waves, are able to erode resistant rocks, whereas weaker waves, operating at lower elevations, are able to erode less resistant rocks. The difference between the level of greatest erosion and the most frequent tidal level may therefore increase with the resistance of the rock (Trenhaile 1987).

Weathering

In some areas cliff erosion largely results from weathering, and removal of the debris by mass movement and weak wave action. Weathering is particularly important in high and low latitudes, where there are suitable climates and weak wave environments. It can also play a significant role in sheltered areas within the vigorous wave environments of the middle latitudes, however, and it prepares rocks for eventual dislodgement and removal by strong waves in more exposed areas.

Frost and related mechanisms

Clay minerals expand and contract in response to tidal- and weather-induced cycles of wetting and drying. This opens up discontinuities in shales and other argillaceous rocks, and also in rocks that are adjacent to them. Thermally induced changes in volume can also reduce the strength of some rocks—diurnal variations in joint widths of up to 0.5 mm have been recorded in the Jurassic rocks of south Wales, where discontinuities open up as limestones expand and mudrocks contract under dry, hot conditions (Williams and Davies 1987).

Clay-rich rocks can also be damaged by wetting and drying resulting from the temperature-dependent adsorption of water (Hudec 1973). This is caused by the attraction of the positively charged ends of water molecules to negatively charged clay surfaces within small rock capillaries. This

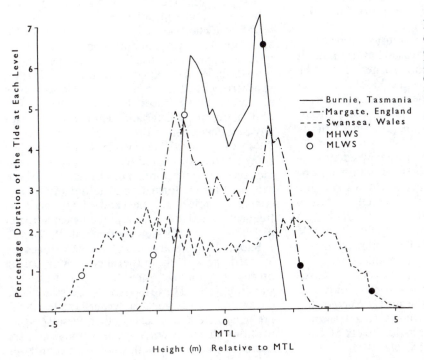

Fig. 11.2. Tidal duration distributions—the amount of time the tidal or still water level is at each intertidal elevation (Carr and Graff 1982)

——— Burnie, Tasmania
—·—· Margate, England
– – – Swansea, Wales
● MHWS
○ MLWS

mechanism may be responsible for much of the field evidence that has traditionally been accorded to frost action. A number of investigators have suggested that the two mechanisms could act together. Matsuoka (1988), for example, proposed that free water is drawn to the freezing front by suction generated by unfreezable adsorbed water trapped in rock pores by expanding ice. The resulting pressures may be more deleterious to rocks than those generated by the approximately 9 percent expansion of water upon freezing.

Although much remains to be determined about the processes responsible for rock breakdown in cool environments, we do have a general sense of the conditions that are most suitable for their operation. Cool coastal regions may be almost optimum environments. Rocks in intertidal and spray and splash zones are able to attain high levels of saturation (Trenhaile and Mercan 1984), and because of the effect of tides they experience many more frost cycles than those further inland. Intertidal rocks are alternately frozen when exposed to freezing air temperatures, and thawed when submerged in water above freezing. Rocks thaw very rapidly when inundated by rising tides, but they may take several hours to freeze once exposed by falling tides (Robinson and Jerwood 1987*a*); it is not known, however, whether they can maintain critical levels of saturation over this period.

The presence of salts in solution appears to inhibit frost action in some cases, but facilitates it in others. Several studies have suggested that the greatest rock deterioration occurs in solutions that contain between 2 and 6 percent of their weight in salt. Frost action may therefore be particularly effective in rocks that are saturated with sea water.

Whether frost or temperature-dependent wetting and drying are effective in coastal environments depends upon the occurrence of suitable rocks, and waves that are strong enough to prevent progressive burial of the cliff under debris (Chapter 12). Frost weathering may assume a dominant role on platforms and cliffs in sheltered sites, but it cannot continue to operate effectively unless waves can remove the debris. Frost cycles occur most frequently in cool, storm-wave environments, and waves and frost also tend to be most effective on the same types of rock. Vigorous wave action may therefore obscure or inhibit the effects of frost action in exposed areas (Trenhaile 1987).

Chemical and salt weathering

Alternate wetting and drying in the spray and intertidal zones creates suitable environments for many chemical and salt weathering processes. These processes assume dominant roles in some warm temperate and tropical regions, but they are probably only important in sheltered areas within cooler, storm-wave environments.

Chemical and salt weathering tend to operate together, and it is very difficult to distinguish the results in the field. The main factor determining the efficacy of chemical weathering is the amount of water available for chemical reactions, and more crucially, to remove the soluble products. High temperatures accelerate most chemical reactions, but this is partly countered by increasing evaporation. Fairly slow rates of chemical weathering in cold regions may reflect the lack of liquid water, rather than the direct effect of low temperatures. Rocks are mechanically weathered by the formation and growth of salt crystals contained within their capillaries. The main processes involve crystal growth from solution, and hydration and temperature-induced expansion. In addition to their erosive role on cliffs and shore platforms, chemical and salt weathering contribute to the development of tafoni and honeycombs, and to the suite of processes collectively referred to as water-layer levelling (Matsukura and Matsuoka 1991).

Tafoni are hollows up to several metres in depth and diameter. They are most common in Mediterranean and foggy, arid coastal regions ranging from the poles to the equator, but they also develop in a variety of other environments. Honeycombs are small, closely spaced depressions up to a few centimetres in diameter and depth. They are usually best developed in the supratidal zone, where they are occasionally reached by splash and heavy spray during storms, and they are especially common on sandstones. Similar mechanisms have been proposed for the formation of tafoni and honeycombs. The role of chemical and salt weathering is frequently cited, but frost and biological agencies may be contributory factors (Trenhaile 1987).

Shore platforms and supratidal rock ledges are lowered, smoothed, and levelled by weathering processes operating around the edges of pools of standing water. The responsible processes, which are collectively referred to as 'water-layer levelling', probably include chemical weathering, salt crystallization, alternate wetting and drying, and the movement of solutions through rock capillaries.

Case hardening often occurs in conjunction with water-layer levelling on platform surfaces, particularly where dissolved ions move along joint planes, depleting or impregnating them with precipitates. Weathering pits with raised rims develop if impregnated joints are more resistant than joint blocks, and small plateaux, sometimes in the form of miniature volcanoes, if the joints are weaker.

The presence of an intertidal saturation level has been a basic tenet of the influential Australasian literature on rock coasts since almost the beginning of this century (Bartrum 1916). This had led to the assumption that there is an abrupt transition from the oxidation zone above the saturation level, where the rock is weathered and weak, into a zone saturated with sea water below, where the rock is largely unweathered and resistant. Present evidence, however, suggests that rocks can only be permanently saturated below the low tidal level, where they are constantly submerged.

There is continuing controversy over the processes operating on limestone coasts. The sharp pinnacles, ridges, grooves, and circular basins that are characteristic of coastal limestones in the spray and splash zones, are similar to the karren formed by fresh water on land. Although surface sea water is normally saturated or supersaturated with calcium carbonate, solution could occur in rock pools at night, when the carbon dioxide produced by faunal respiration is not extracted by algae during the hours of darkness. This lowers the pH of the water and causes calcium carbonate to be transformed into more soluble bicarbonate. Solution can be inhibited or prevented by other biochemical processes, however, including dissolved organic substances coating rock surfaces and building complexes with calcium ions.

Solution is of fairly minor importance in some areas, even in tidal pools where the biomass is large relative to the volume of the water. Nevertheless, it may be responsible for about 10 percent of the total erosion on the coast of Aldabra Atoll. Bioerosional mechanisms have been found to be more important than solution in western Ireland and on Grand Cayman Island, however, and chemical solution may be absent in the rock pools of the northern Adriatic (Schneider 1976; Trudgill 1976, 1987; Spencer 1988a). Although chemical solution is possible in sea water, these studies provide support for the contention that marine karren and other characteristic features of limestone coasts are primarily bioerosional.

Bioerosion

Bioerosion is the removal of the substrate by direct organic activity. It is probably most important in tropical regions, where waves are generally fairly weak and an enormously varied marine biota live on coral, aeolianite, and other calcareous substrates (Spencer 1988a, b; Fischer 1990). Some invertebrates employ purely chemical or mechanical processes to break down rocks. Microflora and fauna that lack hard parts may use only chemical mechanisms, but many other fauna secrete fluids that chemically weaken the rock, before mechanically abrading them with teeth, valvular edges, and other hard parts.

Microflora, including algae, fungi, and lichen, are pioneer colonizers in the inter- and supratidal zones. Algae may be the most important bioerosional agent on rocky coasts. A variety of species is involved, although the minute Cyanophyta or blue-green algae are particularly important in the upper intertidal and supratidal zones. The water in contact with the rock under algal mats receives the products of metabolism and organic waste, and extreme chemical conditions can cause algal colonies to be etched into the substrate.

In addition to their role as rock borers and their effect on water chemistry, algae and other microflora permit subsequent occupation by gastropods, chitons, echinoids, and other grazing organisms. Grazers effectively abrade rock surfaces as they feed on epilithic and the ends of endolithic microflora. In the mid-tidal zone of Aldabra Atoll grazers may be responsible for about one-third of the surface erosion where sand is available for abrasion, and as much as two-thirds where sand is absent (Trudgill 1976).

At least twelve faunal phyla contain members that bore into the substratum, particularly in the lower portions of the intertidal zone. They include *Lithotrya* and other boring barnacles, sipunculoid and polychaete worms, gastropods, echinoids, *Lithophaga* and other bivalve molluscs, and Clionid sponges. Borers directly remove rock material, but they also make the remaining rock more susceptible to wave action and weathering. The carbonate rocks of the tropics favour chemical borers, but mechanical borers are active on a variety of substrates in temperate and boreal waters (Warme 1975).

There is a large body of information on bioerosional rates, but its reliability and applicability vary enormously (Trenhaile 1987; Spencer 1988*a*, *b*). The overall erosion rate, which is usually between about 0.5 to 1 mm yr^{-1} on vertical and horizontal limestone surfaces, may reflect the maximum boring rate of endolithic microflora (Schneider and Torunski 1983).

Mass Movement

Fresh rock surfaces and debris at the foot of cliffs testify to the importance of rock-falls on many coasts. Although they are more frequent than deep-seated slides, most falls are much smaller (Trenhaile 1987; Moon and Healy 1994). Falls occur in well-fractured rocks, especially where notches are cut into the cliff foot by waves, or, in the tropics, by solution or bioerosion. Rock columns defined by joints or bedding planes also topple or overturn by forward tilting. It is the most common of the larger failure mechanisms in Glamorgan, Wales, where pressure release joints and other weathering agencies help to reduce rock cohesion (Davies *et al*. 1991). Rock- and slab-falls, sags, and topples are essentially surficial failures induced by frost and other types of weathering, basal erosion, and hydrostatic pressures exerted by water in rock clefts. Many falls are caused by the reduction in confining pressures resulting from cliff erosion and retreat, and the formation of tension cracks parallel to the erosion surface.

Translational slides usually occur where there are seaward-dipping rocks, alternations of permeable and impermeable strata, massive rocks overlying incompetent materials, or argillaceous and other easily sheared rocks with low bearing strength. Slumps or rotational slides are common in thick, fairly homogeneous deposits of clay, shale, or marl. Sliding takes place in rocks that have been weakened by alternate wetting and drying, clay mineral swelling, or deep chemical weathering. Deep-seated events are triggered by groundwater buildup and basal undercutting. Slides tend to occur during or shortly after snowmelt, or prolonged and/or intense precipitation. Water from septic systems, irrigation, runoff disruption, beach depletion through the building of coastal structures, and other human activities are playing increasing roles in some areas (Griggs and Trenhaile 1994). The damming of rivers has also reduced the bed load reaching the coast, depleting the beaches and exposing the cliffs to more vigorous wave action (Kuhn and Osborne 1987).

Landforms

Coastal scenery is the product of a combination of elements, including: the morphology of the hinterlands; present and past climates; wave and tidal environments; changes in relative sea level; and the structure and lithology of the rocks. These elements provide infinite variation to the basic geotectonic form (Chapter 1).

Bays and Headlands

It is generally assumed that headlands consist of rocks that are more resistant than those in the adjacent bays, although there have been few detailed investigations of variations in rock structure, or direct measurements of rock strength. The occurrence of small bays and headlands often reflects the rather subtle influence of rock structure, including variations in joint density, orientation of discontinuities, and thickness, strike, and dip of the beds. Large bays and prominent headlands are more likely to develop where there are differences in the lithology of the rocks, although they would develop in fairly homogeneous rocks if low cliffs around stream outlets retreated more rapidly than the higher cliffs on the adjacent interfluves.

Because of the effect of wave refraction, the plan shape of crenulated coasts may attain an equilibrium state. This would occur when resistant rocks on exposed headlands are eroded by higher waves at the same rate as weaker rocks are eroded by lower waves in the sheltered bays. The plan shape would then be maintained through time as the coast retreated landwards. Much of the erosive work could have been accomplished in previous interglacial stages, when sea level was similar to today's. It is therefore possible that only minor modification of inherited coasts was required to attain equilibrium at present sea level, although we lack reliable, long-term data on cliff erosion rates to determine whether this has actually occurred.

Cliffs

There are cliffs around about 80 percent of the world's oceanic coasts (Emery and Kuhn 1982). Cliff morphology reflects the interplay of numerous factors, and they are difficult to classify on the basis, for example, of climate or wave regime. Nevertheless, some types are more characteristic of particular morphogenic regions than others.

Cliff morphology

Steep or undercut cliffs are characteristic of wave-dominated environments, whereas convex cliffs are more typical of areas where waves are weak and the climate is conducive to subaerial weathering. Coastal slopes therefore tend to be greater in the vigorous, storm wave environments of the mid-latitudes, than in high latitudes or the tropics, where frost and chemical weathering, respectively, are more important.

Wave action is generally weak in high latitudes because of sea ice, and in some places coastal configuration. These coasts have few true marine cliffs, and most steep slopes are the product of glacial erosion. Nevertheless, wave action is required to remove the weathered debris and prevent progressive burial of the cliff (Howarth and Bones 1972).

In the humid tropics, steep marine cliffs are discouraged by fairly weak wave action, protective coral or algal reefs, and extensive coastal plains. Chemical weathering plays an important role in cliff development, and the weathered material is very susceptible to slumping, mudflows, and other mass movements. Coastal slopes in the hot, wet tropics are often covered in vegetation for all but the lowest few metres, and steep cliffs may be restricted to headlands and other exposed areas. Strong waves are generated on some tropical coasts by onshore monsoons, tropical cyclones, and Trade Winds, however, and the lack of true marine cliffs in the humid tropics may therefore have been exaggerated (Guilcher 1985c). Nevertheless, steep, bare cliffs are much more common in the arid tropics, particularly in exposed areas where spray, high evaporation and salt crystallization, alternate wetting and drying, corrosion, and other weathering processes help to steepen coastal slopes and inhibit the growth of vegetation.

There are numerous exceptions to the morphogenic classification of marine cliffs. For example, despite strong waves, contemporary wave action has accomplished only minor modification of glacially sculptured granites in Maine and eastern Canada. On the other hand, despite weak waves, steep, bare cliffs are common in the limestones of the Canadian Arctic, and throughout the wet and dry tropics.

Geological factors are at least as important as climate and wave regime in determining the relative efficacy of marine and subaerial processes, and therefore the shape of cliff profiles (Emery and

Kuhn 1982) (Fig. 11.3). Steep cliffs usually develop in homogeneous rocks in marine-dominated environments, and convex cliffs where subaerial processes dominate. Where both process suites are effective, subaerial processes produce a convex slope in the upper portion of cliffs, while marine processes cut steep cliff faces at their base. Geological differences superimpose marked variations on these basic morphogenic forms. For example, weaker material, and consequently accelerated rates of subaerial weathering, encourage formation of a convex slope in the upper part of a cliff, despite the dominance of marine processes. Alternatively, resistant cap rocks inhibit subaerial weathering and formation of a convex slope, allowing a steep cliff face to develop in fairly weak materials where subaerial processes are dominant.

The shape and gradient of cliff faces are strongly influenced by structural weaknesses, stratigraphic variations, and the attitude or orientation of the bedding (Emery and Kuhn 1980; Ellenberg and

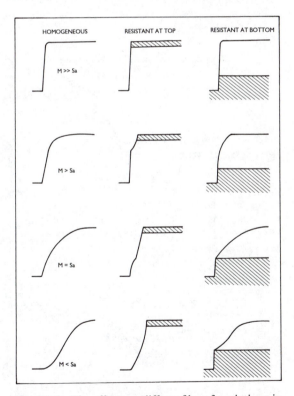

Fig. 11.3. The effect on cliff profiles of variations in rock strength and the relative efficacy of marine (M) and subaerial (Sa) processes. The more resistant rocks are shaded (Emery and Kuhn 1982)

Sturm 1986; Trenhaile 1987). Very steep cliffs generally develop in rocks that are either horizontally or vertically bedded, and more moderate slopes in seaward or landward dipping rocks. Dip slope cliffs in seaward-dipping rocks can be quite smooth, but an irregular cliff face may develop in landward-dipping rock, which exposes different beds to marine and subaerial erosion. The relationship between cliff profiles and the dip of bedding and/ or joint planes is very complex, however, and vertical and seaward-sloping cliffs can be formed in horizontal, or seaward- or landward-dipping rocks (Terzaghi 1962).

Composite cliffs consist of two or more major slope elements. They include multi-storied cliffs, with two or more steep faces separated by more gentle slopes, and bevelled (hog's back, slope over wall) cliffs with a convex or straight seaward-facing slope above a steep, wave-cut face (Wood 1959; Orme 1962; Fleming 1965). The upper and lower surfaces of bevelled cliffs can develop contemporaneously in homogeneous rocks where marine and subaerial processes are of comparable importance, or where weaker rocks overlie more resistant materials. Explanations are less simple for the origin of high bevelled and multi-storied cliffs that have developed in resistant rocks over very long periods of time. They are widely distributed throughout western Britain and Ireland, in Brittany and southern France, and on many oceanic islands in the southern hemisphere. In south-western Britain raised beaches containing middle and upper Pleistocene material suggest that the composite cliffs behind are very old, and have been only slightly modified by contemporary wave action (Andrews *et al.* 1979; K. H. Davies 1983).

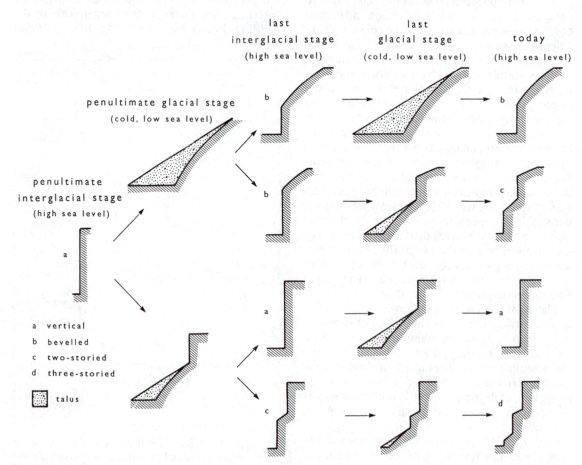

Fig. 11.4. The development of vertical, bevelled, and multi-storied cliffs over two glacial-interglacial cycles (Griggs and Trenhaile 1994)

Marine cliffs experienced marked changes in climate and sea level during the Quaternary. Wave-cut cliffs of the last interglacial stage were abandoned during the subsequent glacial stage, and gradually replaced by convex slopes, developing and extending upwards beneath accumulating talus. Frost and periglacial mass movement were probably dominant processes at this time over much of the ice-free middle latitudes. When the sea rose to its present position, marine processes removed the debris and, depending on the relative strength of the waves and the rock, either trimmed the base of the convex slopes to form composite cliffs, or completely removed it to form steep, wave-cut cliffs. Bevelled profiles developed where the debris reached the top of the cliff during the last glacial stage, whereas multi-storied profiles developed where the debris only rose part of the way up the cliff face (Griggs and Trenhaile 1994) (Figs 11.4 and 11.5).

Although there are usually shore platforms or beaches at the foot of cliffs, some plunge directly into deep water. Plunging cliffs developed where erosion was very slow, and relative sea level therefore rose much faster than sediment could accumulate at the cliff foot. Once sea level became stable, erosion was inhibited by the lack of abrasives at the water level, and reflection of the incoming, non-breaking waves. Nevertheless, hydraulic quarrying caused by the rise and fall of standing waves, and the compression of air in rock clefts, eventually produce caves, notches, and narrow platforms. This may allow more erosive, breaking waves to attack the cliff and destroy the plunging condition. Plunging cliffs are particularly common around the basaltic islands of the southern hemisphere. This is because the rocks are very resistant, and the small size of the islands prevents large amounts of material being transported alongshore, and accumulating at the cliff foot. Coral growth provides further protection from wave attack around some tropical islands.

▲ *bevelled cliffs*

▨ *ice-covered area during the last glaciation*

▢ *additional area covered by ice during the penultimate glaciation*

Fig. 11.5. The relationship between the occurrence of bevelled cliffs and ice limits in the British Isles (Griggs and Trenhaile 1994)

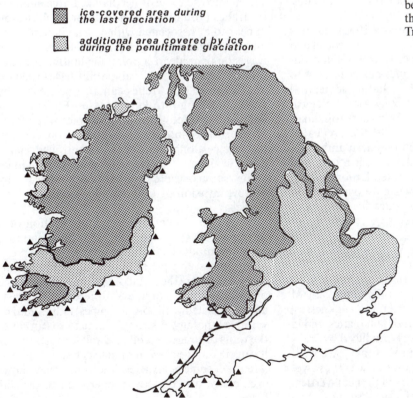

Cliff erosion

Japanese workers have studied the short-term effects of coastal dynamics on cliff erosion. Despite a tendency for erosion to increase with decreasing compressive strength, however, the data were generally inconclusive. Variations in cliff recession rates were partly explained by differences in the frequency of waves that exceed the critical, or minimum, height capable of causing erosion (Sunamura 1992). Although this work tried to quantify important aspects of coastal erosion, many factors other than compressive strength and wave height need to be considered. In south-central England, for example, an easterly decrease in compressive strength is countered by the fracture pattern, which favours increasing rates of cliff erosion to the west, as the mode of failure changes from block detachment, to wedge failures, to topples (Allison 1989).

The erosion of sea cliffs is episodic, site-specific, and closely related to prevailing meteorological conditions. Cliff recession rates have been measured in many parts of the world, although much of the data are little more than rough estimates. Reported rates vary from virtually nothing up to 100 m yr^{-1} (Kirk 1977; Sunamura 1992), but it is difficult to identify patterns that can be correlated with variations in rock type or wave and tidal conditions. A variety of techniques has been used to determine rates of erosion, including sequential terrestrial and aerial photography, morphological evidence, direct measurement using cliff-top stakes or metal pins driven into the cliff face, repeated survey, erosion of ancient and modern anthropogenic structures, old maps, dated inscriptions, and the use of the microerosion meter. Unfortunately, the most accurate techniques have usually been used only to measure erosion rates over very short periods of time.

Caves, Arches, Stacks, and Related Features

Small bays, narrow inlets, caves, arches, stacks, and related features are generally the result of accelerated erosion along structural weaknesses, particularly bedding, joint, and fault planes, and in the fractured and crushed rock produced by faulting. These features form in rocks that have well-defined and well-spaced planes of weakness, yet are strong enough to stand as high, near-vertical slopes, and as the roofs of caves, tunnels, and arches. They are therefore uncommon in weak or thinly bedded rocks with dense joint systems (Trenhaile 1987).

Narrow, steep-sided gorges, or geos, often develop along vertical joint planes or faults in low dipping rocks, and they can also be formed by the erosion of dykes and the collapse of lava tunnels in igneous rocks. The occurrence and shape of caves are usually determined by structural and other weaknesses associated with joints, faults, breccias, schistosity planes, unconformities, irregular sedimentation, and the internal structure of lava flows (Zezza 1981; Davies and Williams 1986). True marine caves tend to be quite small, but large karstic caves have been inherited by the sea through cliff recession. Many cliffs and caves in resistant rocks are not truly contemporary, though they are being modified by marine processes today. Caves in south-western Britain, off northern France, and in southern Australia, for example, are at least as old as the last interglacial stage (Goede *et al.* 1979; Keen *et al.* 1981; K. H. Davies 1983).

It has been assumed that arches and longer tunnels are formed by the coalescence of caves driven into each side of headlands. Large numbers of arches around La Jolla, California, however, appear to be associated with coastal indentations and surge channels, which pile up the approaching waves on one side of a point, facilitating penetration of the weaker rocks at the cliff base (Shepard and Kuhn 1983). Arches can also be created by waves breaking into karstic caves. Fountains of spray are emitted through blowholes when large breakers surge into tunnel-like caves connected to the surface along joint- or fault-controlled shafts. Some of the most spectacular systems are in limestone regions, where they have been inherited by marine invasion of karstic sinks, sinkholes, and tunnels.

Stacks develop where a resistant section of a retreating coast is separated from the mainland. Most are produced by the dissection of the coast along joints and other planes of weakness, although some are the result of folding, submergence of tropical tower karst, and differences in resistance associated with solution pipes, induration, or variable lithology. A strong joint pattern can cause the roofs of arches to collapse and form stacks, or it can encourage erosion of stacks to form arches. Although some stacks develop from arches, however, many form directly from erosion of the cliff face.

Shore Platforms

Shore platforms are erosional rock surfaces created by the retreat of coastal cliffs. Local factors create a wide range of platform morphology, but two main types have frequently been distinguished. Subhorizontal, supra-, inter-, or subtidal platforms terminate abruptly seawards in a low tide cliff, whereas sloping platforms, with gradients between about 1° and 5°, pass below the low tidal level without a major break in slope or abrupt terminus. The classical literature emphasized the occurrence of subhorizontal platforms in Australasia and Hawaii, whereas sloping platforms were usually described from Britain, the north-eastern United States, and elsewhere in the northern Atlantic.

Theories

Most workers accept that mechanical wave action is the main erosive mechanism operating on the sloping platforms in the North Atlantic and other vigorous wave environments. It has been argued, however, that waves cannot cut horizontal platforms because of variations in their strength and the level at which they operate (Gill 1967; Sanders 1970; Hills 1971). Horizontal platforms have therefore been attributed to cliff weathering or the modification of wave-cut ramps by weathering processes. Nevertheless, critical examination of the various weathering theories shows that most accord a major role to wave erosion and transportation (Trenhaile 1987) (Fig. 11.6):

a) Dana (1849) suggested that horizontal platforms are cut in weathered rock by waves, at 'the level of greatest wear'. Bartrum's (1924) supratidal 'storm wave' platform model was similar, although cliff weathering was not emphasized.

b) Old Hat platforms were thought to develop in very sheltered areas, at the level of permanent seawater saturation. Above this level, weak waves wash away the fine, weathered debris, exposing the top of the resistant, unaltered rock below (Bartrum 1916).

The Old Hat hypothesis has been modified to account for horizontal platforms in more vigorous wave environments. It has been proposed that

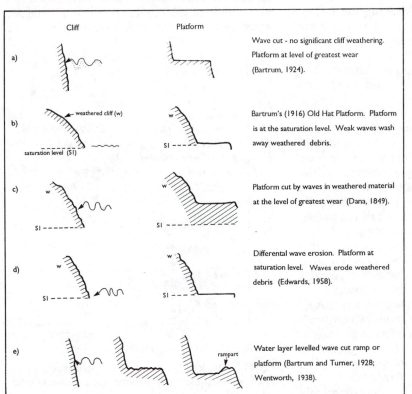

Fig. 11.6. Theories for subhorizontal platform development (Trenhaile 1987)

platforms can be formed by differential wave erosion of cliffs consisting of weak, weathered rocks above the saturation level, and more resistant, unweathered rock below (Edwards 1958; Gill 1967; Bradley and Griggs 1976). Field evidence is lacking for Old Hat platforms, *in sensu stricto*, and in any case, the validity of this theory, and most of its variants, depends upon the existence of a permanent intertidal level of saturation: the evidence presently available suggests that it does not exist (Trenhaile and Mercan 1984).

c) The third group of theories attributes horizontal platforms to the lowering and water-layer levelling of rough, sloping or subhorizontal, wavecut platforms. The inability of these wetting and drying processes to operate effectively at the front of platforms, where the rock is kept wet by spray and splash, was thought to produce residual ridges or ramparts, rising a metre or so above the lowered platform surfaces behind (Bartrum 1935; Hills 1971). Ramparts are often discontinuous, multiple, and not at the seaward edge, however, and some workers therefore consider that they are simply outcrops of locally more resistant rock, which have no genetic significance (Gill 1972*a*).

d) There is ongoing debate over the role of frost action in the formation of strandflats and shore platforms in cool climatic regions (Trenhaile 1983*a*, 1987) (Chapter 12).

e) The role of weathering in the formation of horizontal platforms in Mediterranean and tropical regions is discussed later.

Differences in the morphology of platforms in the North Atlantic and Australasia were traditionally attributed to differences in climate and wave regime. Most of the century-old debate on the origin and development of shore platforms has been conducted in almost total ignorance of the fundamental role of tidal range. There is a strong, positive relationship between tidal range and mean platform gradient in the storm wave environments of the North Atlantic, and it also appears to exist in other regions, irrespective of different climates and wave regimes (Fig. 11.7). It is tempting therefore, to suggest that the very low gradients in Australasia and elsewhere are the result of very small tidal ranges, rather than different formative mechanisms (Trenhaile 1974, 1978, 1987).

The apparent absence of subhorizontal platforms in the wave-dominated North Atlantic provided support for the contention that waves only produce sloping ramps (Hills 1971). Most of this area is macrotidal, however, and it has recently been shown that subhorizontal platforms do exist in several micro- and mesotidal areas in eastern Canada (Trenhaile 1987). Where the tidal range is similar, the gross morphology of these platforms is indistinguishable from those in Australasia. The presence of platforms with seaward slopes of up to 4.5° in the Bay of Fundy, where the maximum spring tidal range is more than 14.5 m, provides additional evidence for the importance of tidal range. Furthermore, although Davies (1972) commented that the platforms studied by Edwards (1958) in north-western Australia 'slope uncharacteristically', their gradient is perfectly in accordance with the 10 m-high tidal range in this region.

The evidence from eastern Canada suggests that horizontal platforms can be cut by waves where there is a small tidal range, thereby supporting the views of several, predominantly Australasian, workers who have insisted that wave-cut shore platforms can be horizontal (Bartrum 1924, 1935; Edwards 1951; Cotton 1963). Although this does not rule out the possibility that they can also be produced by chemical, salt, or frost weathering, or by coastal ice, any attempt to accord a primary role

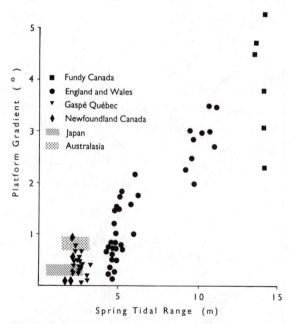

Fig. 11.7. Platform gradient plotted against tidal range. Each point represents the regional average of many surveyed profiles (Trenhaile 1987)

to these mechanisms must explain how they are controlled by the tidal range.

Geological effects

Geological factors are responsible for marked variations in the shape of individual platform profiles, and they can account for different types of platform within areas that have fairly constant morphogenic conditions. In Japan, for example, where the mean tidal range is generally less than 2 m, horizontal platforms are replaced by gently sloping, largely submarine platforms where:

$$\tau_o > 0.005\, S_s$$

where τ_o is the shear stress on the bed and S_s is the shear strength of the rock (Tsujimoto 1987).

Geology influences the development and morphology of shore platforms in many ways (Trenhaile 1980, 1987):

a) The nature, intensity, and efficacy of the erosional mechanisms are governed by the structure, lithology, and mineralogy of the rocks.

b) Geological factors help to determine whether the foot of the cliff and the back of the platform are abraded or protected by accumulating debris and beach sediment.

c) The surface roughness of shore platforms in sedimentary rocks is strongly influenced by the dip and strike, bed thickness, joint density, and variations in the strength of the beds. Platform surfaces are usually coincident with the more resistant members of horizontal or gently dipping strata, but corrugated (washboard) relief often develops in steeply dipping rocks.

d) The age of coastal features, and the possibility that platforms are at least partly inherited from periods when sea level was similar to today's, increase with the strength of the rocks.

Rock strength needs to be determined in relation to the processes acting on it. Compressive strength appears to provide an adequate measure of the susceptibility of different rocks to mechanical wave erosion in some areas (Sunamura 1973; Williams and Davies 1980). Nevertheless, many other factors need to be considered. On a platform in southern Japan, for example, mudstones are eroded more rapidly than interbedded tuffs, despite the fact that they have higher abrasion hardness and greater compressive and impact strength. On the other hand, although the mudstones are probably more resistant to mechanical wave erosion, they have low resistance to flaking, as a result of water absorption or adsorption (Suzuki *et al.* 1970).

Platform morphology

Relationships between platform morphology and other morphogenic, topographic, and geologic factors are less clear than between platform gradient and tidal range (Trenhaile 1987). Mathematical modelling suggests that platform gradient decreases as the strength of the waves increases, and as the rocks become weaker (Trenhaile and Layzell 1981; Trenhaile 1983b). Although there is greater exposure to wave action on headlands, the rocks are generally less resistant in bays, and headland platforms can therefore be either steeper or more gently sloping according to the relative importance of these opposing influences (Hills 1972; Kirk 1977; Trenhaile 1987).

Mathematical modelling suggests that platform width increases with increasing tidal range and wave strength and decreasing rock strength, but the field evidence is often contradictory (So 1965; Takahashi 1977; Trenhaile 1987). Platform gradient and width tend to be positively correlated over areas large enough to encompass significant variations in tidal range, but negatively correlated at the local scale, where tidal range is essentially constant and variations in rock strength and exposure are of much greater importance.

Rock dip determines the degree of protection afforded to weaker beds by the stronger, upstanding members of an alternating sedimentary sequence. Trenhaile (1987) ranked eight combinations of rock strike and dip according to their relative susceptibility to erosion, and their effect on platform width and surface roughness (Fig. 11.8). Fastest erosion was attributed to horizontal rocks and moderately dipping beds that strike perpendicularly to the rock face, and slowest erosion to vertical strata that strike obliquely or parallel to the cliff face. Although bed thickness, joint density, cliff height, and variations in the direction and force of the waves also exert an important influence, the model provides a satisfactory explanation for variations in platform width in several areas. Changes in strike and dip, relative to the orientation of the cliff and the path of the refracted waves, provide an additional explanation, along with rock strength and wave intensity, for variations in platform width along crenulated coasts.

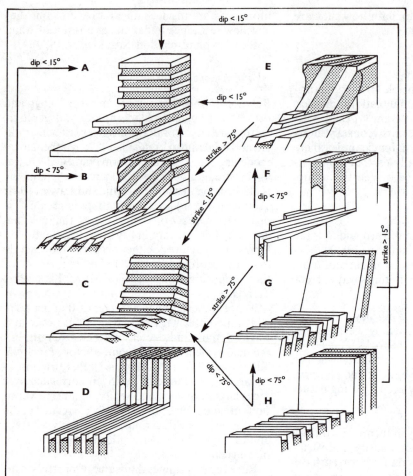

Fig. 11.8. Model of the effect of variable rock dip and strike on erosion rates and platform morphology (Trenhaile 1987)

Recession of low cliffs produces less debris than high cliffs, and they retreat more rapidly in some areas. Although an inverse relationship between cliff height and platform width might therefore be expected, attempts to confirm this relationship are inconclusive. Platform width is also affected by the presence of abrasives, and the degree of protection afforded by superficial deposits at the cliff foot. Robinson (1977c) found that platforms in north-eastern England become progressively narrower as sandy beaches, bare rock, boulder beaches, and talus cones occurred at the cliff base. The effect of superficial deposits on platform width is less clear elsewhere, however, possibly in part because of differences in the amount, grain size, and mobility of the deposits.

The cliff-platform junction is usually close to the high tidal level, but there is considerable variation as a result of geological influences and variations in exposure to wave attack. Junctions in southern Britain tend to be close to the high spring tidal level in resistant rocks, but frequently below the high neap level in weak rocks. Junction elevations are higher in bays in some areas, and on headlands in others, probably reflecting the conflicting influences of rock hardness and wave exposure (Trenhaile 1987).

Attempts to use elevated platforms to identify palaeo-sea levels have been confounded by the lack of consensus on the way in which platforms are formed, and consequently on the elevations at which they develop. In Australasia, and elsewhere in warm meso- and microtidal environments, contemporary shore platforms have been reported at all elevations ranging from the low tidal level to well above the high tidal level. Some workers believe that platforms develop close to the high tidal level while others support the low tidal level, and there is further disagreement over the elevation to which they are ultimately reduced (Bartrum 1916; Fairbridge 1952; Gill 1972b). Differential erosion forms narrow ledges well above the high tidal level. Their origin is usually fairly obvious, but geological control of similar, though larger, features within the intertidal zone is generally more difficult to identify. The genetic significance of platform elevation in microtidal environments is questionable, given that slight variations in lithology, structure, wave exposure, and secondary lowering have a strong influence on the relationship with tidal levels. Furthermore, the possible effect of high Holocene sea levels on platform development in the southern hemisphere has not been adequately considered. The traditional concern with platforms close to, or above, the high tidal level, obscures that fact that most subhorizontal platforms in Australasia, eastern Canada, and elsewhere, are close to the midtidal level. This is consistent with the occurrence of slight, secondary flattening about the midtidal level in the macrotidal environments of western Europe, even though these sloping platforms extend continuously from about the high to below the low tidal level (So 1965; Trenhaile 1972).

Platforms along crenulated coasts are usually higher on headlands, although there are exceptions. Only vigorous waves, which operate at higher elevations, are able to carry out much erosion in resistant rocks, and platform elevation therefore generally increases with the hardness of the rocks (Trenhaile 1987). High platforms in resistant rocks, however, tend to be narrow. They also tend to have gentle gradients, possibly because larger waves operate over a narrower range of elevations than the smaller waves that cut platforms in weaker rocks, at lower elevations. This could explain why high tidal platforms in resistant rocks are horizontal on the Victorian Otway Coast, whereas those in weaker rocks are gently sloping and lower (Gill and Lang 1983; Trenhaile 1987).

Models of platform development

Early models of platform development were descriptive and usually structured within a cycle of erosion. More recently, it has been proposed that platform width is maintained through a balance in the rates of erosion at the high and low tidal levels.

This is consistent with the feedback mechanism that adjusts wave strength and erosion rates at the high tidal level in response to changes in platform width and gradient (Trenhaile 1972, 1987).

There have been several attempts to model wave erosion on rock coasts using plaster and cement blocks in wave tanks. In addition to scaling problems, however, the use of materials that erode much more rapidly than natural rock raises questions about the relevance of the results to natural environments. Most recent attempts to model rock-coast erosion have therefore been mathematical. Most models have been concerned with abrasion and other forms of submarine erosion in tideless seas (Scheidegger 1970; Sunamura 1977). Trenhaile (1983b), however, modelled intertidal platform development according to the relative rates of erosion at the high and low tidal levels. There was a gradual decline in the rate of platform development in each of the model runs, and the eventual attainment of an equilibrium state, when the rates of erosion at the high and low tidal levels had become equal. A slightly modified form of this model was used to examine the effect of different Holocene relative sea level curves on platform development, and middle and upper Pleistocene changes in sea level on the erosional development of continental shelves and elevated coastal terraces (Trenhaile and Byrne 1986; Trenhaile 1989).

The shape of platform profiles, and the close relationship that exists between platform gradient and tidal range, must reflect the way in which wave energy is distributed between the high and low tidal levels. To examine this relationship, the long-term distribution of the water level within the tidal range (Fig. 11.2) was incorporated into a model that considered erosion rates at various levels within the intertidal zone. This model simulated typical platform morphology in Britain, Canada, and Australasia (Trenhaile and Layzell 1981; Trenhaile 1983b). Simulated platforms attained equilibrium when erosion occurred at the same rate at all points along their profiles. This was accomplished by platform gradient varying along a profile in such a way as to compensate for differences in the frequency of tidally controlled wave attack.

Inheritance

In many areas erosional processes may be modifying platforms and other elements of rock coasts that were inherited or partly inherited from the past. This hypothesis is consistent with the palaeo-sea-level record, which shows that interglacial sea levels were similar to today's on a number of occasions during the middle and late Pleistocene (Figs. 2.20 and 2.21). Inheritance is most likely to have occurred on tectonically stable coasts, and in resistant rocks where contemporary rates of erosion are too low to account for the formation of wide shore platforms since the sea reached its present level. On the southern coast of New South Wales U/Th dating of ferruginous and calcareous crusts, and thermoluminescent dating of associated sediments show that a subhorizontal platform, which is awash at high tide, was formed during a period of higher sea level in the last interglacial stage. It was then buried under soil during the last glacial stage, and then exhumed and partly modified by wave erosion in the Holocene. Inclined abrasion ramps, which can rise to more than 10 m above present sea level, are probably polygenic, having developed as sea level rose and fell during the Cenozoic (Bryant *et al.* 1990; Young and Bryant 1993).

There is less justification for assuming that inheritance has necessarily occurred in weaker rock areas. Even if the platforms were inherited from ancient surfaces, in, or slightly above, the modern intertidal zone, till covers, raised beaches, structural remnants, and other evidence would probably have been removed by fairly rapid wave erosion at the present level of the sea. Most workers believe that platforms formed in fairly weak rocks are postglacial features (Hills 1971; Gill 1972*b*; Sunamura 1973; Takahashi 1977; Kirk 1977). Relationships between platform morphology and aspects of the morphogenic environment are often lacking in resistant rocks. Their occurrence in weaker rocks, however, suggests that even if these platforms were partly inherited, they have been substantially modified by contemporary processes. Platforms that may have been partly inherited from the Micmac surface in the St Lawrence Estuary, for example, are now completely adjusted to the contemporary environment (Brodeur and Allard 1983).

Rates of erosion

Many estimates have been made of rates of platform development, but there have been few reliable measurements. Reported backwasting rates on cliffs and platforms range from virtually nothing up to 50–70 m yr^{-1}, and platform downwasting rates from 0.1 to 35 mm yr^{-1} (Sunamura 1973, 1992; Kirk 1977). Downwasting rates often refer to the contributions of single erosional agents, rather than to the total rate of lowering. The micro-erosion meter provides reliable data on absolute rates of platform downwasting (Trudgill 1976; Kirk 1977; Robinson 1977*a*, *b*, *c*; Gill and Lang 1983), but it cannot measure the quarrying of large rock fragments or joint blocks. Microerosion meter measurements are therefore unrepresentative of the total rate of erosion where this mechanism is important, as in most wave-dominated areas.

Limestone Coasts

Many of the processes and landforms of limestone coasts are similar to those of other rock types. Mechanical wave erosion and physical weathering, for example, usually dominate on exposed coasts of all lithologies in cool, storm wave environments. A number of workers have suggested, however, that the susceptibility of calcareous rocks to bio-erosion and possibly chemical solution, distinguish limestone coasts from those in other types of rock. The most characteristic features of limestone coasts develop in lower latitudes, where there are generally weaker waves, higher temperatures, an enormously varied marine biota, and reef limestones, aeolianites, and other young calcareous rocks. Nevertheless, abrasion and other wave mechanisms may account for a large proportion of the erosion of limestone coasts in tropical regions (Trudgill 1976). Furthermore, features generally considered to be typical of limestone coasts, including aveoles, karren (lapies), and shallow pools, are also produced by salt weathering in basalts, granites, and other rocks, particularly, but not exclusively, in low latitudes.

Guilcher (1953) considered that four main types of limestone coast can be distinguished on the basis of temperature and tidal regime. Deep notches and protruding visors are produced by corrosional processes in warm seas, for example, but they must also reflect the generally low tidal range of tropical and Mediterranean regions. Exposure to wave action, rock structure, and other factors create considerable differences in limestone coasts between and within climatic regions. The relief or surface area of coastal karren in south-western Britain may be related to the way that wave energy is expended within the intertidal zone. The relief increases as erosion lowers platform surfaces towards the midtidal level, and decreases as the platform surface is reduced below this zone. As tidal range increases, wave energy becomes less concentrated between the neap tidal levels, and the degree of microrelief about the mid-tidal level therefore decreases (Ley 1979).

Plates-formes à vasques are limestone, and particularly aeolianite, platforms in intertropical and Mediterranean climatic regions (Battistini and Guilcher 1982; Dalongeville and Guilcher 1982). These intertidal platforms are in the form of a terrace-like series of wide, flat-bottomed pools or vasques, and intervening narrow, winding, and lobed ridges (Guilcher 1953; Miller and Mason 1994). They are covered by water at high tide and washed by breaking waves at low tide, with the return flow cascading into successively lower pools. The rims of the pools can be built by organisms, including calcareous algae, vermetids, or even serpulids; residual corrosion features consisting of the pinnacles of lapies; or a combination of the two.

Organogenic formations develop on calcareous coasts in warm Mediterranean and tropical climates (Molinier and Picard 1954; Pérès and Picard 1964; Trenhaile 1987) (Fig. 11.9):

a) Corniche are organic protrusions, 0.5–2 m wide, that grow out from steep rock surfaces at about mean sea level. They are generally largest and best-developed on calcareous rocks, where the physical and chemical effects of spray produce notches in the supralittoral zone. Corniche have been studied in most detail in southern France and in other parts of the north-western Mediterranean, where they consist of the calcareous alga *Tenarea tortuosa* and other diverse Melobesieae algae, although Serpulid worms or sessile Vermetid gastropod tubes can assume a similar role. Despite diagenesis of the interior, they cannot resist the impact of very strong waves, and are therefore best developed in inlets and other sheltered sites.

b) Trottoir are narrow, erosional rock platforms, often with tiers of vasques, at, or a little below, mean sea level. They are thought to result from corrosional erosion in the spray zone, and possibly the protection afforded to the wave battered seaward edges of platforms by Vermetid tubes, and in places, calcareous algae and other organic encrustations. Trottoir are common in the warmer southern and eastern Mediterranean, and in tropical seas. There are trottoir, or surf platforms (ledges),

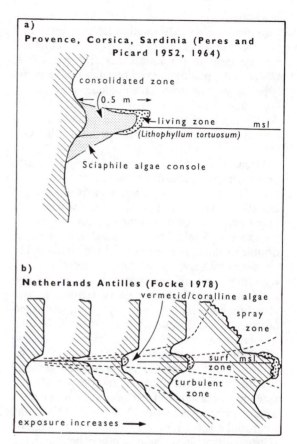

Fig. 11.9. (a) Corniche and (b) trottoir cross-sections

on both sides of the tropical Atlantic. They are up to 10 m wide and 2 m above mean sea level on Curaçao and other islands in the southern Netherlands Antilles, but they only occur in the most exposed areas, where there are thick, well-lithified Vermetid and coralline algae encrustations. In less exposed areas, organic accretions are restricted to a narrow zone in the middle of notches, at about the mid-tidal level. Notch profiles predominate in sheltered areas, without any significant organic accumulation (Focke 1978). As with other types of shore platform, increasing platform width reduces the rate of cliff erosion, and they may therefore eventually attain a state of equilibrium with the processes operating on them (Focke 1978; Woodroffe *et al.* 1983*b*).

Wave-cut notches at the foot of cliffs in cool, wave-dominated areas are usually poorly defined if the rocks are fairly homogeneous, and locally restricted where they reflect lithological or structural weaknesses. Although notches develop in a variety of climates and rock types, they attain their best development in limestones in tropical seas, where cliffs, platforms, and reef flat boulders are often deeply undercut. These notches are typically from about 1–5 m in depth, but much greater depths can be attained where the rocks lack joints and other discontinuities that promote cliff collapse.

Deep, narrow notches usually develop in warm seas where low waves and tidal ranges concentrate the erosive processes within a narrow range of elevations. As tidal range increases, a single notch may be replaced by one at the high and another at the low tidal levels. The presence in some areas of double or multiple notches has been attributed to changes in sea level or intermittent tectonic activity. Other explanations must be considered, however, including the effect of variations in rock structure and lithology, and organisms that operate most efficiently at different elevations. There is no general agreement over the level, or levels, at which notches develop today. The problem is compounded by very small tidal ranges, the general lack of precise measurement, and the paucity and often poor reliability of bench-mark data in many areas. Some investigators believe that the main intertidal notch forms at about the high tidal level (Christiansen 1963), whereas others believe that it is close to the mid-tidal level (Trudgill 1976; Woodroffe *et al.* 1983*b*). Another type of notch forms just below the intertidal zone (Focke 1978).

Debate over the origin of notches in warm seas parallels that over limestone coastal features in general. It was once thought that mechanical wave erosion was responsible, but most workers now accept that chemical or biochemical corrosion, or biological grazing and boring, are usually the dominant mechanisms, particularly in sheltered locations. Nevertheless, abrasion and other forms of mechanical wave erosion are responsible for notch formation in some areas (Vita-Finzi and Cornelius 1973; Tjia 1985). On Aldabra Atoll, for example, abrasion is responsible for about one-third of the erosion occurring in notches where sand is available, whereas grazers carry out between one-third and one-half of the erosion where sand is absent (Trudgill 1976).

Terrestrial karstic features may be inherited and modified by marine processes, generally as a result of coastal recession or changes in relative sea level.

In Victoria, Australia, speleothems give the cliff a fluted appearance, and blowholes have developed in sinkholes (dolines) connected to the sea through narrow, wave-modified conduits. Blowholes near Greymouth, New Zealand, provide spectacular examples of the same phenomenon, and there are also long, narrow geos created by collapse of the seaward portions of the roofs of subterranean stream courses. Five types of marine sinkhole have been described in Asturias, Spain, including those that are now continuously under water, and a variety of partially or periodically inundated forms connected through conduits to the sea (Schulke 1968).

Submerged sinkholes and other elements of karstic landscapes can exert a strong influence on the nature of limestone coasts. Much of the tortuous coast of north-western former Yugoslavia and the sheltered limestone coast of Florida has been produced by the invasion of karstic terrain, and tower karst has been submerged in parts of Java, Vietnam, and Malaysia. Cave collapse and marine invasion have also produced most of the harbours and bays on Bermuda, and they have formed semicircular coves on Malta (Paskoff and Sanlaville 1978). In the south of France partially submerged caves communicate with the sea through narrow conduits or siphons.

Calanques are coastal inlets, often of a gorge-like nature. Those in Provence are well known, but similar features have been described in Corsica, Croatia, Slovenia, and other parts of the Mediterranean. Most workers accept that true calanques (calanques-rias) are karstic dry stream valleys deepened during glacial low sea level stages, and then partially drowned by the sea during the Holocene transgression. Opinions differ, however, on their precise mode of origin (Chardonnet 1948; Corbel 1956; Trenhaile 1987).

Elevated Marine Erosion Surfaces

Gently sloping erosional terraces occupy the hinterlands of many rock coasts. A series of terraces can often be recognized, although the older surfaces may have been greatly dissected by subaerial agencies. Terraces can have caves and stacks on their surface, together with shallow-water marine and terrestrial deposits. There are wide planation surfaces, up to hundreds of metres above sea level, along tectonically stable, plate-imbedded coasts. Those at lower elevations are generally considered to be marine in origin, in contrast to dissected subaerial surfaces in higher upland areas. Flights of emergent marine terraces, ranging from a few metres to several kilometres in width, extend up to hundreds of metres above sea level along tectonically active collision coasts. In most areas, the lowest terraces were cut during periods of high sea level in the last interglacial stage, with previous interglacials being represented by progressively higher and older surfaces.

Although terrace gradients in stable and tectonically mobile regions are usually similar to contemporary shore platforms, some are much wider than the intertidal platforms presently being eroded by the sea. Most workers have assumed that wide surfaces were cut by waves or currents down to considerable depths beneath the water level. This ignores the fact that the most effective erosional processes generally operate at, rather than below, the water level. The effect of middle and upper Pleistocene sea level changes on marine planation surfaces was therefore examined using a mathematical model that was largely concerned with wave erosion within the intertidal zone (Trenhaile 1989). Depending upon rock hardness and wave energy, simulated surfaces of between 1 and 3 km in width were produced at the end of one glacial–interglacial cycle, and much wider terraces at the end of five cycles. The width, gradient (0.5–2°), and concave shape of the erosion surfaces were similar to those of elevated platforms around the world (Fig. 11.10).

The model provides a possible explanation for the way in which very wide marine erosion surfaces are formed, and it supports the contention that they were produced intertidally, particularly when sea level was rising rapidly during periods of deglaciation. Young marine terraces in tectonically mobile areas were uplifted to their present positions, but elevated terraces in stable regions must have formed in the Tertiary, during long periods of high, stable sea level or slow transgression.

COHESIVE CLAY COASTS

Cohesive clay coasts are much more resistant to erosion than cohesionless silt, sand, or coarse clastic coasts, and they have many of the characteristics of weak rock coasts (Bishop *et al.* 1992; Nairn 1992). There are fundamental differences in the erosional behaviour of cohesionless and cohesive coastal sediments. The erosional resistance of cohesionless sediments is determined by the density and volume of the individual particles, and the density and viscosity of the eroding fluid. The resistance of fine-grained cohesive sediments, however, is determined by the forces that exist between the particles, which increase as the distance between them decreases. The forces between clay particles are also influenced by their mineralogy, and the physiochemical environment. Depending on the pore-water chemistry, clay particles can repulse or attract each other, and consequently can be more or less resistant to erosion (Lefebvre *et al.* 1985). The erosive resistance of a cohesive material is therefore complex, and depends upon its compressive or shear strength, clay content, plasticity, consolidation pressure, the structure of the clay material, and the properties of the pore water and the eroding fluid. Consolidation under thick ice is partly responsible for the erosive resistance of cohesive shores that are glacial in origin. The resistance of clays to erosion can also vary according to their water content. In western Wales and along the northern shore of Lake Erie fine-textured tills are resistant to erosion when saturated with water, but they lose all their cohesion when desiccated, and slaking occurs when the material is subsequently wetted by breaking waves (Gelinas and Quigley 1973; Williams and Jones 1991).

Cohesive sediment erosion is also influenced by its structure, the degree and type of fracturing, and the presence of small sandy seams. Pitting and flaking removes material in very small pieces, but separation can occur where fractures and sandy

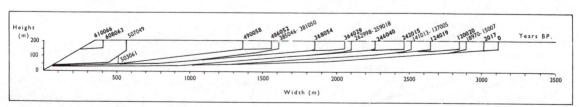

Fig. 11.10. Simulated development of a wide marine erosion surface over five glacial-interglacial cycles (Trenhaile 1989)

planes allow larger units to be removed by spalling (Kamphuis 1990*b*). In eastern Canada, the resistance of intact and unweathered structured clays inhibits erosion at the clay particle level, but erosion occurs through the detachment of larger fragments composed of sand or silt grains, nodules, lenses, or clay aggregates. The removal of clay aggregates in this region appears to reflect the presence of defects in the clay matrix, including microfissures or planes of weakness associated with bedding (Lefebvre *et al.* 1985).

Erosion rates tend to be greater for sand and gravel than for silt and clay, although this is partly countered by the tendency for coarse sediment to form protective beaches at the cliff base. In fine-grained cohesive sediments, eroded material loses its cohesion and the fine debris generally disappears offshore as suspended load. No permanent or continuous beach therefore develops as a result of cliff erosion, unless longshore transport is interrupted by some type of barrier. As the eroded material provides no protection to the cliff, its height has no effect on the rate of erosion. Nevertheless, the erosion of glacial tills, which may alternate in cliffs with water-bearing, glacio-fluvial sands, often releases some granular material,

which accumulates at the cliff foot or on the submarine platform. A small amount of sand or gravel can therefore be swept back and fore by the waves, abrading the underlying surface and accelerating rates of erosion. Larger amounts of material tend to form protective beaches at the foot of cliffs, however, and variations in the erosion rate may then be partially attributed to fluctuations in the elevation, width, steepness, and volume of the beach profile, and consequently in the assessibility of the cliff base to wave action (Jones and Williams 1991; Everts 1991; Shih and Komar 1994).

Lateral and vertical unloading resulting from marine erosion, causes clays to expand and soften, especially if they are over-consolidated and dilatant. Further weakening occurs in fissured or jointed clays through the opening of discontinuities. In temperate regions, erosion of cohesive coasts by marine and subaerial processes is often markedly seasonal in nature, reflecting variations in wave energy, temperature, precipitation, and groundwater pressures (Brunsden 1984; Kostrzewski and Zwolinski 1987).

Shallow sliding occurs as mudslides or, if the water content is very high, as mudflows. Fine-grained, cohesionless soils do not require high

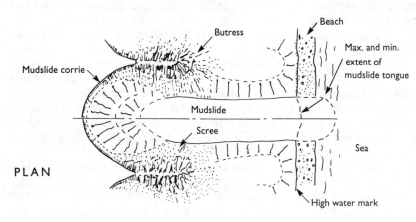

Fig. 11.11. Typical mudslide in London Clay cliff owing to moderate toe erosion (Hutchinson 1973)

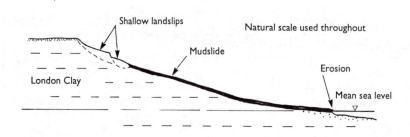

moisture content for flow initiation, but it must be above the liquid limit for flows to occur in sediments with a high proportion of clay. Danish mudslides have water contents as high as 60 percent in spring, and 55–60 per cent has been recorded in active mudslides in Normandy (Auger and Mary 1968; Prior and Renwick 1980). Mudslides have a bimodal shape, consisting of fairly steep feeder flows and gently sloping accumulation flows composed of single or overlapping lobes of clay and debris (Fig. 11.11). The toes have steeper slopes than the accumulation areas, and shallow rotational sliding produces miniature slump blocks. The feeder zones behave like ordinary translational slides that start to move with the seasonal rise in groundwater levels. Surface and subsurface movements of the accumulation flow mainly occur by shearing at its boundaries, and sediment can be incorporated in the mudslides from below. Falls and slides from the bowl-shaped head zone and from the sides also provide debris to the mudslides, and they may help to trigger rapid surges, usually following periods of heavy rainfall (Prior and Renwick 1980; Brunsden 1984; Grainger and Kalaugher 1987). Clay cliffs are also susceptible to deep-seated rotational landslides, which take place where basal erosion is rapid enough to remove mudflow debris and steepen coastal slopes (Bromhead 1978; Hutchinson 1973).

Variations in groundwater level, which change the strength and stability of clay materials, is a crucial factor in the failure of cohesive coastal slopes (McGreal and Craig 1977). Slope failures also occur where grains are removed by seepage or piping of outflowing groundwater, resulting in back sapping and collapse of the free face, often by toppling (Hutchinson 1980; Eyles *et al.* 1986; Nott 1990). Seepage is most common in coarse silt to fine sand, although it can also occur in clays, and it often develops at the base of water-bearing fine, predominantly cohesionless material, underlain by an aquiclude.

There is a strong relationship between landslide and flow activity in cohesive sediments and the occurrence of clays with a high proportion of swelling minerals. These minerals increase the frequency and mobility of slope movements, and allow them to occur on more gentle slopes. Montmorillonite clays, for example, swell to several times their dry volume when hydrated. This increases their plasticity and explains the seasonal nature of many flows, which tend to be most active

in the wet winter months. Alternate freezing and thawing can also weather swelling clays with natural fissures. Danish mudslides occur in clays dominated by swelling montmorillonite minerals, and the most mobile and furthest advanced portion of a mudflow in north-eastern Ireland has the highest montmorillonite and illite-to-kaolinite ratios (Prior and Eve 1975). On St Lucia and Barbados slides in materials with the highest proportion of swelling clay minerals and the highest plasticity cause the greatest distortion and disintegration of the original soil structure, and they have the greatest tendency to flow. The plastic properties of the soils, however, also appear to be sensitive to the types of exchangeable metal cations held in association with the clay minerals. This is especially important in montmorillonite-rich clays, where large amounts of sodium increase their plasticity (Prior and Ho 1972).

Cliffs

Hutchinson's (1973) form/process classification of cliffs in the London Clay of south-eastern England is considered to be generally representative of slopes in stiff fissured clays. Three types were distinguished, based on the relative rates of basal marine erosion and subaerial weathering.

Type 1 occurs where the rate of basal erosion is broadly in balance with weathering and the rate of sediment supply to the toe of the slope by shallow mudsliding. The slope undergoes parallel retreat and erosion only removes slide material, as opposed to *in situ* clay. Removal of the mudslide debris stimulates further sliding, and unloading, softening, and weathering of the exposed clays. London Clay cliffs under equilibrium conditions of moderate toe erosion, with typical cliff edge recession rates between 0.25 and 0.8 m yr^{-1}, develop a series of continuous embayments, each occupied by a shallow mudslide.

Type 2 is found where basal erosion is more rapid than weathering. Waves remove all the material supplied by mudslides and erosion, and undercutting of the *in situ* clay steepens the profile. This eventually causes a deep-seated failure, which is characteristically a rotational slide involving basal failure (Gens *et al.* 1988). The sea removes the slump debris and toe erosion then steepens the slope until another failure occurs. The cliff therefore undergoes cyclical degradation, with slope steepness varying between upper and lower values corresponding to the toe erosion and degradation- or landslide-dominant modes, respectively (Fig. 11.12). This cliff type typically occurs where cliff edge recession rates are between 0.9 and 2.1 m yr^{-1}.

Type 3 coastal slopes develop where there is no basal erosion. When a cliff is abandoned by the sea or coastal defences are constructed at its base, debris is carried to its foot by a series of shallow rotational slides. Abandoned cliffs, therefore, often have a steeper upper slope on which landsliding is initiated, and a flatter lower slope where colluvium accumulates. There is a change in the character of the slips in the degradation zone when the upper slope has been reduced to about 13°. Successive rotational slides then become dominant, and they further reduce the gradient until it is at the ultimate angle of stability against landsliding, which is about 8° in London Clay. The slides then become quiescent and are gradually converted into a series

of undulations, and then into a smooth slope by hill wash and soil creep (Fig. 11.13).

Hutchinson's three cliff types also develop along the northern shore of Lake Erie, in response to long-term changes in lake level. Type 1 cliffs form during periods of normal lake levels, when there are moderate rates of cliff retreat; type 2, or weathering-limited slopes, during periods of high lake levels and rapid retreat; and type 3, transport-limited slopes, during periods of low water levels and slow basal erosion. Variations in wave intensity and the type of glacial sediments in the cliffs, however, also account for the simultaneous distribution of the three cliff types along this coast (Gelinas and Quigley 1973; Quigley and Gelinas 1976). Four types of cyclical instability occur in northern Lake Erie, according to variations in cliff height, wave height, and rates of cliff retreat (Quigley *et al.* 1977). Hutchinson's classification is also generally applicable to cohesive cliffs in Denmark, although some significant differences result from glacio-isostatic recovery and changes in relative sea level (Prior 1977; Prior and Renwick 1980).

Despite the general applicability of Hutchinson's model to a wide range of cohesive sediments, alternate models of cliff retreat are relevant to specific areas. In the glacial sediments of north-eastern Ireland, for example, shallow planar landslides were found to be of greater significance than larger but less-frequent rotational landslides, over a two-year observation period. Although the cliff foot is only subjected to marine attack during extreme tidal and meteorological conditions, it is sufficient periodically to remove the debris produced by landslides, blockfalls, and flows, and then directly erode the cliff base. Although lower gradients may be produced by slope failure, they are steepened by marine action, and cliff retreat in this area usually results in little long-term reduction in slope angle (McGreal 1979).

Hutchinson attributed the transition from mudsliding to deep-seated rotational sliding to increases in the rate of erosion at the foot of the slope, but an alternative hypothesis has been provided by Bromhead (1979), who considered that the correlation of slide processes with erosion rates is largely fortuitous. He suggested that the transition largely depends upon the nature of the material at the crest of the slope, and to a lesser degree on its groundwater hydrology. Mudsliding is the dominant form of mass movement if the crest consists of similar material to the rest of the slope.

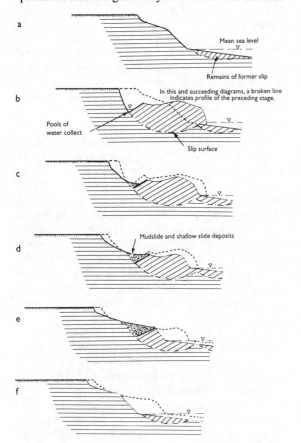

Fig. 11.12. Stages in the cyclic behaviour of London Clay cliffs with strong wave erosion (Hutchinson 1973)

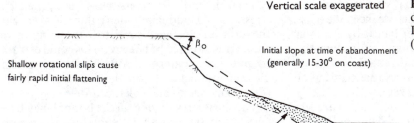

Vertical scale exaggerated

Fig. 11.13. Stages in the degradation of abandoned London Clay cliffs (Hutchinson 1973)

Shallow rotational slips cause fairly rapid initial flattening

β_o

Initial slope at time of abandonment (generally 15-30° on coast)

Colluvium

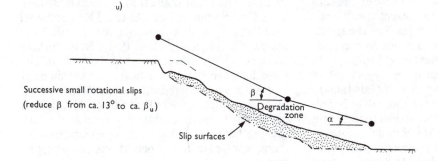

Successive small rotational slips (reduce β from ca. 13° to ca. β_u)

β

Degradation zone

α

Slip surfaces

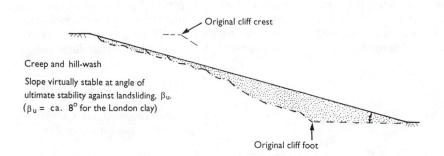

Original cliff crest

Creep and hill-wash

Slope virtually stable at angle of ultimate stability against landsliding, β_u. (β_u = ca. 8° for the London clay)

Original cliff foot

Stronger or better-drained material at the crest, however, inhibits mudslide activity and the accumulation of a protective debris barrier at the cliff foot. This allows wave action to erode and oversteepen the cliff, ultimately precipitating a deep-seated rotational landslide. The slides tend to be of the multiple rotational type when the cliffs are capped with a thick, hard, and jointed caprock.

Shore Platforms

Clay cliffs are often fronted by wide shore platforms. If the platforms are cut from stony glacial tills, however, they may be strewn with lag boulders that retard or eventually prevent wave erosion (Boyd 1986; Healy *et al.* 1990). The formation of shore platforms in abraded glacial till has been modelled for the western Baltic (Healy *et al.* 1987), but there have been few attempts to determine the efficacy of the erosional processes responsible for their development. Platforms consisting of weak landslide material and the planed-off remnants of former landslides are softer, more broken, and consequently more readily eroded than those that developed *in situ*. Major reductions in pore water pressures occur in shore platforms that have experienced recent unloading in front of eroding cliffs, a process that has been termed undrained unloading. Swelling then occurs as groundwater conditions within the platforms slowly recover, reducing the strength of the material as the water content increases. Differential swelling opens up joints and fissures and further weakens the mass. Corrasion, the detachment of blocks by hydraulic quarrying, frost action, alternate wetting and drying, salt crystallization, and faunal boring are among the most important erosive mechanisms on clay platforms (Hutchinson 1986).

Cliff erosion rates of about 1 to 2 m yr^{-1} are typical of the overconsolidated glacial tills and glaciolacustrine clays on the Great Lakes. Continued cliff erosion requires continuous lowering of the subaqueous profile, however, to prevent the dissipation of wave energy before it reaches the shore. Coastal retreat is therefore matched by a corresponding landward shift in nearshore profiles, with generally little change in their shape (Fig. 11.14). In western Lake Ontario subaqueous erosion rates in overconsolidated till range from more than 30 mm yr^{-1} in depths of less than 3 m, to less than 15 mm yr^{-1} in depths of more than 6 m. Rapid decreases in the erosion rate with depth produce concave submarine profiles. Although abrasion by coarse sand and gravel plays an important role in profile evolution, abrasives are largely restricted to shallow water. It is difficult therefore, to attribute downcutting, in water several metres in depth, to abrasion or the erosion of individual grains by wave-induced shear stresses. Cyclical loading and unloading associated with the oscillatory nature of wave-generated stresses could gradually soften a thin till layer near the surface, reducing its strength to the point where shear stresses could be effective. This could explain the presence of patches of soft till, usually less than 20 mm in thickness, on the nearshore profile (Davidson-Arnott and Ollerhead 1995).

The shear stress acting on the bottom may be a very important cause of subaqueous erosion (Coakley *et al.* 1986). Experimental data suggested that the critical shear stress for erosion in western Lake Ontario ranges from 0.5 to 2.3 Pa, compared with the 2–7 Pa values obtained for clay tills from northern Lake Erie (Zeman 1986). Bottom shear stresses, which were computed from hindcaste wave data, exceeded the critical stress for about 61 hours per year. Actual rates of bottom erosion, however, which are similar to those obtained at comparable depths several kilometres to the east, suggest that the hindcaste techniques tend to overestimate wave height and period, and consequently rates of profile lowering.

Other workers have proposed that subaqueous till profiles cannot be eroded by waves in the absence of abrasive material. Recent experimental work, for example, suggests that erosion rates are very low outside the surf zone in the absence of sand. Nevertheless, the erosion of sandless areas

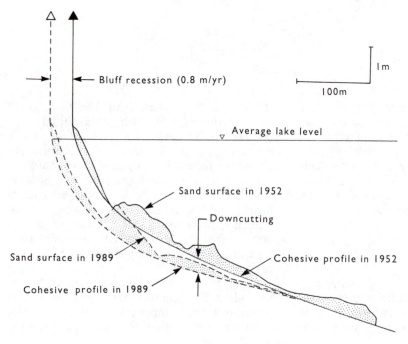

Bluff recession (0.8 m/yr)

1 m

100 m

Average lake level

Sand surface in 1952

Downcutting

Sand surface in 1989

Cohesive profile in 1952

Cohesive profile in 1989

Fig. 11.14. Cohesive submarine profile retreat at Scarborough Bluffs, Lake Ontario (Bishop *et al.* 1992)

under plunging breakers, where turbulent jets are directed down to the bottom, can be similar, or even higher, than in sandy areas within the surf zone (Skafel and Bishop 1994; Skafel 1995). On the north shore of Lake Erie till is generally much too strong to be eroded by wave-generated shear stresses. The critical shear stress to initiate erosion of the softest tills is as low as 6.7 Pa, but for most tills critical stresses must be greater than 20 Pa. Therefore even the largest waves on Lake Erie can only initiate erosion of weaker formations in the breaker zone. Much lower shear stresses are needed to initiate erosion of hard glacial till when granular abrasives are present. Abrasion occurs when glacial till releases even small amounts of sand and gravel, or when the site is adjacent to a source of granular material. Planing of the high points in the fore-shore by abrasion, and protection of the low points by thin veneers of sand, produce a foreshore profile that resembles a sandy beach (Kamphuis 1990*b*). In an experimental study, using intact blocks of cohesive till from the northern coast of Lake Erie, it was found that little or no erosion occurs when sand is absent. Clear water required a critical shear stress of about 7 Pa to initiate erosion, but this was reduced to about 0.8 Pa, corresponding to the threshold of sand movement, when sand was present. On the other hand, too much sand may not be mobile, and it will then protect the underlying surface (Bishop *et al.* 1992).

12 Coasts in Cold Environments

Although some aspects of the morphology and processes operating on cold coasts are similar to those in other environments, they also have some unique characteristics that are attributable to perennial or seasonal sea ice, permafrost, ground ice, frost action, isostasy, glacial history, and other zonal factors.

SEA ICE

Ice reduces the effect of waves in cold environments. Offshore ice prevents wave generation or it limits the fetch, and its presence on the beach or in the nearshore zone reduces the effect of any wave action that does occur. In the Arctic beach sediment transport by waves is inhibited by a thin layer of ice in early autumn, before the ice-foot has developed. The formation of frozen crusts in the lower intertidal zone during low tidal periods can also immobilize sediments where the climate is not severe enough for extensive ice formation. In high energy environments, however, crusts can be removed by waves, scouring by tidal currents, or lifting by anchored ice blocks.

The length of the ice-free season is crucial in determining the role of ice in coastal environments, but other factors also have to be considered, including the surface roughness, age, extent, and continuity of the ice, which influence its strength, thickness, response to winds and ocean currents, and ability to suppress wave activity. Most floating pack ice in the Antarctic is young. Ice disperses fairly easily towards higher latitudes, and it is less likely to be ridged or hummocked than in the Arctic, where constricted waters cause spectacular floe collisions and pressure ridging. First-year ice in Antarctica may be up to 3 m in thickness, but it is seldom greater than 2 m in the Arctic. Older

Antarctic ice can be more than 4 m in thickness, but it is usually less than 3 m in the Arctic. Glaciers and ice shelves supply large amounts of very thick ice, especially in the Antarctic, where there are thick, fast-moving ice streams (John and Sugden 1975).

Ice is a dominant factor in the Antarctic, northern Greenland, and the Canadian High Arctic Islands, where there is perennial pack ice—floating ice that is not attached to the land (Figs. 12.1 and 12.2). It is much less significant on coasts that are only affected during the winter. Although seasonal ice protects beaches from the most severe storms of the year, shore ice may simply displace severe winter wave erosion from the beach to the shoreface, while ice-rafting removes large amounts of sediment (Barnes *et al.* 1993). Ice-pushed and ice-deposited ridges, ice-melt depressions, and other characteristic features of polar and subpolar coasts also form at lower latitudes during winter, but they tend to be smaller and are quickly obliterated by waves after breakup (Davis *et al.* 1976; Knight and Dalrymple 1976; Owens 1976). Nevertheless, the yearly freeze-up and breakup cycle may be preserved in beach sediments in areas with long open-water seasons, where there is sufficient protection from vigorous wave action (Reinson and Rosen 1982).

The ice-foot consists of shore-fast ice, extending well above the normal high tidal level in exposed areas to well below the high-water mark. Tidal cracks separate it from sea ice (Fig. 12.3). Different types of ice-foot can develop, according to the tidal range, shore gradient, and wave and ice conditions. Although originally restricted to areas with a significant tidal range, the term has also been used to describe shore-fast ice in polar microtidal regions, and in the Great Lakes and other tideless seas (Bégin and Allard 1981; Barnes *et al.* 1993). Kaimoos— flat-topped ramparts of alternating layers of ice and sediment—are formed in the

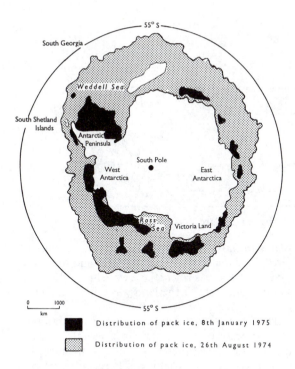

Fig. 12.1. Pack ice distribution in Antarctica (Hansom and Kirk 1989)

upper portion of microtidal beaches by the freezing of swash and wind-blown sand (Rex 1964) (Fig. 12.4).

Several types of ice contribute to the formation of drift ice on the surface of the sea (Gordon and Desplanque 1981). Frazil or slush ice is an unconsolidated mass of water and ice crystals lacking a definite structure. It develops during the initial stages of freezing, although it can also form from snowfall, or from the collision and disintegration of drift-ice fragments. Pan ice is formed through the consolidation of frazil ice. The diameter of individual pans ranges from less than 1 m up to about 10 m, and they can be from a few tens of millimetres up to about 150 mm in thickness. Frequent collisions tend to give them a rounded shape, and they often have a smooth upper surface and a slightly raised rim. The direct accretion of sea water on pans produces thicker ice cakes, that are also circular and rimmed. The coalescence of a large number of pans forms circular ice floes, typically ranging between about 50 and 100 m in diameter, although tabular floes can also develop through the breakup of winter fast ice. Composite ice is composed of a mixture of types. It develops and becomes more abundant as winter progresses, and lateral compression causes pans, floes, and cakes to ride up over each other.

Sediments on cold coasts are seasonally deranged and rearranged by ice and wave action (Hansom and Kirk 1989) (Fig. 12.5). Four periods can be distinguished in an 'ice year' in Arctic Canada (Taylor and McCann 1983):

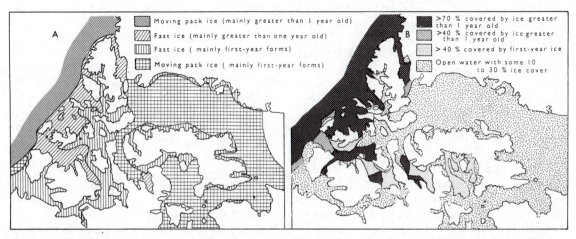

Fig. 12.2. (A) The distribution of sea ice cover in the Canadian Arctic in mid-winter, and (B) in mid-August (Taylor and McCann 1983)

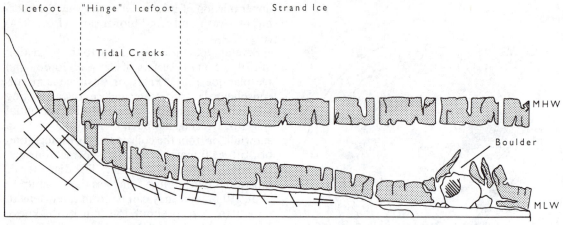

Fig. 12.3. The ice-foot and floating ice at high and low tide (Fournier and Allard 1992)

a) The shore begins to freeze in autumn. Narrow intertidal ice-foots develop from icy slush and frozen swash, spray, and interstitial water. Stranded blocks of ice or floes from the breakup of winter fast ice may also be incorporated.

b) The *in situ* freezing of sea water in winter produces an extensive, unbroken sheet of fast ice, that completely covers the littoral zone. Although it is attached to the land, alternate floating and grounding, as it rises and falls with the tide, is important in areas with a significant tidal range.

c) Breakup begins with melting of the snow cover and enlargement of tidally induced cracks in the ice. Dirty, fractured intertidal ice melts quite rapidly, creating a wide break between the more persistent ice-foot and offshore sea ice.

d) Waves then operate on the coast during the period of fairly open water. The importance of wave action varies from one year to the next, according to differences in the duration of open water, the amount of mobile pack ice offshore, and storm frequency and intensity.

CLIFFS, BEACHES, AND TIDAL FLATS

Postglacial isostatic rebound has been faster than sea level rise in glaciated areas where there was considerable depression of the land, and former beach ridges, barrier spits, barrier beaches, deltas, and other coastal features have been raised to considerable elevations above present sea level. Contemporary coastal features are therefore quite young in areas of rapid uplift. Some coasts in high latitudes are fairly stable today, however, including those

that still have a significant ice cover, including Greenland and probably Antarctica, and those in ice-marginal areas where there was either little isostatic depression, or the land was quickly able to regain its preglacial elevation.

The Antarctic has few major streams, and those that do occur only flow for short periods during the summer. Nevertheless, because of the lack of vegetation, highly variable discharge, little evaporation, no transpiration, and the presence of permafrost, those that do exist are locally important erosional and transportational agents (Hansom and Kirk 1989). Some of the world's largest rivers discharge into the Arctic Ocean. The flow of Arctic rivers is highly seasonal, and a very large proportion of their water and sediment is carried during the spring thaw. As this often precedes the breakup period, some of the sediment may be deposited on to the ice surface. River water draining through holes and cracks in the ice may scour the bottom (strudel scour), forming depressions up to several metres in depth and 20 m in diameter (Reimnitz and Barnes 1987; Forbes and Taylor 1994). Glacially fed rivers have high sediment loads, and moving ice directly supplies rock debris to beaches along glacial termini. Much of the debris floats away on icebergs, but some smaller floating fragments, or brash ice, can be incorporated into beaches, especially where a cliffed ice front is grounded in the shore zone (John and Sugden 1975).

Glacial ice plays an important role in determining the form and sedimentary characteristics of beaches along their termini. A pebble beach is exposed, in places, at the foot of the high wall of ice that extends for over 11,000 km around the

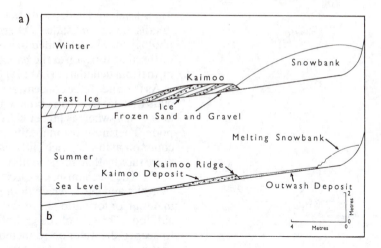

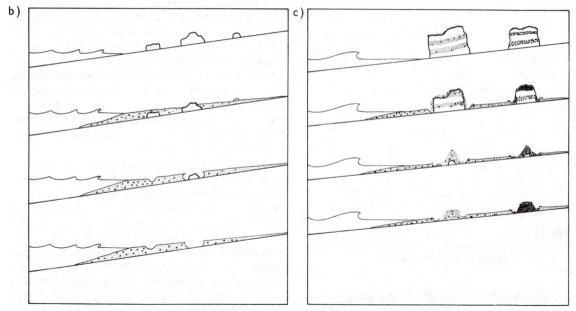

Fig. 12.4. (a) Development of a kaimoo ridge and micro-outwash deposit; (b) the formation of sea ice kettles in kaimoo sediments; and (c) formation of sand and gravel cones from stranded blocks of sea ice or kaimoo fragments (Greene 1970)

Antarctic. Most of this sediment may have been produced by frost riveting, however, rather than by glacial action (Silva 1972). In 1975 the Cook Glacier in South Georgia formed a 30 m-high ice cliff at the high-water mark. By 1982, however, substantial ablation and retreat had transformed it into a gentle, debris-covered ramp, which was located some distance from the high-water mark. Supraglacial morainic till, subglacial till, and glacio-fluvial sands and gravels are the main sources of beach material. Superficial pitting of the beach results from differential melting of buried glacial ice and seasonal snowbanks, and before ice retreat, from locally derived icebergs and brash ice (Gordon and Hansom 1985). A 500 km stretch of coast in south-eastern Alaska, which has more than 100 glaciers, a stormy climate, high relief, and earthquake activity, is one of the most dynamic on Earth. About half the coast is experiencing erosion, and as much as 4 km has been lost

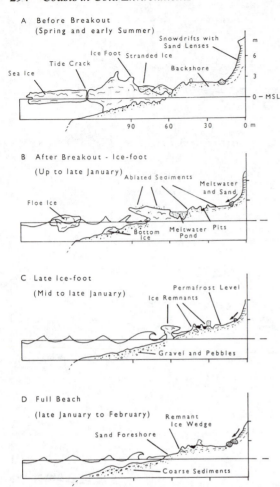

A Before Breakout
(Spring and early Summer)

Snowdrifts with Sand Lenses

Ice Foot Stranded Ice

Tide Crack

Sea Ice

Backshore

m

6

3

0 — MSL

90   60   30   0 m

B After Breakout - Ice-foot
(Up to late January) Ablated Sediments

Meltwater and Sand

Floe Ice

Bottom Ice Meltwater Pits

Meltwater Pond

C Late Ice-foot
(Mid to late January)

Permafrost Level

Ice Remnants

Gravel and Pebbles

D Full Beach
(late January to February) Remnant Ice Wedge

Sand Foreshore

Coarse Sediments

Fig. 12.5. Changes in Antarctic beach profiles as they become less dominated by ice and begin to adjust to wave action (Hansom and Kirk 1989)

in one area in this century. Some glaciers have retreated by over 40 km during this time, and barrier spits, fed by alongshore transport and deposition of glacial sediment across river mouths, have increased in length by as much as 10 km. Glacial retreat has exposed new bays and fiords during the last 200 years, while rapid sedimentation has filled in others. The average sedimentation rate in Lituya Bay, for example, which has three glaciers in its upper reaches, has been about 1 m yr^{-1} (Molnia 1985).

Cliffs consisting of unconsolidated, ice-rich sediments or massive ground ice experience rapid thermal erosion. This results from melting of ice by warmer water, formation of a deep notch at the

water level, and collapse of the overlying material in falls, flows, and slides. Differences in cliff morphology may be attributed to variations in the rate of thermal abrasion at the base, relative to the rate of thermal denudation (Are 1988) (Table 12.1).

As ice tends to be concentrated near the surface, cliffs that are less than about 3 m in height tend to be ice-rich, whereas higher cliffs are relatively ice-poor. Thermo-erosional falls are therefore more common in low, ice-rich cliffs, which tend to erode more rapidly than higher cliffs (Owens and Harper 1983). The mechanism is especially effective in ice-rich silts and clays, which do not provide suitable material for formation of protective beaches at the cliff foot. Thawing of the sea bottom may also help to deepen the water body, thereby increasing wave energy reaching the coast. Thermo-erosional falls are particularly common where notches intersect ice wedges, and the fallen blocks therefore represent a portion of an ice-wedge polygon. Coastal flows can also be initiated by removal of insulating vegetation or unfrozen soil by wave erosion or ice push. Bimodal flows (thaw slumps, ground ice slumps) are similar in appearance to temperate region earthflows, but as they result from melting of large amounts of ground ice their terminal lobes or tongues are much smaller, relative to the size of their bowl-shaped scars.

Coasts are transgressing over thermokarst topography in Alaska and the Canadian Beaufort Sea, and along the northern coast of Russia. Breached thermokarst lakes form lagoons and complex embayments that are partly enclosed by barrier beaches, islands, and spits (Fig. 12.6). The sediment available for beach nourishment depends upon the quantity that is coarse enough to remain in the littoral system, and the proportion of ground ice in the cliffs. Barrier islands usually develop in the southern Beaufort Sea through the breaching of barrier spits, but they can also form from spits attached to small, landward migrating remnants of tundra islands. Although the barrier islands are migrating shorewards at rates of between 2.2 and 3.9 m yr^{-1}, lagoonal coasts are retreating at similar or even higher rates, and barriers therefore seldom weld to the shoreline (Ruz *et al.* 1992).

The occurrence of quite complex barriers and other constructional features testifies to the efficacy of wave action on some of the more open coasts of the Arctic and Antarctic. Nevertheless, ice inhibits sediment movement in cold regions, and there is generally only slight reworking of

Table 12.1. Cliff morphology and rates of thermal erosion

Rate of thermal erosion	Morphology
Higher than the rate of thermal denudation[1]	Vertical cliffs with or without wave-cut notches. Wave action is strong enough to quickly remove the debris produced by thermal denudation in the intervals between successive collapses.
Approximately equal to rates of thermal denudation	Active cliff remains sloping or truncated-sloping[2] for several years; rubble produced by thermal denudation is periodically removed by the sea.
Less than the rate of thermal denudation	On high cliffs consisting of an ice complex, cliffs consist of one or sometimes a series of terraces separated by vertical slopes.
Much less than the rate of thermal denudation	Short stretches of coast involved in thermal denudation, separated by inactive stretches. Inactive cliffs slope seawards, initially at the angle of repose of the thermo-denudational debris.

Notes: [1] Thermal denudation involves the seasonal thawing and downslope movement of sediment. [2] Truncated-sloping cliffs consist of an upper, sloping section and a lower vertical, and sometimes undercut, scarp.

Source: Are 1988.

beach sediments, and small annual changes in beach plan and profile. The longshore movement of sediment is also generally much less efficient than on ice-free coasts with similar wind regimes and fetches. It has been estimated, for example, that net longshore transport along the northern coast of Alaska and the Yukon is only from about 8,000 to 18,000 m^3 per year, which is a small fraction of the amount that is moved in similar ice-free environments (Hill 1990; Dingler 1990). Beach and other coastal sediments therefore tend to have local sources, the material usually being delivered to the coastal zone by streams, mass movement, and other terrestrial mechanisms.

There are marked differences in the length of the open water season, tidal range, and other aspects of coastal environments within high latitudes, which can range from ice- to wave-dominated (Taylor and McCann 1983; Hansom and Kirk 1989). Even where wave energy is limited, ice-rich sediments can be rapidly eroded during short ice-free periods. In October 1963 a storm blowing over the ice-free Beaufort Sea generated a 3.5 m-high surge on the coast of Alaska, and moved more sediment in a few hours than is normally transported in twenty years (Hume and Schalk 1976). In 1970 another storm inundated low-lying tundra and deltas up to 5 km inland (Reimnitz and Maurer 1979). Coastal changes can be very rapid

in ice-rich sediments during open water periods. A maximum erosion rate of at least 55 m per year was recorded on an island in the Laptev Sea between 1944 and 1946 (Are 1988). Rapid, short-term rates of erosion are not necessarily representative of long-term rates, however, which are dependent on debris removal. Therefore, although local short-term rates may be much higher, cliffs on south-western Banks Island, and on the mainland coast of the Beaufort and Chukchi Seas in the western Arctic, are generally retreating at rates of between a few centimetres to more than 2 m per year (Hume *et al.* 1972; Harper 1990; Barnes and Rollyson 1991). Sand and gravel barriers have rapidly developed in the western Arctic. Two barrier spits on south-western Banks Island lengthened by 400 and 600 m in less than thirty years, and in 1972 an island on the coast of north-eastern Alaska lost 40 m to erosion at its eastern end, while a spit at its western end was extended by more than 200 m (Clark *et al.* 1984).

In the Canadian Beaufort Sea there is little relationship between erosion rates and wave energy levels, or the amount of massive ice in coastal cliffs. Human influences are also generally quite slight on high-latitude coasts, although the removal of large amounts of beach material for runways, roads, and aggregate may have exacerbated the erosional problem at Barrow and Wainwright in Alaska

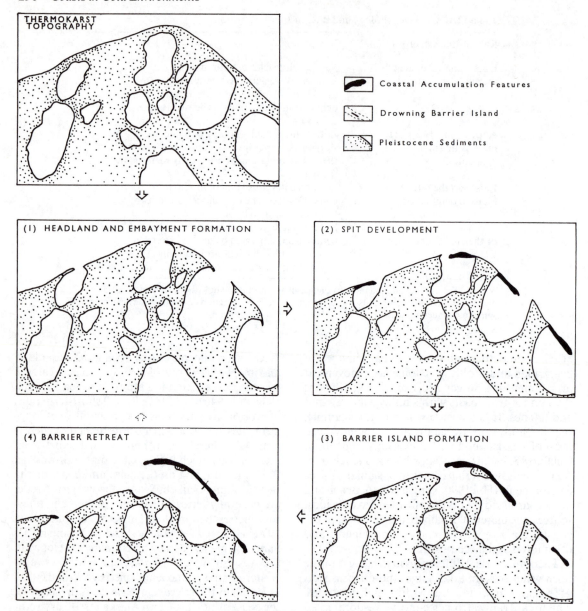

Fig. 12.6. Model of the evolution of a thermokarst coast in the southern Beaufort Sea (Ruz *et al.* 1992)

(Hume *et al.* 1972; Walker 1991). Erosion of coastal cliffs and nearshore zones may therefore partly depend upon deepening of submarine profiles by sea-ice gouging of the bed, and resultant modification of the shoreface as it attempts to establish a profile of equilibrium (Hequette and Barnes 1990). The development of offshore profiles may also be affected by the melting of ice-bonded sediments

that were submerged as a result of shoreline retreat (Dyke 1991).

Although cold region beaches share some characteristics with those in ice-free areas, there are many important differences (Table 12.2). The narrow and poorly developed beaches of cold environments generally consist of coarse sand and cobble, although grounding and melting of ice create

Table 12.2. Some characteristics of modern and raised Arctic and Antarctic beaches

1. Beaches may rest on ice, including buried glacial ice and the ice-foot.
2. Beaches may be pitted as a result of the melting of buried glacial ice, the ice-foot, or sea ice floes.
3. There may be ice-pushed and ice-deposited ridges.
4. Beaches may be truncated, terminating abruptly where there was ice or snowdrift ice slabs when they were formed.
5. Some sediments are ice-rafted.
6. Beach gravel can be poorly rounded.
7. Beaches may be modified subaerially by the effects of frost, gelifluction, and other periglacial processes, including the formation of patterned ground.
8. Bedrock surfaces can be striated by drift ice.
9. Gaps are formed in beach ridges by meltwater streams.
10. Beaches can be associated with ice-contact features and glacio-marine deposits.
11. Elevated beach ridges may contain highly polished ventifacts, although they are not common on active beaches.
12. Beaches can contain cold water fossils, including preserved soft parts.

Source: Nichols 1961.

chaotic assemblages of all grades of sediment. Material indiscriminately eroded, transported, and dumped by sea ice during the winter season can be sorted by waves, winds, and currents during the summer. Nevertheless, beaches are usually less well sorted and rounded than in more vigorous wave environments, and there is generally less variation in grain size and sorting in the cross-shore and longshore directions (Taylor and McCann 1983). If a beach is underlain by frozen ground, ice and storm waves are usually able only to move the unfrozen surface layers. Large storm waves occasionally remove the unfrozen sediment and expose the underlying frost table, however, resulting in melting and erosion of the ice-bonded sediments below. These low-frequency, high-intensity events, which occur about every two or three years in the eastern Canadian Arctic, have a catastrophic effect on beaches.

In addition to its role in limiting wave action, ice has a direct role in transporting sediment on to the land. Large floes or pack ice, driven onshore by winds, waves, and currents, form irregular ice-pushed ridges, striations, and scour marks in the intertidal and supratidal zones of most cold coasts (Gilbert 1990*a*; Dionne 1992; Forbes and Taylor 1994). Ice-pushed ridges are particularly common on headlands, exposed tidal flats, at the mouths of wide, shallow rivers, and on coasts with long fetches that experience moving pack ice for much of the year. High tidal range, boulder barricades, and rocky outcrops inhibit their development in some areas. Bulldozing by grounded ice, ice-pushed boulders, and the scouring action of floating ice also produce a variety of scour marks, pits,

shoals, and ridges on the sea floor (Reimnitz and Barnes 1987; Rearic *et al.* 1990).

The distinction can be made between ice ride-up and pile-up (Reimnitz *et al.* 1990). Ice ride-up occurs when a sheet of ice scrapes or slides up a beach as a smooth, essentially unbroken, sheet, sometimes continuing for hundreds of metres inland. Layers of sediment frozen to the base of the wedge-shaped ice front may be removed, and carried on to the land to form ice-pushed ridges. These ridges are usually quite low, and they can be quickly removed by storm waves. Deeper bulldozing of unfrozen sediments beyond the shore may form larger, more permanent ridges and mounds, ranging up to more than 2 m in height. Ride-up is most common during the early period of ice growth, when the beach and submerged bottom are ice bonded, although it can also occur in the spring. Ice pile-up, which is also most common in Autumn involves slow buckling of a sheet of ice as it advances shorewards, possibly beginning in the region of the tidal crack between bottom-fast and floating fast ice. This produces piles of ice rubble that can grow up to more than 20 m in height, on, or partly seawards of, the beach. When the ice melts during summer, shoreface sediment, which was excavated by the thrusting ice and incorporated into the rubble pile, is deposited as a blanket of sand and coarse clasts. These melt-lag deposits form mounds that are generally up to a few metres high, although they can occasionally range up to as much as 18 m in height.

Buried snow and ice melting on beaches during the thaw season produces kettles or pits, elongated depressions, sand and coarse clastic mounds, conical hummocks, ridges, and sedimentary structures (Moign and Guilcher 1967; Taylor and McCann 1983) (Fig. 12.4). Pitting can result from melting glacial ice, ice-foots, partially buried sea ice driven onshore during storms, and smaller brash ice buried at the storm-wash limit or in overwash deposits. Small deltas, outwash deposits, and erosional scars are also formed where meltwater from melting snow and ice runs across beaches in small braided channels. Ice-melt features only persist, however, above the high tidal level (Taylor and McCann 1976; Martini 1981; Dionne 1988*a*).

Sea ice may contain largely coarse material, which may have been lifted from the bed, or fine-grained sediment that is usually in a layer near the top of the ice cover. Coarse debris is probably entrained through ice gouging, flotation of anchor ice, grounded ice freezing to the bottom and subsequently floating in shallow water, or an influx of ice-bearing sediments from rivers, rock-falls from sea cliffs, and aeolian activity. Although coarse debris is usually concentrated at the ice base, it can be at the surface along fissures, in ice-foot pustules, and in ice blocks that were turned upside down by ice pressures or agitated water, before being incorporated into the ice cover during freeze-up. Ice blocks can also be lifted, or pushed to the surface of an ice cover at high tide, during freeze-up and breakup. The floes then release their sediment load onto the ice surface as they melt (Dionne 1993*a*).

Fine-grained sediment can be deposited directly on to the ice surface by wind, gelifluction, and small streams, and by water flushed up through cracks in the ice as it is submerged by the highest spring tides (Dionne 1993*a*; Gilbert 1990*b*). Fine-grained sediment can be entrained by ice in shallow water through bottom freezing, and the freezing of wave spray containing suspended sediment. It is generally agreed, however, that the most important mechanisms in deeper water, or in areas with a small tidal range involve the formation of slush or frazil ice during freeze-up, when turbulence in the water column keeps sediment in suspension (Osterkamp and Gosink 1984) (Table 12.3).

Fine-grained sediment is resuspended during fall storms and entrained in the minute, disc-like ice crystals, or frazil, that form in turbulent, slightly supercooled water. These crystals float to the surface, mixing with other types of ice and snow and, under suitable meteorological conditions, they rapidly freeze to form a thick layer of sediment-rich ice. Sand and coarser sediment can also be entrained when turbulence carries sticky frazil ice to the bottom, where it adheres to the substrate. This anchor ice is eventually released from the bottom and incorporated in the ice canopy (Reimnitz *et al.* 1987; Kempema *et al.* 1989). Concentrations of fine-grained sediment in sea ice are often far higher than in sea water, possibly because of surface effects and sediment rejection by growing frazil ice crystals as they congeal into an ice cover. This forces sediment particles into flocs in the interstices between crystals and in association with air bubbles (Osterkamp and Gosink 1984).

Sediment frozen to the base of grounded ice during low tide is lifted from the bed when it floats at high tide, and repetition of this process builds a series of thin laminae into the ice base. Although the mechanism may entrain sediment ranging up to small boulders, it is more effective for fine-grained material. The ability of ice floes to lift

Table 12.3. Sediment entrainment processes

Largely coarse-grained material (may also entrain fine-grained sediment)
  1. By anchor ice, with subsequent flotation and incorporation into the overlying ice cover.
  2. By discharge of sediment-laden anchor ice from rivers.
  3. By ice gouging at the sea bed.
  4. By seabed freezing and subsequent flotation.

Fine-grained sediment
  5. By frazil ice during initial ice cover formation under conditions of high turbulence.
  6. Entrainment and deposition under the ice cover by frazil ice formed in leads or in a broken ice cover.
  7. By 'katabatic' flow of dense, cold brine, formed nearshore under the ice and draining downslope (offshore) where suspension of sediments and frazil ice formation could occur.
  8. By anchor ice formation of very short duration and with subsequent flotation.
  9. By trapping in frazil ice when sea water is forced, or filtered, through it.

Source: Osterkamp and Gosink 1984.

boulders depends upon their thickness and size. It has been estimated that an ice pan with a volume of 7 m^3 is needed to transport a boulder 1 m in diameter, while 452 m^3 is required for one 4 m in diameter (Gilbert and Aitken 1981). The main constraint on the ability to transport large, isolated boulders, however, may involve the entrainment rather than flotation mechanisms. It is neither clear whether freezing at the base of grounded ice can entrain larger boulders, nor whether pressures generated by repeated grounding and floating of closely packed ice floes during the early stages of breakup can thrust boulders on to the ice surface. Boulders often protrude through circular, cone-shaped deformation structures (*pustules de pied de glace*, cratered pustules, ballycatters) formed in intertidal ice as it rises and falls with the tides. The boulders may be temporarily lifted as these structures develop, but as they are focus points for the initial melting of the ice, it is likely that they are deposited *in situ* during the early stages of breakup (McCann *et al.* 1981*a*). Movement of boulders with diameters greater than about 0.5–1 m may therefore be by rolling or sliding, rather than by flotation (McCann and Dale 1986).

Most estimates of ice sediment concentrations in the shore zone exceed 10,000 t km^2. The ice-foot on Lake Michigan, for example, has a sediment content of up to 59,000 t km^2, and in the southern part of the lake longshore transport by mobile slush and brash ice is equal to about 13 per cent of the calculated ice-free wave-induced transport (Miner and Powell 1991; Kempema and Reimnitz 1991). Several million tons of sediment is ice-rafted each year along the St Lawrence Estuary, and it is an important component of the sediment budget in the turbidity zone in the middle portion of the estuary. Although most of the sediment is fine-grained, rock fragments up to a metric ton in weight are found on the ice surface. About 20 per cent of the entrained sediment load in this area stays within the tidal zone when the ice breaks up, about 40 percent is returned to the offshore zone, and 40 percent escapes to deeper water. The general absence of permanent sedimentation in this part of the estuary is the result of a rough balance between the amount of sediment moved out of the turbidity zone each year by ice-rafting, and the amount brought in by normal estuarine processes. Dionne (1984) estimated that the ice at Mont-magny, east of Quebec City, contained up to 60,000 t km^2 of mud, whereas concentrations of predominantly coarse clastic material in the rocky, upper intertidal zone in the Upper St Lawrence were between 30,000 and 35,000 t km^2 (Dionne 1993*a*).

The sediment content of the ice is about 90,000 t km^2 in Pangnirtung Fiord on Baffin Island, and about 64,000 t km^2 in the intertidal zone in Frobisher Bay (Gilbert 1983; McCann and Dale 1986). Although sediment loadings are similar to those in the St Lawrence Estuary, offshore ice remains in place in Frobisher Bay while the intertidal ice is melting, and the entrained sediment is therefore recycled within the intertidal flats. High sediment loadings in southern Baffin Island and the St Lawrence reflect high tidal ranges, which facilitate sediment entrainment by bottom freezing. In northern Alaska, however, where there is a low tidal range and sediment is entrained in the ice by wave turbulence, sediment concentrations are less than 1,000 t km^2 (Barnes *et al.* 1982).

There are often large numbers of boulders in the intertidal zones of cold coasts, arranged in the form of mounds, ridges, fields, pavements, or barricades. Boulder fields consist of randomly distributed rocks in loose or dense concentrations, perched on, or embedded in, wide, flat, intertidal zones (Hansom 1986; Forbes and Taylor 1994) (Fig. 12.7). Small ridges and depressions around boulders show that they are being pushed or rolled seawards by the ice on a tidal flat on the northern shore of the Gulf of St Lawrence, over distances ranging from a few metres up to several kilometres (Dionne 1981). Scattered boulders, which commonly have long axes of between 2 and 4 m, lie fully exposed over the surface of tidal flats in Ungava Bay, in northern Quebec. These boulders appear to have moved downslope during coastal emergence, probably by ice entrainment and seaward movement during breakup (Lauriol and Gray 1980). Boulder fields in south-eastern Baffin Island also probably developed through progressive downslope movement of boulders (McCann *et al.* 1981*b*). They are often distributed in linear patterns in this area, as nets or garlands, with diameters between 8 and 16 m, enclosing almost boulder-free areas of nearly flat sand. Garlands may result from individual ice pans, or thicker portions of the coastal ice sheet, pushing boulders aside as they rise and fall with the tide. They are also perpetuated by the tendency for isolated boulders to be moved more easily than those that are clustered in groups. About 21 percent of the boulders moved during a four-year period, about half of them for distances of less than 15 m. Almost 80 per cent of these boulders were located more than 1 m away from their nearest neighbour. Although there was a tendency for seaward migration towards the boulder barricade, there was also a longshore component near the shore. Preferential movement of isolated boulders projecting above the intertidal surface could reflect the greater ease with which they are frozen into the ice in winter, compared with boulders that are grouped together. The absence of furrows or tracks to the lee of boulders, or ridges in front of them, suggest that they are not pushed or rolled by ice pans, unlike areas where greater fetch allows floes to be driven onshore (Gilbert and Aitken 1981).

Well-developed boulder pavements are intertidal surfaces consisting of a compact arrangement of faceted stones, with their larger, flatter, and often striated surfaces facing upwards. Boulder pavements can develop on shore platforms, glacial tills, or tidal flats. They result from grounding of floating ice, which bulldozes, packs, and compacts ice-pushed or ice-rafted boulders into a smooth, highly polished mosaic (Fig. 12.8). They could also develop through the winnowing of stony glacial till by waves or running water, and subsequent embedding of the boulder lag by grounding ice.

Pavements form where there is a supply of boulders, a gently sloping intertidal surface, and the frequent, predominantly onshore, movement of floating ice (Hansom 1983*b*, 1986; Eyles 1994).

Boulder pavements on an island off southern Alaska cannot be attributed to the effects of sea ice. Although the consistent orientation of striations on the boulders are indicative of glacial ice, the shape and orientation of the boulders themselves are not consistent with subglacial transport and deposition. Furthermore, the sediments above and below the pavements appear to be glaciomarine in origin, rather than having been deposited directly by grounded glacial ice. An alternative explanation for pavement formation in this area combines elements of both the subglacial and cold coastal models. The first stage involves formation of a subaqueous boulder lag surface by wave and current winnowing of glacio-marine deposits, followed by increasing ice volume and falling sea level, with abrasion of the lags occurring beneath partially floating ice (Eyles 1988).

There are roughly circular structures with depressed centres and elevated borders on boulder pavements in James Bay and the sub-Antarctic (Martini 1981). In the Antarctic stone circles, nets, and polygons are produced by large ice blocks, brought in by high tides and stranded far up-slope.

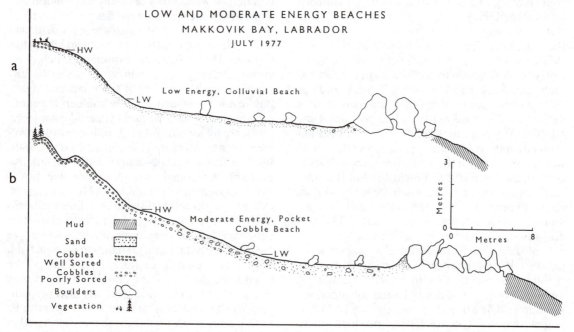

Fig. 12.7. Low- to moderate-energy beaches with boulder barricades and boulder fields in Labrador (Rosen 1980)

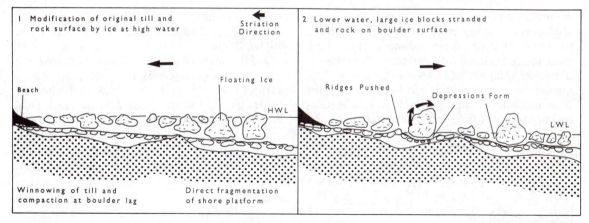

Fig. 12.8. Compaction and polishing of boulder pavements, and the formation of depressions, by stranded ice blocks (Hansom and Kirk 1989)

Rotation of the blocks concentrates finer rock fragments beneath them, while larger material is expelled to the borders. Cross-shore transport of ice blocks forms long, fine-grained furrows with elevated, coarse-grained rims. The furrows and circular structures are quickly destroyed by waves, however, and brash ice reworks the coarser material until the stable pavement organization is restored (Hansom 1983*b*). In the Gulf of Alaska pavements are either flat or they have a nucleated form consisting of irregularly spaced hummocks of smaller clasts distributed around a core of larger boulders. The hummocks may be oriented like small drumlins, with their stoss slopes facing seawards. Although ice block rotation could help to form them, the hummocks partly reflect boulders shielding smaller clasts from strong currents, thereby preventing their movement (Eyles 1994).

Boulders can also be concentrated in a partly submerged row or barricade near the low tidal level. Barricades are separated from the shore by a gently sloping nearshore zone that is frequently occupied by a boulder field. Where there are double barricades (Lauriol and Gray 1980; Gilbert and Aitken 1981), it is assumed that the inner one marks the former edge of the flats, whereas the outer one developed later, following a drop in relative sea level and progradation. Boulder barricades are particularly well developed in Labrador (Rosen 1980) (Fig. 12.9), but they also occur elsewhere in eastern Canada as far south as the St Lawrence Estuary, and in Iceland, the Baltic Sea, and Fennoscandia. Although high tidal range is conducive to bottom freezing, entrainment, and lifting of boul-

ders, the occurrence of barricades in microtidal regions suggests that meteorologically induced changes in water level can perform the same function.

Boulder barricades develop in a number of ways. The boulders may be pushed or dropped where wind- or tide-driven slabs of ice are prevented from advancing further landwards by the ice-foot or shore-fast ice. There is a conspicuous, intermittent line of large boulders in front of the littoral berm in the South Shetland Islands. These boulders were probably deposited in the intertidal zone or at greater depths by icebergs, and subsequently pushed up the beach by the impact of waves laden with brash-ice. On the other hand, barricades in south-eastern Baffin Island consist of boulders that may have been transported seawards by ice, and deposited at the landward edge of the solid sea-ice cover, which persists while the intertidal ice is breaking up (McCann *et al.* 1981*b*; Gilbert and Aitken 1981). The fact that high boulder barricades are only found in estuarine areas in macrotidal Ungava Bay suggests that they result from the interplay of tidal and river currents in this area. In winter, evacuation of fluvial bed load is hindered by stable ice in the intertidal zone, and barricades then develop as boulders are pushed up by ice floes in the river and the sea (Lauriol and Gray 1980).

In ice-free regions tidal flats consist of fine-grained sediment deposited by tidal currents, but in cold regions there is also a large component of coarse, ice-rafted material, ranging from pebbles to boulders. Tidal flats in high latitudes are commonly erosional surfaces, cut into unconsolidated deposits along emergent, glaciated coasts, with very little or no net sedimentation taking place today. Seasonal frost and permafrost tend to dominate at higher latitudes where ice is of little importance, whereas ice is often more important than frost at lower latitudes (Table 12.4). The relative importance of frost and ice varies spatially and temporally, however, according to hydrological, glaciological, and geological influences (Dionne 1988*a*, 1989).

Tidal flats in cold regions are characterized by a variety of ridges, grooves, furrows, small craters, various drag, roll, and skip marks, and other ice-made features (Dionne 1988*a*; Martini 1991). Tidal flats are polished and striated by flat-bottomed ice

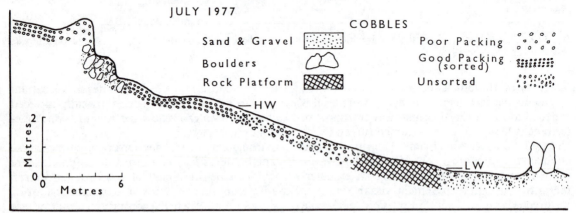

Fig. 12.9. An ice push ridge and boulder barricade in Labrador (Rosen 1980)

Table 12.4. Classification of cold-region tidal flats

Tidal flat type	Ice season (months per year)	Sample locations
Little ice influence	< 1	Nova Scotia, New England, Wadden Sea, North Sea
Moderate ice influence	1–3	Bay of Fundy, southern Gulf of St Lawrence, Baltic Sea
Largely influenced by ice	3–5	St Lawrence Estuary, northern Gulf of St Lawrence, possibly Bothnic Gulf
Ice-dominated and influenced by frost	5–7	James, Hudson, and Ungava Bays, Foxe Basin, Frobisher and Cumberland Bays, possibly White and Barents Seas
Largely frost-dominated, little-to-moderate ice influence	> 7	Canadian Arctic Archipelago, Beaufort Sea, Northern Alaska, Northern Bering Sea, possibly the Kara, Laptev, and Siberia Seas

Source: Dionne 1988*a*.

cakes moved by ebbing tides, but grooves are eroded where the keels of ice cakes or ice floes are dragged along the bottom. Grooves can be straight, sinuous, curved, or crooked, and they can be up to 2 km in length, although they are generally less than 1 m in width and 0.3 m in depth. Many are double tracks, but single furrows are also quite common, and they are often accompanied by small lateral ridges, and a larger frontal ridge. In the St Lawrence the horns of lunate ridges generally point towards the land, reflecting the dominant influence of seaward-moving ice floes during breakup. In James Bay, however, most grooves and ice-push features reflect the dominance of landward-moving ice, possibly because of the formation of thick ice in this micro- to mesotidal environment, ice piling up on the coast during storms, and the obstruction of seaward-moving shore ice by pack ice (Martini 1981). Large ice-pushed boulders create similar furrows up to about 40 m in length, with widths that are roughly determined by the size of the boulders. In the St Lawrence Estuary ice floes dragged or pushed a boulder weighing 176 t about 3 m seawards, and boulders of between 20 and 50 t have moved 20 to 40 m seawards (Dionne 1988b). Tidal flats also contain ice-scoured depressions that create a chaotic micro-relief. Circular and subcircular depressions, sometimes with a raised rim, develop where ice floes are alternately raised and lowered by waves and tides. Similar features are formed when a block of ice pulls away a layer of sediment or a boulder frozen into its base, and on sandy tidal flats where meltwater drains through holes in the ice cover. The melting of small pieces of buried ice also creates small kettle holes in sandy and muddy tidal flats.

Sedimentary processes on Arctic and subarctic tidal flats are affected by permafrost and seasonal frost. Cryogenic processes heave ice-drifted boulders and concentrate them on the surface of tidal flats in subarctic Quebec, and to a lesser extent along the St Lawrence Estuary. Tidal flats may be raised by frost mounds or the formation of segregation ice in the underlying permafrost, whereas melting ice lenses create fairly large, circular depressions. Hydrological pressures generated by ice melting beneath the surface are also responsible for a variety of injection features, including miniature mud volcanoes and small polygonal ridges (inverted mud cracks).

Winter ice does not completely protect coastal marshes from wave, tide, or current action. Although ice is frozen to the surface of the upper marsh, it is only frozen intermittently and discontinuously to the lower marsh surface. Turbid water penetrating beneath the ice at high tide can therefore cause local erosion or sedimentation in the lower marsh. Tens of centimetres of mud are deposited in this way each year in the St Lawrence Estuary, and it is easily incorporated into the ice cover. A large volume of mud and fine sand is also spread over the ice surface when turbid water is flushed through cracks in the ice cover. This sediment is then incorporated into the ice through freezing (Dionne 1988a).

Ice dragged by waves or tidal currents removes thin layers of marsh surface, forming shallow depressions. This may create patterned, jigsaw-like marshes with pools comprising between one-half and one-third of the surface area. Long, narrow, and straight furrows and pools develop, however, where tidal and longshore currents are funnelled through narrow inlets and straits (Martini 1981). Winter ice along the walls of drainage channels restricts tidal flows, thereby increasing the energy available for channel incision. This may help to produce deep drainage channels in macrotidal regions. On the other hand, scouring by drift ice moved through the channels by tides may help to increase their width (Bartsch-Winkler and Ovenshine 1984; Dionne 1988a).

Although ice and ice-boulder scour occur during breakup, most erosion takes place when ice blocks, lifted by spring tides, tear large pieces of turf from the marsh surface in the upper inter- and supratidal zones. The mechanism is very effective in the St Lawrence, but because of lower tidal range, somewhat less so in Hudson and James Bays (Martini 1981). Turf slabs are deposited in deeper water and strewn about the marsh surface and on bare tidal flats (Hardwick-Witman 1986). Clumps deposited on the lower marsh form a nucleus for the deposition of fine-grained sediments, and spotty vertical and lateral accretion then causes marshes to grow irregularly, producing a chaotic micro-relief.

Sedimentation rates are probably lower on salt marshes in cold than in temperate regions because of seasonal ice cover, offshore transport of mud by ice floes, and the cutting down of marsh grasses by ice, which reduces deposition in the spring. In the St Lawrence Estuary, for example, average annual sedimentation rates on seven high tidal marshes for the last thousand years range from only 0.8 to 2.75

mm yr^{-1} (Dionne 1986). Patterns of sedimentation are also modified by more-random deposition by floating ice. Marshes in cold regions usually contain a fairly high proportion of coarse sediment, including many large, ice-rafted boulders. Along the St Lawrence Estuary, for example, marshes are strewn with ice-transported boulders and other coarse material of varied composition. Many of the boulders are of local origin, but some have been picked up from sites along the estuary as much as 100 to 200 km away. Marsh pools or salt pans can also be formed by ice erosion, melting ground ice, or permafrost degradation. In northern Quebec pans have been attributed to the effects of ground ice, thermokarst, frost heave, and thermal erosion, reflecting the presence of permafrost or seasonally frozen ground ice (Dionne 1988a, 1989; Fournier *et al.* 1987) (Fig. 12.10).

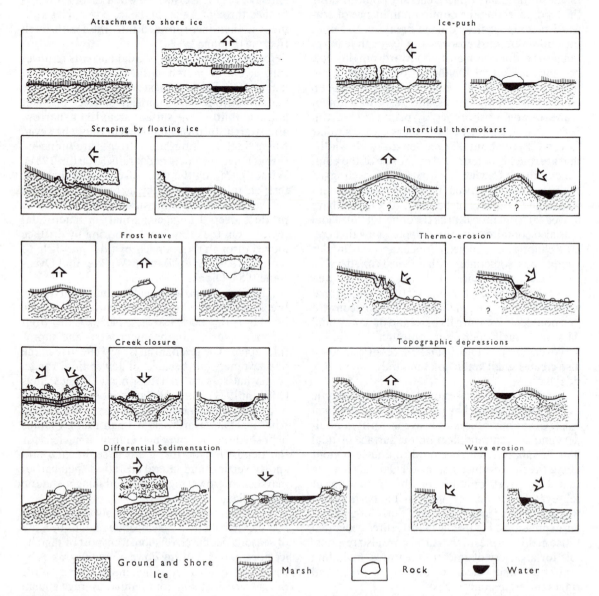

Fig. 12.10. Modes of formation of marsh pans in northern Quebec. The direction of the arrows distinguishes movement by heaving, pushing, scraping, or collapsing (Fournier *et al.* 1987)

ROCK COASTS

Cold coasts are generally low wave-energy environments, because of ice, and in some cases, the shape of the water body. Weathering is therefore important in preparing resistant rock surfaces for erosion. Although large amounts of debris accumulate at the foot of cliffs, however, it is not yet clear whether it reflects rapid weathering in high latitudes, or the inability of weak waves to remove it.

The long-term effectiveness of frost, temperature-dependent wetting and drying (Chapter 11), and other weathering mechanisms depends upon removal of the debris, which prevents slopes becoming buried under talus (Howarth and Bones 1972). Data on rates of cliff retreat in cold coastal regions are often contradictory and of questionable reliability. Estimates generally range between 0.0001 and 1 m per century in northern Fennoscandia and Spitsbergen, but much higher rates of 4–50 m per century have been cited from Gottland, Sweden, Spitsbergen, and the South Shetland Islands (Trenhaile 1987).

Cold coastal regions may provide optimum conditions for frost action, including:

a) high levels of saturation in cliffs and shore platforms through tidal inundation, groundwater flow, snow and ice melt, and wave spray and splash;
b) salts in sea water, which appear to increase its severity; and
c) frequent, rapid, tidally induced changes in intertidal temperatures.

Tidally induced frost action occurs in the intertidal zone when air temperature is below freezing, while the temperature of the water is slightly above. Intertidal rocks freeze when exposed to air by the falling tide, and thaw when covered by the rising tide. The mechanism may be particularly effective because of the frequency of the freeze–thaw cycles, especially in semi-diurnal tidal environments. In coastal Maine, for example, there are on average 133 freeze–thaw cycles each year in the intertidal zone, compared with only thirty to forty inland (Kennedy and Mather 1953). Intertidal frost action may also be particularly effective because of rapid changes in temperature induced by sudden emergence and submergence of the rocks. Laboratory experiments confirm that rock temperatures rapidly increase when inundated in sea water, but they suggest that they decline very slowly when

exposed to air (Robinson and Jerwood 1987*b*). High saturation levels are conducive to effective frost weathering in the intertidal zone. Saturation levels in several types of rock were greater than 90 percent when first exposed by the falling tide on the midtidal, subhorizontal platforms of Gaspé, Quebec, and between 75 and 80 percent on various portions of the sloping platforms of the Bay of Fundy (Trenhaile and Mercan 1984). High saturation levels are also attained in sorption-sensitive rocks during periods of high humidity. The breakup of argillaceous, fine-grained rocks can then be accomplished by temperature-dependent wetting and drying, as they adsorb water and expand as temperatures rise, and desorb and contract as temperatures fall (Hudec 1973).

Tidally induced frost action is inhibited at high latitudes by low water temperatures, but water and air temperatures suggest that it may occur at various times of the year in cool temperate regions (Trenhaile 1987). Normal frost action, resulting from changes in air temperature, may also be most effective in the mid-latitudes than in higher latitudes, where there are less frequent fluctuations about the freezing point. On the other hand, the length and intensity of the freezing periods, which favour the growth of ice crystals and lenses, are greatest at high latitudes.

It is difficult to assess the role of chemical and salt weathering in cold environments. Chemical weathering is generally considered to be slow in Antarctica, and tafoni and honeycombs have been attributed to frost action and salt crystallization from sea spray. Salt crystallization is also thought to be partly responsible for the accumulation of loose flakes of granite around rock outcrops, and the formation of weathering pits and cavernous forms in the Canadian Arctic (Watts 1985). Corrosional features have developed in a wide range of non-calcareous rocks in south-eastern Canada, and in Finland small weathering pits in granite and gneiss have been attributed to alternate wetting and drying, combined with marine abrasion and frost action (Uusinoka and Matti 1979). Nevertheless, although bioerosion and chemical and salt weathering occur in cold regions, their overall effect is generally thought to be quite weak.

Shore Platforms

The possible role of frost action in the development of shore platforms and ledges is still a contentious issue (Trenhaile 1983*a*, 1987). It is thought to have played an important role in their formation in the Younger Dryas in Norway and in sheltered areas of western Scotland (Andersen 1968; Dawson 1980). It has been suggested that elevated platforms in southern Britain were cut by periglacial shore erosion at the end of interglacial stages, when there was high sea level, rapid southward excursions of polar water, and low air temperatures in the north Atlantic (Dawson 1986). Frost is essential for shore platform development in Spitsbergen (Guilcher 1974), and it is active on the platforms in the St Lawrence Estuary and Hudson Bay (Allard and Tremblay 1983; Dionne and Brodeur 1988).

Tidally induced frost action damaged Chalk shore platforms during severe winters in temperate southern England (Williams and Robinson 1981). Most cracking and surface spalling occurs in the upper portion of platforms, which are exposed to freezing air temperatures for a much longer period than the lower parts. Once the falling tide exposes the rock, at least five to six hours is necessary for it to cool to the freezing point of sea water and to dissipate the released latent heat, before destructive freezing can occur. Limited exposure to low temperatures may therefore prevent widespread freezing on the lower portion of shore platforms. Tidally induced frost action was simulated in the laboratory, using air and water temperatures measured in the field. The results confirmed that low temperatures occur often enough in this area for frost weathering to be a significant factor in

platform development (Robinson and Jerwood 1987*a*, *b*) (Fig. 12.11).

It has been suggested that narrow, rugged ledges can be produced by frost erosion associated with the ice-foot. Shore platforms are best developed where frost and fairly strong wave action can work together, however, and although frost weakens rocks, it is less clear how it could operate as a planating agent. If it is responsible for platform formation, as distinct from contributing to cliff recession, then its efficacy must markedly decrease at a specific datum that is coincident with the surface of the resulting platform. A possible mechanism has been proposed by Matthews *et al.* (1986), who developed a semi-quantitative model to account for the formation of narrow rock platforms in southern Norway (Fig. 12.12). These platforms developed around an ice-dammed lake that existed for only about a hundred years during the 'Little Ice Age'. They proposed that frost shattering was most effective near the inner edge of the platform surface and at the cliff base. Their model emphasizes deep penetration of the annual freeze–thaw cycle, the movement of unfrozen lake water towards the freezing plane, and the growth of segregation ice in fissures and cracks at the interface between lake ice and bedrock. Ice-push and ice-pull processes are primarily responsible for removal of the debris. It remains to be determined whether this model is applicable to platform formation in cold regions where there are tides, stronger wave action, salt water, and in places an ice-foot and permafrost. Another possible reason

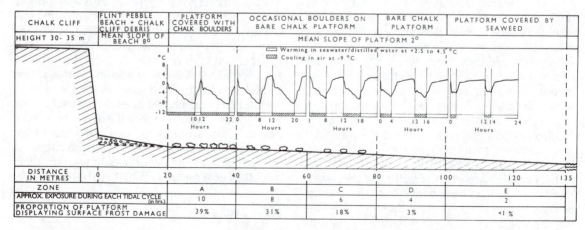

Fig. 12.11. Idealized profile of a Chalk shore platform in south-eastern England. The boxes show rock temperatures recorded in simulated freeze–thaw cycles during tidal cycles in each intertidal zone (Robinson and Jerwood 1987*b*).

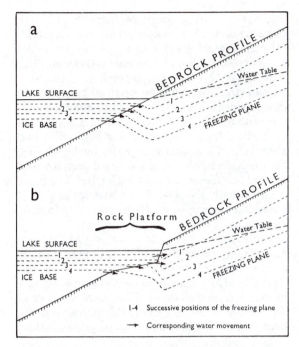

Fig. 12.12. Model of the development of a rock platform by frost action on a lake: (a) thickening of the lake ice, frost penetration into the bedrock, and the movement of unfrozen water to the freezing platform during the annual freezing cycle; (b) the same processes superimposed on a typical lake platform in Norway (Matthews *et al.* 1986)

for frost planation may be provided by the tendency, as noted in the Chalk of south-eastern England, for tidally induced frost action preferentially to attack upstanding protrusions, therefore flattening platform surfaces. This may reflect the greater surface-to-mass ratio of the protrusions, which allows them to cool in air more rapidly than the flatter portions of the platform surface, and their greater exposure to wind chill (Robinson and Jerwood 1987*a*, *b*).

Ice-foot collapse can dislodge material from the cliff face, and despite the fact that much of the ice melts *in situ*, debris can be removed as it floats away (N. Nielsen 1979; Allard and Tremblay 1983). Drift ice also transports loose rock debris, and striations and small grooves can be cut by rock fragments frozen into the base of the ice (Dionne 1985). Nevertheless, until recently most workers believed that drift ice is an ineffective erosional agent in the coastal zone (Moign 1976; Martini 1981): this is being reassessed. Shore ice, pack ice, and rock-

shod glacial ice are thought to abrade, quarry, and transport debris across well-developed shore platforms in the Antarctic Peninsula (Hansom and Kirk 1989). The formation of subhorizontal shore platforms in the South Shetland Islands has been attributed to fast ice freezing to the underlying bedrock, quarrying by grounded ice and stranded ice rocked by the tides, and abrasion by rocks frozen into the ice base (Hansom 1983*a*; Hansom and Kirk 1989). Sea ice pressure, and less frequently wave action, also assist gelifraction and frost wedging in quarrying gneissic joint blocks in macrotidal Ungava Bay. Fresh water ice riveting and ice-push, involving wind-driven floating ice loaded with rock fragments, appear to be responsible for quarrying thin flakes or scales of rock. The mobile portions of the ice-foot and ice floes at freeze-up and breakup override the rocky intertidal slopes and push, slide, and carry loose blocks. Although much of the ice-foot melts in place and does not contribute to debris removal, it may facilitate deep frost penetration by providing a thermal barrier to sporadic warming by tidal water (Fournier and Allard 1992). Frost weathering and the effects of ice abrasion, dislodgement, and quarrying are also considered to be the main processes responsible for the formation of wide, subhorizontal platforms in the upper St Lawrence Estuary. Pressures exerted by ice cakes and floes cause direct plucking of the rock surface by the ice, or indirect plucking or ploughing by large erratic boulders (Dionne and Brodeur 1988; Dionne 1993*b*). There is clear evidence in this area of the erosive efficacy of shore ice, and probable evidence of its ability to planate rocky intertidal zones. Nevertheless, the occurrence of weak slates and shales, high tidal range, strong currents, and large erratic blocks provides particularly suitable conditions for frost and ice action, and it remains to be determined whether these cold region mechanisms assume similar roles in less favourable places.

Mechanical wave erosion is important on cold coasts that have a significant ice-free period and a stormy wave climate. The relationship between platform gradient and tidal range, which exists in many storm- and swell-wave environments (Chapter 11) (Fig. 11.7), does not appear to be affected by the presence of coastal ice and frost during the winter months. Platform gradient in the Bay of Fundy, where the tidal range is more than 13.5 m, is between 2° and 4.5°, whereas horizontal platforms have developed in Gaspé, Quebec, and western

Newfoundland, where the tidal range is from 2 to 3.5 m. This suggests that the morphology of platforms in cold regions is determined, as elsewhere, by the tidal distribution of wave energy. Platform morphology may also be determined, however, by the way tides direct the work of frost and/or ice in the intertidal zone. In macrotidal Ungava Bay the greatest erosion of frost loosened rocks by waves and ice is at the mean high water neap level—which corresponds to the maximum tidal duration value (Chapter 2). The lack of a second erosional maximum at the mean low water neaps may be attributed to the presence of mud and boulders, lower fetch, and the shorter period of exposure for frost action at this level (Fournier and Allard 1992).

Strandflats

The term 'strandflat' has become almost synonymous with the extensive coastal plains on the western and north-western coasts of Norway, although there are similar low, quasi-horizontal coastal platforms in Antarctica, Iceland, Spitsbergen, and a number of other high-latitude areas (Nansen 1922; Guilcher *et al.* 1986). Large portions of the Norwegian strandflat, which is up to 60 km in width, developed in resistant crystalline and metamorphic rocks. It extends as a single, fairly well defined platform over low islands, peninsulas, and skerries, and is dissected in places by deep fiords, although it may consist of a series of partially submerged terraces in areas of high relief. The inner edge is usually about 40–50 m above sea level, and it terminates abruptly in places at the steep front of mountains.

There is little agreement on the origin of strandflats (Trenhaile 1983*a*, 1987). Although various combinations of frost, shore and glacial ice, and wave action have been promoted, it is possible only to speculate on their origin. The role of shore ice as an effective erosional agent in the coastal zone has been disputed, but recent work suggests that its

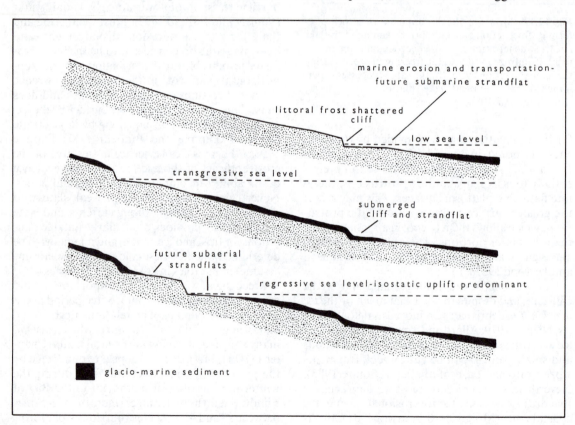

Fig. 12.13. The origin of strandflats in Spitsbergen (Guilcher 1974)

role in the formation of shore platforms and strandflats in high latitudes should be reassessed. Larsen and Holtedahl (1985), for example, proposed that the Norwegian strandflat was formed during glacial stages by sea-ice planation and transportation, in combination with frost shattering. Many investigators, however, have emphasized the combined role of frost and wave action, although their relative importance has not been determined (Guilcher 1974) (Fig. 12.13).

While some investigators stressed the work of frost, and relegated the role of waves to the removal of loose debris, others have accorded an important role to mechanical wave erosion. This is consistent with the assumption that waves that are strong enough to remove the coarse products of frost action in crystalline rocks, should also be capable of dislodging frost-weakened rocks in cliffs and shore platforms. It is also supported by the occurrence of the best-developed strandflats on capes and other areas exposed to strong waves. Although a variety of mechanisms prepares rock surfaces for wave erosion in different morphogenic environments, frost action may therefore be an essential precursor to erosion and removal of resistant rocks by waves in high latitudes.

References

Aagaard, T. 1985: Observations on beach cusps. *Geografisk Tidsskr*. 85, pp. 27–31.

Aagaard, T. 1988a: Rhythmic beach and nearshore topography: examples from Denmark. *Geografisk Tidsskr*. 88, pp. 55–60.

Aagaard, T. 1988b: Nearshore bar morphology on the low-energy coast of northern Zealand, Denmark. *Geografiska Annlr*. A 70, pp. 59–67.

Aagaard, T. 1990: Infragravity waves and nearshore bars in protected, storm-dominated coastal environments. *Marine Geol*. 94, pp. 181–203.

Aagaard, T. 1991: Multiple-bar morphodynamics and its relation to low-frequency edge waves. *J. Coastal Res*. 7, pp. 801–13.

Aagaard, T. and Greenwood, B. 1994: Suspended sediment transport and the role of infragravity waves in a barred surf zone. *Marine Geol*. 118, pp. 23–48.

Aagaard, T., Nielsen, N., and Nielsen, J. 1994: Cross-shore structure of infragravity standing wave motion and morphological adjustment: an example from northern Zealand, Denmark. *J. Coastal Res*. 10, pp. 716–31.

Acevedo, R. and Morelock, J. 1988: Effects of terrigenous sediment influx on coral reef zonation in southwestern Puerto Rico. *Proc. 6th Int. Coral Reef Symp*., Townsville 2, pp. 189–94.

Adegbehin, J. O. and Nwaigbo, L. C. 1990: Mangrove resources in Nigeria: use and management perspectives. *Nature and Resources* 26, pp. 13–21.

Adey, W. H. 1975: The algal ridges and coral reefs of St. Croix: their structure and Holocene development. *Atoll Res. Bull*. 187.

Adey, W. H. 1978: Coral reef morphogenesis: a multidimensional model. *Science* 202, pp. 831–7.

Adey, W. H. and Burke, R. B. 1976: Holocene bioherms (algal ridges and bank-barrier reefs) of the eastern Caribbean. *Geol. Soc. Amer. Bull*. 87, pp. 95–109.

Adey, W. H. and Burke, R. B. 1977: Holocene bioherms of Lesser Antilles—geological control of development. In *Reefs and Related Carbonates—Ecology and Sedimentology*, S. H. Frost, M. P. Weiss, and J. B. Saunders (eds.), *Amer. Assoc. Petrol. Geol. Studies Geol*. 4, pp. 67–81.

Adey, W. H. and Macintyre, I. G. 1973: Crustose coralline algae: a re-evaluation in the geological sciences. *Geol. Soc. Amer. Bull*. 84, pp. 883–904.

Adey, W. H., Macintyre, I. G., and Stuckenrath, R. 1977: Relict barrier reef system off St. Croix: its implications with respect to late Cenozoic coral reef development in the western Atlantic. *Proc. 3rd Int. Coral Reef Symp., Miami* 2, pp. 15–21.

Ahnert, F. 1960: Estuarine meanders in the Chesapeake Bay area. *Geogr. Rev*. 50, pp. 390–401.

Ahrens, J. P., Aff, M., and Titus, M. F. 1985: Wave runup formulas for smooth slopes. *J. Waterw. Port Coastal Ocean Div. Amer. Soc. Civ. Engnr*. 111, pp. 128–33.

Alexander, C. R., Nittrouer, C. A., Demaster, D. J., Park, Y.-A., and Park, S.-C. 1991: Macrotidal mudflats of the southwestern Korean coast: a model for interpretation of intertidal deposits. *J. Sed. Petrol*. 61, pp. 805–24.

Alino, P. M., Viva Banzon, P., Yap, H. T., Gomez, E. D., Morales, J. T., and Bayoneto, R. P. 1985: Recovery and recolonization on a damaged backreef area at Cangaluyan Is. (Northern Philippines). *Proc. 5th Int. Coral Reef Congr., Tahiti* 4, pp. 279–84.

Aliotta, S. and Farinati, E. 1990: Stratigraphy of Holocene sand-shell ridges in the Bahia Blanca Estuary, Argentina. *Marine Geol*. 94, pp. 353–60.

Allard, M. and Tremblay, G. 1983: Les Processus d'érosion littorale périglaciaire de la région de Poste-de-la-Baleine et des Îles Manitounuk sur la côte est de la mer d'Hudson, Canada. *Z. Geomorph*. Suppl. Band 47, pp. 27–60.

Allen, G. P. 1991: Sedimentary processes and facies in the Gironde Estuary: a recent model for macrotidal estuarine systems. In *Clastic Tidal Sedimentology*, D. G. Smith, G. E. Reinson, B. A. Zaitlin, and R. A. Rahmani (eds.), *Can. Soc. Petrol. Geol. Mem*. 16, pp. 29–40.

Allen, G. P. and Posamentier, H. W. 1993: Sequence stratigraphy and facies model of an incised valley fill: the Gironde Estuary, France. *J. Sed. Petrol*. 63, pp. 378–91.

Allen, G. P., Salomon, J. C., Bassoullet, P., Du Penhoat, Y., and De Grandpré, C. 1980: Effects of tides on mixing and suspended sediment transport in macrotidal estuaries. *Sed. Geol*. 26, pp. 69–90.

Allen, J. R. 1988: Nearshore sediment transport. *Geogr. Rev*. 78, pp. 148–57.

Allen, J. R., Bauer, B. O., Psuty, N. P., and Carter, R. W. G. 1991: Process variation across a barred, tidal nearshore. *Coastal Sediments '91*, N. C. Kraus, K. J. Gingerich, and D. L. Kriebel (eds.), Amer. Soc. Civ. Engnr., pp. 498–511.

Allen, J. R. L. 1970: Sediments of the modern Niger Delta: a summary and review. In *Deltaic Sedimentation: Modern and Ancient*, J. P. Morgan (ed.), *Soc. Econ. Paleont. Mineral.* Spec. Publ. 15, pp. 138–51.

Allen, J. R. L. 1982: Simple models for the shape and symmetry of tidal sand waves: (1) statically stable equilibrium forms. *Marine Geol.* 48, pp. 31–49.

Allen, J. R. L. 1989: Evolution of salt-marsh cliffs in muddy and sandy systems: a qualitative comparison of British west-coast estuaries. *Earth Surface Processes and Landforms* 14, pp. 85–92.

Allen, J. R. L. 1994: A continuity-based sedimentological model for temperate-zone tidal salt marshes. *J. Geol. Soc. London* 151, pp. 41–9.

Allen, J. R. L. and Pye, K. 1992: Coastal saltmarshes: their nature and importance. In *Saltmarshes: Morphodynamics, Conservation and Engineering Significance*, J. R. L. Allen and K. Pye (eds.), Cambridge Univ. Press, Cambridge, pp. 1–18.

Allison, R. J. 1989: Rates and mechanisms of change in hard rock coastal cliffs. *Z. Geomorph.* Suppl. Band 73, pp. 125–38.

Andersen, B. C. 1968: Glacial geology of western Troms, north Norway. *Norges Geol. Unders.* 256, pp. 1–160.

Anderson, F. E. 1983: The northern muddy intertidal: seasonal factors controlling erosion and deposition—a review. *Can. J. Fish. Aquat. Sci.* 40 (Suppl. 1), pp. 143–59.

Anderson, F. E., Black, L., Watling, L. E., Mook, W., and Mayer, L. M. 1981: A temporal and spatial study of mudflat erosion and deposition. *J. Sed. Petrol.* 51, pp. 729–36.

Anderson, R. S. 1989: Saltation of sand: a qualitative review with biological analogy. In *Coastal Sand Dunes*, C. H. Gimingham, W. Ritchie, B. B. Willetts, and A. J. Willis (eds.), *Proc. Roy. Soc. Edinb.* B 96, pp. 149–65.

Anderson, R. S. 1990: Eolian ripples as examples of self-organization in geomorphological systems. *Earth-Sci. Rev.* 29, pp. 77–96.

Anderson, R. S. and Haff, P. K. 1988: Simulation of eolian saltation. *Science* 241, pp. 820–3.

Anderson, R. S. and Haff, P. K. 1991: Wind modification and bed response during saltation of sand in air. In *Aeolian Grain Transport 1 Mechanics*, O. E. Barndorff-Nielsen and B. B. Willetts (eds.), Acta Mechanica Suppl. 1, Springer-Verlag, Wien, pp. 21–51.

Anderson, R. S., Sørensen, M., and Willetts, B. B. 1991: A review of recent progress in our understanding of aeolian sediment transport. In *Aeolian Grain Transport 1 Mechanics*, O. E. Barndorff-Nielsen and B. B. Willetts (eds.), Acta Mechanica Suppl. 1, Springer-Verlag, Wien, pp. 1–19.

Andrews, J. T. (ed.) 1985: *Quaternary Environments*, Allen & Unwin, London.

Andrews, J. T., Bowen, D. Q., and Kidson, C. 1979: Amino acid ratios and the correlation of raised beach deposits in south-west England and Wales. *Nature* 281, pp. 556–8.

Anthony, E. J. 1989: Chenier plain development in northern Sierra Leone, West Africa. *Marine Geol.* 90, pp. 297–309.

Anthony, E. J. 1991: Beach-ridge plain development: Sherbro Island, Sierra Leone. *Z. Geomorph.* Suppl. Band 81, pp. 85–98.

Antia, E. E. 1989: A short-term study of the effects of the changing coastal conditions on some geomorphic elements on Nigerian beaches. *Z. Geomorph.* Suppl. Band 73, pp. 1–16.

Are, F. E. 1988: Thermal abrasion of sea coasts. *Polar Geogr. Geol.* 12, pp. 1–157.

Arens, S. M., Van Kaam-Peters, H. M. E., and Van Boxel, J. H. 1995: Air flow over foredunes and implications for sand transport. *Earth Surface Processes and Landforms* 20, pp. 315–32.

Armon, J. W. 1979: Landward sediment transfers in a transgressive barrier island system, Canada. In *Barrier Islands*, S. P. Leatherman (ed.), Academic Press, New York, pp. 65–80.

Arnskov, M. M., Fredsøe, J., and Sumer, B. M. 1993: Bed shear stress measurements over a smooth bed in three-dimensional wave-current motion. *Coastal Engnr.* 20, pp. 277–316.

Asano, T. 1992: Observations of granular-fluid mixture under an oscillatory sheet flow. *Proc. 23rd Coastal Engnr. Conf.* pp. 1896–1909.

Asano, T., Deguchi, H., and Kobayashi, N. 1992: Interaction between water waves and vegetation. *Proc. 23rd Coastal Engnr. Conf.* pp. 2710–23.

Ashley, G. M. and Renwick, W. H. 1983: Channel morphology and processes at the riverine-estuarine transition, the Raritan River, New Jersey. In *Modern and Ancient Fluvial Systems*, J. D. Collinson and J. Lewin (eds.), *Int. Assoc. Sedimentologists* (Blackwell, Oxford) Spec. Publ. 6, pp. 207–18.

Ashley, G. M. and Zeff, M. L. 1988: Tidal channel classification for a low-mesotidal salt marsh. *Marine Geol.* 82, pp. 17–32.

Atwater, B. F. 1987: Evidence for great Holocene earthquakes along the outer coast of Washington State. *Science* 236, pp. 942–4.

Aubrey, D. G. and Gaines, A. G. 1982: Rapid formation and degradation of barrier spits in areas with low rates of littoral drift. *Marine Geol.* 49, pp. 257–78.

Aubrey, D. G. and Speer, P. E. 1985: A study of non-linear tidal propagation in shallow inlet/estuarine systems. Part 1: observations. *Estuarine, Coastal and Shelf Sci.* 21, pp. 185–205.

Auger, P. and Mary, G. 1968: Glissements et coulées boueuses en Basse-Normandie. *Rev. Géogr. Phys. Géol. Dyn.* 10, pp. 213–24.

Augustinus, P. G. E. F. 1980: Actual development of the

chenier coast of Suriname (South America). *Sed. Geol.* 26, pp. 91–113.

Augustinus, P. G. E. F. 1989: Cheniers and chenier plains: a general introduction. *Marine Geol.* 90, pp. 219–29.

Augustinus, P. G. E. F., Hazelhoff, L., and Kroon, A. 1989: The chenier coast of Suriname: modern and geological development. *Marine Geol.* 90, pp. 269–81.

Avoine, J. 1986: Sediment exchanges between the Seine Estuary and its adjacent shelf. *J. Geol. Soc., London* 144, pp. 135–48.

Baba, J. and Komar, P. D. 1981*a*: Measurements and analysis of settling velocities of natural quartz sand grains. *J. Sed. Petrol.* 51, pp. 631–40.

Baba, J. and Komar, P. D. 1981*b*: Settling velocities of irregular grains at low Reynolds numbers. *J. Sed. Petrol.* 51, pp. 121–8.

Bagnold, R. A. 1941: *The Physics of Blown Sand and Desert Dunes*, Methuen, London.

Bagnold, R. A. 1963: Beach and nearshore processes. Part 1. Mechanics of marine sedimentation. In *The Sea*, M. N. Hill (ed.), Interscience, New York, 3, pp. 507–28.

Bagnold, R. A. 1966: An approach to the sediment transport problem from general physics. *U. S. Geol. Surv. Prof. Paper* 422-I.

Bagnold, R. A. 1974: Fluid forces on a body in shear flow; experimental use of 'stationary flow'. *Proc. Roy. Soc. London* A 340, pp. 147–71.

Bailard, J. A. 1981: An energetics total load sediment transport model for a plane sloping beach. *J. Geophys. Res.* 86, pp. 10938–54.

Bailard, J. A. 1982: Modeling on-offshore sediment transport in the surf zone. *Proc. 18th Coastal Engnr. Conf.* pp. 1419–38.

Bakker, T. W. M. 1990: The geohydrology of coastal dunes. In *Dunes of the European Coasts*, W. Bakker, P. D. Jungerius, and J. A. Klijn (eds.), *Catena* Suppl. 18, pp. 109–19.

Bale, A. J., Morris, A. W., and Howland, R. J. M. 1985: Seasonal sediment movement in the Tamar Estuary. *Oceanologica Acta* 8, pp. 1–6.

Ball, M. C. 1980: Patterns of secondary successions in a mangrove forest of southern Florida. *Oecologia* 44, pp. 226–35.

Bao-can, W. and Eisma, D. 1988: Mudflat deposition along the Wenzhou coastal plain in southern Zhejiang, China. In *Tide-influenced Sedimentary Environments and Facies*, P. L. De Boer, A. Van Gelder, and S. D. Nio (eds.), D. Reidel, Dordrecht, Netherlands, pp. 265–74.

Bardach, J. E. 1989: Global warming and the coastal zone. *Climatic Change* 15, pp. 117–50.

Barkaszi, S. F. and Dally, W. R. 1992: Fine-scale measurement of sediment suspension by breaking waves at SUPERTANK. *Proc. 23rd Coastal Engnr. Conf.* pp. 1910–23.

Barnes, P. W., Kempema, E. W., Reimnitz, E., McCormick, M., Weber, W. S., and Hayden, E. C. 1993: Beach profile modification and sediment transport by ice: an overlooked process on Lake Michigan. *J. Coastal Res.* 9, pp. 65–86.

Barnes, P. W., Reimnitz, E., and Fox, D. 1982: Ice rafting of fine-grained sediment, a sorting and transport mechanism, Beaufort Sea, Alaska. *J. Sed. Petrol.* 52, pp. 493–502.

Barnes, P. W. and Rollyson, B. P. 1991: Erosion and accretion along the Arctic coast of Alaska: the influence of ice and climate. *Coastal Sediments '91*, N. C. Kraus, K. J. Gingerich, and D. L. Kriebel (eds.), Amer. Soc. Civ. Engnr., pp. 1518–31.

Barnett, P. J. 1990: Relationships of bluff composition to bluff profile and coastal outline, north shore Lake Erie. *Proc. Can. Coastal Conf.*, M. H. Davies (ed.), National Res. Council Can. (ACOS), Ottawa, pp. 197–211.

Barnett, T. P. and Kenyon, K. E. 1975: Recent advances in the study of wind waves. *Rept. Progr. Phys.* 38, pp. 667–729.

Barnett, T. P. and Sutherland, A. J. 1968: A note on an overshoot effect in wind-generated waves. *J. Geophys. Res.* 73, pp. 6879–85.

Bartrum, J. A. 1916: High water rock platforms: a phase of shoreline erosion. *Trans. N.Z. Inst.* 48, pp. 132–4.

Bartrum, J. A. 1924: The shore platform of the west coast near Auckland: its storm wave origin. *Rept. Austr. Assoc. Sci.* 16, pp. 493–5.

Bartrum, J. A. 1935: Shore platforms. *Proc. Austr. N.Z. Assoc. Adv. Sci.* 22, pp. 135–43.

Bartsch-Winkler, S. and Ovenshine, A. T. 1984: Macrotidal subarctic environment of Turnagain and Knik Arms, Upper Cook Inlet, Alaska: sedimentology of the intertidal zone. *J. Sed. Geol.* 54, pp. 1221–38.

Barua, D. K. 1990: Suspended sediment movement in the estuary of the Ganges–Brahmaputra–Meghna river system. *Marine Geol.* 91, pp. 243–53.

Barusseau, J. P., Diop, E. H. S., and Saos, J. L. 1985: Evidence of dynamics reversal in tropical estuaries, geomorphological and sedimentological consequences (Salum and Casamance Rivers, Senegal). *Sedimentology* 32, pp. 543–52.

Basco, D. R. 1985: A qualitative description of wave breaking. *J. Waterw. Port Coastal Ocean Div. Amer. Soc. Civ. Engnr.* 111, pp. 171–88.

Basco, D. R. and Yamashita, T. 1986: Toward a simple model of the wave breaking transition region in surf zones. *Proc. 20th Coastal Engnr. Conf.* pp. 955–70.

Bassett, J. A. and Curtis, T. G. F. 1985: The nature and occurrence of sand-dune machair in Ireland. *Proc. Roy. Irish Acad.* B 85, pp. 1–20.

Bates, C. C. 1953: Rational theory of delta formation. *Amer. Assoc. Petrol. Geol. Bull.* 37, pp. 2119–61.

Battistini, R. and Guilcher, A. 1982: Les Plates-formes littorales à vasques en roches calcaires: répartition dans le monde, mer Méditerranée non comprise. *Karst Littoraux*. Comité National Francais de Geographie, Actes du Colloqium de Perpignan, 15–17 Mai 1982, 1, pp. 1–11.

Battjes, J. A. 1974: Surf similarity. *Proc. 14th Coastal Engnr. Conf.* pp. 466–79.

Battjes, J. A. 1988: Surf-zone dynamics. *Annual Rev. Fluid Mech.* 20, pp. 257–93.

Bauer, B. O. 1991: Aeolian decoupling of beach sediments. *Ann. Assoc. Amer. Geogr.* 81, pp. 290–303.

Bauer, B. O. and Allen, J. R. 1995: Beach steps: an evolutionary perspective. *Marine Geol.* 123, pp. 143–66.

Bauer, B. O. and Greenwood, B. 1988: Surf-zone similarity. *Geogr. Rev.* 78, pp. 137–47.

Bauer, B. O. and Greenwood, B. 1990: Modification of a linear bar–trough system by a standing edge wave. *Marine Geol.* 92, pp. 177–204.

Bauer, B. O., Sherman, D. J., Nordstrom, K. F., and Gares, P. A. 1990: Aeolian transport measurement and prediction across a beach and dune at Castroville, California. In *Coastal Dunes: Form and Process*, K. F. Nordstrom, N. P. Psuty, and R. W. G. Carter (eds.), Wiley, Chichester, pp. 39–55.

Bayliss-Smith, T. P., Healy, R., Lailey, R., Spencer, T., and Stoddart, D. R. 1979: Tidal flows in salt marsh creeks. *Estuarine and Coastal Marine Sci.* 9, pp. 235–55.

Beach, R. A. and Sternberg, R. W. 1987: The influence of infragravity motions on suspended sediment transport in the inner surf zone. *Coastal Sediments '87*, N. C. Kraus (ed.), Amer. Soc. Civ. Engnr., pp. 913–28.

Beach, R. A. and Sternberg, R. W. 1991: Infragravity driven suspended sediment transport in the swash, inner and outer-surf zone. *Coastal Sediments '91*, N. C. Kraus, K. J. Gingerich, and D. L. Kriebel (eds.), Amer. Soc. Civ. Engnr., pp. 114–28.

Beach, R. A. and Sternberg, R. W. 1992: Suspended sediment transport in the surf zone: response to incident wave and longshore current interaction. *Marine Geol.* 108, pp. 275–94.

Bégin, Y. and Allard, M. 1981: La Dynamique glacielle a l'embouchure de la Grande Riviere de la Baleine au Quebec subarctique. *Proc. Workshop on Ice Action on Shores*, J.-C. Dionne (ed.), National Res. Council Can. (ACROSES) Ottawa, pp. 117–31.

Belknap, D. F. and Kraft, J. C. 1985: Influence of antecedent geology on stratigraphic preservation potential and evolution of Delaware's barrier system. *Marine Geol.* 63, pp. 235–62.

Belly, P.-Y. 1964: Sand movement by wind. U.S. Army Corps of Engnr., *Coastal Engineering Research Center Techn. Memo.* 1.

Bender, L. C. and Wong, K.-C. 1993: The effect of wave–current interaction on tidally forced estuarine circulation. *J. Geophys. Res.* 98, pp. 16521–8.

Bender, M. L., Fairbanks, R. G., Taylor, F. W., Matthews, R. K., Goddard, J. G., and Broecker, W. S. 1979: Uranium-series dating of the Pleistocene reef tracts of Barbados, West Indies. *Geol. Soc. Amer. Bull.* 90, pp. 577–94.

Best, J. L. and Leeder, M. R. 1993: Drag reduction in turbulent muddy seawater flows and some sedimentary consequences. *Sedimentology* 40, pp. 1129–37.

Bijker, E. W. 1966: The increase of bed shear in a current due to wave action. *Proc. 10th Coastal Engnr. Conf.* pp. 746–65.

Bird, E. C. F. 1969: *Coasts*, MIT Press, Cambridge, Mass.

Bird, E. C. F. 1985a: *Coastal Changes: A Global Review*, Wiley, Chichester.

Bird, E. C. F. 1985b: The study of coastal changes. *Z. Geomorph.* Suppl. band 57, pp. 1–9.

Bird, E. C. F. 1986: Mangroves and intertidal morphology in Westernport Bay, Victoria, Australia. *Marine Geol.* 69, pp. 251–71.

Bird, E. C. F. 1990: Classification of European dune coasts. In *Dunes of the European Coasts*, W. Bakker, P. D. Jungerius, and J. A. Klijn (eds.), *Catena* Suppl. 18, pp. 15–24.

Bird, E. C. F. and Guilcher, A. 1982: Observations préliminaires sur les récifs frangeants actuels du Kenya et sur les formes littorales associées. *Rev. Géomorph. Dyn.* 31, pp. 113–25.

Bird, E. C. F. and Jones, D. J. B. 1988: The origin of foredunes on the coast of Victoria, Australia. *J. Coastal Res.* 4, pp. 181–92.

Birkemeier, W. A. 1985: Time scales of nearshore profile changes. *Proc. 19th Coastal Engnr. Conf.* pp. 1507–21.

Bishop, C. T. 1983: A shore protection alternative: artificial headlands. *Proc. Can. Coastal Conf.*, B. J. Holden (ed.), National Res. Council Can. (ACROSES), Ottawa, pp. 305–19.

Bishop, C. T. and Donelan, M. A. 1989: Wave prediction models. *In Applications in Coastal Modeling*, V. C. Lakhan and A. S. Trenhaile (eds.), Elsevier, Amsterdam, pp. 75–105.

Bishop, C. T., Skafel, M. G., and Nairn, R. 1992: Cohesive profile erosion by waves. *Proc. 23rd Coastal Engnr. Conf.* pp. 2976–89.

Bluck, B. J. 1967: Sedimentation of beach gravels: examples from south Wales. *J. Sed. Petrol.* 37, pp. 128–56.

Boczar-Karakiewicz, B., Bona, J. L., and Cohen, D. L. 1987a: Interaction of shallow-water waves and bottom topography. In *Dynamical Problems in Continuum Physics*, J. L. Bona, C. Dafermos, J. L. Ericksen, and D. Kinderlehrer, Springer-Verlag, New York, pp. 131–76.

Boczar-Karakiewicz, B. and Davidson-Arnott, R. G. D. 1987: Nearshore bar formation by non-linear wave processes—a comparison of model results and field data. *Marine Geol.* 77, pp. 287–304.

Boczar-Karakiewicz, B., Drapeau, G., and Tessier, B. 1987b: Barres sableuses littorales des Îles-De-La Madeleine: Mecanisme de leur formation et mathematique. *Proc. Can. Coastal Conf.*, Y. Ouellet (ed.), National Res. Council Can. (ACROSES), Ottawa, pp. 115–29.

Bodge, K. R. 1988: Longshore current and transport across non-singular equilibrium beach profiles. *Proc. 21st Coastal Engnr. Conf.* pp. 1396–1410.

Bodge, K. R. 1989: A literature review of the distribution of longshore sediment transport across the surf zone. *J. Coastal Res.* 5, pp. 307–28.

Bodge, K. R. 1992: Representing equilibrium beach profiles with an exponential expression. *J. Coastal Res.* 8, pp. 47–55.

Bodge, K. R. and Dean, R. G. 1987: Short-term impoundment of longshore transport. *Coastal Sediments '87*, N. C. Kraus (ed.), Amer. Soc. Civ. Engnr., pp. 468–83.

Bodge, K. R. and Kraus, N. C. 1991: Critical examination of longshore transport rate magnitude. *Coastal Sediments '91*, N. C. Kraus, K. J. Gingerich, and D. L. Kriebel (eds.), Amer. Soc. Civ. Engnr., pp. 139–55.

Bogen, J. 1983: Morphology and sedimentology of deltas in fjord and fjord valley lakes. *Sed. Geol.* 36, pp. 245–67.

Boon, J. D. 1975: Tidal discharge asymmetry in a salt marsh drainage system. *Limnol. Oceanogr.* 20, pp. 71–80.

Boon, J. D. and Byrne, R. J. 1981: On basin hypsometry and the morphodynamic response of coastal inlet systems. *Marine Geol.* 40, pp. 27–48.

Boorman, L. A. 1977: Sand dunes. In *The Coastline*, R. S. K. Barnes (ed.), Wiley, London, pp. 161–97.

Boorman, L. A. 1989: The influence of grazing on British sand dunes. In *Perspectives in Coastal Dune Management*, F. Van Der Meulen, P. D. Jungerius, and J. Visser (eds.), SPB Academic Publishing, The Hague, The Netherlands, pp. 121–4.

Boothroyd, J. C. 1985: Tidal inlets and tidal deltas. In *Coastal Sedimentary Environments*, R. A. Davis (ed.), Springer-Verlag, New York, pp. 445–532.

Boothroyd, J. C., Friedrich, N. E., and McGinn, S. R. 1985: Geology of microtidal coastal lagoons: Rhode Island. *Marine Geol.* 63, pp. 35–76.

Boothroyd, J. C. and Hubbard, D. K. 1975: Genesis of bedforms in mesotidal estuaries. In *Estuarine Research*, L. E. Cronin (ed.), Academic Press, New York, ii. 217–34.

Borówka, R. K. 1990a: The Holocene development and present morphology of the Łeba Dunes, Baltic coast of Poland. In *Coastal Dunes: Form and Process*, K. F. Nordstrom, N. P. Psuty, and R. W. G. Carter (eds.), Wiley, Chichester, pp. 289–313.

Borówka, R. K. 1990b: Coastal dunes in Poland. In *Dunes of the European Coasts*, W. Bakker, P. D. Jungerius, and J. A. Klijn (eds.), *Catena* Suppl. 18, pp. 25–30.

Bosence, D. W. J. 1983: Coralline algal reef frameworks. *J. Geol. Soc. London* 140, pp. 365–76.

Bosence, D. W. J. 1984: Construction and preservation of two modern coralline algal reefs, St. Croix, Caribbean. *Palaeont.* 27, pp. 549–74.

Boss, S. K. and Liddell, W. D. 1987: Patterns of sediment composition of Jamaican fringing reef facies. *Sedimentology* 34, pp. 77–87.

Boss, S. K. and Neumann, A. C. 1993: Impacts of Hurricane Andrew on carbonate platform environments, northern Great Bahama Bank. *Geology* 21, pp. 897–900.

Bosscher, H. and Schlager, W. 1992: Computer simulation of reef growth. *Sedimentology* 39, pp. 503–12.

Bouchon, C. and Laborel, J. 1988: The coral communities of the Grand Cul-De-Sac Marin of Guadeloupe Island (French West Indies). *Proc. 6th Int. Coral Reef Symp., Townsville* 3, pp. 333–8.

Bourman, R. P. 1986: Aeolian sand transport along beaches. *Austr. Geogr.* 17, pp. 30–4.

Bowden, K. F. 1980: Physical factors: salinity, temperature, circulation, and mixing processes. In *Chemistry and Biogeochemistry of Estuaries*, E. Olausson and I. Cato (eds.), Wiley, Chichester, pp. 37–70.

Bowen, A. J. 1972: Edge waves and the littoral environment. *Proc. 13th Coastal Engnr. Conf.* pp. 1313–20.

Bowen, A. J. 1980: Simple models of nearshore sedimentation: beach profiles and longshore bars. In *The Coastline of Canada*, S. B. McCann (ed.), *Geol. Surv. Can. Paper* 80–10, pp. 1–11.

Bowen, A. J. and Guza, R. T. 1978: Edge waves and surf beat. *J. Geophys. Res.* 83, pp. 1913–20.

Bowen, A. J. and Holman, R. A. 1989: Shear instabilities of the mean longshore current 1. Theory. *J. Geophys. Res.* 94, pp. 18023–30.

Bowen, A. J. and Huntley, D. A. 1984: Waves, long waves and nearshore morphology. *Marine Geol.* 60, pp. 1–13.

Bowen, A. J. and Inman, D. L. 1971: Edge waves and crescentic bars. *J. Geophys. Res.* 76, pp. 8662–71.

Bowman, D. and Goldsmith, V. 1983: Bar morphology of dissipative beaches: an empirical model. *Marine Geol.* 51, pp. 15–33.

Boyd, G. L. 1986: A geomorphic model of bluff erosion on the Great Lakes. *Proc. Symp. Cohesive Shores*, M. G. Skafel (ed.), National Res. Council Can. (ACROSES), Ottawa, pp. 60–8.

Boyd, R., Bowen, A. J., and Hall, R. K. 1987: An evolutionary model for transgressive sedimentation on the eastern shore of Nova Scotia. In *Glaciated Coasts*, D. M. Fitzgerald and P. S. Rosen (eds.), Academic Press, London, pp. 87–114.

Boyd, R., Dalrymple, R., and Zaitlin, B. A. 1992: Classification of clastic coastal depositional environments. *Sed. Geol.* 80, pp. 139–50.

Bradbury, R. H. (ed.) 1990: *Acanthaster and the Coral Reef: a Theoretical Perspective*, Springer-Verlag, Berlin, *Lecture Notes in Biomath.* 88.

Bradley, W. C. and Griggs, G. B. 1976: Form, genesis, and deformation of central California wave-cut platforms. *Geol. Soc. Amer. Bull.* 87, pp. 433–49.

Braithwaite, C. J. R. 1982*a*: Patterns of accretion of reefs in the Sudanese Red Sea. *Marine Geol.* 46, pp. 297–325.

Braithwaite, C. J. R. 1982*b*: Progress in understanding reef structure. *Progr. Phys. Geogr.* 6, pp. 505–23.

Brenninkmeyer, B. M., James, C. P., and Wood, L. J. 1977: Bore–bore interaction on the foreshore. *Coastal Sediments '77*, Amer. Soc. Civ. Engnr., pp. 622–38.

Bressolier, C., Froidefond, J. -M., and Thomas, Y.-F. 1990: Chronology of coastal dunes in the south-west of France. In *Dunes of the European Coasts*, W. Bakker, P. D. Jungerius, and J. A. Klijn (eds.), *Catena* Suppl. 18, pp. 101–7.

Bressolier, C. and Thomas, Y-F. 1977: Studies on wind and plant interactions on French Atlantic coastal dunes. *J. Sed. Petrol.* 47, pp. 331–8.

Bricker-Urso, S., Nixon, S. W., Cochran, J. K., Hirschberg, D. J., and Hunt, C. 1989: Accretion rates and sediment accumulation in Rhode Island salt marshes. *Estuaries* 12, pp. 300–17.

Brinke, W. B. M. ten 1987: Quantifying mud exchange between the eastern Scheldt Basin and the North Sea. *Coastal Sediments '91*, N. C. Kraus, K. J. Gingerich, and D. L. Kriebel (eds.), Amer. Soc. Civ. Engnr., pp. 760–74.

Brodeur, D. and Allard, M. 1983: Les Plates-formes littorales de l'île aux Coudres, moyen estuaire du Saint-Laurent, Québec. *Géogr. Phys. Quat.* 37, pp. 179–95.

Bromhead, E. N. 1978: Large landslides in London Clay at Herne Bay, Kent. *Quart. J. Engnr. Geol.* 11, pp. 291–304.

Bromhead, E. N. 1979: Factors affecting the transition between the various types of mass movement in coastal cliffs consisting largely of over-consolidated clay with special reference to southern England. *Quart. J. Engnr. Geol.* 12, pp. 291–300.

Brousse, R. 1985: The age of the islands in the Pacific Ocean: volcanism and coral-reef build up. *Proc. 5th Int. Coral Reef Congr., Tahiti* 6, pp. 389–400.

Brown, B. E. and Howard, L. S. 1985: Assessing the effects of 'stress' on reef corals. In *Advances in Marine Biology* 22, J. H. S. Blaxter, F. S. Russell, and M. Yonge (eds.), Academic Press, London, pp. 1–63.

Brugmans, F. 1983: Wind ripples in an active drift sand area in the Netherlands: a preliminary report. *Earth Surface Processes and Landforms* 8, pp. 527–34.

Brunsden, D. 1984: Mudslides. In *Slope Instability*, D. Brunsden and D. B. Prior (eds.), Wiley, Chichester, pp. 363–418.

Bruun, P. 1988: The Bruun Rule of erosion by sea-level rise: a discussion of large-scale two- and three-dimensional usages. *J. Coastal Res.* 4, pp. 627–48.

Bruun, P. 1989: The coastal drain: what can it do or not do? *J. Coastal Res.* 5, pp. 123–5.

Bryant, E. A. 1988: Storminess and high tide beach change, Stanwell Park, Australia, 1943–1978. *Marine Geol.* 79, pp. 171–87.

Bryant, E. A., Young, R. W., and Price, D. M. 1992: Evidence of tsunami sedimentation on the south-eastern coast of Australia. *J. Geol.* 100, pp. 753–65.

Bryant, E. A., Young, R. W., Price, D. M., and Short, S. A. 1990: Thermoluminescence and uranium-thorium chronologies of Pleistocene coastal landforms of the Illawarra Region, New South Wales. *Austr. Geogr.* 21, pp. 101–11.

Buddemeier, R. W. and Hopley, D. 1988: Turn-ons and turn-offs: causes and mechanisms of the initiation and termination of coral reef growth. *Proc. 6th Int. Coral Reef Symp., Townsville* 1, pp. 253–61.

Buigues, D. 1985: Principal facies and their distribution at Mururoa Atoll (French Polynesia). *Proc. 5th Int. Coral Reef Congr., Tahiti* 3, pp. 249–55.

Bujalesky, G. G. and González-Bonorino, G. 1991: Gravel spit stabilized by unusual(?) high-energy wave climate in bay side, Tierra Del Fuego. *Coastal Sediments '91*, N. C. Kraus, K. J. Gingerich, and D. L. Kriebel (eds.), Amer. Soc. Civ. Engnr., pp. 960–74.

Bunkley-Williams, L. and Williams, E. H. 1990: Global assault on coral reefs. *Natural Hist.* April (4), pp. 46–54.

Bunt, J. S., Williams, W. T., and Bunt, E. D. 1985: Mangrove species distribution in relation to tides at the seafront and up rivers. *Austr. J. Marine and Freshwater Res.* 36, pp. 481–92.

Bunt, J. S., Williams, W. T., and Duke, N. C. 1982: Mangrove distributions in north-east Australia. *J. Biogeogr.* 9, pp. 111–20.

Burban, P.-Y., Lick, W., and Lick, J. 1989: The flocculation of fine-grained sediments in estuarine waters. *J. Geophys. Res.* 94, pp. 8323–30.

Burke, H. W. 1952: Contributions by the Japanese to the study of coral reefs. *U.S. Geol Surv. Pacific Geol. Surv. Rept.* 517, Memo. for Record.

Butterfield, G. R. 1991: Grain transport rates in steady and unsteady turbulent airflows. In *Aeolian Grain Transport 1 Mechanics*, O. E. Barndorff-Nielsen and B. B. Willetts (eds.), Acta Mechanica Suppl. 1, Springer-Verlag, Wien, pp. 97–122.

Byrne, J. V., Leroy, D. O. S., and Riley, C. M. 1959: The chenier plain and its stratigraphy, southwestern Louisiana. *Trans. Gulf Coast Assoc. Geol. Soc.* 9, pp. 237–60.

Byrne, M.-L. and McCann, S. B. 1993: The internal structure of vegetated coastal sand dunes, Sable Island, Nova Scotia. *Sed. Geol.* 84, pp. 199–218.

Bythell, J. C., Gladfelter, E. H., and Bythell, M. 1993: Chronic and catastrophic natural mortality of three common Caribbean reef corals. *Coral Reefs* 12, pp. 143–52.

Cabioch, G., Thomassin, A., and LeColle, J. F. 1989: Age d'émersion des récifs frangeants holocènes autour de la 'Grande Terre' de Nouvelle-Calédonie (SO Pacifique); nouvelle interprétation de la courbe des niveaux marins depuis 8 000 ans B.P. *C. R. Acad. Sci.* 308, series II, pp. 419–25.

Caldwell, N. E. and Williams, A. T. 1986: Spatial and seasonal beach profile characteristics. *Geol. J.* 21, pp. 127–38.

Campbell, K. M., Coleman, C. J., Entsminger, L. D., and Tanner, W. F. 1977: Langmuir circulation as a factor in the formation of depositional beach cusps. In *Coastal Sedimentology*, W. F. Tanner (ed.), Dept. Geol., Florida State Univ., Tallahassee, Florida, pp. 245–62.

Carr, A. P. and Graff, J. 1982: The tidal immersion factor and shore platform development. *Trans. Inst. Brit. Geogr.* 7, pp. 240–5.

Carter, R. W. G. 1986: The morphodynamics of beach-ridge formation: Magilligan, Northern Ireland. *Marine Geol.* 73, pp. 191–214.

Carter, R. W. G. 1988: *Coastal Environments*, Academic Press, London.

Carter, R. W. G., Bauer, B. O., Sherman, D. J., Davidson-Arnott, R. G. D., Gares, P. A., Nordstrom, K. F., and Orford, J. D. 1992a: Dune development in the aftermath of stream outlet closure: examples from Ireland and California. In *Coastal Dunes*, R. W. G. Carter, T. G. F. Curtis, and M. J. Sheehy-Skeffington (eds.), Proc. 3rd European Dune Congr., Balkema, Rotterdam, pp. 57–69.

Carter, R. W. G., Forbes, D. L., Jennings, S. C., Orford, J. D., Shaw, J., and Taylor, R. B. 1989: Barrier and lagoon coast evolution under differing relative sea-level regimes: examples from Ireland and Nova Scotia. *Marine Geol.* 88, pp. 221–42.

Carter, R. W. G., Hesp, P. A., and Nordstrom, K. F. 1990c: Erosional landforms in coastal dunes. In *Coastal Dunes: Form and Process*, K. F. Nordstrom, N. P. Psuty, and R. W. G. Carter (eds.), Wiley, Chichester, pp. 217–50.

Carter, R. W. G., Jennings, S. C., and Orford, J. D. 1990a: Headland erosion by waves. *J. Coastal Res.* 6, pp. 517–29.

Carter, R. W. G., Nordstrom, K. F., and Psuty, N. P. 1990d: The study of coastal dunes. In *Coastal Dunes: Form and Process*, K. F. Nordstrom, N. P. Psuty, and R. W. G. Carter (eds.), Wiley, Chichester, pp. 1–16.

Carter, R. W. G. and Orford, J. D. 1980: Gravel barrier genesis and management: a contrast. *Coastal Zone '80*, B. L. Edge (ed.), Amer. Soc. Civ. Engnr., pp. 1304–20.

Carter, R. W. G. and Orford, J. D. 1988: Conceptual model of coarse clastic barrier formation from multiple sediment sources. *Geogr. Rev.* 78, pp. 221–39.

Carter, R. W. G. and Orford, J. D. 1991: The sedimentary organisation and behaviour of drift-aligned gravel barriers. *Coastal Sediments '91*, N. C. Kraus, K. J. Gingerich, and D. L. Kriebel (eds.), Amer. Soc. Civ. Engnr., pp. 934–48.

Carter, R. W. G. and Orford, J. D. 1993: The morphodynamics of coarse clastic beaches and barriers: a short- and long-term perspective. *J. Coastal Res.* Spec. Issue 15, pp. 158–79.

Carter, R. W. G., Orford, J. D., Forbes, D. L., and Taylor, R. B. 1987: Gravel barriers, headlands and lagoons: an evolutionary model. *Coastal Sediments '87*, N. C. Kraus (ed.), Amer. Soc. Civ. Engnr., pp. 1776–92.

Carter, R. W. G., Orford, J. D., Forbes, D. L., and Taylor, R. B. 1990b: Morphosedimentary development of drumlin-flank barriers with rapidly rising sea level, Story Head, Nova Scotia. *Sed. Geol.* 69, pp. 117–38.

Carter, R. W. G., Orford, J. D., Jennings, S. C., Shaw, J., and Smith, J. P. 1992b: Recent evolution of a paraglacial estuary under conditions of rapid sea-level rise: Chezzetcook Inlet, Nova Scotia. *Proc. Geol. Assoc.* 103, pp. 167–85.

Carter, R. W. G. and Wilson, P. 1990: The geomorphological, ecological and pedological development of coastal foredunes at Magilligan Point, Northern Ireland. In *Coastal Dunes: Form and Process*, K. F. Nordstrom, N. P. Psuty, and R. W. G. Carter (eds.), Wiley, Chichester, pp. 129–57.

Casapieri, P. 1984: Environmental impact of pollution controls on the Thames Estuary, United Kingdom. In *The Estuary as a Filter*, V. S. Kennedy (ed.), Academic Press, New York, pp. 489–504.

Castaing, P. and Allen, G. P. 1981: Mechanisms controlling seaward escape of suspended sediment from the Gironde: a macrotidal estuary in France. *Marine Geol.* 40, pp. 101–18.

Chafetz, H. S. and Kocurek, G. 1981: Coarsening-upward sequences in beach cusp accumulations. *J. Sed. Petrol.* 51, pp. 1157–61.

Chamberlain, J. A. 1978: Mechanical properties of coral skeleton: compressive strength and its adaptive significance. *Paleobiol.* 4, pp. 419–35.

Chapman, V. J. 1974: *Salt Marshes and Salt Deserts of the World*, J. Cramer, Lehre, 2nd edn.

Chapman, V. J. 1975: *Mangrove Vegetation*, J. Cramer, Lehre.

Chapman, V. J. 1976: *Coastal Vegetation*, Pergamon Press, Oxford, 2nd edn.

Chapman, V. J. 1977a: Introduction. In *Wet Coastal Ecosystems*, V. J. Chapman (ed.), Elsevier, Amsterdam, pp. 1–29.

Chapman, V. J. 1977b: Wet coastal formations of Indo-Malesia and Papua-New Guinea. In *Wet Coastal Ecosystems*, V. J. Chapman (ed.), Elsevier, Amsterdam, pp. 261–70.

Chapman, V. J. and Ronaldson, J. W. 1958: The mangrove and salt marsh flats of the Auckland Isthmus. *N.Z. Dept. Sci. and Industr. Res. Bull.* 125, pp. 1–79.

Chappell, J. 1974: Geology of coral terraces, Huon Peninsula, New Guinea: a study of Quaternary tectonic movements and sea-level changes. *Geol. Soc. Amer. Bull.* 85, pp. 553–70.

Chappell, J. 1980*a*: Inshore–nearshore morphodynamics—a predictive model. *Proc. 17th Coastal Engnr. Conf.* pp. 963–77.

Chappell, J. 1980*b*: Coral morphology, diversity and reef growth. *Nature* 286, pp. 249–52.

Chappell, J. 1983: Sea level changes and coral reef growth. In D. J. Barnes (ed.), *Perspectives on Coral Reefs*, Brian Clouston, Manuka A.C.T., *Austr. Inst. Marine Sci. Contrib.* 200, pp. 46–55.

Chappell, J. 1993: Contrasting Holocene sedimentary geologies of lower Daly River, northern Australia, and lower Sepik-Ramu, Papua New Guinea. *Sed. Geol.* 83, pp. 339–58.

Chappell, J. and Eliot, I. G. 1979: Surf-beach dynamics in time and space—an Australian case study, and elements of a predictive model. *Marine Geol.* 32, pp. 231–50.

Chappell, J., Eliot, I. G., Bradshaw, M. P., and Lonsdale, E. 1979: Experimental control of beach face dynamics by water-table pumping. *Engnr. Geol.* 14, pp. 29–41.

Chappell, J. and Grindrod, J. 1984: Chenier plain formation in northern Australia. In *Coastal Geomorphology in Australia*, B. G. Thom (ed.), Academic Press, Sydney, pp. 197–231.

Chappell, J. and Shackleton, N. J. 1986: Oxygen isotopes and sea level. *Nature* 324, pp. 137–40.

Chardonnet, J. 1948: Les Calanques provencales, origin et divers types. *Ann. Géogr.* 57, pp. 289–97.

Chase, J. 1975: Wind-driven circulation in a Spanish estuary. *Estuarine and Coastal Marine Sci.* 3, pp. 303–10.

Chepil, W. S. 1945: Dynamics of wind erosion. III The transport capacity of the wind. *Soil Sci.* 60, pp. 475–80.

Chepil, W. S. 1959: Equilibrium of soil grains at the threshold of movement by wind. *Soil Sci. Soc. Amer. Proc.* 23, pp. 422–8.

Chesher, R. 1985: Practical problems in coral reef utilization and management: a Tongan case study. *Proc. 5th Coral Reef Congr., Tahiti* 4, pp. 213–17.

Chevalier, J. P. 1973: Geomorphology and geology of coral reefs in French Polynesia. In *Biology and Geology of Coral Reefs*, O. A. Jones and R. Endean (eds.), Academic Press, New York, 1, pp. 113–41.

Chevalier, J. P. 1977: Origin of the reef formations of Moorea Island (Archipelago of La Societe). *Proc. 3rd Int. Coral Reef Symp., Miami* 2, pp. 283–7.

Choi, D. R. 1981: Quaternary reef foundations in the southernmost Belize Shelf, British Honduras. *Proc. 4th Int. Coral Reef Symp., Manila* 1, pp. 635–42.

Chough, S. K. and Orton, G. J. 1995: Preface (to special issue on coarse-grained deltas). *Sedimentology* 98, pp. 1–2.

Christensen, B. 1983: Mangroves—what are they worth? *Unasylva* 35, pp. 2–10.

Christiansen, Ch., Dalsgaard, K., Møller, J. T., and Bowman, D. 1990: Coastal dunes in Denmark. Chronology in relation to sea level. In *Dunes of the European Coasts*, W. Bakker, P. D. Jungerius, and J. A. Klijn (eds.), *Catena* Suppl. 18, pp. 61–70.

Christiansen, S. 1963: Morphology of some coral cliffs, Bismarck Archipelago. *Geogr. Tidsskr.* 62, pp. 1–23.

Chubb, J. 1957: The pattern of some Pacific islands chains. *Geol. Mag.* 94, pp. 221–8.

Clark, J. A. 1980: A numerical model of world-wide sea level changes on a viscoelastic Earth. In *Earth Rheology, Isostasy, and Eustasy*, N.-A. Morner (ed.), Wiley, Chichester, pp. 525–34.

Clark, J. A. and Lingle, C. S. 1979: Predicted relative sea level changes (18,000 years BP to present) caused by late-glacial retreat of the Antarctic ice sheet. *Quat. Res.* 11, pp. 279–98.

Clark, J. A. and Primus, J. A. 1987: Sea-level changes resulting from future retreat of ice sheets: an effect of CO_2 warming of the climate. In *Sea-level Changes*, M. J. Tooley and I. Shennan (eds.), *Inst. Brit. Geogr.* (Blackwell, Oxford) Spec. Publ. 27, pp. 356–70.

Clark, M. J., French, H. M., and Harry, D. G. 1984: Reconnaisance techniques for the estimation of Arctic coastal sediment budget and process. In *Coastal Research: UK Perspectives*, M. W. Clark (ed.), Geo Books, Norwich, pp. 1–15.

Clarke, T. L., Lesht, B., Young, R. A., Swift, D. J. P., and Freeland, G. L. 1982: Sediment resuspension by surface wave action: an examination of possible mechanisms. *Marine Geol.* 49, pp. 43–59.

Clayton, K. M. 1990: Sea-level rise and coastal defences in the UK. *Quart. J. Engnr. Geol.* 23, pp. 283–7.

Cleary, W. J., Hosier, P. E., and Wells, G. R. 1979: Genesis and distribution of marsh islands within southeastern North Carolina lagoons. *J. Sed. Petrol.* 49, pp. 703–10.

Clemmensen, L. B. 1986: Storm-generated eolian sand shadows and their sedimentary structures, Vejers Strand, Denmark. *J. Sed. Geol.* 56, pp. 520–7.

Clifton, H. E. 1969: Beach lamination: nature and origin. *Marine Geol.* 7, pp. 553–9.

Clifton, H. E. 1976: Wave-formed sedimentary structures—a conceptual model. In *Beach and Nearshore Sedimentation*, R. A. Davis and R. L. Ethington (eds.), *Soc. Econ. Paleont. Mineral.* Spec. Publ. 24, pp. 126–48.

Clifton, H. E. and Dingler, J. R. 1984: Wave-formed structures and paleo-environmental reconstruction. *Marine Geol.* 60, pp. 165–98.

Climap Project Members 1976: The surface of the ice-age Earth. *Science* 191, pp. 1131–7.

Cloud, P. E. 1966: Beach cusps: response to Plateau's rule. *Science* 154, pp. 890–1.

Coakley, J. P. 1989: The origin and evolution of a complex cuspate foreland: Pointe-Aux-Pins, Lake Erie, Ontario. *Géogr. Phys. Quat.* 43, pp. 65–76.

Coakley, J. P., Rukavina, N. A., and Zeman, A. J. 1986: Wave-induced subaqueous erosion of cohesive tills: preliminary results. *Proc. Symp. Cohesive Shores,*

M. G. Skafel (ed.), National Res. Council Can. (ACROSES), Ottawa, pp. 120–36.

Coastal Engineering Research Center 1973: *Shore Protection Manual*, US Army Corps of Engnr., Washington DC, 1st edn.

Coastal Engineering Research Center 1984: *Shore Protection Manual*, US Army Corps of Engnr., Washington DC, 4th edn.

Coleman, J. M. 1982: *Deltas: Processes of Deposition and Models for Exploration*, Int. Human Resources Development Corporation: Boston, 2nd edn.

Coleman, J. M., Gagliano, S. M., and Smith, W. G. 1970: Sedimentation in a Malaysian high tide tropical delta. In *Deltaic Sedimentation: Modern and Ancient*, J. P. Morgan (ed.), Soc. Econ. Paleont. Mineral. Spec. Publ. 15, pp. 185–97.

Coleman, J. M. and Roberts, H. H. 1989: Deltaic coastal wetlands. In *Coastal Lowlands*, W. J. M. Van der Linden, S. A. P. L. Cloetingh, J. P. K. Kaasschieter, W. J. E. Van de Graaf, J. Vandenberghe, and J. A. M. Van der Gun (eds.), Kluwer, Dordrecht, Proc. Symp. Coastal Lowlands, Roy. Geol. Mining Soc. Netherlands, pp. 1–24.

Coleman, J. M. and Wright, L. D. 1975: Modern river deltas: variability of processes and sand bodies. In *Deltas—Models for Exploration*, M. L. Broussard (ed.), Houston Geol. Soc., Texas, pp. 99–149.

Coles, S. L. 1985: The effects of elevated temperature on reef coral planula settlement as related to power station entrainment. *Proc. 5th Int. Coral Reef Congr., Tahiti* 4, pp. 171–6.

Colgan, M. W. 1987: Coral reef recovery on Guam (Micronesia) after catastrophic predation by *Acanthaster planci. Ecology* 68, pp. 1592–605.

Collins, L. B., Zhu, Z. R., Wyrwoll, K.-H., Hatcher, B. G., Playford, P. E., Chen, J. H., Eisenhauer, A., and Wasserburg, G. J. 1993b: Late Quaternary evolution of coral reefs on a cool-water carbonate margin: the Abrolhos carbonate platforms, southwest Australia. *Marine Geol.* 110, pp. 203–12.

Collins, L. B., Zhu, Z. R., Wyrwoll, K.-H., Hatcher, B. G., Playford, P. E., Eisenhauer, A., Chen, J. H., Wasserburg, G. J., and Bonani, G. 1993a: Holocene growth history of a reef complex on a cool-water carbonate margin: Easter Group of the Houtman Abrolhos, eastern Indian Ocean. *Marine Geol.* 115, pp. 29–46.

Collins, L. M., Collins, J. N., and Leopold, L. B. 1987: Geomorphic processes of an estuarine marsh: preliminary results and hypotheses. In *International Geomorphology*, V. Gardiner (ed.), Wiley, Chichester, 1, pp. 1049–72.

Colquhoun, D. J., Pierce, J. W., and Schwartz, M. L. 1968: Field and laboratory observations on the genesis of barrier islands. *Geol. Soc. Amer. Annual Meeting, Abstr.* pp. 59–60. Repr. in 1973 in *Barrier Islands*, M. L. Schwartz (ed.), Benchmark Papers in Geology 9, Dowden, Hutchinson, and Ross, Stroudsburg, Penn., p. 290.

Conley, D. C. and Inman, D. L. 1992: Field observations of the fluid–granular boundary layer under near-breaking waves. *J. Geophys. Res.* 97, pp. 9631–43.

Cook, P. G. 1986: A review of coastal dune building in eastern Australia. *Austr. Geogr.* 17, pp. 133–43.

Cook, P. J. and Polach, H. A. 1973: A chenier sequence at Broad Sound, Queensland, and evidence against a Holocene high sea level. *Marine Geol.* 14, pp. 253–68.

Cooper, J. A. G. 1993a: Sedimentation in a river dominated estuary. *Sedimentology* 40, pp. 979–1017.

Cooper, J. A. G. 1993b: Sedimentation in the cliff-bound, microtidal Mtamvuna Estuary, South Africa. *Marine Geol.* 112, pp. 237–56.

Cooper, W. S. 1958: Coastal sand dunes of Oregon and Washington. *Geol. Soc. Amer. Mem.* 72.

Cooper, W. S. 1967: Coastal sand dunes of California. *Geol. Soc. Amer. Mem.* 104.

Corbel, J. 1956: Un karst Méditerranéen de basse altitude: le massif des calanques et la formation de son relief. *Rev. Géogr. Lyon* 31, pp. 129–36.

Cornaglia, P. 1889: On beaches. *Accademia Nazionale dei Lincei, Atti. Cl. Sci. Fis. Mat. e Nat. Mem.* 5, Ser. 4, pp. 284–304. English translation (1977) by W. N. Felder, in *Beach Processes and Coastal Hydrodynamics*, J. S. Fisher and R. Dolan (eds.), Benchmark Papers in Geology 39, Dowden, Hutchinson, and Ross, Stroudsburg, Penn., pp. 11–26.

Cornish, V. 1898: On sea-beaches and sandbanks. *Geogr. J.* 11, pp. 528–43.

Costa, R. G. and Mehta, A. J. 1990: Flow-fine sediment hysteresis in sediment-stratified coastal waters. *Proc. 22nd Coastal Engnr. Conf.* pp. 2047–60.

Cotton, C. A. 1954: Deductive morphology and genetic classification of coasts. *Sci. Monthly* 78 (3), pp. 163–81.

Cotton, C. A. 1963: Levels of planation of marine benches. *Z. Geomorph.* 7, pp. 97–110.

Coudray, J., Thomassin, B. A., and Vasseur, P. 1985: Comparative geomorphology of New Caledonia and Mayotte barrier reefs (Indo-Pacific Province). *Proc. 5th Int. Coral Reef Congr., Tahiti* 6, pp. 427–32.

Coutellier, V. and Stanley, D. J. 1987: Late Quaternary stratigraphy and paleogeography of the eastern Nile Delta, Egypt. *Marine Geol.* 77, pp. 257–75.

Craft, C. B., Seneca, E. D., and Broome, S. W. 1993: Vertical accretion in microtidal regularly and irregularly flooded estuarine marshes. *Estuarine, Coastal and Shelf Sci.* 37, pp. 371–86.

Craik, W., Kenchington, R., and Kelleher, G. 1990: Coral-reef management. In *Coral Reefs* (Ecosystems of the World 25), Z. Dubinsky (ed.), Elsevier, Amsterdam, pp. 453–67.

Crossland, C. J. 1988: Latitudinal comparisons of coral reef structure and function. *Proc. 6th Int. Coral Reef Symp., Townsville* 1, pp. 221–6.

Dabrio, C. J. and Polo, M. D. 1981: Flow regime and bedforms in a ridge and runnel system, S. E. Spain. *Sed. Geol.* 28, pp. 97–110.

Dahl, A. L. 1985: Status and conservation of South Pacific coral reefs. *Proc. 5th Int. Coral Reef Congr., Tahiti.* 6, pp. 509–13.

Dahl, A. L. and Salvat, B. 1988: Are human impacts, either through traditional or contemporary uses, stabilizing or destabilizing to reef community structure? *Proc. 6th Int. Coral Reef Symp., Townsville* 1, pp. 63–9.

Dally, W. R. 1987: Longshore bar formation—surf beat or undertow? *Coastal Sediments '87* N. C. Kraus (ed.), Amer. Soc. Civ. Engnr., pp. 71–86.

Dalongeville, M. and Guilcher, A. 1982: Les Platesformes à vasques en Méditerranée notamment leur extension vers le nord. *Karst Littoraux.* Comité National Francais de Geographie, Actes du Colloqium de Perpignan, 15–17 Mai 1982, 1, pp. 13–22.

Dalongeville, R. (ed.) 1984: Le beach rock. Colloque tenu à Lyon en novembre 1983. *Trav. Maison de l'Orient* 8, Lyons.

Dalrymple, R. A. 1975: A mechanism for rip current generation on an open coast. *J. Geophys. Res.* 80, pp. 3485–7.

Dalrymple, R. A. (ed.) 1985: *Physical Modelling in Coastal Enginering*, Balkema, Rotterdam.

Dalrymple, R. A. and Lanan, G. A. 1976: Beach cusps formed by intersecting waves. *Geol. Soc. Amer. Bull.* 87, pp. 57–60.

Dalrymple, R. W., Knight, R. J., Zaitlin, B. A., and Middleton, G. V. 1990: Dynamics and facies model of a macrotidal sand-bar complex, Cobequid Bay–Salmon River Estuary (Bay of Fundy). *Sedimentology* 37, pp. 577–612.

Dalrymple, R. W., Zaitlin, B. A., and Boyd, R. 1992: Estuarine facies models: conceptual basis and stratigraphic implications. *J. Sed. Petrol.* 62, pp. 1130–46.

Dana, J. D. 1849: *Geology*, Putnam, New York, Rept. US Exploration Exped. 1838–42, 10, pp. 35–8.

D'Anglejan, B. F. 1980: Effects of seasonal changes on the sedimentary regime of a subarctic estuary, Rupert Bay (Canada). *Sed. Geol.* 26, pp. 51–68.

D'Anglejan, B. F. and Ingram, R. G. 1976: Time–depth variations in tidal flux of suspended matter in the Saint Lawrence Estuary. *Estuarine and Coastal Marine Sci.* 4, pp. 401–16.

Daniel, J. R. K. 1989: The chenier plain coastal system of Guyana. *Marine Geol.* 90, pp. 283–7.

Dankers, N., Binsbergen, M., Zegers, K., Laane, R., and van der Loeff, M. R. 1984: Transportation of water, particulate and dissolved organic and inorganic matter between a salt marsh and the Ems-Dollard Estuary, The Netherlands. *Estuarine, Coastal and Shelf Sci.* 19, pp. 143–65.

Darienzo, M. E., Peterson, C. D., and Clough, C. 1994: Stratigraphic evidence for great subduction-zone earthquakes at four estuaries in northern Oregon, U.S.A. *J. Coastal Res.* 10, pp. 850–76.

Davidson-Arnott, R. G. D. 1988: Controls on formation and form of barred nearshore profiles. *Geogr. Rev.* 78, pp. 185–93.

Davidson-Arnott, R. G. D. and Law, M. N. 1990: Seasonal patterns and controls on sediment supply to coastal foredunes, Long Point, Lake Erie. In *Coastal Dunes: Form and Process*, K. F. Nordstrom, N. P. Psuty, and R. W. G. Carter (eds.), Wiley, Chichester, pp. 177–200.

Davidson-Arnott, R. G. D. and McDonald, R. A. 1989: Nearshore water motion and mean flows in a multiple parallel bar system. *Marine Geol.* 86, pp. 321–38.

Davidson-Arnott, R. G. D. and Ollerhead, J. 1995: Nearshore erosion on a cohesive shoreline. *Marine Geol.* 122, pp. 349–65.

Davidson-Arnott, R. G. D. and Pember, G. F. 1980: Morphology and sedimentology of multiple parallel bar systems, southern Georgian Bay, Ontario. In *The Coastline of Canada*, S. B. McCann (ed.), *Geol. Surv. Can. Paper* 80–10, pp. 417–28.

Davidson-Arnott, R. G. D. and Pyskir, N. M. 1988: Morphology and formation of an Holocene coastal dune field, Bruce Peninsula, Ontario. *Géogr. Phys. Quat.* 42, pp. 163–70.

Davidson, M. A., Russell, P. E., Huntley, D. A., and Hardisty, J. 1993: Tidal asymmetry in suspended sand transport on a macrotidal intermediate beach. *Marine Geol.* 110, pp. 333–53.

Davidson, M. A., Russell, P. E., Huntley, D. A., Hardisty, J., and Cramp, A. 1992: An overview of the British beach and nearshore dynamics (B-Band) programme. *Proc. 23rd Coastal Engnr. Conf.* pp. 1987–2000.

Davies, A. G. 1983: Wave interactions with rippled sand beds. In *Physical Oceanography of Coastal and Shelf Seas*, B. Johns (ed.), Elsevier, Amsterdam, pp. 1–65.

Davies, A. G. 1985: Field observations of the threshold of sediment motion by wave action. *Sedimentology* 32, pp. 685–704.

Davies, J. L. 1957: The importance of cut and fill in the development of sand beach ridges. *Austr. J. Sci.* 20, pp. 105–11.

Davies, J. L. 1960: Beach alignment in southern Australia. *Austr. Geogr.* 8, pp. 42–4.

Davies, J. L. 1972: *Geographical Variation in Coastal Development*, Oliver & Boyd, Edinburgh.

Davies, K. H. 1983: Amino acid analysis of Pleistocene marine molluscs from the Gower Peninsula. *Nature* 302, pp. 137–9.

Davies, M. H. 1984: The search for a coastal sediment transport formula. *Coastal Engineering in Canada 1984*, Queen's Univ., Kingston, Ontario, pp. 139–60.

Davies, P. and Williams, A. T. 1986: Cave development in lower Lias coastal cliffs, the Glamorgan Heritage Coast, Wales, UK. In *Iceland Coastal and River*

Symp., Proc., G. Sigbjarnarson (ed.), National Energy Authority, Reykjavik, pp. 75–92.

Davies, P., Williams, A. T., and Bomboe, P. 1991: Numerical modelling of lower Lias rock failures in the coastal cliffs of south Wales. *Coastal Sediments '91*, N. C. Kraus, K. J. Gingerich, and D. L. Kriebel (eds.), Amer. Soc. Civ. Engnr., pp. 1599–612.

Davies, P. J. 1977: Modern reef growth—Great Barrier Reef. *Proc. 3rd Int. Coral Reef Symp., Miami* 2, pp. 325–30.

Davies, P. J. 1988: Evolution of the Great Barrier Reef—reductionist dream or expansionist vision. *Proc. 6th Int. Coral Reef Symp., Townsville* 1, pp. 9–17.

Davies, P. J. and Kinsey, D. W. 1977: Holocene reef growth—One Tree Island, Great Barrier Reef. *Marine Geol.* 24, pp. M1–M11.

Davies, P. J., Marshall, J. F., and Hopley, D. 1985: Relationships between reef growth and sea level in the Great Barrier Reef. *Proc. 5th Coral Reef Congr., Tahiti* 3, pp. 95–103.

Davies, P. J., Marshall, J. F., Thom, B. G., Harvey, N., Short, A., and Martin, K. 1977: Reef development—Great Barrier Reef. *Proc. 3rd Int. Coral Reef Symp., Miami* 2, pp. 331–7.

Davies, P. J. and Montaggioni, L. (seminar convenors) 1985: Reef growth and sea-level change: the environmental signature. *Proc. 5th Int. Coral Reef Congr., Tahiti* 3, pp. 477–515.

Davis, R. A. 1985: Beach and nearshore zone. In *Coastal Sedimentary Environments*, R. A. Davis (ed.), Springer-Verlag, New York, pp. 379–444.

Davis, R. A. 1989: Morphodynamics of the west-central Florida barrier system: the delicate balance between wave- and tide-domination. In *Coastal Lowlands*, W. J. M. Van der Linden, S. A. P. L. Cloetingh, J. P. K. Kaasschieter, W. J. E. Van de Graaf, J. Vandenberghe, and J. A. M. Van der Gun (eds.), Kluwer, Dordrecht, Proc. Symp. Coastal Lowlands, Roy. Geol. Mining Soc. Netherlands, pp. 225–35.

Davis, R. A., Burke, R. B., and Brame, J. W. 1979: Origin and development of barrier islands on west-central Peninsula of Florida. *Amer. Assoc. Petrol. Geol. Bull.* 63, pp. 438–9.

Davis, R. A., Fox, W. T., Hayes, M. O., and Boothroyd, J. C. 1972: Comparison of ridge and runnel systems in tidal and non-tidal environments. *J. Sed Petrol.* 42, pp. 413–21.

Davis, R. A., Goldsmith, V., and Goldsmith, Y. E. 1976: Ice effects on beach sedimentation from Massachusetts and Lake Michigan. *Rev. Géogr. Montréal* 30, pp. 201–6.

Davis, R. A. and Hayes, M. O. 1984: What is a wave-dominated coast? *Marine Geol.* 60, pp. 313–29.

Dawson, A. G. 1980: Shore erosion by frost: an example from the Scottish late-glacial. In *Studies in the Late-Glacial of North-West Europe*, J. J. Lowe, J. M. Gray, and J. E. Robinson (eds.), Pergamon, Oxford, pp. 45–53.

Dawson, A. G. 1986: Quaternary shore platforms and sea-levels in southern Britain. In *Essays for Professor R. E. H. Mellor*, W. Ritchie, J. C. Stone, and A. S. Mather (eds.), Dept. Geogr., Univ. Aberdeen, pp. 377–82.

Dawson, A. G., Long, D., and Smith, D. E. 1988: The Storegga Slides: evidence from eastern Scotland for a possible tsunami. *Marine Geol.* 82, pp. 271–6.

Day, J. W., Hall, C. A. S., Kemp, W. M., and Yanez-Arancibia, A. 1989: Intertidal wetlands: salt marshes and mangrove swamps. In *Estuarine Ecology*, J. W. Day, C. A. S. Hall, W. M. Kemp, and A. Yanez-Arancibia (eds.), Wiley, New York, pp. 188–225.

Day, M. 1983: Doline morphology and development in Barbados. *Ann. Assoc. Amer. Geogr.* 73, pp. 206–19.

Dean, R. G. 1973: Heuristic models of sand transport in the surf zone. *Engineering Dynamics in the Surf Zone*, Proc. First Austr. Conf. Coastal Engnr., Inst. Engnr., Australia, pp. 208–14.

Dean, R. G. 1991: Equilibrium beach profiles: characteristics and applications. *J. Coastal Res.* 7, pp. 53–84.

Dean, R. G., Healy, T. R., and Dommerholt, A. P. 1993: A 'blind-folded' test of equilibrium beach profile concepts with New Zealand data. *Marine Geol.* 109, pp. 253–66.

Dean, R. G. and Maurmeyer, E. M. 1980: Beach cusps at Port Reyes and Drakes Bay beaches, California. *Proc. 17th Coastal Engnr. Conf.* pp. 863–84.

Dean, R. G., Srinivas, R., Parchure, T. M. 1992: Longshore bar generation mechanisms. *Proc. 23rd Coastal Engnr. Conf.* pp. 2001–14.

Defant, A. 1961: *Physical Oceanography*, Pergamon Press, Oxford, vol. 2.

Deigaard, R., Fredsøe, J., and Hedegaard, I. B. 1986: Mathematical model for littoral drift. *J. Waterw. Port Coastal Ocean Div. Amer. Soc. Civ. Engnr.* 112, pp. 351–69.

Delesalle, B. 1985: Mataiva Atoll, Tuamotu Archipelago. *Proc. 5th Int. Coral Reef Congr., Tahiti* 1, pp. 269–321.

Demarest, J. M. and Leatherman, S. P. 1985: Mainland influence on coastal transgression: Delmarva Peninsula. *Marine Geol.* 63, pp. 19–33.

De Raeve, F. 1989: Sand dune vegetation and management dynamics. In *Perspectives in Coastal Dune Management*, F. Van Der Meulen, P. D. Jungerius, and J. Visser (eds.), SPB Academic Publishing, The Hague, pp. 99–109.

De Silva, M. W. R. N. 1985: Status of the coral reefs of Sri Lanka. *Proc. 5th Int. Coral Reef Congr., Tahiti* 6, pp. 515–18.

Devall, M. S., Thien, L. B., and Platt, W. J. 1990: The ecology of *Ipomoea pes-caprae*, a pantropical strand plant. *Proc. Symp. Coastal Sand Dunes*, R. G. D. Davidson-Arnott (ed.), National Res. Council Can. (ACOS), Ottawa, pp. 231–49.

D.G.I. 1986: Executive summary (with 1985 observations) of the Danish report—full scale test of the coastal drains system for beach protection at Torsminde, Denmark. *Danish Geotechn. Inst.* Lyngby, Denmark (ref. 17083322 KH/KIB).

Dibajnia, M. and Watanabe, A. 1992: Sheet flow under nonlinear waves and currents. *Proc. 23rd Coastal Engnr. Conf.* pp. 2015–28.

Dietz, R. S., Holden, J. C., and Sproll, W. P. 1970: Geotectonic evolution and subsidence of Bahama Platform. *Geol. Soc. Amer. Bull.* 81, pp. 1915–28.

Dijkema, K. S. 1987: Geography of salt marshes in Europe. *Z. Geomorph.* 31, pp. 489–99.

Dinesen, Z. D. 1988: Complementary management of marine parks and island national parks in the Great Barrier Reef region. *Proc. 6th Int. Coral Reef Symp., Townsville* 2, pp. 363–8.

Dingler, J. R. 1979: The threshold of grain motion under oscillatory flow in a laboratory wave channel. *J. Sed. Petrol.* 49, pp. 287–94.

Dingler, J. R. 1990: Wave-generated littoral transport near Nome, Alaska. *Proc. Can. Coastal Conf.*, M. H. Davies (ed.), National Res. Council Can. (ACOS), Ottawa, pp. 111–22.

Dingler, J. R., Hsu, S. A., and Reiss, T. E. 1992: Theoretical and measured aeolian sand transport on a barrier island, Louisiana, USA. *Sedimentology* 39, pp. 1031–43.

Dingler, J. R. and Inman, D. L. 1976: Wave-formed ripples in nearshore sands. *Proc. 15th Coastal Engnr. Conf.* pp. 2109–21.

Dionne, J.-C. 1981: A boulder-strewn tidal flat, north shore of the Gulf of St. Lawrence, Québec. *Géogr. Phys. Quat.* 35, pp. 261–7.

Dionne, J.-C. 1984: An estimate of ice-drifted sediments based on the mud content of the ice cover at Montmagny, middle St. Lawrence Estuary. *Marine Geol.* 57, pp. 149–66.

Dionne, J.-C. 1985: Drift-ice abrasion marks along rocky shores. *J. Glaciol.* 31, pp. 237–41.

Dionne, J.-C. 1986: Érosion récente des marais intertidaux de L'Estuaire du Saint-Laurent, Québec. *Géogr. Phys. Quat.* 40, pp. 307–23.

Dionne, J.-C. 1988a: Characteristic features of modern tidal flats in cold regions. In *Tide-influenced Sedimentary Environments and Facies*, P. L. de Boer (ed.), D. Reidel, Dordrecht, The Netherlands, pp. 301–32.

Dionne, J.-C. 1988b: Ploughing boulders along shorelines, with particular reference to the St. Lawrence Estuary. *Geomorphology* 1, pp. 297–308.

Dionne, J.-C. 1989: The role of ice and frost in tidal marsh development—a review with particular reference to Québec, Canada. In *Zonality of Coastal Geomorphology and Ecology*, E. C. F. Bird and D. Kelletat (eds.), *Essener Geographische Arbeiten* 18, pp. 171–210.

Dionne, J.-C. 1992: Canadian landform examples—25 Ice-push features. *Can. Geogr.* 36, pp. 86–91.

Dionne, J.-C. 1993a: Sediment load of shore ice and ice rafting potential, Upper St. Lawrence Estuary, Québec, Canada. *J. Coastal Res.* 9, pp. 628–46.

Dionne, J.-C. 1993b: Influence glacielle dans le faconnement d'une plate-forme rocheuse intertidale, Estuaire du Saint-Laurent, Québec. *Rev. Géomorph. Dyn.* 42, pp. 1–10.

Dionne, J.-C. and Brodeur, D. 1988: Frost weathering and ice action in shore platform development with particular reference to Québec, Canada. *Z. Geomorph.* Suppl. Band 71, pp. 117–30.

Dodd, N. 1994: On the destabilization of a longshore current on a plane beach: bottom shear stress, critical conditions, and onset of instability. *J. Geophys. Res.* 99, pp. 811–24.

Dodd, N., Oltman-Shay, J., and Thornton, E. B. 1992: Shear instabilities in the longshore current: a comparison of observation and theory. *J. Phys. Oceanogr.* 22, pp. 62–82.

Doering, J. C. and Bowen, A. J. 1988: Wave-induced flow and nearshore suspended sediment. *Proc. 21st Coastal Engnr. Conf.* pp. 1452–63.

Dolan, R. 1973: Barrier islands: natural and controlled. In *Coastal Geomorphology*, D. R. Coates (ed.), State Univ. New York, Binghampton, New York, pp. 263–78.

Dolan, R. and Ferm, J. C. 1968: Crescentic landforms along the Atlantic coast of the United States. *Science* 159, pp. 627–9.

Dolan, R. and Godfrey, P. 1973: Effects of Hurricane Ginger on the barrier islands of North Carolina. *Geol. Soc. Amer. Bull.* 84, pp. 1329–34.

Dolan, R. and Hayden, B. 1983: Patterns and prediction of shoreline change. In *CRC Handbook of Coastal Processes and Erosion*, P. D. Komar (ed.), CRC Press, Boca Raton, Florida, pp. 123–49.

Dolan, R., Hayden, B., and Felder, W. 1979: Shoreline periodicities and edge waves. *J. Geol.* 87, pp. 175–85.

Dolotov, Y. S. 1992: Possible types of coastal evolution associated with the expected rise of the world's sea level caused by the 'Greenhouse Effect'. *J. Coastal Res.* 8, pp. 719–26.

Donaldson, A. C., Martin, R. H., and Kanes, W. H. 1970: Holocene Guadalupe Delta of Texas Gulf Coast. In *Deltaic Sedimentation: Modern and Ancient*, J. P. Morgan (ed.), *Soc. Econ. Paleont. Mineral.* Spec. Publ. 15, pp. 107–37.

Done, T. J. 1983: Coral zonation: its nature and significance. In D. J. Barnes (ed.), *Perspectives on Coral Reefs*, Brian Clouston, Manuka A.C.T., *Austr. Inst. Marine Sci. Contrib.* 200, pp. 107–47.

Doody, J. P. 1992: Nature conservation on the coast—the role of coastal zone management. In *Coastal Dunes*, R. W. G. Carter, T. G. F. Curtis, and M. J. Sheehy-Skeffington (eds.), Proc. 3rd European Dune Congr., Balkema, Rotterdam, pp. 495–501.

Doornkamp, J. C. and King, C. A. M. 1971: *Numerical Analysis in Geomorphology*, Edward Arnold, London, pp. 258–68.

Douglas, B. C. 1992: Global sea level acceleration. *J. Geophys. Res.* 97, pp. 12699–706.

Douglass, S. L. and Weggel, J. R. 1988: Laboratory experiments on the influence of wind on near-shore wave breaking. *Proc. 21st Coastal Engnr. Conf.* pp. 632–43.

Downing, J. P. 1984: Suspended sand transport on a dissipative beach. *Proc. 19th Coastal Engnr. Conf.* pp. 1765–81.

Drapeau, G., Long, B., and Kamphuis, J. W. 1990: Evaluation of radioactive sand tracers to measure longshore sediment transport rates. *Proc. 22nd Coastal Engnr. Conf.* pp. 2710–23.

Dronkers, J. 1986: Tidal asymmetry and estuarine morphology. *Netherlands J. Sea Res.* 20, pp. 117–31.

Duane, D. B. and James, W. R. 1980: Littoral transport in the surf zone elucidated by an eulerian sediment tracer experiment. *J. Sed. Petrol.* 50, pp. 929–42.

Dubois, R. N. 1972: Inverse relation between foreshore slope and mean grain size as a function of the heavy mineral content. *Geol. Soc. Amer. Bull.* 83, pp. 871–6.

Dubois, R. N. 1978: Beach topography and beach cusps. *Geol. Soc. Amer. Bull.* 89, pp. 1133–9.

Dubois, R. N. 1992: A re-evaluation of Bruun's Rule and supporting evidence. *J. Coastal Res.* 8, pp. 618–28.

Duffy, W., Belknap, D. F., and Kelley, J. T. 1989: Morphology and stratigraphy of small barrier-lagoon systems in Maine. *Marine Geol.* 88, pp. 243–62.

Duke, N. C. 1992: Mangrove floristics and biogeography. In *Tropical Mangrove Ecosystems*, A. I. Robertson and D. M. Alongi (eds.), Amer. Geophys. Union, Washington, DC, pp. 63–100.

Dyer, K. R. 1972: Sedimentation in estuaries. In *The Estuarine Environment*, R. S. K. Barnes and J. Green (eds.), Applied Science Publ., London, pp. 10–32.

Dyer, K. R. 1973: *Estuaries: A Physical Introduction*, Wiley, London.

Dyer, K. R. 1980: Velocity profiles over a rippled bed and the threshold of movement of sand. *Estuarine and Coastal Marine Sci.* 10, pp. 181–99.

Dyer, K. R. 1982: The initiation of sedimentary furrows by standing internal waves. *Sedimentology* 29, pp. 885–9.

Dyer, K. R. 1986: *Coastal and Estuarine Sediment Dynamics*, Wiley, Chichester.

Dyer, K. R. 1989: Sediment processes in estuaries: future research requirements. *J. Geophys. Res.* 94, pp. 14327–39.

Dyer, K. R. and Evans, E. M. 1989: Dynamics of turbidity maximum in a homogeneous tidal channel. *J. Coastal Res.* Spec. Issue 5, pp. 23–30.

Dyer, K. R. and New, A. L. 1986: Intermittency in estuarine mixing. In *Estuarine Variability*, D. A. Wolfe (ed.), Academic Press, New York, pp. 321–39.

Dyke, L. D. 1991: Temperature changes and thaw of permafrost adjacent to Richards Island, Mackenzie Delta, N.W.T. *Can. J. Earth Sci.* 28, pp. 1834–42.

Dyke, P. P. G., Moscardini, A. O., and Robson, E. H. (eds.) 1985: *Offshore and Coastal Modeling*, Springer-Verlag, Berlin.

Easton, W. H. and Olson, E. A. 1976: Radiocarbon profile of Hanauma Reef, Oahu, Hawaii. *Geol. Soc. Amer. Bull.* 87, pp. 711–9.

Edwards, A. B. 1951: Wave action in shore platform formation. *Geol. Mag.* 88, pp. 41–9.

Edwards, A. B. 1958: Wave cut platforms at Yampi Sound in the Buccaneer Archipelago. *J. Roy. Soc. West. Austr.* 41, pp. 17–21.

Eisma, D., Bernard, P., Cadee, G. C., Ittekkot, V., Kalf, J., Laane, R., Martin, J. M., Mook, W. G., Van Put, A., and Schuhmacher, T. 1991: Suspended-matter particle size in some west-European estuaries; part 1: particle-size distribution. *Netherlands J. Sea Res.* 28, pp. 193–214.

Eldredge, L. C. and Kropp, R. K. 1985: Volcanic ashfall effects on intertidal and shallow-water coral reef zones at Pagan (Mariana Islands). *Proc. 5th Coral Reef Congr., Tahiti* 4, pp. 195–200.

Elgar, S., Herbers, T. H. C., Okihiro, M., Oltman-Shay, J., and Guza, R. T. 1992: Observations of infragravity waves. *J. Geophys. Res.* 97, pp. 15573–7.

Eliot, I. G. and Clarke, D. J. 1982: Temporal and spatial variability of the sediment budget of the subaerial beach at Warilla, New South Wales. *Austr. J. Marine and Freshwater Res.* 33, pp. 945–69.

Eliot, I. G. and Clarke, D. J. 1986: Minor storm impact on the beachface of a sheltered sandy beach. *Marine Geol.* 73, pp. 61–83.

Ellenberg, L. and Sturm, M. 1986: *Kliffs*, Peter Fieber, Braunschweig.

Elliott, A. J. 1978: Observations of the meteorologically induced circulation in the Potomac Estuary. *Estuarine and Coastal Marine Sci.* 6, pp. 285–99.

Ellison, J. C. 1993: Mangrove retreat with rising sea-level, Bermuda. *Estuarine, Coastal and Shelf Sci.* 37, pp. 75–87.

Ellison, J. C. and Stoddart, D. R. 1991: Mangrove ecosystem collapse during predicted sea-level rise: Holocene analogues and implications. *J. Coastal Res.* 7, pp. 151–65.

Emanuel, K. A. 1987: The dependence of hurricane intensity on climate. *Nature* 326, pp. 483–5.

Emery, K. O. and Kuhn, G. G. 1980: Erosion of rock coasts at La Jolla, California. *Marine Geol.* 37, pp. 197–208.

Emery, K. O. and Kuhn, G. G. 1982: Sea cliffs: their processes, profiles, and classification. *Geol. Soc. Amer. Bull.* 93, pp. 644–54.

Emery, K. O., Tracey, J. I., and Ladd, H. S. 1954: Geology of Bikini and nearby atolls. *U.S. Geol. Surv. Prof. Paper* 260A, pp. 1–265.

Endean, R. and Cameron, A. M. 1990: *Acanthaster planci* population outbreaks. In *Coral Reefs* (Ecosystems of the World 25), Z. Dubinsky (ed.), Elsevier, Amsterdam, pp. 419–37.

Endean, R., Cameron, A. M., and DeVantier, L. M. 1988: *Acanthaster planci* predation on massive corals: the myth of rapid recovery of devastated reefs. *Proc. 6th Int. Coral Reef Symp., Townsville* 2, pp. 143–8.

Entsminger, L. D. 1977: A study of selected beach cusps on St. Joseph Spit, Florida. In *Coastal Sedimentology*, W. F. Tanner (ed.), Dept. Geol., Florida State Univ., Tallahassee, Florida, pp. 229–43.

Escher, B. G. 1937: Experiments on the formation of beach cusps. *Leidsche Geologische Mededeelingen.* 9, pp. 79–104.

Evans, D. V. 1988: Mechanisms for the generation of edge waves over a sloping beach. *J. Fluid Mech.* 186, pp. 379–91.

Evans, G., Schmidt, V., Bush, P., and Nelson, H. 1969: Stratigraphy and geologic history of the sabkha, Abu Dhabi, Persian Gulf. *Sedimentology* 12, pp. 145–59.

Everts, C. H. 1987: Continental shelf evolution in response to a rise in sea level. In *Sea-level Fluctuation and Coastal Evolution*, D. Nummedal, O. H. Pilkey, and J. D. Howard (eds.), *Soc. Econ. Paleont. Mineral.* Spec. Publ. 41, pp. 49–57.

Everts, C. H. 1991: Seacliff retreat and coarse sediment yields in southern California. *Coastal Sediments '91*, N. C. Kraus, K. J. Gingerich, and D. L. Kriebel (eds.), Amer. Soc. Civ. Engnr., pp. 1586–98.

Everts, C. H., Dill, R. F., Jones, A., Lorensen, T., Kelly, K., and Wilkins, R. B. 1987: Littoral sand losses to Scripps submarine canyon. *Coastal Sediments '87*, N. C. Kraus (ed.), Amer. Soc. Civ. Engnr., pp. 1549–62.

Ewing, J. A. 1983: Wind waves: a review of research during the last twenty-five years. *Geophys. J. Roy. Astron. Soc.* 74, pp. 313–29.

Eyles, C. H. 1988: A model for striated boulder pavement formation on glaciated, shallow-marine shelves: an example from the Yakataga Formation, Alaska. *J. Sed. Petrol.* 58, pp. 62–71.

Eyles, C. H. 1994: Intertidal boulder pavements in the northeastern Gulf of Alaska and their geological significance. *Sed. Geol.* 88, pp. 161–73.

Eyles, N., Buergin, R., and Hincenbergs, A, 1986: Sedimentological controls on piping structures and the development of scalloped slopes along an eroding shoreline; Scarborough Bluffs, Ontario. *Proc. Symp. Cohesive Shores*, M. G. Skafel (ed.), National Res. Council Can. (ACROSES), Ottawa, pp. 69–86.

Fagerstrom, J. A. 1987: *The Evolution of Reef Communities*, Wiley, New York.

Fairbridge, R. W. 1952: Marine erosion. *Proc. 7th Pacific Sci. Congr.* pp. 347–58.

Fairbridge, R. W. 1980: The estuary: its definition and geodynamic cycle. In *Chemistry and Biogeochemistry of Estuaries*, E. Olausson and I. Cato (eds.), Wiley, Chichester, pp. 1–35.

Falqués, A. and Iranzo, V. 1994: Numerical simulation of vorticity waves in the nearshore. *J. Geophys. Res.* 99, pp. 825–41.

Fanos, A. M. 1995: The impact of human activities on the erosion and accretion of the Nile Delta coast. *J. Coastal Res.* 11, pp. 821–33.

Fenster, M. and Dolan, R. 1994: Large-scale reversals in shoreline trends along the U.S. mid-Atlantic coast. *Geology* 22, pp. 543–46.

Fenton, J. D. 1990: Nonlinear wave theories. In *The Sea (Ocean Engnr. Sci.)*, B. Le Méhauté and D. M. Hanes (eds.), Wiley, New York, 9, pp. 3–25.

Field, M. E. and Duane, D. B. 1977: Post-Pleistocene history of the United States inner continental shelf: significance to origin of barrier islands: reply. *Geol. Soc. Amer. Bull.* 88, pp. 735–6.

Fieller, N. R. J., Gilbertson, D. D., and Olbricht, W. 1984: A new method for environmental analysis of particle size distribution data from shoreline sediments. *Nature* 311, pp. 648–51.

Filion, L. 1984: A relationship between dunes, fire and climate recorded in the Holocene deposits of Québec. *Nature* 309, pp. 543–6.

Finkelstein, K. 1982: Morphological variations and sediment transport in crenulate-bay beaches, Kodiak Island, Alaska. *Marine Geol.* 47, pp. 261–81.

Fischer, R. 1990: Biogenetic and nonbiogenetically determined morphologies of the Costa Rican Pacific coast. *Z. Geomorph.* 34, pp. 313–21.

Fisher, J. J. 1968a: Origin of barrier island chain shorelines: middle Atlantic States. *Geol. Soc. Amer.* Spec. Paper 115, pp. 66–7.

Fisher, J. J. 1968b: Barrier island formation: discussion. *Geol. Soc. Amer. Bull.* 79, pp. 1421–6.

Fisher, J. J. and Simpson, E. J. 1979: Washover and tidal sedimentation rates as environmental factors in development of a transgressive barrier shoreline. In *Barrier Islands*, S. P. Leatherman (ed.), Academic Press, New York, pp. 127–48.

Fisher, J. S., Leatherman, S. P., and Perry, F. C. 1974: Overwash processes on Assateague Island. *Proc. 14th Coastal Engnr. Conf.* pp. 1194–211.

Fisher, J. S. and Stauble, D. K. 1977: Impact of Hurricane Belle on Assateague Island washover. *Geology* 5, pp. 765–8.

Fleming, C. A. 1965: Two-storied cliffs at the Auckland Islands. *Trans. Roy. Soc. N.Z. Geol.* 3, pp. 171–4.

Flemming, N. C. 1964: Tank experiments on the sorting of beach material during cusp formation. *J. Sed. Petrol.* 34, pp. 112–22.

Flemming, N. C. 1965: Form and relation to present sea level of Pleistocene marine erosion features. *J. Geol.* 73, pp. 799–811.

Fletcher, C. H. 1992: Sea-level trends and physical consequences: applications to the U.S. shore. *Earth-Sci. Rev.* 33, pp. 73–109.

Flick, R. E., Guza, R. T., and Inman, D. L. 1981: Elevation and velocity measurements of laboratory shoaling waves. *J. Geophys. Res.* 86, pp. 4149–60.

Focke, J. W. 1978: Limestone cliff morphology on Curacao (Netherlands Antilles), with special attention to the origin of notches and vermetid/coralline algal surf benches (corniches, trottoirs). *Z. Geomorph.* 22, pp. 329–49.

Fonseca, M. S. 1989: Sediment stabilization by *Halophila decipiens* in comparison to other seagrasses. *Estuarine, Coastal and Shelf Sci.* 29, pp. 501–7.

Forbes, D. L., Orford, J. D., Carter, R. W. G., Shaw, J., and Jennings, S. C. 1995: Morphodynamic evolution, self-organisation, and instability of coarse-clastic barriers on paraglacial coasts. *Marine Geol.* 126, pp. 63–85.

Forbes, D. L. and Taylor, R. B. 1987: Coarse-grained beach sedimentation under paraglacial conditions, Canadian Atlantic coast. In *Glaciated Coasts*, D. M. Fitzgerald and P. S. Rosen (eds.), Academic Press, San Diego, pp. 51–86.

Forbes, D. L. and Taylor, R. B. 1994: Ice in the shore zone and the geomorphology of cold coasts. *Progr. Phys. Geogr.* 18, pp. 59–89.

Forbes, D. L., Taylor, R. B., Orford, J. D., Carter, R. W. G., and Shaw, J. 1991: Gravel-barrier migration and overstepping. *Marine Geol.* 97, pp. 305–13.

Fosberg, F. R. 1966: Vegetation as a geological agent in tropical deltas. In *Scientific Problems of the Humid Tropical Zone Deltas and their Implications*, UNESCO, Paris, Proc. Dacca Symp. Feb/Mar. 1964, pp. 227–33.

Fournier, A. and Allard, M. 1992: Periglacial shoreline erosion of a rocky coast: George River Estuary, northern Quebec. *J. Coastal Res.* 8, pp. 926–42.

Fournier, A., Allard, M., and Seguin, M. 1987: Typologie morpho-génétique des marelles du marais littoral de la baie de Kangiqsualujjuaq, estuaire du George, Québec nordique. *Géogr. Phys. Quat.* 41, pp. 47–64.

Fowler, R. E. and Dalrymple, A. 1990: Wave group forced nearshore circulation. *Proc. 22nd Coastal Engnr. Conf.* pp. 729–42.

Fox, W. T. 1985: Modeling coastal environments. In *Coastal Sedimentary Environments*, R. A. Davis (ed.), Springer-Verlag, New York, pp. 665–705.

Fox, W. T. and Davis, R. A. 1973: Simulation model for storm cycles and beach erosion on Lake Michigan. *Geol. Soc. Amer. Bull.* 84, pp. 1769–90.

Fredsøe, J. 1993: Modelling of non-cohesive sediment transport processes in the marine environment. *Coastal Engnr.* 21, pp. 71–103.

Fredsøe, J., Andersen, O. H., and Silberg, S. 1985: Distribution of suspended sediment in large waves. *J. Waterw. Port Coastal Ocean Div. Amer. Soc. Civ. Engnr.* 111, pp. 1041–59.

Freitas, C. R. de 1991: The greenhouse crisis: myths and misconceptions. *Area* 23, pp. 11–18.

French, J. R. 1993: Numerical simulation of vertical marsh growth and adjustment to accelerated sea-level rise, north Norfolk, UK. *Earth Surface Processes and Landforms* 18, pp. 63–81.

French, J. R. and Spencer, T. 1993: Dynamics of sedimentation in a tide-dominated backbarrier salt marsh, Norfolk, UK. *Marine Geol.* 110, pp. 315–31.

French, J. R. and Stoddart, D. R. 1992: Hydrodynamics of salt marsh creek systems: implications for marsh morphological development and material exchange. *Earth Surface Processes and Landforms* 17, pp. 235–52.

French, P. W., Allen, J. R. L., and Appleby, P. G. 1994: 210-lead dating of a modern period saltmarsh deposit from the Severn Estuary (southwest Britain), and its implications. *Marine Geol.* 118, pp. 327–34.

Friedrichs, C. T. and Aubrey, D. G. 1988: Non-linear tidal distortion in shallow well-mixed estuaries: a synthesis. *Estuarine, Coastal and Shelf Sci.* 27, pp. 521–45.

Frihy, O. E., Lotfy, M. F., Komar, P. D. 1995: Spatial variations in heavy minerals and patterns of sediment sorting along the Nile Delta, Egypt. *Sed. Geol.* 97, pp. 33–41.

Froidefond, J.-M. and Prud'homme, R. 1991: Coastal erosion and aeolian sand transport on the Aquitaine coast, France. In *Aeolian Grain Transport 2, The Erosional Environment*, O. E. Barndorff-Nielsen and B. B. Willetts (eds.), Acta Mechanica Suppl. 2, Springer-Verlag, Wien, pp. 147–59.

Fryberger, S. G., Krystinik, L. F., and Schenk, C. J. 1990: Tidally flooded back-barrier dunefield, Guerrero Negro area, Baja California, Mexico. *Sedimentology* 37, pp. 23–43.

Fulford, E. T. 1987: Distribution of sediment transport across the surf zone. *Coastal Sediments '87*, N. C. Kraus (ed.), Amer. Soc. Civ. Engnr., pp. 452–67.

Funicelli, N. A. 1984: Assessing and managing effects of reduced freshwater inflow to two Texas estuaries. In *The Estuary as a Filter*, V. S. Kennedy (ed.), Academic Press, New York, pp. 435–46.

Funnell, B. M. and Pearson, I. 1989: Holocene sedimentation of the north Norfolk barrier coast in relation to relative sea-level change. *J. Quat. Sci.* 4, pp. 25–36.

Gabrie, C., Porcher, M., and Masson, M. 1985: Dredging in French Polynesian coral reefs: towards a general policy of resource exploitation and site development. *Proc. 5th Int. Coral Reef Congr., Tahiti* 4, pp. 271–7.

Gagliano, S. M. and Van Beek, J. L. 1975: An approach to multiuse management in the Mississippi delta system. In *Deltas—Models for Exploration*, M. L. Broussard (ed.), Houston Geol. Soc., Houston, Texas, pp. 223–38.

Gallagher, B. 1971: Generation of surf beat by non-linear wave interactions. *J. Fluid Mech.* 49, pp. 1–20.

Gallenne, B. 1974: Study of fine material in suspension

in the estuary of the Loire and its dynamic grading. *Estuarine and Coastal Marine Sci.* 2, pp. 261–72.

Galloway, W. E. 1975: Process framework for describing the morphologic and stratigraphic evolution of deltaic depositional systems. In *Deltas—Models for Exploration*, M. L. Broussard (ed.), Houston Geol. Soc., Houston, Texas, pp. 87–98.

Galvin, C. J. 1972: Waves breaking in shallow water. In *Waves on Beaches*, R. E. Meyer (ed.), Academic Press, New York, pp. 413–56.

Gao, S. and Collins, M. 1994: Tidal inlet equilibrium in relation to cross-sectional area and sediment tranport patterns. *Estuarine, Coastal and Shelf Sci.* 38, pp. 157–72.

Gardner, R. A. M. 1983: Aeolianite. In *Chemical Sediments and Geomorphology*, A. S. Goudie and K. Pye (eds.), Academic Press, London, pp. 265–300.

Gares, P. A. 1990: Eolian processes and dune changes at developed and undeveloped sites, Island Beach, New Jersey. In *Coastal Dunes: Form and Process*, K. F. Nordstrom, N. P. Psuty, and R. W. G. Carter (eds.), Wiley, Chichester, pp. 361–80.

Gares, P. A. 1992: Topographic changes associated with coastal dune blowouts at Island Beach State Park, New Jersey. *Earth Surface Processes and Landforms* 17, pp. 589–604.

Gares, P. A. and Nordstrom, K. F. 1988: Creation of dune depressions by foredune accretion. *Geogr. Rev.* 78, pp. 194–204.

Garrett, C. and Smith, J. 1976: Interaction between long and short surface waves. *J. Phys. Oceanogr.* 6, pp. 925–30.

Garvine, R. W. 1984: Radial spreading of buoyant plumes in coastal waters. *J. Geophys. Res.* 89, pp. 1989–96.

Gehrels, W. R., Belknap, D. F., Pearce, B. R., and Gong, B. 1995: Modeling the contribution of M_2 tidal amplication to the Holocene rise of mean high water in the Gulf of Maine and the Bay of Fundy. *Marine Geol.* 124, pp. 71–85.

Geister, J. 1977: The influence of wave exposure on the ecological zonation of Caribbean coral reefs. *Proc. 3rd Int. Coral Reef Symp., Miami* 1, pp. 23–9.

Gelfenbaum, G. 1983: Suspended-sediment response to semidiurnal and fortnightly tidal variations in a mesotidal estuary: Columbia River, U.S.A. *Marine Geol.* 52, pp. 39–57.

Gelinas, P. J. and Quigley, R. M. 1973: The influence of geology on erosion rates along the north shore of Lake Erie. *Proc. 16th Conf. Great Lakes Res.* pp. 421–30.

Gell, R. A. 1978: Shelly beaches on the Victorian coast. *Proc. Roy. Soc. Vict.* 90, pp. 257–69.

Gens, A., Hutchinson, J. N., and Cavounidis, S. 1988: Three-dimensional analysis of slides in cohesive soils. *Géotechnique* 38, pp. 1–23.

George, R., Flick, R. E., and Guza, R. T. 1994: Observations of turbulence in the surf zone. *J. Geophys. Res.* 99, pp. 801–10.

Geyer, W. R. 1988: The advance of a salt wedge front: observations and dynamical model. In *Physical Processes in Estuaries*, J. Dronkers and van W. Leussen (eds.), Springer-Verlag, Berlin, pp. 181–95.

Gibbs, R. J. 1977: Distribution and transport of suspended particulate material of the Amazon River in the ocean. In *Estuarine Processes*, M. Wiley (ed.), Academic press, New York, ii. 35–47.

Gibbs, R. J., Tshudy, D. M., Konwar, L., and Martin, J. M. 1989: Coagulation and transport of sediments in the Gironde Estuary. *Sedimentology* 36, pp. 987–99.

Gibeaut, J. C. and Davis, R. A. 1993: Statistical geomorphic classification of ebb-tidal deltas along the west-central Florida coast. *J. Coastal Res.* Spec. Issue 18, pp. 165–84.

Giglioli, M. E. C. and Thornton, I. 1965: The mangrove swamps of Keneba, lower Gambia River basin. *J. Applied Ecol.* 2, pp. 81–103.

Gilbert, R. 1983: Sedimentary processes of Canadian Arctic fjords. *Sed. Geol.* 36, pp. 147–75.

Gilbert, R. 1990a: A distinction between ice-pushed and ice-lifted landforms on lacustrine and marine coasts. *Earth Surface Processes and Landforms* 15, pp. 15–24.

Gilbert, R. 1990b: Rafting in glacimarine environments. In *Glacimarine Environments: Processes and Sediments*, J. A. Dowdeswell and J. D. Scourse (eds.), Geol. Soc. London Spec. Publ. 53, pp. 105–20.

Gilbert, R. and Aitken, A. E. 1981: The role of sea ice in biophysical processes on intertidal flats at Pangnirtung (Baffin Island), N.W.T. *Proc. Workshop on Ice Action on Shores*, J.-C. Dionne (ed.), National Res. Council Can. (ACROSES) Ottawa, pp. 89–103.

Gilbert, R., Aitken, A. E., and Lemmen, D. S. 1993: The glacimarine sedimentary environment of Expedition Fiord, Canadian High Arctic. *Marine Geol.* 110, pp. 257–73.

Gill, E. D. 1967: The dynamics of the shore platform process and its relation to changes in sea-level. *Proc. Roy. Soc. Vict.* 80, pp. 183–92.

Gill, E. D. 1972a: Ramparts on shore platforms. *Pacific Geol.* 4, pp. 121–33.

Gill, E. D. 1972b: The relationship of present shore platforms to past sea levels. *Boreas* 1, pp. 1–25.

Gill, E. D. and Lang, J. G. 1983: Microerosion meter measurements of rock wear on the Otway Coast of southeast Australia. *Marine Geol.* 52, pp. 141–56.

Ginsburg, R. N. and Schroeder, J. H. 1973: Growth and submarine fossilization of algal cup reefs, Bermuda. *Sedimentology* 20, pp. 575–614.

Glaeser, J. D. 1978: Global distribution of barrier islands in terms of tectonic setting. *J. Geol.* 86, pp. 283–97.

Gleason, M. G. 1993: Effects of disturbance on coral communities: bleaching in Moorea, French Polynesia. *Coral Reefs* 12, pp. 193–201.

Gleason, M. L., Elmer, D. A., Pien, N. C., and Fisher, J. S. 1979: Effects of stem density upon sediment retention by salt marsh cord grass, *Spartina alterniflora* Loisel. *Estuaries* 2, pp. 271–3.

Gleason, R. and Hardcastle, P. J. 1973: The significance of wave parameters in the sorting of beach pebbles. *Estuarine and Coastal Marine Sci.* 1, pp. 11–18.

Glenn, S. M. and Grant, W. D. 1987: A suspended sediment stratification correction for combined wave and current flows. *J. Geophys. Res.* 92, pp. 8244–64.

Glynn, P. W. 1993: Coral reef bleaching: ecological perspectives. *Coral Reefs* 12, pp. 1–17.

Glynn, P. W., Cortes, J., Guzman, H. M., and Richmond, R. H. 1988: El Nino (1982–83) associated coral mortality and relationship to sea surface temperature deviations in the tropical eastern Pacific. *Proc. 6th Int. Coral Reef Symp., Townsville* 3, pp. 237–43.

Glynn, P. W. and Macintyre, I. G. 1977: Growth rate and age of coral reefs on the Pacific coast of Panama. *Proc. 3rd Int. Coral Reef Symp., Miami* 2, pp. 251–9.

Godfrey, P. J., Leatherman, S. P., and Zaremba, R. 1979: A geobotanical approach to classification of barrier beach systems. In *Barrier Islands*, S. P. Leatherman (ed.), Academic Press, New York, pp. 99–126.

Godin, G. 1985: Modification of river tides by the discharge. *J. Waterw. Port Coastal Ocean Div. Amer. Soc. Civ. Engnr.* VIII, pp. 257–74.

Goede, A., Harmon, R., and Kiernan, K. 1979: Sea caves of King Island. *Helictite* 17, pp. 51–64.

Goldsmith, V. 1985: Coastal dunes. In *Coastal Sedimentary Environments*, R. A. Davis (ed.), Springer-Verlag, New York, pp. 303–78.

Goldsmith, V., Rosen, P., and Gertner, Y. 1990: Eolian transport measurements, winds, and comparison with theoretical transport in Israeli coastal dunes. In *Coastal Dunes: Form and Process*, K. F. Nordstrom, N. P. Psuty, and R. W. G. Carter (eds.), Wiley, Chichester, pp. 79–101.

Gordon, D. C. and Desplanque, C. 1981: Ice dynamics in the Chignecto Bay region of the Bay of Fundy. *Proc. Workshop on Ice Action on Shores*, J.-C. Dionne (ed.), National Res. Council Can. (ACROSES) Ottawa, pp. 35–52.

Gordon, J. E. and Hansom, J. D. 1985: Beach forms and changes associated with retreating glacier ice, South Georgia. *Geografiska Annlr.* A 68, pp. 15–24.

Goreau, T. F., Goreau, N. I., and Goreau, T. J. 1979: Corals and coral reefs. *Sci. Amer.* 241, pp. 124–36.

Gornitz, V. 1991: Global coastal hazards from future sea level rise. *Palaeogeogr., Palaeoclim., Palaeoecol. (Global and Planetary Change Section)*, 89, pp. 379–98.

Gorycki, M. A. 1973: Sheetflow structure: mechanism of beach cusp formation and related phenomena. *J. Geol.* 81, pp. 109–17.

Goudie, A. S., Warren, A., Jones, D. K. C., and Cooke, R. U. 1987: The character and possible origins of the aeolian sediments of the Wahiba Sand Oman. *Geogr. J.* 153, pp. 231–56.

Gould, H. R. 1970: The Mississippi Delta complex. In *Deltaic Sedimentation: Modern and Ancient*, J. P. Morgan (ed.), *Soc. Econ. Paleont. Mineral.* Spec. Publ. 15, pp. 3–30.

Gourlay, M. R. 1988: Coral cays: products of wave action and geological processes in a biogenic environment. *Proc. 6th Int. Coral Reef Symp., Townsville* 2, pp. 491–6.

Gourlay, M. R. 1992: Wave set-up, wave run-up and beach water table: interaction between surf zone hydraulics and groundwater hydraulics. *Coastal Engnr.* 17, pp. 93–144.

Gourlay, M. R. 1994: Wave transformation on a coral reef. *Coastal Engnr.* 23, pp. 17–42.

Grainger, P. and Kalaugher, P. G. 1987: Intermittent surging movements of a coastal landslide. *Earth Surface Processes and Landforms* 12, pp. 597–603.

Grant, D. R. 1980: Quaternary sea-level change in Atlantic Canada as an indication of crustal delevelling. In *Earth Rheology, Isostasy and Eustasy*, N.-A. Morner (ed.), Wiley, Chichester, pp. 201–14.

Grant, W. D. and Madsen, O. S. 1986: The continental-shelf bottom boundary layer. *Ann. Rev. Fluid Mech.* 18, pp. 265–305.

Graus, R. R., Chamberlain, J. A., and Boker, A. M. 1977: Structural modification of corals in relation to waves and currents. In *Reefs and Related Carbonates—Ecology and Sedimentology*, S. H. Frost, M. P. Weiss, and J. B. Saunders (eds.), *Amer. Assoc. Petrol. Geol. Studies in Geol.* 4, pp. 135–53.

Gray, A. J. 1992: Saltmarsh plant ecology: zonation and succession revisited. In *Saltmarshes: Morphodynamics, Conservation and Engineering Significance*, J. R. L. Allen and K. Pye (eds.), Cambridge Univ. Press, Cambridge, pp. 63–79.

Greeley, R. and Iversen, J. D. 1985: *Wind as a Geological Process*, Cambridge Univ. Press, Cambridge.

Greene, H. G. 1970: Microrelief of an arctic beach. *J. Sed. Petrol.* 40, 419–27.

Greensmith, J. T. and Tucker, E. V. 1975: Dynamic structures in the Holocene chenier plain setting of Essex, England. In *Nearshore Sediment Dynamics and Sedimentation*, J. Hails and A. Carr (eds.), Wiley, London, pp. 251–72.

Greenwood, B. 1982: Bars. In *The Encyclopedia of Beaches and Coastal Environments*, M. L. Schwartz (ed.), Hutchinson Ross, Stroudsburg, Penn., pp. 135–9.

Greenwood, B. and Davidson-Arnott, R. G. D. 1979: Sedimentation and equilibrium in wave-formed bars: a review and case study. *Can. J. Earth Sci.* 16, pp. 312–32.

Greenwood, B. and Mittler, P. R. 1984: Sediment flux and equilibrium slopes in a barred nearshore. *Marine Geol.* 60, pp. 79–98.

Greenwood, B. and Osborne, P. D. 1990: Vertical and horizontal structure in cross-shore flows: an example of undertow and wave set-up on a barred beach. *Coastal Engnr.* 14, pp. 543–80.

Greenwood, B. and Osborne, P. D. 1991: Equilibrium slopes and cross-shore velocity asymmetries in a storm-dominated, barred nearshore system. *Marine Geol.* 96, pp. 211–35.

Greenwood, B., Osborne, P. D., and Bowen, A. J. 1991: Measurements of suspended sediment transport: prototype shorefaces. *Coastal Sediments '91*, N. C. Kraus, K. J. Gingerich, and D. L. Kriebel (eds.), Amer. Soc. Civ. Engnr., pp. 284–99.

Greenwood, B., Osborne, P. D., Bowen, A. J., Hazen, D. G., and Hay, A. E. 1990a: Nearshore sediment flux and bottom boundary dynamics. The Canadian coastal sediment transport programme (C-Coast). *Proc. 22nd Coastal Engnr. Conf.* pp. 2227–40.

Greenwood, B., Osborne, P. D., Bowen, A. J., Hazen, D. G., and Hay, A. E. 1990b: C-coast: the Canadian coastal sediment transport programme. *Proc. Can. Coastal Conf.*, M. H. Davies (ed.), National Res. Council Can. (ACOS), Ottawa, pp. 319–36.

Greenwood, B. and Sherman, D. J. 1984: Waves, currents, sediment flux and morphological response in a barred nearshore system. *Marine Geol.* 60, pp. 31–61.

Grigg, R. W. 1992: Coral reef environmental science: truth versus the Cassandra syndrome. *Coral Reefs* 11, pp. 183–6.

Grigg, R. W. and Dollar, S. J. 1990: Natural and anthropogenic disturbance on coral reefs. In *Coral Reefs* (Ecosystems of the World 25), Z. Dubinsky (ed.), Elsevier, Amsterdam, pp. 439–52.

Grigg, R. W. and Epp, D. 1989: Critical depth for the survival of coral islands: effects on the Hawaiian Archipelago. *Science* 243, pp. 638–41.

Griggs, G. B. 1987: The production, transport, and delivery of coarse-grained sediment by California's coastal streams. *Coastal Sediments '87*, N. C. Kraus (ed.), Amer. Soc. Civ. Engnr., pp. 1825–38.

Griggs, G. B. and Trenhaile, A. S. 1994: Coastal cliffs and platforms. In *Coastal Evolution*, R. W. G. Carter and C. D. Woodroffe (eds.), Cambridge Univ. Press, Cambridge, pp. 425–50.

Grindrod, J. 1985: The palynology of mangroves on a prograded shore, Princess Charlotte Bay, north Queensland, Australia. *J. Biogeogr.* 12, pp. 323–48.

Grindrod, J. and Rhodes, E. G. 1984: Holocene sea-level history of a tropical estuary: Missionary Bay, north Queensland. In *Coastal Geomorphology in Australia*, B. G. Thom (ed.), Academic Press, Sydney, pp. 151–78.

Gruszczynski, M., Rudowski, S., Semil, J., Słominski, J., and Zrobek, J. 1993: Rip currents as a geological tool. *Sedimentology* 40, pp. 217–36.

Guilcher, A. 1953: Essai sur la zonation et la distribution des formes littorales de dissolution du calcaire. *Ann. Géogr.* 62, pp. 161–79.

Guilcher, A. 1971: Mayotte barrier reef and lagoon, Comoro Islands, as compared with other barrier reefs, atolls and lagoons in the world. In *Symposia of the Zoological Soc. London*, D. R. Stoddart and M. Yonge (eds.), Academic Press, London, 28, pp. 65–86.

Guilcher, A. 1974: Les Rasas: un problème de morphologie littorale generale. *Ann. Géogr.* 83, pp. 1–33.

Guilcher, A. 1976: Double and multiple barrier reefs in the world. *Studia Societatis Scientiarum Torunensis Torun-Polonia* VIII C Nr 4–6, pp. 85–99.

Guilcher, A. 1985a: Red Sea coasts. In *The World's Coastline*, E. C. F. Bird and M. L. Schwartz (eds.), Van Nostrand Reinhold, New York, pp. 713–17.

Guilcher, A. 1985b: Nature and human change of sedimentation in lagoons behind barrier reefs in the humid tropics. Proc. 5th Int. Coral Reef Congr., Tahiti 4, pp. 207–12.

Guilcher, A. 1985c: Retreating cliffs in the humid tropics: an example from Paraiba, northeastern Brazil. *Z. Geomorph.* Suppl. Band 57, pp. 95–103.

Guilcher, A. 1988: *Coral Reef Geomorphology*, Wiley, Chichester.

Guilcher, A. 1989: Are mangroves good tests in coastal geomorphological zonality? *Essener Geogr. Arbeiten* 18, pp. 125–48.

Guilcher, A., Bodéré, J.-C., Coudé, A., Hansom, J. D., Moign, A., and Peulvast, J.-P. 1986: Le problème des strandflats en cinq pays de hautes latitudes. *Rev. Géogr. Phys. Géol. Dyn.* 27, pp. 47–79.

Guillén, J. and Palanques, A. 1993: Longshore bar and trough systems in a microtidal, storm-wave dominated coast: the Ebro Delta (northwestern Mediterranean). *Marine Geol.* 115, pp. 239–52.

Guillou, H., Brousse, R., Gillot, P. Y., and Guille, G. 1993: Geological reconstruction of Fangataufa Atoll, south Pacific. *Marine Geol.* 110, pp. 377–91.

Guza, R. T. and Bowen, A. J. 1975: The resonant instabilities of long waves obliquely incident on a beach. *J. Geophys. Res.* 80, pp. 4529–34.

Guza, R. T. and Davis, R. E. 1974: Excitation of edge waves by waves incident on a beach. *J. Geophys. Res.* 79, pp. 1285–91.

Guza, R. and Inman, D. 1975: Edge waves and beach cusps. *J. Geophys. Res.* 80, pp. 2997–3012.

Guza, R. T. and Thornton, E. B. 1978: Variability of longshore currents. *Proc. 16th Coastal Engnr. Conf.* pp. 756–75.

Guza, R. T. and Thornton, E. B. 1980: Local and shoaled comparisons of sea surface elevations, pressures, and velocities. *J. Geophys. Res.* 85, pp. 1524–30.

Guza, R. T. and Thornton, E. B. 1981: Wave set-up on a natural beach. *J. Geophys. Res.* 86, pp. 4133–7.

Guza, R. T. and Thornton, E. B. 1982: Swash oscillations on a natural beach. *J. Geophys. Res.* 87, pp. 483–91.

Guza, R. T. and Thornton, E. B. 1985: Observations of surf beat. *J. Geophys. Res.* 90, pp. 3161–72.

Guza, R. T. and Thornton, E. B. 1989: Run-up and surf beat. In *Nearshore Sediment Transport*, R. J. Seymour (ed.), Plenum Press, New York, pp. 173–81.

Haff, P. K. and Anderson, R. S. 1993: Grain scale simulations of loose sedimentary beds: the example of grain-bed impacts in aeolian saltation. *Sedimentology* 40, pp. 175–98.

Hale, P. B. and McCann, S. B. 1982: Rhythmic topography in a mesotidal low-wave-energy environment. *J. Sed. Petrol.* 52, pp. 415–29.

Hallam, A. 1981: Plate tectonics, biogeography, and evolution. *Nature* 293, pp. 31–2.

Hallermeier, R. J. 1981: Terminal settling velocity of commonly occurring sand grains. *Sedimentology* 28, pp. 859–65.

Hallermeier, R. J. 1982*a*: Bedload and wave thrust computations of alongshore sand transport. *J. Geophys. Res.* 87, pp. 5741–51.

Hallermeier, R. J. 1982*b*: Oscillatory bedload transport: data review and simple formulation. *Continental Shelf Res.* 1, pp. 159–90.

Hallermeier, R. J., Nosek, K. B., and Andrassy, C. J. 1990: Evaluation of empirical model for wave runup elevations. *Proc. 22nd Coastal Engnr. Conf.* pp. 41–54.

Hallock, P. 1988: The role of nutrient availability in bioerosion: consequences to carbonate buildups. *Palaeogeogr., Palaeoclim., Palaeoecol.* 63, pp. 275–91.

Hallock, P. and Schlager, W. 1986: Nutrient excess and the demise of coral reefs and carbonate platforms. *Palaios* 1, pp. 389–98.

Hamm, L., Madsen, P. A., and Peregrine, D. H. 1993: Wave transformation in the nearshore zone: a review. *Coastal Engnr.* 21, pp. 5–39.

Hammond, F. D. C., Heathershaw, A. D., and Langhorne, D. N. 1984: A comparison between Shields' threshold criterion and the movement of loosely packed gravel in a tidal channel. *Sedimentology* 31, pp. 51–62.

Hammond, R. R. and Wallace, W. J. 1982: Seabed drifter movement in San Diego Bay and adjacent waters. *Estuarine, Coastal and Shelf Sci.* 14, pp. 623–34.

Hammond, T. M. and Collins, M. B. 1979: On the threshold of transport of sand-sized sediment under the combined influence of unidirectional and oscillatory flow. *Sedimentology* 26, pp. 795–812.

Hands, E. B. 1983: The Great Lakes as a test model for profile responses to sea level changes. In *CRC Handbook of Coastal Processes and Erosion*, P. D. Komar (ed.), CRC Press, Boca Raton, Florida, pp. 167–89.

Hanes, D. M. 1991: Suspension of sand due to wave groups. *J. Geophys. Res.* 96, pp. 8911–15.

Hansen, D. V. and Rattray, M. 1966: New dimensions in estuary classification. *Limnol. Oceanogr.* 11, pp. 319–26.

Hanslow, D. and Nielsen, P. 1993: Shoreline set-up on natural beaches. *J. Coastal Res.* Spec. Issue 15, pp. 1–10.

Hansom, J. D. 1983*a*: Shore-platform development in the South Shetland Islands, Antarctica. *Marine Geol.* 53, pp. 211–29.

Hansom, J. D. 1983*b*: Ice-formed intertidal boulder pavements in the sub-Antarctic. *J. Sed. Geol.* 53, pp. 135–45.

Hansom, J. D. 1986: Intertidal forms produced by floating ice in Vestfirdir, Iceland. *Marine Geol.* 71, pp. 289–98.

Hansom, J. D. and Kirk, R. M. 1989: Ice in the intertidal zone: examples from Antarctica. In *Zonality of Coastal Geomorphology and Ecology*, E. C. F. Bird and D. Kelletat (eds.), *Essener Geographische Arbeiten* 18, pp. 211–36.

Haq, B. U., Hardenbol, J., and Vail, P. R. 1987: Chronology of fluctuating sea levels since the Triassic. *Science* 235, pp. 1156–66.

Hardisty, J. 1983: An assessment and calibration of formulations for Bagnold's bedload equation. *J. Sed. Petrol.* 53, pp. 1007–10.

Hardisty, J. 1986: A morphodynamic model for beach gradients. *Earth Surface Processes and Landforms* 11, pp. 327–33.

Hardisty, J. 1989: Morphodynamic and experimental assessments of wave theories in intermediate water depths. *Earth Surface Processes and Landforms* 14, pp. 107–18.

Hardisty, J. 1990*a*: *Beaches Form and Process*, Unwin Hyman, London.

Hardisty, J. 1990*b*: A note on suspension transport in the beach gradient model. *Earth Surface Processes and Landforms* 15, pp. 91–6.

Hardisty, J., Collier, J., and Hamilton, D. 1984: A calibration of the Bagnold beach equation. *Marine Geol.* 61, pp. 95–101.

Hardisty, J. and Whitehouse, R. J. S. 1988: Evidence for a new sand transport process from experiments on Saharan dunes. *Nature* 332, pp. 532–4.

Hardwick-Witman, M. N. 1986: Aerial survey of a salt marsh: ice rafting to the lower intertidal zone. *Estuarine, Coastal and Shelf Sci.* 22, pp. 379–83.

Harper, J. R. 1990: Morphology of the Canadian Beaufort Sea coast. *Marine Geol.* 91, pp. 75–91.

Harris, P. T. 1988: Large-scale bedforms as indicators of mutually evasive sand transport and the sequential infilling of wide-mouthed estuaries. *Sed. Geol.* 57, pp. 273–98.

Harrison, P. L. and Wallace, C. C. 1990: Reproduction, dispersal and recruitment of Scleractinian corals. In *Coral Reefs* (Ecosystems of the World 25), Z. Dubinsky (ed.), Elsevier, Amsterdam, pp. 133–207.

Hart, B. S. and Long, B. F. 1990: Recent evolution of the Outardes Estuary, Quebec, Canada: consequences of dam construction on the river. *Sedimentology* 37, pp. 495–507.

Hart, B. S., Prior, D. B., Barrie, J. V., Currie, R. G., and Luternauer, J. L. 1992: A river mouth submarine channel and failure complex, Fraser Delta, Canada. *Sed. Geol.* 81, pp. 73–87.

Hartmann, D. and Bowman, D. 1993: Efficiency of the log-hyperbolic distribution—a case study: pattern of sediment sorting in a small tidal-inlet—Het Zwin, The Netherlands. *J. Coastal Res.* 9, pp. 1044–53.

Hartnall, T. J. 1984: Salt-marsh vegetation and micro-relief development on the New Marsh at Gibraltar Point, Lincolnshire. In *Coastal Research: UK Perspectives*, M. W. Clark (ed.), Geo Books, Norwich, pp. 37–58.

Hasselmann, K. and 15 others 1973: Measurements of wind-wave growth and swell decay during the Joint North Sea wave Project (JONSWAP). *Deutsche Hydrograph. Z.* A8 (no. 12).

Hasselmann, K., Ross, D. B., Muller, P., and Sell, W. 1976: A parametric wave prediction model. *J. Phys. Oceanogr.* 6, pp. 200–28.

Hattori, M. and Kawamata, R. 1980: Onshore–offshore transport and beach profile change. *Proc. 17th Coastal Engnr. Conf.* pp. 1175–93.

Hay, A. E. and Bowen, A. J. 1993: Spatially correlated depth changes in the nearshore zone during Autumn storms. *J. Geophys. Res.* 98, pp. 12387–404.

Hay, A. E. and Sheng, J. 1992: Vertical profiles of suspended sand concentration and size from multi-frequency acoustic backscatter. *J. Geophys. Res.* 97, pp. 15661–77.

Hayes, M. O. 1975: Morphology of sand accumulation in estuaries: an introduction to the symposium. In *Estuarine Research*, L. E. Cronin (ed.), Academic Press, New York, ii. 3–22.

Hayes, M. O. 1979: Barrier island morphology as a function of tidal and wave regime. In *Barrier Islands*, S. P. Leatherman (ed.), Academic Press, New York, pp. 1–27.

Hayes, M. O. 1980: General morphology and sediment patterns in tidal inlets. *Sed. Geol.* 26, pp. 139–56.

Hayes, M. O. 1991: Geomorphology and sedimentation patterns of tidal inlets: a review. *Coastal Sediments '91*, N. C. Kraus, K. J. Gingerich, and D. L. Kriebel (eds.), Amer. Soc. Civ. Engnr., pp. 1343–55.

Hazen, D. G., Greenwood, B., and Bowen, A. J. 1990: Nearshore current patterns on barred beaches. *Proc. 22nd Coastal Engnr. Conf.* pp. 2061–72.

Healy, T. 1991: Coastal erosion and sea level rise. *Z. Geomorph.* Suppl. Band 81, pp. 15–29.

Healy, T., Sneyd, A. D., and Werner, F. 1987: First approximation sea-level dependent mathematical model for volume eroded and submarine profile development in a semi-enclosed sea: Kiel Bay, western Baltic. *Math. Geol.* 19, pp. 41–56.

Healy, T., Sterr, H., and Werner, F. 1990: Lag boulders on a till submarine abrasion platform in a semi-enclosed sea, Kieler Bucht, western Baltic. *Z. Geomorph.* 34, pp. 323–34.

Heaps, N. S. 1983: Storm surges, 1967–1982. *Geophys. J. Roy. Astron. Soc.* 74, pp. 331–76.

Heathershaw, A. D. and Davies, A. G. 1985: Resonant wave reflection by transverse bedforms and its relation to beaches and offshore bars. *Marine Geol.* 62, pp. 321–38.

Hegge, B. J. and Masselink, G. 1991: Groundwater-table responses to wave run-up: an experimental study from Western Australia. *J. Coastal Res.* 7, pp. 623–34.

Hendry, M. and Digerfeldt, G. 1989: Palaeogeography and palaeoenvironments of a tropical coastal wetland and offshore shelf during Holocene submergence, Jamaica. *Palaeogeogr., Palaeoclim., Palaeoecol.* 73, pp. 1–10.

Hequette, A. and Barnes, P. W. 1990: Coastal retreat and shoreface profile variations in the Canadian Beaufort Sea. *Marine Geol.* 91, pp. 113–32.

Hesp, P. A. 1981: The formation of shadow dunes. *J. Sed. Petrology* 51, pp. 101–12.

Hesp, P. A. 1983: Morphodynamics of incipient foredunes in New South Wales, Australia. In *Eolian Sediments and Processes*, M. E. Brookfield and T. Ahlbrandt (eds.), Elsevier, Amsterdam, pp. 325–42.

Hesp, P. A. 1984: Foredune formation in southeast Australia. In *Coastal Geomorphology in Australia*, B. G. Thom (ed.), Academic Press, Sydney, pp. 69–97.

Hesp, P. A. 1988: Morphology, dynamics and internal stratification of some established foredunes in southeast Australia. *Sed. Geol.* 55, pp. 17–41.

Hesp, P. A. 1989: A review of biological and geomorphological processes involved in the initiation and development of incipient foredunes. In *Coastal Sand Dunes*, C. H. Gimingham, W. Ritchie, B. B. Willetts, and A. J. Willis (eds.), *Proc. Roy. Soc. Edinb.* B 96, pp. 181–201.

Hesp, P., Illenberger, W., Rust, I., McLachlan, A., and Hyde, R. 1989: Some aspects of transgressive dune-field and transverse dune geomorphology and dynamics, south coast, South Africa. *Z. Geomorph.* Suppl. Band 73, pp. 111–23.

Hesp, P. A. and Thom, B. G. 1990: Geomorphology and evolution of active transgressive dunefields. In *Coastal Dunes: Form and Process*, K. F. Nordstrom, N. P. Psuty, and R. W. G. Carter (eds.), Wiley, Chichester, pp. 253–88.

Hicks, D. M. and Inman, D. L. 1987: Sand dispersion from an ephemeral river delta on the central California coast. *Marine Geol.* 77, pp. 305–18.

Hill, P. R. 1990: Coastal geology of the King Point area, Yukon Territory, Canada. *Marine Geol.* 91, pp. 93–111.

Hills, E. S. 1971: A study of cliffy coastal profiles based on examples in Victoria, Australia. *Z. Geomorph.* 15, pp. 137–80.

Hills, E. S. 1972: Shore platforms and wave ramps. *Geol. Mag.* 109, pp. 81–8.

Hine, A. C. 1979: Mechanisms of berm development and resulting beach growth along a barrier spit complex. *Sedimentology* 26, pp. 333–51.

Hino, M. 1974: Theory on formation of rip-current and cuspidal coast. *Proc. 14th Coastal Engnr. Conf.* pp. 910–19.

Hoekstra, P. 1993: Late Holocene development of a tide-induced elongate delta, the Solo delta, east Java. *Sed. Geol.* 83, pp. 211–33.

Hoffmeister, J. E. and Ladd, H. S. 1944: The antecedent platform theory. *J. Geol.* 52, pp. 388–402.

Holland, K. T. and Holman, R. A. 1993: The statistical distribution of swash maxima on natural beaches. *J. Geophys. Res.* 98, pp. 10271–8.

Holman, R. A. 1983: Edge waves and the configuration of the shoreline. In *CRC Handbook of Coastal Processes and Erosion*, P. D. Komar (ed.), CRC Press, Boca Raton, Florida, pp. 21–33.

Holman, R. A. and Bowen, A. J. 1979: Edge waves on complex beach profiles. *J. Geophys. Res.* 84, pp. 6339–46.

Holman, R. A. and Bowen, A. J. 1982: Bars, bumps and holes: models for the generation of complex beach topography. *J. Geophys. Res.* 87, pp. 457–68.

Holman, R. A. and Bowen, A. J. 1984: Longshore structure of infragravity wave motions. *J. Geophys. Res.* 89, pp. 6446–52.

Holman, R. A., Howd, P., Oltman-Shay, J., and Komar, P. D. 1990: Observations of the swash expression of far infragravity wave motions. *Proc. 22nd Coastal Engnr. Conf.* pp. 1242–53.

Holman, R. A., Huntley, D. A., and Bowen, A. J. 1978: Infragravity waves in storm conditions. *Proc. 16th Coastal Engnr. Conf.* pp. 268–84.

Holman, R. A. and Sallenger, A. H. 1985: Setup and swash on a natural beach. *J. Geophys. Res.* 90, pp. 945–53.

Holman, R. A. and Sallenger, A. H. 1993: Sand bar generation: a discussion of the Duck experiment series. *J. Coastal Res.* Spec. Issue 15, pp. 76–92.

Holthus, P. F. 1985: A reef resource conservation and management plan for Ponape Island (Caroline Archipelago, Micronesia). *Proc. 5th Int. Coral Reef Congr., Tahiti* 4, pp. 231–6.

Hopley, D. 1974: Coastal changes produced by tropical cyclone Althea in Queensland; December 1971. *Austr. Geogr.* 12, pp. 445–56.

Hopley, D. 1982: *The Geomorphology of the Great Barrier Reef*, Wiley, New York.

Hopley, D. 1983: Morphological classifications of shelf reefs: a critique with special reference to the Great Barrier Reef. In D. J. Barnes (ed.), *Perspectives on Coral Reefs*, Brian Clouston, Manuka A.C.T., *Austr. Inst. Marine Sci. Contrib.* 200, pp. 180–99.

Hopley, D. 1989: Coral reefs: zonation, zonality and gradients. *Essener Geogr. Arbeiten* 18, pp. 79–123.

Hopley, D. and Barnes, R. 1985: Structure and development of a windward fringing reef, Orpheus Island, Palm Group, Great Barrier Reef. *Proc. 5th Int. Coral Reef Congr., Tahiti* 3, pp. 141–6.

Hopley, D., McLean, R. F., Marshall, and Smith, A. S. 1978: Holocene–Pleistocene boundary in a fringing reef: Hayman Island, north Queensland. *Search* 9, pp. 323–5.

Horikawa, K. 1988 (ed.): *Nearshore Dynamics and Coastal Processes*, Univ. Tokyo Press, Tokyo.

Horikawa, K., Hotta, S., and Kraus, N. C. 1986: Literature review of sand transport by wind on a dry sand surface. *Coastal Engnr.* 9, pp. 503–26.

Horn, D. P. 1992: A numerical model for shore-normal sediment size variation on a macrotidal beach. *Earth Surface Processes and Landforms* 17, pp. 755–73.

Horn, D. P. 1993: Sediment dynamics on a macrotidal beach: Isle of Man, U.K. *J. Coastal Res.* 9, pp. 189–208.

Hotta, S. 1988: Sand transport by wind. In *Nearshore Dynamics and Coastal Processes*, K. Horikawa (ed.), Univ. Tokyo Press, Tokyo, pp. 218–38.

Hotta, S., Kubota, S., Katori, S., and Horikawa, K. 1984: Sand transport by wind on a wet sand surface. *Proc. 19th Coastal Engnr. Conf.* pp. 1265–81.

Howarth, P. J. and Bones, J. G. 1972: Relationships between process and geometrical form on high Arctic slopes, south-west Devon Island, Canada. In *Polar Geomorphology*, R. J. Price and D. E. Sugden (eds.), *Inst. Brit. Geogr.* Spec. Publ. 4, pp. 139–53.

Howd, P. A. and Birkemeier, W. A. 1987: Storm-induced morphology changes during DUCK85. *Coastal Sediments '87*, N. C. Kraus (ed.), Amer. Soc. Civ. Engnr., pp. 834–47.

Howd, P. A. Bowen, A. J., and Holman, R. 1992: Edge waves in the presence of strong longshore currents. *J. Geophys. Res.* 97, pp. 11357–71.

Howd, P. A., Bowen, T., Holman, R., and Oltman-Shay, J. 1991: Infragravity waves, longshore currents, and linear sand bar formation. *Coastal Sediments '91*, N. C. Kraus, K. J. Gingerich, and D. L. Kriebel (eds.), Amer. Soc. Civ. Engnr., pp. 72–84.

Howd, P. A., Oltman-Shay, J., Holman, R., and Komar, P. D. 1990: Evolution of infragravity variance during a storm. *Proc. 22nd Coastal Engnr. Conf.* pp. 1118–30.

Hoyt, J. H. 1967: Barrier island formation. *Geol. Soc. Amer. Bull.* 78, pp. 1125–35.

Hoyt, J. H. 1969: Chenier versus barrier, genetic and stratigraphic distinction. *Amer. Assoc. Petrol. Geol. Bull.* 53, pp. 299–306.

Hoyt, W. H. 1991: Migrating east Australian barriers and reefs under greenhouse scenario. *Coastal Sediments '91*, N. C. Kraus, K. J. Gingerich, and D. L. Kriebel (eds.), Amer. Soc. Civ. Engnr., pp. 1329–42.

Hsu, J. R. C., Silvester, R., and Xia, Y-M. 1989a: Static equilibrium bays: new relationships. *J. Waterw. Port Coastal Ocean Div. Amer. Soc. Civ. Engnr.* 115, pp. 285–98.

Hsu, J. R. C., Silvester, R., and Xia, Y-M. 1989b: Applications of headland control. *J. Waterw. Port Coastal Ocean Div. Amer. Soc. Civ. Engnr.* 115, pp. 299–310.

Hsu, S. A. 1973: Computing eolian sand transport from shear velocity measurements. *J. Geol.* 81, pp. 739–43.

Huang, N. E., Tung, C.-C., and Long, S. R. 1990: Wave

spectra. In *The Sea (Ocean Engnr. Sci.)*, B. Le Méhauté and D. M. Hanes, (eds.), Wiley, New York, 9, pp. 197–237.

Hubbard, D. K. 1985: What do we mean by reef growth? *Proc. 5th Int. Coral Reef Congr., Tahiti* 6, pp. 433–8.

Hubbard, D. K., Miller, A. I., and Scaturo, D. 1990. Production and cycling of calcium carbonate in a shelf-edge reef system (St. Croix, U. S. Virgin Islands): applications to the nature of reef systems in the fossil record. *J. Sed. Petrol.* 60, pp. 335–60.

Hubertz, J. M. 1986: Observations of local wind effects on longshore currents. *Coastal Engnr.* 10, pp. 275–88.

Hudec, P. P. 1973: Weathering of rocks in Arctic and Sub-arctic environment. In *Canadian Arctic Geology*, J. D. Aitken and D. J. Glass (eds.), Geol. Soc. Assoc. Can.—Can. Soc. Petrol. Geol. Symp. (Saskatoon, Saskatchewan), pp. 313–35.

Huh, O. K., Roberts, H. H., and Rouse, L. J. 1991: Fine grain sediment transport and deposition in the Atchafalaya and chenier plain sedimentary system. *Coastal Sediments '91*, N. C. Kraus, K. J. Gingerich, and D. L. Kriebel (eds.), Amer. Soc. Civ. Engnr., pp. 817–30.

Hume, J. D. and Schalk, M. 1967: Shoreline processes near Barrow, Alaska: a comparison of the normal and the catastrophic. *Arctic* 20, pp. 86–103.

Hume, J. D. and Schalk, M. 1976: The effects of ice on the beach and nearshore, Point Barrow, Arctic Alaska. *Rev. Géogr. Montréal* 30, pp. 105–14.

Hume, J. D., Schalk, M., and Hume, P. W. 1972: Short-term climate changes and coastal erosion, Barrow, Alaska. *Arctic* 25, pp. 272–8.

Hume, T. M. and Herdendorf, C. E. 1993: On the use of empirical stability relationships for characterising estuaries. *J. Coastal Res.* 9, pp. 413–22.

Hunter, R. E. 1977a: Basic types of stratification in small eolian dunes. *Sedimentology* 24, pp. 361–87.

Hunter, R. E. 1977b: Terminology of cross-stratified sedimentary layers and climbing-ripple structures. *J. Sed. Petrol.* 47, pp. 697–706.

Hunter, R. E., Richmond, B. M., and Alpha, T. R. 1983: Storm-controlled oblique dunes of the Oregon coast. *Geol. Soc. Amer. Bull.* 94, pp. 1450–65.

Huntley, D. A. 1980: Edge waves in a crescentic bar system. In *The Coastline of Canada*, S. B. McCann (ed.), *Geol. Surv. Can. Paper* 80–10, pp. 111–21.

Huntley, D. A. 1988a: Evidence for phase coupling between edge waves modes. *J. Geophys. Res.* 93, pp. 12393–408.

Huntley, D. A. 1988b: A modified inertial dissipation method for estimating seabed stresses at low Reynolds numbers, with application to wave/current boundary layer measurements. *J. Phys. Oceanogr.* 18, pp. 339–46.

Huntley, D. A. and Bowen, A. J. 1975: Comparison of the hydrodynamics of steep and shallow beaches. In *Nearshore Sediment Dynamics and Sedimentation*, J. Hails and A. Carr (eds.), Wiley, London, pp. 69–109.

Huntley, D. A., Guza, R. T., and Thornton, E. B. 1981: Field observations of surf beat 1. Progressive edge waves. *J. Geophys. Res.* 86, pp. 6451–66.

Huntley, D. A. and Hanes, D. M. 1987: Direct measurement of suspended sediment transport. *Coastal Sediments '87*, N. C. Kraus (ed.), Amer. Soc. Civ. Engnr., pp. 723–37.

Huntley, D. A. and Short, A. D. 1992: On the spacing between observed rip currents. *Coastal Engnr.* 17, pp. 211–25.

Hutchinson, J. N. 1973: The response of London Clay cliffs to differing rates of toe erosion. *Geologia Applicata e Idrogeologia* 8, pp. 221–39. Repr. in 1975 as Building Research Establishment (UK) Current Paper CP 27/75.

Hutchinson, J. N. 1980: Various forms of cliff instability arising from coast erosion in south-east England. *Fjellsprengningsteknikk Bergmekanikk/Geoteknikk* Tapir Publishers, Tronheim, pp. 19.1–19.32.

Hutchinson, J. N. 1986: Cliffs and shores in cohesive materials: geotechnical and engineering geological aspects. *Proc. Symp. Cohesive Shores*, M. G. Skafel (ed.), National Res. Council Can. (ACROSES), Ottawa, pp. 1–44.

Illenberger, W. K. 1991: Pebble shape (and size!). *J. Sed. Petrol.* 61, pp. 756–67.

Illenberger, W. K. and Rust, I. C. 1988: A sand budget for the Alexandria coastal dunefield, South Africa. *Sedimentology* 35, pp. 513–21.

Inman, D. and Bagnold, R. 1963: Beach and nearshore processes. Part II. Littoral processes. In *The Sea*, M. N. Hill (ed.), Interscience, New York, 3, pp. 529–53.

Inman, D. L., Ewing, G. C., and Corliss, J. B. 1966: Coastal sand dunes of Guerrero Negro, Baja California, Mexico. *Geol. Soc. Amer. Bull.* 77, pp. 787–802.

Inman, D. L. and Frautschy, J. D. 1966: Littoral processes and the development of shorelines. *Coastal Engnr. Speciality Conf., Santa Barbara*, Amer. Soc. Civ. Engnr. pp. 511–36.

Inman, D. L. and Guza, R. T. 1982: The origin of swash cusps on beaches. *Marine Geol.* 49, pp. 133–48.

Inman, D. L. and Nordstrom, C. E. 1971: On the tectonic and morphologic classification of coasts. *J. Geol.* 79, pp. 1–21.

Innocenti, L. and Pranzini, E. 1993: Geomorphological evolution and sedimentology of the Ombrone River Delta, Italy. *J. Coastal Res.* 9, pp. 481–93.

Ippen, A. T. and Eagleson, P. S. 1955: A study of sediment sorting by waves shoaling on a plane beach. US Army Corps of Engnr., *Beach Erosion Board Techn. Memo.* 63.

Isla, F. I. 1993: Overpassing and armouring phenomena on gravel beaches. *Marine Geol.* 110, pp. 369–76.

Isla, F. I. and Bujalesky, G. G. 1993: Saltation on gravel beaches: Tierra del Fuego, Argentina. *Marine Geol.* 115, pp. 263–70.

Jackson, D. W. T. and Nevin, G. H. 1992: Sand transport in a cliff top dune system at Fonte de Telha, Portugal. In *Coastal Dunes*, R. W. G. Carter, T. G. F. Curtis, and M. J. Sheehy-Skeffington (eds.), Proc. 3rd European Dune Congr., Balkema, Rotterdam, pp. 81–92.

Jackson, N. L. and Nordstrom, K. F. 1993: Depth of activity of sediment by plunging breakers on a steep sand beach. *Marine Geol.* 115, pp. 143–51.

Jacobson, H. A. 1988: Historical development of the saltmarsh at Wells, Maine. *Earth Surface Processes and Landforms* 13, pp. 475–86.

Jaffe, B. E. and Sallenger, A. H. 1992: The contribution of suspension events to sediment transport in the surf zone. *Proc. 23rd Coastal Engnr. Conf.* pp. 2680–93.

Jagger, K. A., Psuty, N. P., and Allen, J. R. 1991: Caleta morphodynamics, Perdido Key, Florida, U.S.A. *Z. Geomorph.* Suppl. Band 81, pp. 99–113.

Jago, C. F. 1980: Contemporary accumulation of marine sand in a macrotidal estuary, southwest Wales. *Sed. Geol.* 26, pp. 21–49.

Jago, C. F. and Hardisty, J. 1984: Sedimentology and morphodynamics of a macrotidal beach, Pendine Sands, SW Wales. *Marine Geol.* 60, pp. 123–54.

James, N. P. 1983: Reef environment. In *Carbonate Depositional Environments*, P. A. Scholle, D. G. Bebout, and C. H. Moore (eds.), *Amer. Assoc. Petrol. Geol. Mem.* 33, pp. 346–440.

Jane, F. W. and White, D. J. B. 1964: The botany and plant ecology of Blakeney Point and Scolt Head Island. In *Blakeney Point and Scolt Head Island*, J. A. Steers (ed.), National Trust, London, pp. 22–39.

Jay, D. A. and Musiak, J. D. 1994: Particle trapping in estuarine tidal flows. *J. Geophys. Res.* 99, pp. 20445–61.

Jelgersma, S., Van der Zijp, M., and Brinkman, R. 1993: Sealevel rise and the coastal lowlands in the developing world. *J. Coastal Res.* 9, pp. 958–72.

Jennings, J. N. 1955: The influence of wave action on coastal outline in plan. *Austr. Geogr.* 6, pp. 36–44.

Jennings, J. N. 1964: The question of coastal dunes in tropical humid climates. *Z. Geomorph.* 8, pp. 150–4.

Jennings, J. N. 1975: Desert dunes and estuarine fill in the Fitzroy Estuary (north-western Australia). *Catena* 2, pp. 215–62.

Jennings, J. N. and Coventry, R. J. 1973: Structure and texture of a gravelly barrier island in the Fitzroy Estuary, Western Australia, and the role of mangroves in the shore dynamics. *Marine Geol.* 15, pp. 145–67.

Jennings, S. C., Carter, R. W. G., and Orford, J. D. 1993: Late Holocene salt marsh development under a regime of rapid relative-sea-level rise: Chezzetcook Inlet, Nova Scotia. Implications for the interpretation of palaeomarsh sequences. *Can. J. Earth Sci.* 30, pp. 1374–84.

Jensen, F. 1994: Dune management in Denmark: application of the Nature Protection Act of 1992. *J. Coastal Res.* 10, pp. 263–9.

Jensen, J. B. and Stecher, O. 1992: Paraglacial barrier-lagoon development in the late Pleistocene Baltic Ice Lake, southwestern Baltic. *Marine Geol.* 107, pp. 81–101.

Jiménez, J. A. and Sánchez-Arcilla, A. 1993: Medium-term coastal response at the Ebro Delta, Spain. *Marine Geol.* 114, pp. 105–18.

Johannes, R. E., Wiebe, W. J., Crossland, C. J., Rimmer, D. W., and Smith, S. V. 1983: Latitudinal limits of coral reef growth. *Marine Ecol.—Progr. Series* 11, pp. 105–11.

John, B. S. and Sugden, D. E. 1975: Coastal geomorphology of high latitudes. *Progr. Phys. Geogr.* 7, pp. 53–132.

Johnson, D. P. and Risk, M. J. 1987: Fringing reef growth on a terrigenous mud foundation, Fantome Island, central Great Barrier reef, Australia. *Sedimentology* 34, pp. 275–87.

Johnson, D. W. 1919: *Shore Processes and Shoreline Development*, Wiley, New York.

Johnston, P. 1995: The role of hydro-isostasy for Holocene sea-level changes in the British Isles. *Marine Geol.* 124, pp. 61–70.

Jolliffe, I. P. 1983: Coastal erosion and flood abatement: what are the options? *Geogr. J.* 149, pp. 62–7.

Jones, D. G. and Williams, A. T. 1991: Statistical analysis of factors influencing cliff erosion along a section of the west Wales coast, U.K. *Earth Surface Processes and Landforms* 16, pp. 95–111.

Jonsson, I. G. 1966: Wave boundary layers and friction factors. *Proc. 10th Coastal Engnr. Conf.* pp. 127–48.

Jonsson, I. G. 1980: A new approach to oscillatory rough turbulent boundary layers. *Ocean Engnr.* 7, pp. 109–52.

Jungerius, P. D. 1989: Geomorphology, soils and dune management. In *Perspectives in Coastal Dune Management*, F. Van Der Meulen, P. D. Jungerius, and J. Visser (eds.), SPB Academic Publishing, The Hague, pp. 91–8.

Jungerius, P. D. 1990: The characteristics of dune soils. In *Dunes of the European Coasts*, W. Bakker, P. D. Jungerius, and J. A. Klijn (eds.), *Catena* Suppl. 18, pp. 155–62.

Jungerius, P. D. and Dekker, L. W. 1990: Water erosion in the dunes. In *Dunes of the European Coasts*, W. Bakker, P. D. Jungerius, and J. A. Klijn (eds.), *Catena* Suppl. 18, pp. 185–93.

Jungerius, P. D. and Van Der Meulen, F. 1989: The development of dune blowouts, as measured with erosion pins and sequential air photographs. *Catena* 16, pp. 369–76.

Jungerius, P. D., Van Der Meulen, F., Loedeman, J. H., and Stuiver, J. 1992: A geometrical approach to monitoring blowout development from aerial photographs using a Geographical Information System (GIS). In *Coastal Dunes*, R. W. G. Carter, T. G. F. Curtis, and M. J. Sheehy-Skeffington (eds.), Proc. 3rd European Dune Congr., Balkema, Rotterdam, pp. 129–38.

Justesen, P. 1988: Turbulent wave boundary layers. *Inst.*

Hydrodyn. Hydraul. Engnr., Techn. Univ. Denmark, Lyngby, Series Paper 43.

Kajiura, K. 1968: A model of the bottom boundary layer in water waves. *Bull. Earthquake Res. Inst. Univ. Tokyo*, 46, pp. 75–123.

Kajiura, K. and Shuto, N. 1990: Tsunamis. In *The Sea (Ocean Engnr. Sci.)*, B. Le Méhauté and D. M. Hanes (eds.), Wiley, New York, 9, pp. 395–420.

Kale, V. S. and Awasthi, A. 1993: Morphology and formation of armored mud balls on Revadanda Beach, western India. *J. Sed. Petrol.* 63, pp. 809–13.

Kamaludin, H. 1993: The changing mangrove shorelines in Kuala Kurau, Peninsular Malaysia. *Sed. Geol.* 83, pp. 187–97.

Kamphuis, J. W. 1975: Friction factor under oscillatory waves. *J. Waterw. Harbors Coastal Div. Amer. Soc. Civ. Engnr.* 101, pp. 135–44.

Kamphuis, J. W. 1990a: Littoral transport rate. *Proc. 22nd Coastal Engnr. Conf.* pp. 2402–15.

Kamphuis, J. W. 1990b: Influence of sand or gravel on the erosion of cohesive sediment. *J. Hydraul. Res.* 28, pp. 43–53.

Kamphuis, J. W. 1991a: Alongshore sediment transport rate. *Waterw. Port Coastal Ocean Div. Amer. Soc. Civ. Engnr.* 112, pp. 624–40.

Kamphuis, J. W. 1991b: Alongshore sediment transport rate distribution. *Coastal Sediments '91*, N. C. Kraus, K. J. Gingerich, and D. L. Kriebel (eds.), Amer. Soc. Civ. Engnr., pp. 170–83.

Kana, T. W. 1978: Surf zone measurements of suspended sediments. *Proc. 16th Coastal Engnr. Conf.* pp. 1725–43.

Kana, T. W. and Ward, L. G. 1980: Nearshore suspended sediment load during storm and post-storm conditions. *Proc. 17th Coastal Engnr. Conf.* pp. 1158–74.

Kaneko, A. 1985: Formation of beach cusps in a wave tank. *Coastal Engnr.* 9, pp. 81–98.

Katoh, K., Tanaka, N., and Irie, I. 1984: Field observation on suspended-load in the surf zone. *Proc. 19th Coastal Engnr. Conf.* pp. 1046–62.

Katoh, K. and Yanagishima, S. 1992: Berm formation and beach erosion. *Proc. 23rd Coastal Engnr. Conf.* pp. 2136–49.

Katori, S., Sakakiyama, T., and Watanabe, A. 1984: Measurement of sand transport in a cross unidirectional-oscillatory flow tank. *Coastal Engnr. Japan*, 27, pp. 193–203.

Kawamura, R. 1951: *Study on Sand Movement by Wind*, Rept. Inst. Sci. Techn., Univ. Tokyo, 5, pp. 95–112. English trans. in Hydraul. Engnr. Lab., Univ. California, Berkeley, (1964), Inst. of Engnr. Res., Techn, Rept. HEL-2-8.

Kawata, Y., Shirai, T., and Tsuchiya, Y. 1992: Field observation on sand ripples under rough sea state. *Proc. 23rd Coastal Engnr. Conf.* pp. 2164–75.

Kawata, Y. and Tsuchiya, Y. 1986: Applicability of sub-

sand system to beach erosion control. *Proc. 20th Coastal Engnr. Conf.* pp. 1255–67.

Kearney, M. S., Grace, R. E., and Stevenson, J. C. 1988: Marsh loss in Nanticoke Estuary, Chesapeake Bay. *Geogr. Rev.* 78, pp. 205–20.

Keen, D. H., Harmon, R. S., and Andrews, J. T. 1981: U series and amino acid dates from Jersey. *Nature* 289, pp. 162–4.

Kemp, P. H. 1975: Wave asymmetry in the nearshore zone and breaker area. In *Nearshore Sediment Dynamics and Sedimentation*, J. Hails and A. Carr (eds.), Wiley, London, pp. 47–67.

Kemp, P. H. and Plinston, D. T. 1974: Internal velocities in the uprush and backwash zone. *Proc. 14th Coastal Engnr. Conf.* pp. 575–85.

Kempema, E. W., Reimnitz, E., and Barnes, P. W. 1989: Sea ice sediment entrainment and rafting in the Arctic. *J. Sed. Petrol.* 59, pp. 308–17.

Kempema, E. W. and Reimnitz, E. 1991: Nearshore sediment transport by slush/brash ice in southern Lake Michigan. *Coastal Sediments '91*, N. C. Kraus, K. J. Gingerich, and D. L. Kriebel (eds.), Amer. Soc. Civ. Engnr., pp. 212–19.

Kendall, C. G. St. C. and Skipwith, P. A. d'E. 1969: Geomorphology of a recent shallow water carbonate province: Khor Al Bazam, Trucial Coast, southwest Persian Gulf. *Geol. Soc. Amer. Bull.* 80, pp. 865–92.

Kennedy, T. B. and Mather, K. 1953: Correlation between laboratory freezing and thawing and weathering at Treat Island, Maine. *Proc. Amer. Concrete Inst.* 50, pp. 141–72.

Kenyon, P. M. and Turcotte, D. L. 1985: Morphology of a delta prograding by bulk sediment transport. *Geol. Soc. Amer. Bull.* 96, pp. 1457–65.

Kerr, R. A. 1987: Refining and defending the Vail sea level curve. *Science* 235, pp. 1141–2.

Khafagy, A. A., Naffaa, M. G., Fanos, A. M., and Dean, R. G. 1992: Nearshore coastal changes along the Nile Delta shores. *Proc. 23rd Coastal Engnr. Conf.* pp. 3260–72.

Kiene, W. E. 1988: A model of bioerosion on the Great Barrier Reef. *Proc. 6th Int. Coral Reef Symp., Townsville* 3, pp. 449–54.

Kim, C. S. and Huntley, D. A. 1985: Long waves in the nearshore zone. *Proc. Can. Coastal Conf.*, D. L. Forbes (ed.), National Res. Council Can, (ACROSES), Ottawa, pp. 37–51.

King, B. A., Blackley, M. W. L., Carr, A. P., and Hardcastle, P. J. 1990: Observations of wave-induced set-up on a natural beach. *J. Geophys. Res.* 95, pp. 22289–97.

King, C. A. M. 1972: *Beaches and Coasts*, Edward Arnold, London, 2nd edn.

King, D. B. 1991: The effect of beach slope on oscillatory flow bedload transport. *Coastal Sediments '77*, Amer. Soc. Civ. Engnr., pp. 734–44.

King, D. B. and Seymour, R. J. 1989: State of the art in oscillatory sediment transport models. In *Nearshore*

Sediment Transport, R. J. Seymour (ed.), Plenum Press, New York, pp. 371–85.

Kinsey, D. W. 1981: The Pacific/Atlantic reef growth controversy. *Proc. 4th Int. Coral Reef Symp., Manila* 1, pp. 493–8.

Kinsey, D. W. 1985: Metabolism, calcification and carbon production 1 system level studies. *Proc. 5th Int. Coral Reef Congr., Tahiti* 4, pp. 505–26.

Kinsey, D. W. and Davies, P. J. 1979: Effects of elevated nitrogen and phosphorus on coral reef growth. *Limnol. Oceanogr.* 24, pp. 935–40.

Kinsey, D. W. and Hopley, D. 1991: The significance of coral reefs as global carbon sinks—response to Greenhouse. *Palaeogeogr., Palaeoclim., Palaeoecol. (Global and Planetary Change Section)* 89, pp. 363–77.

Kinsman, B. 1965: *Wind Waves*, Prentice-Hall, Englewood Cliffs, NJ.

Kirby, J. T., Dalrymple, R. A., and Liu, P. L.-F. 1981: Modification of edge waves by barred-beach topography. *Coastal Engnr.* 5, pp. 35–49.

Kirby, R. 1988: High concentration suspension (fluid mud) layers in estuaries. In *Physical Processes in Estuaries*, J. Dronkers and W. Van Leussen (eds.), Springer-Verlag, Berlin, pp. 463–87.

Kirk, R. M. 1975: Aspects of surf and runup processes on mixed sand and gravel beaches. *Geografiska Annlr.* A 57, pp. 117–33.

Kirk, R. M. 1977: Rates and forms of erosion on intertidal platforms at Kaikoura Peninsula, South Island, New Zealand. *NZ J. Geol. Geophys.* 20, pp. 571–613.

Kirk, R. M. 1980: Mixed sand and gravel beaches: morphology, processes and sediments. *Progr. Phys. Geogr.* 4, pp. 189–210.

Kjerfve, B. 1989: Estuarine geomorphology and physical oceanography. In *Estuarine Ecology*, J. W. Day, C. A. S. Hall, W. M. Kemp, and A. Yanez-Arancibia (eds.), Wiley, New York, pp. 47–78.

Kjerfve, B. and Magill, K. E. 1989: Geographic and hydrodynamic characteristics of shallow coastal lagoons. *Marine Geol.* 88, pp. 187–99.

Klein, G. deVries 1985: Intertidal flats and intertidal sand bodies. In *Coastal Sedimentary Environments*, R. A. Davis (ed.), Springer-Verlag, New York, pp. 187–224.

Klijn, J. A. 1990: The younger dunes in the Netherlands; chronology and causation. In *Dunes of the European Coasts*, W. Bakker, P. D. Jungerius, and J. A. Klijn (eds.), *Catena* Suppl. 18, pp. 89–100.

Knight, R. J. 1980: Linear sand bar development and tidal current flow in Cobequid Bay, Bay of Fundy, Nova Scotia. In *The Coastline of Canada*, S. B. McCann (ed.), *Geol. Surv. Can. Paper* 80–10, pp. 123–52.

Knight, R. J. and Dalrymple, R. W. 1976: Winter conditions in a macrotidal environment, Cobequid Bay, Nova Scotia. *Rev. Géogr. Montréal* 30, pp. 65–85.

Knighton, A. D., Woodroffe, C. D., and Mills, K. 1992: The evolution of tidal creek networks, Mary River, Northern Australia. *Earth Surface Processes and Landforms* 17, pp. 167–90.

Kobayashi, N. 1982: Sediment transport on a gentle slope due to waves. *J. Waterw. Port Coastal Ocean Div. Amer. Soc. Civ. Engnr.* 108, pp. 254–71.

Kobayashi, N. 1988: Review of wave transformation and cross-shore sediment transport processes in surf zones. *J. Coastal Res.* 4, pp. 435–45.

Kobayashi, N., DeSilva, G. S., Watson, K. D. 1989: Wave transformation and swash oscillation on gentle and steep slopes. *J. Geophys. Res.* 94, pp. 951–66.

Kocurek, G. and Nielson, J. 1986: Conditions favourable for the formation of warm-climate aeolian sand sheets. *Sedimentology* 33, pp. 795–816.

Kocurek, G., Townsley, E., Yeh, K., Havholm, K., and Sweet, M. L. 1992: Dune and dune-field development on Padre Island, Texas, with implications for interdune deposition and water-table-controlled accumulation. *J. Sed. Petrol.* 62, pp. 622–35.

Kolb, C. R. and Dornbusch, W. K. 1975: The Mississippi and Mekong Deltas—a comparison. In *Deltas—Models for Exploration*, M. L. Broussard (ed.), Houston Geol. Soc., Houston, Texas, pp. 193–207.

Kolb, C. R. and Van Lopik, J. R. 1966: Depositional environments of the Mississippi River deltaic plain—southeastern Louisiana. In *Deltas in their Geologic Framework*, M. L. Shirley and J. A. Ragsdale (eds.), Houston Geol. Soc., Houston, Texas, pp. 17–61.

Komar, P. D. 1973*a*: Observations of beach cusps at Mono Lake, California. *Geol. Soc. Amer. Bull.* 84, pp. 3593–600.

Komar, P. D. 1973*b*: Computer models of delta growth due to sediment input from rivers and longshore transport. *Geol. Soc. Amer. Bull.* 84, pp. 2217–26.

Komar, P. D. 1975: Nearshore currents: generation by obliquely incident waves and longshore variations in breaker height. In *Nearshore Sediment Dynamics and Sedimentation*, J. Hails and A. Carr (eds.), Wiley, London, pp. 17–45.

Komar, P. D. 1976: *Beach Processes and Sedimentation*, Prentice-Hall, Englewood Cliffs, NJ.

Komar, P. D. 1977: Selective longshore transport rates of different grain-size fractions within a beach. *J. Sed. Petrol.* 47, pp. 1444–53.

Komar, P. D. 1978: Relative quantities of suspension versus bed-load transport on beaches. *J. Sed. Petrol.* 48, pp. 921–32.

Komar, P. D. 1980*a*: Settling velocities of circular cylinders at low Reynolds numbers. *J. Geol.* 88, pp. 327–36.

Komar, P. D. 1980*b*: K-type coefficients in the littoral drift system: variable or constant? *Marine Geol.* 34, pp. M99–101.

Komar, P. D. 1981: The applicability of the Gibbs equation for grain settling velocities to conditions other than quartz grains in water. *J. Sed. Petrol.* 51, pp. 1125–32.

Komar, P. D. 1983*a*: Nearshore currents and sand transport on beaches. In *Physical Oceanography of Coastal and Shelf Seas*, B. Johns (ed.), Elsevier, Amsterdam, pp. 67–109.

Komar, P. D. 1983*b*: Rhythmic shoreline features and their origins. In *Megageomorphology*, R. Gardner and H. Scoging (eds.), Clarendon—Oxford Univ. Press, Oxford, pp. 92–112.

Komar, P. D. 1985: Computer models of shoreline configuration: headland erosion and the graded beach revisited. In *Models in Geomorphology*, M. J. Woldenberg (ed.), Allen & Unwin, pp. 155–70.

Komar, P. D. and Enfield, D. B. 1987: Short-term sea-level changes and coastal erosion. In *Sea-level Fluctuation and Coastal Evolution*, D. Nummedal, O. H. Pilkey, and J. D. Howard (eds.), *Soc. Econ. Paleont. Mineral.* Spec. Publ. 41, pp. 17–27.

Komar, P. D. and Gaughan, M. K. 1972: Airy wave theory and breaker height prediction. *Proc. 13th Coastal Engnr. Conf.* pp. 405–18.

Komar, P. D. and Holman, R. A. 1986: Coastal processes and the development of shoreline erosion. *Annual Rev. Earth Planet. Sci.* 14, pp. 237–65.

Komar, P. D. and Inman, D. L. 1970: Longshore sand transport on beaches. *J. Geophys. Res.* 75, pp. 5914–27.

Komar, P. D., Lanfredi, N., Baba, M., Dean, R. G., Dyer, K., Healy, T., Ibe, A. C., Terwindt, J. H. J., and Thom, B. G. 1991: The response of beaches to sea-level changes: a review of predictive models. *J. Coastal Res.* 7, pp. 895–921.

Komar, P. D. and Li, Z. 1986: Pivoting analysis of the selective entrainment of sediments by shape and size with application to gravel threshold. *Sedimentology* 33, pp. 425–36.

Komar, P. D. and McDougal, W. G. 1994: The analysis of exponential beach profiles. *J. Coastal Res.* 10, pp. 59–69.

Komar, P. D. and Miller, M. C. 1973: The threshold of sediment movement under oscillatory water waves. *J. Sed. Petrol.* 43, pp. 1101–10.

Komar, P. D. and Miller, M. C. 1974: Sediment threshold under oscillatory waves. *Proc. 14th Coastal Engnr. Conf.* pp. 756–75.

Komar, P. D. and Miller, M. C. 1975: On the comparison between the threshold of sediment motion under waves and unidirectional currents with a discussion of the practical evaluation of the threshold. *J. Sed. Petrol.* 45, pp. 362–7.

Komar, P. D. and Reimers, C. E. 1978: Grain shape effects on settling rates. *J. Geol.* 86, pp. 193–209.

Komar, P. D. and Wang, C. 1984: Processes of selective grain transport and the formation of placers on beaches. *J. Geol.* 92, pp. 637–55.

Kostaschuk, R. A. and Atwood, L. A. 1990: River discharge and tidal controls on salt-wedge position and implications for channel shoaling: Fraser River, British Columbia. *Can. J. Civ. Engnr.* 17, pp. 452–9.

Kostaschuk, R. A., Church, M. A., and Luternauer, J. L. 1991: Acoustic images of turbulent flow structures in Fraser River estuary, British Columbia. *Current Res. Geol. Surv. Can. Paper* 91-1E, pp. 83–90.

Kostaschuk, R. A., Church, M. A., and Luternauer, J. L. 1992*a*: Sediment transport over salt-wedge intrusions: Fraser River estuary, Canada. *Sedimentology* 39, 305–17.

Kostaschuk, R. A., Luternauer, J. L., and Church, M. A. 1989: Suspended sediment hysteresis in a salt-wedge estuary: Fraser River, Canada. *Marine Geol.* 87, pp. 273–85.

Kostaschuk, R. A., Luternauer, J. L., McKenna, G. T., and Moslow, T. F. 1992*b*: Sediment transport in a submarine channel system: Fraser River Delta, Canada. *J. Sed. Petrol.* 62, pp. 273–82.

Kostaschuk, R. A. and McCann, S. B. 1983: Observations on delta-forming processes in a fjord-head delta, British Columbia, Canada. *Sed. Geol.* 36, pp. 269–88.

Kostrzewski, A. and Zwolinski, Z. 1987: Operation and morphologic effects of present-day morphogenetic processes modelling the cliffed coast of Wolin Island, N.W. Poland. In *International Geomorphology*, V. Gardiner (ed.), Wiley, Chichester, pp. 1231–52.

Kraft, J. C., Allen, E. A., Belknap, D. F., John, C. J., and Maurmeyer, E. M. 1979: Processes and morphologic evolution of an estuarine and coastal barrier system. In *Barrier Islands*, S. P. Leatherman (ed.), Academic Press, New York, pp. 149–83.

Kraft, J. C., Allen, E. A., and Maurmeyer, E. M. 1978: The geological and paleogeomorphological evolution of a spit system and its associated coastal environments: Cape Henlopen Spit, Delaware. *J. Sed. Petrol.* 48, pp. 211–26.

Kranenburg, C. 1988: On internal waves in partially mixed and stratified tidal flows. In *Physical Processes in Estuaries*, J. Dronkers and W. Van Leussen (eds.), Springer-Verlag, Berlin, pp. 213–38.

Kraus, N. C. and Dean, J. L. 1987: Longshore sediment transport rate distributions. *Coastal Sediments '87*, N. C. Kraus (ed.), Amer. Soc. Civ. Engnr., pp. 881–96.

Kraus, N. C. and Horikawa, K. 1990: Nearshore sediment transport. In *The Sea* Vol. 9 Part B (*Ocean Engnr. Sci.*), B. Le Méhauté and D. M. Hanes (eds.), Wiley, New York, pp. 775–813.

Kraus, N. C., Isobe, M., Igarashi, H., Sasaki, T. O., and Horikawa, K. 1982: Field experiments on longshore sand transport in the surf zone. *Proc. 18th Coastal Engnr. Conf.* pp. 969–98.

Kraus, N. C., Larson, M., and Kriebel, D. L. 1991: Evaluation of beach erosion and accretion predictors. *Coastal Sediments '91*, N. C. Kraus, K. J. Gingerich, and D. L. Kriebel (eds.), Amer. Soc. Civ. Engnr., pp. 572–87.

Kroon, A. and Hoekstra, P. 1990: Eolian transport on a natural beach. *J. Coastal Res.* 6, pp. 367–79.

Kuhn, G. G. and Osborne, R. H. 1987: Sea cliff erosion

in San Diego County, California. *Coastal Sediments '87*, N. C. Kraus (ed.), Amer. Soc. Civ. Engnr., pp. 1839–54.

Kuo, C. T. and Chen, W. J. 1990: Bottom shear stress and friction factor due to the asymmetric wave action. *Proc. 22nd Coastal Engnr. Conf.* pp. 637–46.

Laborel, J. 1974: West African reef corals: an hypothesis on their origin *Proc. 2nd Int. Coral Reef Symp., Brisbane* 1, pp. 425–43.

Laboute, P. 1985: Evaluation des dégâts causés par les passages des cyclones de 1982–1983 en Polynésie Française sur les pentes externes des atolls de Tikehau et de Takapoto (Archipel des Tuamotu). *Proc. 5th Int. Coral Reef Congr., Tahiti* 3, pp. 323–9.

Ladd, H. S. 1977: Types of coral reefs and their distribution. In *Biology and Geology of Coral Reefs*, O. A. Jones and R. Endean (eds.), Academic Press, New York, 4, pp. 1–19.

Ladd, H. S., Newman, W. A., and Sohl, N. F. 1974: Darwin Guyot, the Pacific's oldest atoll. *Proc. 2nd Int. Coral Reef Symp., Brisbane* 2, pp. 513–22.

Ladd, H. S., Tracey, J. I, and Gross, M. G. 1967: Drilling on Midway Atoll, Hawaii. *Science* 156, pp. 1088–94.

Lakhan, V. C. 1989: Modeling and simulation of the coastal system. In *Applications in Coastal Modeling*, V. C. Lakhan and A. S. Trenhaile (eds.), Elsevier, Amsterdam, pp. 17–41.

Lakhan, V. C. and Trenhaile, A. S. 1989: Models and the coastal system. In *Applications in Coastal Modeling*, V. C. Lakhan and A. S. Trenhaile (eds.), Elsevier, Amsterdam, pp. 1–16.

Lambeck, K. and Nakada, M. 1992: Constraints on the age and duration of the last interglacial period and on sea-level variations. *Nature* 357, pp. 125–8.

Lancaster, N. 1982: Dunes on the Skeleton Coast, Namibia (South West Africa): geomorphology and grain size relationships. *Earth Surface Processes and Landforms* 7, pp. 575–87.

Langbein, W. B. 1963: The hydraulic geometry of a - shallow estuary. *Bull. Int. Assoc. Sci. Hydrol.* 8, pp. 84–94.

Lange, W. P. de and Lange, P. J. de 1994: An appraisal of factors controlling the latitudinal distribution of mangrove (*Avicannia marina* var. *resinifera*) in New Zealand. *J. Coastal Res.* 10, pp. 539–48.

Lanyon, J. A., Eliot, I. G., and Clarke, D. J. 1982: Groundwater-level variation during semidiurnal spring tidal cycles on a sandy beach. *Austr. J. Marine and Freshwater Res.* 33, pp. 377–400.

Larsen, E. and Holtedahl, H. 1985: The Norwegian strandflat: a consideration of its age and origin. *Norsk Geologisk Tidsskr.* 65, pp. 247–54.

Larsen, L. H., Sternberg, R. W., Shi, N. C., Marsden, M. A. H., and Thomas, L. 1981: Field investigations of the threshold of grain motion by ocean waves and currents. *Marine Geol.* 42, pp. 105–32.

Lauriol, B. and Gray, J. T. 1980: Processes responsible for the concentration of boulders in the intertidal zone in Leaf Basin, Ungava. In *The Coastline of Canada*, S. B. McCann (ed.), *Geol. Surv. Can. Paper* 80–10, pp. 281–92.

Lavelle, J. W. and Mofjeld, H. O. 1987: Do critical stresses for incipient motion and erosion really exist? *J. Hydraul. Engnr. Amer. Soc. Civ. Engnr.* 113, pp. 370–85.

Law, J. 1990: Seasonal variations in coastal dune form. *Proc. Symp. Coastal Sand Dunes*, R. G. D. Davidson-Arnott (ed.), National Res. Council Can. (ACOS), Ottawa, pp. 69–88.

Law, M. N. and Davidson-Arnott, R. G. D. 1990: Seasonal controls on aeolian processes on the beach and foredune. *Proc. Symp. Coastal Sand Dunes*, R. G. D. Davidson-Arnott (ed.), National Res. Council Can. (ACOS), Ottawa, pp. 49–68.

Leatherman, S. P. 1979: Barrier dune systems: a reassessment. *Sed. Geol.* 24, pp. 1–16.

Leatherman, S. P. 1985: Geomorphic and stratigraphic analysis of Fire Island, New York. *Marine Geol.* 63, pp. 173–95.

Leatherman, S. P. 1987: Reworking of glacial outwash sediments along outer Cape Cod: development of Provincetown Spit. In *Glaciated Coasts*, D. M. Fitzgerald and P. S. Rosen (eds.), Academic Press, London, pp. 307–25.

Leatherman, S. P. 1990: Modelling shore response to sea-level rise on sedimentary coasts. *Progr. Phys. Geogr.* 14, pp. 447–64.

Leatherman, S. P., Williams, A. T., and Fisher, J. S. 1977: Overwash sedimentation associated with a large-scale northeaster. *Marine Geol.* 24, pp. 109–21.

LeBlond, P. H. 1979: An explanation of the logarithmic spiral plan shape of headland-bay beaches. *J. Sed. Petrol.* 49, pp. 1093–100.

Lee, A. C. and Osborne, R. H. 1995: Quartz grain-shape of southern California beaches. *J. Coastal Res.* 11, pp. 1336–45.

Lee, C.-B. 1985: Sedimentary processes of fine sediments and the behaviour of associated metals in the Keum Estuary, Korea. In *Marine and Estuarine Geochemistry*, A. C. Sigleo and A. Hattori (eds.), Lewis Publishers, Chelsea, Michigan, pp. 209–25.

Lee, P. Z.-F. 1994: The submarine equilibrium profile. *J. Coastal Res.* 10, pp. 1–17.

Leeder, M. R. 1982: *Sedimentology: Process and Product*, George Allen & Unwin, London.

Lees, B. G. 1987: Age structure of the Point Stuart chenier plain: a reassessment. *Search* 18, pp. 257–9.

Lees, B. G. 1992: The development of a chenier sequence on the Victoria Delta, Joseph Bonaparte Gulf, northern Australia. *Marine Geol.* 103, pp. 215–24.

Lees, B. G., Hayne, M., and Price, D. 1993: Marine transgression and dune initiation on western Cape York, northern Australia. *Marine Geol.* 114, pp. 81–9.

Lee-Young, J. S. and Sleath, J. F. A. 1988: Initial motion in combined wave and current flows. *Proc. 21st Coastal Engnr. Conf.* pp. 1140–51.

Lefebvre, G., Rohan, K., and Douville, S. 1985: Erosivity of natural intact structured clay: evaluation. *Can. Geotechn. J.* 22, pp. 508–17.

Lennon, G. W., Bowers, D. G., Nunes, R. A., Scott, B. D., Ali, M., Boyle, J., Wenju, C., Herzfeld, M., Johansson, G., Nield, S., Petrusevics, P., Stephenson, P., Suskin, A. A., and Wijffels, S. E. A. 1987: Gravity currents and the release of salt from an inverse estuary. *Nature* 327, pp. 695–7.

Leonard, J. E. and Cameron, B. W. 1981: Origin of high-latitude carbonate beach: Mt. Desert Island, Maine. *Northeastern Geol.* 3, pp. 178–83.

Leont'ev, I. O. 1985: Sediment transport and beach equilibrium profile. *Coastal Engnr.* 9, pp. 277–91.

Lessa, G. and Masselink, G. 1995: Morphodynamic evolution of a macrotidal barrier estuary. *Marine Geol.* 129, pp. 25–46.

Lessios, H., Robertson, D., and Cubit, J. 1984: Spread of *Diadema* mass mortality through the Caribbean. *Science* 226, pp. 335–7.

Lettau, K. and Lettau, H. H. 1978: Experimental and micrometeorological field studies of dune migration. In H. H. Lettau and K. Lettau (eds.), *Exploring the World's Driest Climate*, Inst. of Environmental Studies (IES) Rept. 101, Univ. Wisconsin–Madison, pp. 110–47.

Letzsch, W. S. and Frey, R. W. 1980: Deposition and erosion in a Holocene salt marsh, Sapelo Island, Georgia. *J. Sed. Petrol.* 50, pp. 529–42.

Levin, D. R. 1993: Tidal inlet evolution in the Mississippi River delta plain. *J. Coastal Res.* 9, pp. 462–80.

Lewis, W. V. 1938: The evolution of shoreline curves. *Proc. Geol. Assoc.* 49, pp. 107–27.

Ley, R. G. 1979: The development of marine karren along the Bristol Channel coastline. *Z. Geomorph.* Suppl. Band 32, pp. 75–89.

Li, M. Z. 1994: Direct skin friction measurements and stress partitioning over movable sand ripples. *J. Geophys. Res.* 99, pp. 791–99.

Li, M. Z. and Komar, P. D. 1992a: Longshore grain sorting and beach placer formation adjacent to the Columbia River. *J. Sed. Petrol.* 62, pp. 429–41.

Li, M. Z. and Komar, P. D. 1992b: Selective entrainment and transport of mixed size and density sands: flume experiments simulating the formation of black-sand placers. *J. Sed. Petrol.* 62, pp. 584–90.

Li, Y., Wolanski, E., and Xie, Q. 1993: Coagulation and settling of suspended sediment in the Jiaojiang River Estuary, China. *J. Coastal Res.* 9, pp. 390–402.

Li, Z. and Komar, P. D. 1986: Laboratory measurements of pivoting angles for applications to selective entrainment of gravel in a current. *Sedimentology* 33, pp. 413–23.

Lick, W., Huang, H., and Jepsen, R. 1993: Flocculation of fine-grained sediments due to differential settling. *J. Geophys. Res.* 98, pp. 10279–88.

Lighty, R. G., Macintyre, I. G., and Stuckenrath, R. 1978: Submerged early Holocene barrier reef southeast Florida shelf. *Nature* 276, pp. 59–60.

Lincoln, J. M. and FitzGerald, D. M. 1988: Tidal distortions and flood dominance at five small tidal inlets in southern Maine. *Marine Geol.* 82, pp. 133–48.

Lippmann, T. C. and Holman, R. A. 1990: The spatial and temporal variability of sand bar morphology. *J. Geophys. Res.* 95, pp. 11575–90.

Lippmann, T. C., Holman, R. A., and Hathaway, K. K. 1993: Episodic, nonstationary behavior of a double bar system at Duck, North Carolina, U.S.A., 1986–1991. *J. Coastal Res.* Spec. Issue 15, pp. 49–75.

List, J. H. 1992: Breakpoint-forced and bound long waves in the nearshore: a model comparison. *Proc. 23rd Coastal Engnr. Conf.* pp. 860–7.

Littler, M. M. and Littler, D. S. 1985: Factors controlling relative dominance of primary producers on biotic reefs. *Proc. 5th Int. Coral Reef Congr., Tahiti* 4, pp. 35–9.

Liu, J. T. and Zarillo, G. A. 1993: Simulation of grain-size abundances on a barred upper shoreface. *Marine Geol.* 109, pp. 237–51.

Logan, B. W., Read, J. F., Hagan, G. M., Hoffman, P., Brown, R. G., Woods, P. J., and Gebelein, C. D. 1974: Evolution and diagenesis of Quaternary carbonate sequences, Shark Bay, Western Australia. *Amer. Assoc. Petrol. Geol. Mem.* 22.

Long, A. J. and Hughes, P. D. M. 1995: Mid- and late-Holocene evolution of the Dungeness foreland, UK *Marine Geol.* 124, pp. 253–71.

Longuet-Higgins, M. S. 1953: Mass transport in water waves. *Phil. Trans. Roy. Soc. London* A 245, pp. 535–87.

Longuet-Higgins, M. S. 1969a: Action of a variable stress at the surface of water waves. *The Physics of Fluids* 12, pp. 737–40.

Longuet-Higgins, M. S. 1969b: A nonlinear mechanism for the generation of sea waves. *Proc. Roy. Soc. London* A 311, pp. 371–89.

Longuet-Higgins, M. S. and Stewart, R. W. 1962: Radiation stress and mass transport in gravity waves, with application to 'surf beats'. *J. Fluid Mech.* 13, pp. 481–504.

Loomis, H. G. 1978: Tsunami. In *Geophysical Predictions*, National Acad. Sci. Washington, DC, pp. 155–65.

Lund-Hansen, L. C. and Oehmig, R. 1992: Comparing sieve and sedimentation balance analysis of beach, lake and eolian sediments using log-hyperbolic parameters. *Marine Geol.* 107, pp. 139–47.

Lynch, J. C., Meriwether, J. R., McKee, B. A., Vera-Herrera, F., and Twilley, R. R. 1989: Recent accretion in mangrove ecosystems based on ^{137}Cs and [210]Pb. *Estuaries* 12, pp. 284–99.

Lynts, G. W. 1970: Conceptual model of the Bahamian

Platform for the last 135 million years. *Nature* 225, pp. 1226–8.

Machemehl, J. L., French, T. J., and Huang, N. E. 1975: New method for beach erosion control. *Proc. Civ. Engnr. in the Oceans, Amer. Soc. Civ. Engnr.* pp. 142–60.

Macintyre, I. G. 1988: Modern coral reefs of western Atlantic: new geological perspective. *Amer. Assoc. Petrol. Geol. Bull.* 72, pp. 1360–9.

Macintyre, I. G., Burke, R. B., and Stuckenrath, R. 1977: Thickest recorded Holocene reef section, Isla Pérez core hole, Alacran Reef, Mexico. *Geology* 5, pp. 749–54.

Macintyre, I. G., Burke, R. B., and Stuckenrath, R. 1981: Core holes in the outer fore-reef off Carrie Bow Cay, Belize: a key to the Holocene history of the Belizean barrier reef complex. *Proc. 4th Int. Coral Reef Symp., Manila* 1, pp. 567–74.

Macintyre, I. G. and Glynn, P. W. 1976: Evolution of modern Caribbean fringing reef, Galeta Point, Panama. *Amer. Assoc. Petrol. Geol. Bull.* 60, pp. 1054–72.

Macintyre, I. G., Raymond, B. and Stuckenrath, R. 1983: Recent history of a fringing reef, Bahia Salina Del Sur, Vieques Island, Puerto Rico. *Atoll Res. Bull.* 268.

MacKay, D. K. 1970: The ice regime of the Mackenzie Delta, Northwest Territories. In *Hydrology of Deltas*, Gentbrugge, Paris, Proc. Bucharest Symp. May 1969, IASH/AIHS—UNESCO, ii. 356–62.

Macmillan, D. H. 1966: *Tides*, Elsevier, New York.

Macnae, W. 1968: A general account of the fauna and flora of mangrove swamps and forests in the Indo-West-Pacific region. *Adv. Marine Biol.* 6, pp. 73–270.

Madsen, O. S. and Grant, W. D. 1976: Quantitative description of sediment transport by waves. *Proc. 15th Coastal Engnr. Conf.* pp. 1093–112.

Madsen, P. A. and Svendsen, I. A. 1983: Turbulent bores and hydraulic jumps. *J. Fluid Mech.* 129, pp. 1–25.

Maldonado, A. 1975: Sedimentation, stratigraphy, and development of the Ebro Delta, Spain. In *Deltas— Models for Exploration*, M. L. Broussard (ed.), Houston Geol. Soc., Houston, Texas, pp. 311–38.

Maragos, J. E., Evans, C., and Holthus, P. 1985: Reef corals in Kaneohe Bay six years before and after termination of sewage discharges (Oahu, Hawaiian Archipelago). *Proc. 5th Int. Coral Reef Congr., Tahiti* 4, pp. 189–94.

Marino, J. N. and Mehta, A. J. 1987: Inlet ebb shoals related to coastal parameters. *Coastal Sediments '87*, N. C. Kraus (ed.), Amer. Soc. Civ. Engnr., pp. 1608–23.

Marqués, M. A. and Julià, R. 1988: St. Pere Pescador beach–dune interaction. *J. Coastal Res.* Spec. Issue 3, pp. 57–61.

Marsh, W. M. and Marsh, B. D. 1987: Wind erosion and sand dune formation on high Lake Superior bluffs. *Geografiska Annlr.* A 69, pp. 379–91.

Marshall, J. F. and Davies, P. J. 1984: Facies variation and Holocene reef growth in the southern Great Barrier Reef. In *Coastal Geomorphology in Australia*, B. G. Thom (ed.), Academic Press, Sydney, pp. 123–34.

Martini, I. P. 1981: Ice effect on erosion and sedimentation on the Ontario shores of James Bay, Canada. *Z. Geomorph.* 25, pp. 1–16.

Martini, I. P. 1990: Aeolian features of the recent, subarctic coastal zone of the Hudson Bay Lowland, Ontario, Canada. *Proc. Symp. Coastal Sand Dunes*, R. G. D. Davidson-Arnott (ed.), National Res. Council Can. (ACOS), Ottawa, pp. 137–57.

Martini, I. P. 1991: Sedimentology of subarctic tidal flats of western James Bay and Hudson Bay, Ontario, Canada. In *Clastic Tidal Sedimentology*, D. G. Smith, G. E. Reinson, B. A. Zaitlin, and R. A. Rahmani (eds.), *Can. Soc. Petrol. Geol. Mem.* 16, pp. 301–12.

Mason, C., Sallenger, A. H., Holman, R. A., and Birkemeier, W. A. 1985: DUCK82—a coastal storm processes experiment. *Proc. 19th Coastal Engnr. Conf.* pp. 1913–28.

Mason, O. K. 1993: The geoarchaeology of beach ridges and cheniers: studies of coastal evolution using archaeological data. *J. Coastal Res.* 9, pp. 126–46.

Mason, O. K. and Jordan, J. W. 1993: Heightened North Pacific storminess during synchronous late Holocene erosion of northwest Alaska beach ridges. *Quat. Res.* 40, pp. 55–69.

Massari, F. and Parea, G. C. 1988: Progradational gravel beach sequences in a moderate- to high-energy, microtidal marine environment. *Sedimentology* 35, pp. 881–913.

Masselink, G. 1993: Simulating the effects of tides on beach morphodynamics. *J. Coastal Res.* Spec. Issue 15, pp. 180–97.

Masselink, G. and Hegge, B. 1995: Morphodynamics of meso- and macrotidal beaches: examples from central Queensland, Australia. *Marine Geol.* 129, pp. 1–23.

Masselink, G. and Short, A. D. 1993: The effect of tide range on beach morphodynamics and morphology: a conceptual beach model. *J. Coastal Res.* 9, pp. 785–800.

Matsukura, Y. and Matsuoka, N. 1991: Rates of tafoni weathering on uplifted shore platforms in Nojima-Zaki, Boso Peninsula, Japan. *Earth Surface Processes and Landforms* 16, pp. 51–6.

Matsunaga, N., Takehara, K., and Awaya, Y. 1988: Coherent eddies induced by breakers on a sloping bed. *Proc. 21st Coastal Engnr. Conf.* pp. 234–45.

Matsuoka, N. 1988: Laboratory experiments on frost shattering of rocks. *Sci. Rept. Inst. Geosci. Univ. Tsukuba* (Japan) A 9, pp. 1–36.

Mathew, J. and Baba, M. 1995: Mudbanks of the southwest coast of India. II: Wave–mud interactions. *J. Coastal Res.* 11, pp. 179–87.

Matthews, J. A., Dawson, A. G., and Shakesby, R. A. 1986: Lake shoreline development, frost weathering and rock platform erosion in an alpine periglacial environment, Jotunheimen, southern Norway. *Boreas* 15, pp. 33–50.

Maxwell, W. G. H. 1968: *Atlas of the Great Barrier Reef*, Elsevier, Amsterdam.

May, J. P. and Tanner, W. F. 1973: The littoral power gradient and shoreline changes. In *Coastal Geomorphology*, D. R. Coates (ed.), State Univ. New York, Binghamton, pp. 43–60.

McBride, R. A., Byrnes, M. R., and Hiland, M. W. 1995: Geomorphic response-type model for barrier coastlines: a regional perspective. *Marine Geol.* 126, pp. 143–59.

McCabe, J. C., Dyer, K. R., Huntley, D. A., and Bale, A. J. 1992: The variation of floc sizes within a turbidity maximum at spring and neap tides. *Proc. 23rd Coastal Engnr. Conf.* pp. 3178–88.

McCann, S. B. 1979: Barrier islands in the southern Gulf of St. Lawrence, Canada. In *Barrier Islands*, S. P. Leatherman (ed.), Academic Press, New York, pp. 29–63.

McCann, S. B. and Byrne, M.-L. 1989: Stratification models for vegetated coastal dunes in Atlantic Canada. In *Coastal Sand Dunes*, C. H. Gimingham, W. Ritchie, B. B. Willetts, and A. J. Willis (eds.), *Proc. Royal Soc. Edinb.* B 96, pp. 203–15.

McCann, S. B. and Dale, J. E. 1986: Sea ice breakup and tidal flat processes, Frobisher Bay, Baffin Island. *Physical Geogr.* 7, pp. 168–80.

McCann, S. B., Dale, J. E., and Hale, P. B. 1981a: Ice conditions and effects in a macrotidal subarctic environment, southeast Baffin Island. *Proc. Workshop on Ice Action on Shores*, J.-C. Dionne (ed.), National Res. Council Can. (ACROSES) Ottawa, pp. 105–15.

McCann, S. B., Dale, J. E., and Hale, P. B. 1981b: Subarctic tidal flats in areas of large tidal range, southern Baffin Island, eastern Canada. *Géogr. Phys. Quat.* 35, pp. 183–204.

McClimans, T. A. 1988: Estuarine fronts and river plumes. In *Physical Processes in Estuaries*, J. Dronkers and W. Van Leussen (eds.), Springer-Verlag, Berlin, pp. 55–69.

McCoy, E. D. and Heck, K. L. 1976: Biogeography of corals, seagrasses and mangroves: an alternative to the center of origin concept. *Systematic Zoology* 25, pp. 201–10.

McDonald, R. R. and Anderson, R. S. 1995: Experimental verification of aeolian and lee side deposition models. *Sedimentology* 42, pp. 39–56.

McDougal, W. G. and Hudspeth, R. T. 1983: Wave setup/setdown and longshore current on non-planar beaches. *Coastal Engnr.* 7, pp. 103–17.

McDougal, W. G. and Hudspeth, R. T. 1984: Longshore sediment transport on Dean beach profiles. *Proc. 19th Coastal Engnr. Conf.* pp. 1488–506.

McDowell, D. M. 1983: Sediment transport due to waves and currents. In *Mechanics of Sediment Transport*, B. M. Sumer and A. Muller (eds.), Proc. Euromech 156/Instanbul 1982, Balkema, Rotterdam, pp. 281–5.

McEwan, I. K. 1993: Bagnold's kink, a physical feature of a wind velocity profile modified by blown sand? *Earth Surface Processes and Landforms* 18, pp. 145–56.

McEwan, I. K. and Willetts, B. B. 1991: Numerical model of the saltation cloud. In *Aeolian Grain Transport 1 Mechanics*, O. E. Barndorff-Nielsen and B. B. Willetts (eds.), Acta Mechanica Suppl. 1, Springer-Verlag, Wien, pp. 53–66.

McEwan, I. K., Willetts, B. B., and Rice, M. A. 1992: The grain/bed collision in sand transport by wind. *Sedimentology* 39, pp. 971–81.

McGreal, W. S. 1979: Cliffline recession near Kilkeel N. Ireland: an example of a dynamic coastal system. *Geografiska Annlr.* A 61, pp. 211–19.

McGreal, W. S. and Craig, D. 1977: Mass-movement activity: an illustration of differing responses to groundwater conditions from two sites in Northern Ireland. *Irish Geogr.* 10, pp. 28–35.

McKay, P. J. and Terich, T. A. 1992: Gravel barrier morphology: Olympic National Park, Washington State, U.S.A. *J. Coastal Res.* 8, pp. 813–29.

McKenna Neuman, C. 1993: A review of aeolian transport processes in cold environments. *Progr. Phys. Geogr.* 17, pp. 137–55.

McLaren, P. 1981: An interpretation of trends in grain size measures. *J. Sed. Petrol.* 51, pp. 611–24.

McLean, R. 1967: Plan shape and orientation of beaches along the east coast, South Island. *N. Z. Geogr.* 23, pp. 16–22.

McLean, S. R. 1981: The role of non-uniform roughness in the formation of sand ribbons. *Marine Geol.* 42, pp. 49–74.

McLellan, A. G. 1971: Ambiguous 'glacial' striae formed near waterbodies. *Can. J. Earth Sci.* 8, pp. 477–9.

McManus, J. W., Ferrer, E. M., and Campos, W. L. 1988: A village-level approach to coastal adaptive management and resource assessment (CAMRA). *Proc. 6th Int. Coral Reef Symp., Townsville* 2, pp. 381–6.

McNutt, M. and Menard, H. W. 1978: Lithospheric flexure and uplifted atolls. *J. Geophys. Res.* 83, pp. 1206–12.

McPherson, J. G., Shanmugam, G., and Moiola, R. J. 1987: Fan-deltas and braid deltas: varieties of coarse-grained deltas. *Geol. Soc. Amer. Bull.* 99, pp. 331–40.

Meckel, L. D. 1975: Holocene sand bodies in the Colorado Delta area, northern Gulf of California. In *Deltas—Models for Exploration*, M. L. Broussard (ed.), Houston Geol. Soc., Houston, Texas, pp. 239–65.

Medina, R., Losada, M. A., Losada, I. J., and Vidal, C. 1994: Temporal and spatial relationship between sediment grain size and beach profile. *Marine Geol.* 118, pp. 195–206.

Mehta, A. J. and Dyer, K. R. 1990: Cohesive sediment transport in estuarine and coastal waters. In *The Sea* Vol. 9 Part B (*Ocean Engnr. Sci.*), B. Le Méhauté and D. M. Hanes (eds.), Wiley, New York, pp. 815–39.

Mei, C. C. 1985: Resonant reflection of surface water waves by periodic sandbars. *J. Fluid Mech.* 152, pp. 315–35.

Meistrell, F. J. 1966: The spit-platform concept: laboratory observation of spit development. Unpubl. M.Sc. thesis, Univ. Alberta. Republished in 1972 in *Spits and Bars*, M. L. Schwartz (ed.), Benchmark Papers in Geology, Dowden, Hutchinson and Ross, Stroudsburg, Penn., pp. 225–84.

Meldahl, K. H. 1995: Pleistocene shoreline ridges from tide-dominated and wave-dominated coasts: northern Gulf of California and western Baja California, Mexico. *Marine Geol.* 123, pp. 61–72.

Melville, G. 1984: Headlands and offshore islands as dominant controlling factors during late Quaternary barrier formation in the Forster–Tuncurry area, New South Wales, Australia. *Sed. Geol.* 39, pp. 243–71.

Miall, A. D. 1992: Exxon global cycle chart: an event for every occasion? *Geology* 20, pp. 787–90.

Middleton, G. V. 1976: Hydraulic interpretation of sand size distributions. *J. Geol.* 84, pp. 405–26.

Miles, J. W. 1957: On the generation of surface waves by shear flows. *J. Fluid Mech.* 3, pp. 185–204.

Miller, C. and Barcilon, A. 1978: Hydrodynamic instability in the surf zone as a mechanism for the formation of horizontal gyres. *J. Geophys. Res.* 83, pp. 4107–16.

Miller, D. L. R. and Mackenzie, F. T. 1988: Implications of climate change and associated sea-level rise for atolls. *Proc. 6th Int. Coral Reef Symp., Townsville* 3, pp. 519–22.

Miller, M. C. and Komar, P. D. 1980a: Oscillation sand ripples generated by laboratory apparatus. *J. Sed. Petrol.* 50, pp. 173–82.

Miller, M. C. and Komar, P. D. 1980b: A field investigation of the relationship between oscillation ripple spacing and the near-bottom water orbital motions. *J. Sed. Petrol.* 50, pp. 183–91.

Miller, M. C. and Komar, P. D. 1981: Oscillation ripples—reply. *J. Sed. Petrol.* 51, pp. 678–9.

Miller, W. R. and Mason, T. R. 1994: Erosional features of coastal beachrock and aeolianite outcrops in Natal and Zululand, South Africa. *J. Coastal Res.* 10, pp. 374–94.

Milliman, J. D. 1973: Caribbean coral reefs. In *Biology and Geology of Coral Reefs*, O. A. Jones and R. Endean (eds.), Academic Press, New York, 1, pp. 1–50.

Milliman, J. D. 1974: *Marine Carbonates*, Springer-Verlag, Berlin.

Milliman, J. D. and Meade, R. H. 1983: World-wide delivery of river sediments to the oceans. *J. Geol.* 91, pp. 1–21.

Miner, J. J. and Powell, R. D. 1991: An evaluation of ice-rafted erosion caused by an icefoot complex, southwestern Lake Michigan, U.S.A. *Arctic Alpine Res.* 23, pp. 320–7.

Minoura, K., Nakaya, S., and Uchida, M. 1994: Tsunami deposits in a lacustrine sequence of the Sanriku coast, northeast Japan. *Sed. Geol.* 89, pp. 25–31.

Mocke, G. P. and Smith, G. G. 1992: Wave breaker turbulence as a mechanism for sediment suspension. *Proc. 23rd Coastal Engnr. Conf.* pp. 2279–92.

Mogridge, G. R., Davies, M. H., and Willis, D. H. 1994: Geometry prediction for wave-generated bedforms. *Coastal Engnr.* 22, pp. 255–86.

Moign, A. 1976: L'action des glaces flottantes sur le littoral et les fonds marins du Spitsberg central et nord-occidental. *Rev. Géogr. Montréal* 30, pp. 51–64.

Moign, A. and Guilcher, A. 1967: Une flèche littorale en milieu periglaciaire arctique: la flèche de Sars (Spitsberg). *Norois* 56, pp. 549–68.

Molinier, R. and Picard, J. 1954: Eléments de bionomie marine sur les côtes de Tunisie. *Ann. Station Océanogr. Salammbô* 48.

Molnia, B. F. 1985: Processes on a glacier-dominated coast, Alaska. *Z. Geomorph.* Suppl. Band 57, pp. 141–53.

Montaggioni, L. F. 1988: Holocene reef growth history in mid-plate high volcanic islands. *Proc. 6th Int. Coral Reef Symp., Townsville* 3, pp. 455–60.

Montague, C. L. 1986: Influence of biota on erodibility of sediments. In *Estuarine Cohesive Sediment Dynamics*, A. J. Mehta (ed.), Springer-Verlag, Berlin, pp. 251–69.

Moon, V. G. and Healy, T. 1994: Mechanisms of coastal cliff retreat and hazard zone delineation in soft flysch deposits. *J. Coastal Res.* 10, pp. 663–80.

Morfett, J. C. 1991: Numerical model of longshore transport of sand in surf zone. *Proc. Inst. Civ. Engnr.* 91, part 2, pp. 55–70.

Morgan, J. P. 1970: Depositional processes and products in the deltaic environment. In *Deltaic Sedimentation: Modern and Ancient*, J. P. Morgan (ed.), *Soc. Econ. Paleont. Mineral.* Spec. Publ. 15, pp. 31–47.

Morgan, J. P., Coleman, J. M., and Gagliano, S. M. 1968: Mudlumps: diapiric structures in Mississippi Delta sediments. In *Diapirism and Diapirs*, J. Braunstein and G. D. O'Brien (eds.), *Amer. Assoc. Petrol. Geol. Mem.* 8, pp. 145–61.

Morner, N.-A. 1981: Eustasy, paleoglaciation, and paleoclimatology. *Geol. Rundschau* 70, pp. 691–702.

Morner, N.-A. 1987: Models of global sea-level changes. In *Sea-level Changes*, M. J. Tooley and I. Shennan (eds.), *Inst. Brit. Geogr.* (Blackwell, Oxford) Spec. Publ. 27, pp. 332–55.

Moslow, T. F. and Heron, S. D. 1979: Quaternary evolution of core banks, North Carolina: Cape Lookout to New Drum Inlet. In *Barrier Islands*, S. P. Leatherman (ed.), Academic Press, New York, pp. 211–36.

Moslow, T. F. and Heron, S. D. 1981: Holocene depositional history of a microtidal cuspate foreland cape: Cape Lookout, North Carolina. *Marine Geol.* 41, pp. 251–70.

Moslow, T. F. and Tye, R. S. 1985: Recognition and characterization of Holocene tidal inlet sequences. *Marine Geol.* 63, pp. 129–51.

Moss, A. J. 1972: Bed-load sediments. *Sedimentology* 18, pp. 159–219.

Moutzouris, C. I. 1990: Experimental results on the sediment grain threshold under short-wave action. *Proc. 22nd Coastal Engnr. Conf.* pp. 2552–65.

Muhs, D. R. and Szabo, B. J. 1994: New uranium-series ages of the Waimanalo Limestone, Oahu, Hawaii: implications for sea level during the last interglacial period. *Marine Geol.* 118, pp. 315–26.

Mulrennan, M. E. 1992: Ridge and runnel beach morphodynamics: an example from the central east coast of Ireland. *J. Coastal Res.* 8, pp. 906–18.

Murty, T. S. 1977: *Seismic Sea Waves—Tsunamis*, Dept. of Fisheries and the Environment, Fisheries and Marine Service, Ottawa, Canada, Bull. 198.

Murty, T. S. 1984: *Storm Surges—Meteorological Ocean Tides*. Dept. of Fisheries and Oceans, Ottawa,—Can. Bull. of Fisheries and Aquatic Sci. 212.

Muscatine, L. 1990: The role of symbiotic algae in carbon and energy flux in reef corals. In *Coral Reefs* (Ecosystems of the World 25), Z. Dubinsky (ed.), Elsevier, Amsterdam, pp. 75–87.

Muto, T. and Steel, R. J. 1992: Retreat of the front in a prograding delta. *Geology*, 20, pp. 967–70.

Muzik, K. 1985: Dying coral reefs of the Ryukyu Archipelago (Japan). *Proc. 5th Int. Coral Reef Congr., Tahiti* 6, pp. 483–9.

Nadaoka, K., Hino, M., and Koyano, Y. 1989: Structure of the turbulent flow field under breaking waves in the surf zone. *J. Fluid Mech.* 204, pp. 359–87.

Nagaraja, V. N. 1966: Hydrometeorological and tidal problems of the deltaic areas in India. In *Scientific Problems of the Humid Tropical Zone Deltas and their Implications*, UNESCO, Paris, Proc. Dacca Symp. Feb./Mar. 1964, pp. 115–21.

Naidu, A. S. and Mowatt, T. C. 1975: Depositional environments and sediment characteristics of the Colville and adjacent deltas, northern Arctic Alaska. In *Deltas—Models for Exploration*, M. L. Broussard (ed.), Houston Geol. Soc., Houston, Texas, pp. 283–309.

Nairn, R. B. 1991: Validation of a detailed alongshore transport model. In *Sand Transport in Rivers, Estuaries and the Sea*, R. Soulsby and R. Bettess (eds.), Proc. Euromech 262 Colloq., Balkema, Rotterdam, pp. 233–9.

Nairn, R. B. 1992: Designing for cohesive shores. Paper presented to the Coastal Engineering in Canada Conference, Queen's Univ., Kingston, Ontario.

Nairn, R. B. and Southgate, H. N. 1993: Deterministic profile modelling of nearshore processes. Part 2. Sediment transport and beach profile development. *Coastal Engnr.* 19, pp. 57–96.

Nakaza, E., Tsukayama, S., and Hino, M. 1990: Bore-like surf beat on reef coasts. *Proc. 22nd Coastal Engnr. Conf.* pp. 743–56.

Nansen, F. 1922: *The Strandflat and Isostasy*. Videnskapssel skapets Skrifter 1, Mat.-Naturv. Klasse 1921, no. 11 (Kristiana).

Nasr, D. H. 1985: Coral reef conservation in Sudan. *Proc. 5th Int. Coral Reef Congr., Tahiti* 4, pp. 243–6.

National Research Council (US) 1987: *Responding to Changes in Sea Level*, National Academy Press, Washington, DC.

Neiheisel, J. and Weaver, C. E. 1967: Transport and deposition of clay minerals southeastern United States. *J. Sed. Petrol.* 37, pp. 1084–116.

Nelson, B. W. 1970: Hydrography, sediment dispersal, and recent historical development of the Po River Delta, Italy. In *Deltaic Sedimentation: Modern and Ancient*, J. P. Morgan (ed.), Soc. Econ. Paleont. Mineral. Spec. Publ. 15, pp. 152–84.

Nemec, W. 1990a: Deltas—remarks on terminology and classification. In *Coarse-grained Deltas*, A. Colella and D. B. Prior (eds.), Int. Assoc. Sedimentologists (Blackwell, Oxford) Spec. Publ. 10, pp. 3–12.

Nemec, W. 1990b: Aspects of sediment movement on steep delta slopes. In *Coarse-grained Deltas*, A. Colella and D. B. Prior (eds.), Int. Assoc. Sedimentologists (Blackwell, Oxford) Spec. Publ. 10, pp. 29–73.

Nemec, W. and Steel, R. J. 1988: What is a fan delta and how do we recognise it? In *Fan Deltas: Sedimentology and Tectonic Settings*, W. Nemec and R. J. Steel (eds.), Blackie, Glasgow, pp. 3–13.

Neudecker, S. 1981: Growth and survival of scleractinian corals exposed to thermal effluents at Guam. *Proc. 4th Int. Coral Reef Symp., Manila* 1, pp. 173–80.

Neumann, A. C. and Macintyre, I. 1985: Reef response to sea level rise: keep up, catch-up or give-up. *Proc. 5th Int. Coral Reef Congr., Tahiti* 3, pp. 105–10.

New, A. L. and Dyer, K. R. 1988: Internal waves and mixing in stratified estuarine flows. In *Physical Processes in Estuaries*, J. Dronkers and W. Van Leussen (eds.), Springer-Verlag, Berlin, pp. 239–54.

Newell, N. D., Rigby, J. K., Whiteman, A. J., and Bradley, J. S. 1951: Shoal-water geology and environments, eastern Andros Island, Bahamas. *Bull. Amer. Museum Natural History* 97, pp. 1–30.

Newman, W. S., Marcus, L. F., Pardi, R. R., Paccione, J. A., and Tomecek, S. M. 1980: Eustasy and deformation of the geoid: 1,000–6,000 radiocarbon years BP. In *Earth Rheology, Isostasy, and Eustasy*, N.-A. Morner (ed.), Wiley, Chichester, pp. 555–67.

Nichol, S. L. 1991: Zonation and sedimentology of estuarine facies in an incised valley, wave-dominated, microtidal setting, New South Wales, Australia. In

Clastic Tidal Sedimentology, D. G. Smith, G. E. Reinson, B. A. Zaitlin, and R. A. Rahmani (eds.), *Can. Soc. Petrol. Geol. Mem.* 16, pp. 41–58.

Nichol, S. L. and Boyd, R. 1993: Morphostratigraphy and facies architecture of sandy barriers along the Eastern Shore of Nova Scotia. *Marine Geol.* 114, pp. 59–80.

Nicholls, R. J. and Webber, N. B. 1988: Characteristics of shingle beaches with reference to Christchurch Bay, S. England. *Proc. 21st Coastal Engnr. Conf.* pp. 1922–36.

Nicholls, R. J. and Wright, P. 1991: Longshore transport of pebbles: experimental estimates of K. *Coastal Sediments '91*, N. C. Kraus, K. J. Gingerich, and D. L. Kriebel (eds.), Amer. Soc. Civ. Engnr., pp. 920–33.

Nichols, M. M. 1986: Effects of fine sediment resuspension in estuaries. In *Estuarine Cohesive Sediment Dynamics*, A. J. Mehta (ed.), Springer-Verlag, New York, pp. 5–42.

Nichols, M. M. 1989: Sediment accumulation rates and relative sea-level rise in lagoons. *Marine Geol.* 88, pp. 201–19.

Nichols, M. M. and Biggs, R. B. 1985: Estuaries. In *Coastal Sedimentary Environments*, R. A. Davis (ed.), Springer-Verlag, New York, pp. 77–186.

Nichols, R. L. 1961: Characteristics of beaches formed in polar climates. *Amer. J. Sci.* 259, pp. 694–708.

Nickling, W. G. 1984: The stabilizing role of bonding agents on the entrainment of sediment by wind. *Sedimentology* 31, pp. 111–17.

Nickling, W. G. 1988: The initiation of particle movement by wind. *Sedimentology* 35, pp. 499–511.

Nickling, W. G. and Davidson-Arnott, R. G. D. 1990: Aeolian sediment transport on beaches and coastal sand dunes. *Proc. Symp. Coastal Sand Dunes*, R. G. D. Davidson-Arnott (ed.), National Res. Council Can. (ACOS), Ottawa, pp. 1–35.

Nickling, W. G. and Ecclestone, M. 1981: The effects of soluble salts on the threshold shear velocity of fine sand. *Sedimentology* 28, pp. 505–10.

Niedoroda, A. W. 1973: Sand bars along low energy beaches part 2: transverse bars. In *Coastal Geomorphology*, D. R. Coates (ed.), State Univ. New York, Binghamton, pp. 103–13.

Niedoroda, A. W., Sheppard, D. M., and Devereaux, A. B. 1991: The effect of beach vegetation on aeolian sand transport. *Coastal Sediments '91*, N. C. Kraus, K. J. Gingerich, and D. L. Kriebel (eds.), Amer. Soc. Civ. Engnr., pp. 246–60.

Niedoroda, A. W., Swift, D. J. P., Figueiredo, A. G., and Freeland, G. L. 1985: Barrier island evolution, middle Atlantic shelf, U.S.A. Part II: evidence from the shelf floor. *Marine Geol.* 63, pp. 363–96.

Nielsen, L. H., Johannessen, P. N., and Surlyk, F. 1988: A late Pleistocene coarse-grained spit-platform sequence in northern Jylland, Denmark. *Sedimentology* 35, pp. 915–37.

Nielsen, N. 1979: Ice-foot processes. Observations of erosion on a rocky coast, Disko, west Greenland. *Z. Geomorph.* 23, pp. 321–31.

Nielsen, P. 1979: Some basic concepts of wave sediment transport. *Inst. Hydrodyn. Hydraul. Engnr., Techn. Univ. Denmark, Lyngby, Series Paper* 20, pp. 1–160.

Nielsen, P. 1988: Three simple models of wave sediment transport. *Coastal Engnr.* 12, pp. 43–62.

Nielsen, P. 1990: Tidal dynamics of the water table in beaches. *Water Resources Res.* 26, pp. 2127–34.

Nielsen, P. 1992: *Coastal Bottom Boundary Layers and Sediment Transport* (World Scientific, Singapore).

Nielsen, P. and Hanslow, D. J. 1991: Wave runup distributions on natural beaches. *J. Coastal Res.* 7, pp. 1139–52.

Nieuwenhuyse, A. and Kroonenberg, S. B. 1994: Volcanic origin of Holocene beach ridges along the Caribbean coast of Costa Rica. *Marine Geol.* 120, pp. 13–26.

Nishi, R., Sato, M., and Nakamura, K. 1992: Grain-size distribution of suspended sediments. *Proc. 23rd Coastal Engnr. Conf.* pp. 2293–306.

Nittrouer, C. A., Kuehl, S. A., Demaster, D. J., and Kowsmann, R. O. 1986: The deltaic nature of Amazon shelf sedimentation. *Geol. Soc. Amer. Bull.* 97, pp. 444–58.

Nordstrom, K. F. and Allen, J. R. 1980: Geomorphologically compatible solutions to beach erosion. *Z. Geomorph.* Suppl. Band 34, pp. 142–54.

Nordstrom, K. F. and Jackson, N. L. 1992: Effect of source width and tidal elevation changes on aeolian transport on an estuarine beach. *Sedimentology* 39, pp. 769–78.

Nordstrom, K. F. and Lotstein, E. L. 1989: Perspectives on resource use of dynamic coastal dunes. *Geogr. Rev.* 79, pp. 1–12.

Nossin, J. J. 1965: Analysis of younger beach ridge deposits in eastern Malaya. *Z. Geomorph.* 9, pp. 186–208.

Nott, J. F. 1990: The role of subaerial processes in sea cliff retreat—a south east Australian example. *Z. Geomorph.* 34, pp. 75–85.

Nowell, A. R. M. 1983: The benthic boundary layer and sediment transport. *Rev. Geophys. Space Phys.* 21, pp. 1181–92.

Nummedal, D. 1983: Barrier islands. In *CRC Handbook of Coastal Processes and Erosion*, P. D. Komar (ed.), CRC Press, Boca Raton, Florida, pp. 77–121.

Nunn, P. D. 1993: Role of *Porolithon* algal-ridge growth in the development of the windward coast of Tongatapu Island, Tonga, south Pacific. *Earth Surface Processes and Landforms* 18, pp. 427–39.

Oak, H. L. 1984: The boulder beach: a fundamentally distinct sedimentary assemblage. *Ann. Assoc. Amer. Geogr.* 74, pp. 71–82.

Oberdorfer, J. A. and Buddemeier, R. W. 1988: Climate change: effects on reef island resources. *Proc. 6th Int. Coral Reef Symp., Townsville* 3, pp. 523–7.

O'Brien, M. P. 1969: Equilibrium flow areas of inlets on sandy coasts. *J. Waterw. Harbors Div. Amer. Soc. Civ. Engnr.* 95, pp. 43–52.

O'Connor, B. A. 1987: Short and long term changes in estuary capacity. *J. Geol. Soc. London* 144, pp. 187–95.

O'Connor, B. A. and Yoo, D. 1988: Mean bed friction of combined wave/current flow. *Coastal Engnr.* 12, pp. 1–21.

Oertel, G. F. 1979: Barrier island development during the Holocene recession, southeastern United States. In *Barrier Islands*, S. P. Leatherman (ed.), Academic Press, New York, pp. 273–90.

Offen, G. R. and Kline, S. J. 1975: A proposed model of the bursting process in turbulent boundary layers. *J. Fluid Mech.* 70, pp. 209–28.

Officer, C. B. 1976: *Physical Oceanography of Estuaries*, Wiley, New York.

Officer, C. B. 1981: Physical dynamics of estuarine suspended sediments. *Marine Geol.* 40, pp. 1–14.

Ogden, J. C., Brown, R. A., and Salesky, N. 1973: Grazing by the echinoid *Diadema antillarum* Philippi: formation of halos around West Indian patch reefs. *Science* 182, pp. 715–17.

O'Hare, T. J. and Huntley, D. A. 1994: Bar formation due to wave groups and associated long waves. *Marine Geol.* 116, pp. 313–25.

Oliver, J. and McGinnity, P. 1985: Commercial coral collecting on the Great Barrier Reef. *Proc. 5th Int. Coral Reef Congr., Tahiti* 5, pp. 563–8.

Olson, J. S. and Van Der Maarel, E. 1989: Coastal dunes in Europe: a global view. In *Perspectives in Coastal Dune Management*, F. Van Der Meulen, P. D. Jungerius, and J. Visser (eds.), SPB Academic Publishing, The Hague, The Netherlands, pp. 3–32.

Oltman-Shay, J. and Guza, R. T. 1987: Infragravity edge wave observations on two California beaches. *J. Phys. Oceanogr.* 17, pp. 644–63.

Oltman-Shay, J. and Howd, P. A. 1993: Edge waves on nonplanar bathymetry and alongshore currents: a model and data comparison. *J. Geophys. Res.* 98, pp. 2495–507.

Oost, A. P., de Haas, H., Ijnsen, F., van den Boogert, J. M., and de Boer, P. L. 1993: The 18.6 yr nodal cycle and its impact on tidal sedimentation. *Sed. Geol.* 87, pp. 1–11.

Orford, J. D. 1975: Discrimination of particle zonation on a pebble beach. *Sedimentology* 22, pp. 441–63.

Orford, J. D. 1977: A proposed mechanism for storm beach sedimentation. *Earth Surface Processes* 2, pp. 381–400.

Orford, J. D., Carter, R. W. G., and Jennings, S. C. 1991: Coarse clastic barrier environments: evolution and implications for Quaternary sea level interpretation. *Quat. Int.* 9, pp. 87–104.

Orford, J. D. and Wright, P. 1978: What's in a name?— Descriptive or genetic implications of 'ridge and runnel' topography. *Marine Geol.* 28, pp. M1–M8.

Orme, A. R. 1962: Abandoned and composite sea cliffs in Britain and Ireland. *Irish Geogr.* 4, pp. 279–91.

Orme, A. R. 1990: The instability of Holocene coastal dunes: the case of the Morro Dunes, California. In *Coastal Dunes: Form and Process*, K. F. Nordstrom, N. P. Psuty, and R. W. G. Carter (eds.), Wiley, Chichester, pp. 315–36.

Orme, A. R. and Orme, A. J. 1988: Ridge-and-runnel enigma. *Geogr. Rev.* 78, pp. 169–84.

Orton, G. J. and Reading, H. G. 1993: Variability of deltaic processes in terms of sediment supply, with particular emphasis on grain size. *Sedimentology* 40, pp. 475–512.

Osborne, P. D. and Greenwood, B. 1992a: Frequency dependent cross-shore suspended sediment transport. 1. A non-barred shoreface. *Marine Geol.* 106, pp. 1–24.

Osborne, P. D. and Greenwood, B. 1992b: Frequency dependent cross-shore suspended sediment transport. 2. A barred shoreface. *Marine Geol.* 106, pp. 25–51.

Osborne, P. D. and Greenwood, B. 1993: Sediment suspension under waves and currents: time scales and vertical structure. *Sedimentology* 40, pp. 599–622.

Osborne, P. D., Greenwood, B., and Bowen, A. J. 1990: Cross-shore suspended sediment transport on a non-barred beach: the role of wind waves, infragravity waves and mean flows. *Proc. Can. Coastal Conf.*, M. H. Davies (ed.), National Res. Council Can. (ACOS), Ottawa, pp. 349–61.

Osborne, P. D. and Vincent, C. E. 1993: Dynamics of large and small scale bedforms on a macrotidal shoreface under shoaling and breaking waves. *Marine Geol.* 115, pp. 207–26.

Osborne, R. H., Yeh, C.-C., and Lu, Y. 1991: Grain-shape analysis of littoral and shelf sands, southern California. Coastal Sediments '91, N. C. Kraus, K. J. Gingerich, and D. L. Kriebel (eds.), Amer. Soc. Civ. Engnr., pp. 846–59.

Osterkamp, T. E. and Gosink, J. P. 1984: Observations and analyses of sediment-laden sea ice. In *The Alaskan Beaufort Sea: Ecosystems and Environments*, P. W. Barnes, D. M. Schell, and E. Reimnitz (eds.), Academic Press, Orlando, Florida, pp. 73–93.

Otvos, E. G. 1985: Barrier platforms: northern Gulf of Mexico. *Marine Geol.* 63, pp. 285–305.

Otvos, E. G. and Price, W. A. 1979: Problems of chenier genesis and terminology—an overview. *Marine Geol.* 31, pp. 251–63.

Oueslati, A. 1992: Salt marshes in the Gulf of Gabes (southeastern Tunisia): their morphology and recent dynamics. *J. Coastal Res.* 8, pp. 727–33.

Owens, E. H. 1976: The effects of ice on the littoral zone at Richibucto Head, eastern New Brunswick. *Rev. Géogr. Montréal* 30, pp. 95–104.

Owens, E. H. and Harper, J. R. 1983: Arctic coastal processes: a state-of-knowledge review. *Proc. Can.*

Coastal Conf., B. J. Holden (ed.), National Res. Council Can. (ACROSES), Ottawa, pp. 3–18.

Ozhan, E. 1983: Laboratory study of breaker type effect on longshore sand transport. In *Mechanics of Sediment Transport*, B. M. Sumer and A. Muller (eds.), Proc. Euromech 156/Instanbul 1982, Balkema, Rotterdam, pp. 265–74.

Parkinson, R. W. 1989: Decelerating Holocene sea-level rise and its influence on southwest Florida coastal evolution: a transgressive/regressive stratigraphy. *J. Sed. Petrol.* 59, pp. 960–72.

Parks, J. M. 1989: Beachface dewatering: a new approach to beach stabilization. *The Compass* 66, pp. 65–72.

Parnell, K. E. 1988: Physical process studies in the Great Barrier Reef Marine Park. *Progr. Phys. Geogr.* 12, pp. 209–36.

Paskoff, R. and Kelletat, D. 1991: Introduction: review of coastal problems. *Z. Geomorph.* Suppl. Band 81, pp. 1–13.

Paskoff, R. P. and Sanlaville, P. 1978: Observations géomorphologiques sur les côtes de l'archipel Maltais. *Z. Geomorph.* 22, pp. 310–28.

Paulay, G. and McEdward, L. R. 1990: A simulation model of island reef morphology: the effects of sea level fluctuations, growth, subsidence and erosion. *Coral Reefs* 9, pp. 51–62.

Pedersen, C., Deigaard, R., Fredsøe, J., and Hansen, E. A. 1992: Numerical simulation of sand in plunging breakers. *Proc. 23rd Coastal Engnr. Conf.* pp. 2344–57.

Pejrup, M. 1988: Suspended sediment transport across a tidal flat. *Marine Geol.* 82, pp. 187–98.

Penland, S., Boyd, R., and Suter, J. R. 1988: Transgressive depositional systems of the Mississippi Delta plain: a model for barrier shoreline and shelf sand development. *J. Sed. Petrol.* 58, pp. 932–49.

Penland, S., McBride, R. A., Williams, S. J., Boyd, R., and Suter, J. R. 1991: Effects of sea level rise on the Mississippi River delta plain. *Coastal Sediments '91*, N. C. Kraus, K. J. Gingerich, and D. L. Kriebel (eds.), Amer. Soc. Civ. Engnr., pp. 1248–64.

Penland, S. and Suter, J. R. 1989: The geomorphology of the Mississippi River chenier plain. *Marine Geol.* 90, pp. 231–58.

Peregrine, D. H. 1983: Breaking waves on beaches. *Annual Rev. Fluid Mech.* 15, pp. 149–78.

Peregrine, D. H. and Smith, R. 1979: Nonlinear effects upon waves near caustics. *Phil. Trans. Roy. Soc. London.* A 292, pp. 341–70.

Pérès, J. M. and Picard, J. 1964: Nouveau manuel de bionomie benthique de la mer Méditerranée. *Rec. Travaux Station D'Endoume* 31, pp. 1–137.

Pestrong, R. 1972: Tidal-flat sedimentation at Cooley Landing, southwest San Francisco Bay. *Sed. Geol.* 8, pp. 251–88.

Pethick, J. S. 1969: Drainage in tidal marshes. In *The Coastline of England and Wales*, J. A. Steers (ed.), Cambridge Univ. Press, London, pp. 725–30.

Pethick, J. S. 1974: The distribution of salt pans on tidal salt marshes. *J. Biogeogr.* 1, pp. 57–62.

Pethick, J. S. 1980: Velocity surges and asymmetry in tidal channels. *Estuarine and Coastal Marine Sci.* 11, pp. 331–45.

Pethick, J. S. 1981: Long-term accretion rates on tidal salt marshes. *J. Sed. Petrol.* 51, pp. 571–7.

Pethick, J. S. 1984: *An Introduction to Coastal Geomorphology*, Edward Arnold, London.

Pethick, J. S. 1992: Saltmarsh geomorphology. In *Saltmarshes: Morphodynamics, Conservation and Engineering Significance*, J. R. L. Allen and K. Pye (eds.), Cambridge Univ. Press, Cambridge, pp. 41–62.

Pethick, J. S. 1993: Shoreline adjustments and coastal management: physical and biological processes under accelerated sea-level rise. *Geogr. J.* 159, pp. 162–8.

Phillips, J. D. 1985: Headland-bay beaches revisited: an example from Sandy Hook, New Jersey. *Marine Geol.* 65, pp. 21–31.

Phillips, O. M. 1957: On the generation of waves by turbulent wind. *J. Fluid Mech.* 2, pp. 417–45.

Pichon, M. 1977: Physiography, morphology and ecology of the double barrier reef of north Bohol (Philippines). *Proc. 3rd Int. Coral Reef Symp., Miami* 2, pp. 261–7.

Pierce, J. W. and Colquhoun, D. J. 1970: Holocene evolution of a portion of the North Carolina coast. *Geol. Soc. Amer. Bull.* 81, pp. 3697–714.

Pilkey, O. H. and Wright, H. L. 1988: Seawalls versus beaches. *J. Coastal Res.* Spec. Issue 4, pp. 41–64.

Pilkey, O. H., Young, R. S., Riggs, S. R., Smith, A. W. S., Wu, H., and Pilkey, W. D. 1993: The concept of shoreface profile of equilibrium: a critical review, *J. Coastal Res.* 9, pp. 255–78.

Pillans, B. 1983: Upper Quaternary marine terrace chronology and deformation, south Taranaki, New Zealand. *Geology* 11, pp. 292–7.

Piotrowska, H. 1989: Natural and anthropogenic changes in sand-dunes and the vegetation on the southern Baltic coast. In *Perspectives in Coastal Dune Management*, F. Van Der Meulen, P. D. Jungerius, and J. Visser (eds.), SPB Academic Publishing, The Hague, The Netherlands, pp. 33–40.

Pirazzoli, P. A. 1977: Sea level relative variations in the world during the last 2,000 years. *Z. Geomorph.* 21, pp. 284–96.

Pirazzoli, P. A. 1985: Leeward Islands (Maupiti, Tupai, Bora Bora, Huahine), Society Archipelago. *5th Int. Coral Reef Congr., Tahiti* 1, pp. 17–72.

Pirazzoli, P. A. 1991: *World Atlas of Holocene Sea-Level Changes*, Elsevier, Amsterdam.

Pirazzoli, P. A., Koba, M., Montaggioni, L. F., and Person, A. 1988: Anaa (Tuamotu Islands, central Pacific): an incipient raising atoll? *Marine Geol.* 82, pp. 261–9.

Pirazzoli, P. A. and Montaggioni, L. F. 1985: Lithospheric deformation in French Polynesia (Pacific Ocean) as deduced from Quaternary shorelines. *Proc. 5th Int. Coral Reef Congr., Tahiti* 3, pp. 195–200.

Pirazzoli, P. A., Radtke, U., Hantoro, W. S., Jouannic, C., Hoang, C. T., Causse, C., and Borel Best, M. 1993: A one million-year-long sequence of marine terraces on Sumba Island, Indonesia. *Marine Geol.* 109, pp. 221–36.

Pluis, J. L. A. 1992: Relationships between deflation and near surface wind velocity in a coastal dune blowout. *Earth Surface Processes and Landforms* 17, pp. 663–73.

Pluis, J. L. A. and Winder, B. de 1990: Natural stabilization. In *Dunes of the European Coasts*, W. Bakker, P. D. Jungerius, and J. A. Klijn (eds.), *Catena* Suppl. 18, pp. 195–208.

Porter, J. W. 1985: The maritime weather of Jamaica: its effects on annual carbon budgets of the massive reef-building coral *Montastrea Annularis. Proc. 5th Int. Coral Reef Congr., Tahiti* 6, pp. 363–79.

Postma, G. 1990a: An analysis of the variation in delta architecture. *Terra Nova* 2, pp. 124–30.

Postma, G. 1990b: Depositional architecture and facies of river and fan deltas: a synthesis. In *Coarse-grained Deltas*, A. Colella and D. B. Prior (eds.), *Int. Assoc. Sedimentologists* (Blackwell, Oxford) Spec. Publ. 10, pp. 13–27.

Postma, G. and Nemec, W. 1990: Regressive and transgressive sequences in a raised Holocene gravelly beach, southwestern Crete. *Sedimentology* 37, pp. 907–20.

Postma, H. 1961: Transport and accumulation of suspended matter in the Dutch Wadden Sea. *Netherlands J. Sea Res.* 1, pp. 148–90.

Postma, H. 1980: Sediment transport and sedimentation. In *Chemistry and Biogeochemistry of Estuaries*, E. Olausson and I. Cato (eds.), Wiley, Chichester, pp. 153–86.

Poulos, S., Papadopoulos, A., and Collins, M. B. 1994: Deltaic progradation in Thermaikos Bay, northern Greece and its socio-economical implications. *Ocean and Coastal Management* 22, pp. 229–47.

Powell, K. A., Quinn, P. A., and Greated, C. A. 1992: Shingle beach profiles and wave kinematics. *Proc. 23rd Coastal Engnr. Conf.* pp. 2358–69.

Powell, R. D. and Molnia, B. F. 1989: Glacimarine sedimentary processes, facies and morphology of the south–southeast Alaska shelf and fiords. *Marine Geol.* 85, pp. 359–90.

Powers, M. C. 1953: A new roundness scale for sedimentary particles. *J. Sed. Petrol.* 23, pp. 117–19.

Pratt, R. M. and Dill, R. F. 1974: Deep eustatic terrace levels: further speculations. *Geology* 2, pp. 155–9.

Preobrazhensky, B. V. 1993: *Contemporary Reefs*, Balkema, Rotterdam.

Prior, D. B. 1977: Coastal mudslide morphology and processes on Eocene clays in Denmark. *Geografisk Tidsskr.* 76, pp. 14–33.

Prior, D. B. and Bornhold, B. D. 1990: The underwater development of Holocene fan deltas. In *Coarse-grained Deltas*, A. Colella and D. B. Prior (eds.), *Int. Assoc. Sedimentologists* (Blackwell, Oxford) Spec. Publ. 10, pp. 75–90.

Prior, D. B. and Coleman, J. M. 1979: Submarine landslides—geometry and nomenclature. *Z. Geomorph.* 23, pp. 415–26.

Prior, D. B. and Eve, R. M. 1975: Coastal landslide morphology at Røsnaes, Denmark. *Geografisk Tidsskr.* 74, pp. 12–20.

Prior, D. B. and Ho, C. 1972: Coastal and mountain slope instability on the islands of St. Lucia and Barbados. *Engnr. Geol.* 6, pp. 1–18.

Prior, D. B. and Renwick, W. H. 1980: Landslide morphology and processes on some coastal slopes in Denmark and France. *Z. Geomorph.* Suppl. Band 34, pp. 63–86.

Pritchard, D. W. 1967a: What is an estuary: physical viewpoint. In *Estuaries*, G. H. Lauff (ed.), *Amer. Assoc. Adv. Sci. Publ.* 83, pp. 3–5.

Pritchard, D. W. 1967b: Observations of circulation in coastal plain estuaries. In *Estuaries*, G. H. Lauff (ed.), *Amer. Assoc. Adv. Sci. Publ.* 83, pp. 37–44.

Prost, M. T. 1989: Coastal dynamics and chenier sands in French Guiana. In *Quaternary of South America and Antarctic Peninsula*, J. Rabassa (ed.), Balkema, Rotterdam, pp. 191–218.

Pryor, W. A. 1975: Biogenic sedimentation and alteration of argillaceous sediments in shallow marine environments. *Geol. Soc. Amer. Bull.* 86, pp. 1244–54.

Psuty, N. P. 1967: *The Geomorphology of Beach Ridges in Tabasco, Mexico*, Louisiana State Press, Baton Rouge.

Psuty, N. P. 1990: Foredune mobility and stability, Fire Island, New York. In *Coastal Dunes: Form and Process*, K. F. Nordstrom, N. P. Psuty, and R. W. G. Carter (eds.), Wiley, Chichester, pp. 159–76.

Psuty, N. P. 1992: Spatial variation in coastal foredune development. In *Coastal Dunes*, R. W. G. Carter, T. G. F. Curtis, and M. J. Sheehy-Skeffington (eds.), Proc. 3rd European Dune Congr., Balkema, Rotterdam, pp. 3–13.

Pugh, D. T. 1987: *Tides, Surges and Mean Sea-Level*, Wiley, Chichester.

Purdy, E. G. 1974: Reef configurations: cause and effect. In *Reefs in Time and Space*, L. F. Laporte (ed.), *Soc. Econ. Paleont. Mineral.* Spec. Publ. 18, pp. 9–76.

Purdy, E. G. 1981: Evolution of the Maldive atolls, Indian Ocean. *Proc. 4th Int. Coral Reef Symp., Manila* 1, p. 659.

Putrevu, U. and Svendsen, I. A. 1992: Shear instability of longshore currents: a numerical study. *J. Geophys. Res.* 97, pp. 7283–303.

Putrevu, U. and Svendsen, I. A. 1993: Vertical structure of the undertow outside the surf zone. *J. Geophys. Res.* 98, pp. 22707–16.

Pye, K. 1983: Early post-depositional modification of

aeolian dune sands. In *Eolian Sediments and Processes*, M. E. Brookfield and T. Ahlbrandt (eds.), Elsevier, Amsterdam, pp. 197–221.

Pye, K. 1990: Physical and human influences on coastal dune development between the Ribble and Mersey estuaries, northwest England. In *Coastal Dunes: Form and Process*, K. F. Nordstrom, N. P. Psuty, and R. W. G. Carter (eds.), Wiley, Chichester, pp. 339–59.

Pye, K. 1992: Saltmarshes on the barrier coastline of north Norfolk, eastern England. In *Saltmarshes: Morphodynamics, Conservation and Engineering Significance*, J. R. L. Allen and K. Pye (eds.), Cambridge Univ. Press, Cambridge, pp. 148–78.

Pye, K. and Tsoar, H. 1990: *Aeolian Sand and Sand Dunes*, Unwin Hyman, London.

Qinshang, Y., Shiyuan, X., and Xusheng, S. 1989: Holocene cheniers in the Yangtze Delta, China. *Marine Geol.* 90, pp. 337–43.

Quick, M. C. 1991: Onshore–offshore sediment transport on beaches. *Coastal Engnr.* 15, pp. 313–32.

Quick, M. C. and Ametepe, J. 1991: Relationship between longshore and cross-shore transport. *Coastal Sediments '91*, N. C. Kraus, K. J. Gingerich, and D. L. Kriebel (eds.), Amer. Soc. Civ. Engnr., pp. 184–96.

Quick, M. C. and Har, B. C. 1985: Criteria for onshore–offshore sediment movement on beaches. *Proc. Can. Coastal Conf.*, D. L. Forbes (ed.), National Res. Council Can. (ACROSES), Ottawa, pp. 257–69.

Quigley, R. M. and Gelinas, P. J. 1976: Soil mechanics aspects of shoreline erosion. *Geosci. Can.* 3, pp. 169–73.

Quigley, R. M., Gelinas, P. J., Bou, W. T., and Packer, R. W. 1977: Cyclic erosion—instability relationships: Lake Erie north shore bluffs. *Can. Geotechn. J.* 14, pp. 310–23.

Quinn, T. M. and Merriam, D. F. 1988: Evolution of Florida Bay islands from a supratidal precursor: evidence from westernmost Bob Allen Key and Sid Key. *J. Geol.* 96, pp. 375–81.

Raichlen, F. and Papanicolaou, P. 1988: Some characteristics of breaking waves. *Proc. 21st Coastal Engnr. Conf.* pp. 377–97.

Rampino, M. R. and Sanders, J. E. 1981: Evolution of the barrier islands of southern Long Island, New York. *Sedimentology* 28, pp. 37–47.

Ramsey, K. E., Penland, S., and Roberts, H. H. 1991: Implications of accelerated sea-level rise on Louisiana coastal environments. *Coastal Sediments '91*, N. C. Kraus, K. J. Gingerich, and D. L. Kriebel (eds.), Amer. Soc. Civ. Engnr., pp. 1207–22.

Ranwell, D. S. 1958: Movement of vegetated sand dunes at Newborough Warren, Anglesey. *J. Ecol.* 46, pp. 83–100.

Ranwell, D. S. 1972: *Ecology of Salt Marshes and Sand Dunes*, Chapman & Hall, London.

Rasmussen, K. R. 1989: Some aspects of flow over coastal dunes. In *Coastal Sand Dunes*, C. H. Gimingham, W. Ritchie, B. B. Willetts, and A. J. Willis (eds.), *Proc. Roy. Soc. Edinb.* B 96, pp. 129–47.

Raudkivi, A. J. 1988: The roughness height under waves. *J. Hydraul. Res.* 26, pp. 569–84.

Rearic, D. M., Barnes, P. W., and Reimnitz, E. 1990: Bulldozing and resuspension of shallow-shelf sediment by ice keels: implications for Arctic sediment transport trajectories. *Marine Geol.* 91, pp. 133–47.

Reed, D. J. 1987: Temporal sampling and discharge asymmetry in salt marsh creeks. *Estuarine, Coastal and Shelf Sci.* 25, pp. 459–66.

Reed, D. J. 1988: Sediment dynamics and deposition in a retreating coastal salt marsh. *Estuarine, Coastal and Shelf Sci.* 26, pp. 67–79.

Reed, D. J. 1991: Ponds and bays: natural processes of coastal marsh erosion in the Mississippi Deltaic Plain, Louisiana, U.S.A. *Z. Geomorph.* Suppl. band 81, pp. 41–51.

Reed, D. J. 1995: The response of coastal marshes to sea-level rise: survival or submergence? *Earth Surface Processes and Landforms* 20, pp. 39–48.

Refaat, H. E. A. A. and Tsuchiya, Y. 1992: Formation and reduction processes of river deltas; theory and experiments. *Proc. 23rd Coastal Engnr. Conf.* pp. 2772–85.

Reimnitz, E. and Barnes, P. W. 1987: Sea-ice influence on Arctic coastal retreat. *Coastal Sediments '87*, N. C. Kraus (ed.), Amer. Soc. Civ. Engnr., pp. 1578–91.

Reimnitz, E., Barnes, P. W., and Harper, J. R. 1990: A review of beach nourishment from ice transport of shoreface materials, Beaufort Sea, Alaska. *J. Coastal Res.* 6, pp. 439–70.

Reimnitz, E., Kempema, E. W., and Barnes, P. W. 1987: Anchor ice, seabed freezing, and sediment dynamics in shallow Arctic seas. *J. Geophys. Res.* 92, pp. 14671–8.

Reimnitz, E. and Maurer, D. K. 1979: Effects of storm surges on the Beaufort Sea coast, northern Alaska. *Arctic* 32, pp. 329–44.

Reimold, R. J. 1977: Mangals and salt marshes of eastern United States. In *Wet Coastal Ecosystems*, V. J. Chapman (ed.), Elsevier, Amsterdam, pp. 157–66.

Reinson, G. E. and Rosen, P. S. 1982: Preservation of ice-formed features in a subarctic sandy beach sequence: geologic implications. *J. Sed. Petrol.* 52, pp. 463–71.

Rex, R. W. 1964: Arctic beaches, Barrow Alaska. In *Papers in Marine Geology, Shepard Commemorative Volume*, R. L. Miller (ed.), Macmillan, New York, pp. 384–400.

Rhodes, E. G. 1982: Depositional model for a chenier plain, Gulf of Carpentaria, Australia. *Sedimentology* 29, pp. 201–21.

Ribberink, J. S. and Al-Salem, A. 1990: Bedforms, sediment concentrations and sediment transport in simulated wave conditions. *Proc. 22nd Coastal Engnr. Conf.* pp. 2318–31.

Rice, M. A. 1991: Grain shape effects on aeolian sediment transport. In *Aeolian Grain Transport 1 Mechanics*, O. E. Barndorff-Nielsen and B. B. Willetts (eds.), Acta Mechanica Suppl. 1, Springer-Verlag, Wien, pp. 159–66.

Rice, M. A., Willetts, B. B., and McEwan, I. K. 1995: An experimental study of multiple grain-size ejecta produced by collisions of saltating grains with a flat bed. *Sedimentology* 42, pp. 695–706.

Richards, D. R. and Granat, M. A. 1986: Salinity redistributions in deepened estuaries. In *Estuarine Variability*, D. A. Wolfe (ed.), Academic Press, New York, pp. 463–82.

Richardson, J. F. and Jerónimo, M. A. da S. 1979: Velocity-voidage relations for sedimentation and fluidisation. *Chem. Engnr. Sci.* 34, pp. 1419–22.

Richmond, B. M. and Sallenger, A. H. 1984: Cross-shore transport of bimodal sands. *Proc. 19th Coastal Engnr. Conf.* pp. 1997–2008.

Ridd, P., Sandstrom, M. W., and Wolanski, E. 1988: Outwelling from tropical tidal salt flats. *Estuarine, Coastal and Shelf Sci.* 26, pp. 243–53.

Ritchie, W. 1967: The machair of South Uist. *Scot. Geogr. Mag.* 83, pp. 161–73.

Ritchie, W. 1986: Anomalous east coast machair in the Uists, Outer Hebrides. In *Essays for Professor R. E. H. Mellor*, W. Ritchie, J. C. Stone, and A. S. Mather (eds.), Dept. Geogr., Univ. Aberdeen, pp. 383–9.

Ritchie, W. 1992: Scottish landform examples—4. Coastal parabolic dunes of the Sands of Forvie. *Scot. Geogr. Mag.* 108, pp. 39–44.

Roberts, H. H., Lugo, A., Carter, B., and Simms, M. 1988: Across-reef flux and shallow subsurface hydrology in modern coral reefs. *Proc. 6th Int. Coral Reef Symp., Townsville* 2, pp. 509–15.

Roberts, H. H., Suhayda, J. N., and Coleman, J. M. 1980: Sediment deformation on low-angle slopes: Mississippi River Delta. In *Thresholds in Geomorphology*, D. R. Coates and J. D. Vitek (eds.), Allen & Unwin, London, pp. 131–67.

Robertson-Rintoul, M. J. 1990: A quantitative analysis of the near-surface wind flow pattern over coastal parabolic dunes. In *Coastal Dunes: Form and Process*, K. F. Nordstrom, N. P. Psuty, and R. W. G. Carter (eds.), Wiley, Chichester, pp. 57–78.

Robertson-Rintoul, M. J. and Ritchie, W. 1990: The geomorphology of coastal sand dunes in Scotland: a review. In *Dunes of the European Coasts*, W. Bakker, P. D. Jungerius, and J. A. Klijn (eds.), *Catena* Suppl. 18, pp. 41–9.

Robinson, D. A. and Jerwood, L. C. 1987a: Sub-aerial weathering of chalk shore platforms during harsh winters in southeast England. *Marine Geol.* 77, pp. 1–14.

Robinson, D. A. and Jerwood, L. C. 1987b: Frost and salt weathering of chalk shore platforms near Brighton, Sussex, U.K. *Trans. Inst. Brit. Geogr.* 12, pp. 217–26.

Robinson, L. A. 1977a: Marine erosive processes at the cliff foot. *Marine Geol.* 23, pp. 257–71.

Robinson, L. A. 1977b: Erosive processes on the shore platform of northeast Yorkshire, England. *Marine Geol.* 23, pp. 339–61.

Robinson, L. A. 1977c: The morphology and development of the northeast Yorkshire shore platform. *Marine Geol.* 23, pp. 237–55.

Rodolfo, K. S. 1975: The Irrawaddy Delta: Tertiary setting and modern offshore sedimentation. In *Deltas—Models for Exploration*, M. L. Broussard (ed.), Houston Geol. Soc., Houston, Texas, pp. 339–56.

Roelvink, J. A. and Brøker, I. 1993: Cross-shore profile models. *Coastal Engnr.* 21, pp. 163–91.

Roelvink, J. A. and Stive, M. J. F. 1989: Bar-generating cross-shore flow mechanisms on a beach. *J. Geophys. Res.* 94, pp. 4785–800.

Rogers, C. S. 1985: Degradation of Caribbean and western Atlantic coral reefs and decline of associated fisheries. *Proc. 5th Int. Coral Reef Congr., Tahiti* 6, pp. 491–6.

Rogers, C. S., McLain, L., and Zullo, E. 1988: Damage to coral reefs in Virgin Islands National Park and biosphere reserve from recreational activities. *Proc. 6th Int. Coral Reef Symp., Townsville* 2, pp. 405–10.

Rosati, J. D., Gingerich, K. J., Kraus, N. C., McKee Smith, J., and Beach, R. A. 1991: Longshore sand transport rate distributions measured in Lake Michigan. *Coastal Sediments '91*, N. C. Kraus, K. J. Gingerich, and D. L. Kriebel (eds.), Amer. Soc. Civ. Engnr., pp. 156–69.

Rosen, P. S. 1975: Origin and processes of cuspate spit shorelines. In *Estuarine Research II*, L. E. Cronin (ed.), Academic Press, New York, pp. 77–92.

Rosen, P. S. 1979: Aeolian dynamics of a barrier island system. In *Barrier Islands*, S. P. Leatherman (ed.), Academic Press, New York, pp. 81–98.

Rosen, P. S. 1980: Coastal environments of the Makkovik Region, Labrador. In *The Coastline of Canada*, S. B. McCann (ed.), *Geol. Surv. Can. Paper* 80–10, pp. 267–80.

Rosen, P. S. and Leach, K. 1987: Sediment accumulation forms, Thompson Island, Boston Harbor, Massachusetts. In *Glaciated Coasts*, D. M. Fitzgerald and P. S. Rosen (eds.), Academic Press, San Diego, pp. 233–50.

Roux, J. P. le 1992: Settling velocity of spheres: a new approach. *Sed. Geol.* 81, pp. 11–16.

Roy, P. S. 1984: New South Wales estuaries: their origin and evolution. In *Coastal Geomorphology in Australia*, B. G. Thom (ed.), Academic Press, Sydney, pp. 99–121.

Roy, P. S. and Connell, J. 1991: Climatic change and the future of atoll states. *J. Coastal Res.* 7, pp. 1057–75.

Roy, P. S., Thom, B. G., and Wright, L. D. 1980: Holocene sequences on an embayed high-energy coast: an evolutionary model. *Sed. Geol.* 26, pp. 1–19.

Ruddiman, W. F., Raymo, M. E., Martinson, D. G., Clement, B. M., and Backman, J. 1989: Pleistocene evolution: northern hemisphere ice sheets and North Atlantic Ocean. *Paleoceanogr.* 4, pp. 353–412.

Ruddiman, W. F., Raymo, M., and McIntyre, A. 1986: Matuyama 41,000-year cycles: North Atlantic Ocean and northern hemisphere ice sheets. *Earth Planetary Sci. Letters* 80, pp. 117–29.

Rumpel, D. A. 1985: Successive aeolian saltation: studies of idealized collisions. *Sedimentology* 32, pp. 267–80.

Russell, P., Davidson, M., Huntley, D., Cramp, A., Hardisty, J., and Lloyd, G. 1991: The British beach and nearshore dynamics (B-Band) programme. *Coastal Sediments '91*, N. C. Kraus, K. J. Gingerich, and D. L. Kriebel (eds.), Amer. Soc. Civ. Engnr., pp. 371–84.

Russell, R. J. 1942: Geomorphology of the Rhone Delta. *Ann. Assoc. Amer. Geogr.* 32, pp. 149–254.

Russell, R. J. and McIntire, W. G. 1965: Beach cusps. *Geol. Soc. Amer. Bull.* 76, pp. 307–20.

Rutin, J. 1992: Geomorphic activity of rabbits on a coastal sand dune, De Blink Dunes, The Netherlands. *Earth Surface Processes and Landforms* 17, pp. 85–94.

Ruz, M.-H. 1989: Les Crêtes littorales sableuses dunifiées du sud-est de l'Irlande comparées à celles du massif Armoricain (France). *Z. Geomorph. Suppl.* Band 73, pp. 73–86.

Ruz, M.-H. and Allard, M. 1995: Sedimentary structures of cold-climate coastal dunes, eastern Hudson Bay, Canada. *Sedimentology* 42, pp. 725–34.

Ruz, M.-H., Héquette, A., and Hill, P. R. 1992: A model of coastal evolution in a transgressed thermokarst topography, Canadian Beaufort Sea. *Marine Geol.* 106, pp. 251–78.

Safriel, U. N. 1974: Vermetid gastropods and intertidal reefs in Israel and Bermuda. *Science* 186, pp. 1113–15.

Sallenger, A. H. 1979: Beach-cusp formation. *Marine Geol.* 29, pp. 23–37.

Sallenger, A. H., Holman, R. A., and Birkemeier, W. A. 1985: Storm-induced response of a nearshore-bar system. *Marine Geol.* 64, pp. 237–57.

Sallenger, A. H. and Howd, P. A. 1989: Nearshore bars and the break-point hypothesis. *Coastal Engnr.* 12, pp. 301–13.

Salomon, J. C. and Allen, G. P. 1983: Rôle sedimentologique de la marée dans les estuares a fort marnage. Compagnie Français des Petroles. Notes and memoires 18, pp. 35–44.

Salvat, B. 1981: Preservation of coral reefs: scientific whim or economic necessity? Past, present and future. *Proc. 4th Int. Coral Reef Symp., Manila* 1, pp. 225–9.

Salvat, B. and Soegiarto, A. (chairmen) 1981: Section 1 Reef and man. Sub section 3 Resource management and marine parks. *Proc. 4th Int. Coral Reef Symp., Manila* 1, pp. 217–332.

Sanders, N. K. 1968: The development of Tasmanian shore platforms. Unpubl. Ph.D. thesis, Univ. Tasmania.

Sanders, N. K. 1970: The production of horizontal high-tidal shore platforms. *Austr. Nat. Hist.* 16, pp. 315–19.

Sanford, L. P. and Boicourt, W. C. 1990: Wind-forced salt intrusion into a tributary estuary. *J. Geophys. Res.* 95, pp. 13357–71.

Sanford, L. P. and Halka, J. P. 1993: Assessing the paradigm of mutually exclusive erosion and deposition of mud, with examples from upper Chesapeake Bay. *Marine Geol.* 114, pp. 37–57.

Santema, P. 1966: The effect of tides, coastal currents, waves and storm surges on the natural conditions prevailing in deltas. In *Scientific Problems of the Humid Tropical Zone Deltas and their Implications*, UNESCO, Paris, Proc. Dacca Symp. Feb./Mar. 1964, pp. 109–13.

Sarma, G. V. S., Satyakumar, P., Varma, K. U. M., Rao, E. N. D., and Subba Rao, M. 1993: Clay minerals of the Sarada–Varaha Estuary, east coast of India. *J. Coastal Res.* 9, pp. 885–94.

Sarre, R. D. 1987: Aeolian sand transport. *Progr. Phys. Geogr.* 11, pp. 157–82.

Sarre, R. D. 1989: The morphological significance of vegetation and relief on coastal foredune processes. *Z. Geomorph.* Suppl. Band 73, pp. 17–31.

Sarre, R. D. and Chancey, C. C. 1990: Size segregation during aeolian saltation on sand dunes. *Sedimentology* 37, pp. 357–65.

Sato, M., Kuroki, K., and Shinohara, T. 1992: A field experiment on the formation of beach cusps. *Proc. 23th Coastal Engnr. Conf.* pp. 2205–18.

Sauer, J. D. 1959: *Coastal Pioneer Plants of the Caribbean and Gulf of Mexico*. Geogr. Branch, Off. Naval Res. Washington, DC, Project NR 388–047.

Sauer, J. D. 1962: Effects of recent tropical cyclones on the coastal vegetation of Mauritius. *J. Ecol.* 50, pp. 275–90.

Sawaragi, T. and Deguchi, I. 1980: On–offshore sediment transport rate in the surf zone. *Proc. 17th Coastal Engnr. Conf.* pp. 1194–214.

Schaffer, H. A. and Jonsson, I. G. 1992: Edge waves revisited. *Coastal Engnr.* 16, pp. 349–68.

Scheidegger, A. E. 1970: *Theoretical Geomorphology*, Springer-Verlag, New York.

Schneider, J. 1976: Biological and inorganic factors in the destruction of limestone coasts. *Contrib. Sedimentology* 6, pp. 1–112.

Schneider, J. and Torunski, H. 1983: Biokarst on limestone coasts, morphogenesis and sediment production. *Marine Ecol.* 4, pp. 45–63.

Scholl, D. W. 1968: Mangrove swamps: geology and sedimentology. In *The Encyclopedia of Geomorphology*, R. W. Fairbridge (ed.), Dowden, Hutchinson, and Ross, Stroudsburg, Penn., pp. 683–8.

Schou, A. 1952: Direction determining influence of the wind on shoreline simplification and coastal dune forms. *Proc. 17th Congr. Int. Geogr. Union*, Washington, DC, pp. 370–3.

Schroeder, J. H. 1981: Man versus reef in the Sudan: threats, destruction, protection. *Proc. 4th Int. Coral Reef Symp., Manila* 1, pp. 253–7.

Schroeder, J. H. and Nasr, D. H. 1983: The fringing reefs of Port Sudan, Sudan: 1 morphology—sedimentology—zonation. *Essener Geogr. Arb.* 6, pp. 29–44.

Schubel, J. R. and Carter, H. H. 1984: The estuary as a filter for fine-grained suspended sediment. In *The Estuary as a Filter*, V. S. Kennedy (ed.), Academic Press, New York, pp. 81–105.

Schubel, J. R., Shen, H.-T., and Park, M.-J. 1986: Comparative analysis of estuaries bordering the Yellow Sea. In *Estuarine Variability*, D. A. Wolfe (ed.), Academic Press, New York, pp. 43–62.

Schuhmacher, H. and Zibrowius, H. 1985: What is hermatypic? A redefinition of ecological groups in corals and other organisms. *Coral Reefs* 4, pp. 1–9.

Schulke, H. 1968: Quelques types de dépressions fermeés littorales et supralittorales lieés à l'action destructive de la mer (Bretange, Corse, Asturies). *Norois* 57, pp. 23–42.

Schwartz, M. L. 1972: Theoretical approach to the origin of beach cusps. *Geol. Soc. Amer. Bull.* 83, pp. 1115–16.

Schwiderski, E. W. 1979: Global ocean tides, part II: the semidiurnal principal lunar tide (M_2), atlas of tidal charts and maps. *Naval Surface Weapons Center* (Dahlgren, Virginia), Rept. No. NSWC TR 79–414.

Scoffin, T. P. 1970: The trapping and binding of subtidal carbonate sediments by marine vegetation in Bimini Lagoon, Bahamas. *J. Sed. Petrol.* 40, pp. 249–73.

Scoffin, T. P. 1993: The geological effects of hurricanes on coral reefs and the interpretation of storm deposits. *Coral Reefs* 12, pp. 203–21.

Scoffin, T. P. and Dixon, J. E. 1983: The distribution and structure of coral reefs: one hundred years since Darwin. *Biol. J. Linnean Soc.* 20, pp. 11–38.

Scoffin, T. P., Stearn, C. W., Boucher, D., Frydl, P., Hawkins, C. M., Hunter, I. G., and MacGeachy, J. K. 1980: Calcium carbonate budget of a fringing reef on the west coast of Barbados, part II. *Bull. Mar. Sci.* 30, pp. 475–508.

Scoffin, T. P. and Stoddart, D. R. 1983: Beachrock and intertidal cements. In *Chemical Sediments and Geomorphology*, A. S. Goudie and K. Pye (eds.), Academic Press, London, pp. 401–25.

Scott, G. A. J. and Rotondo, G. M. 1983a: A model for the development of types of atolls and volcanic islands on the Pacific lithospheric plate. *Atoll Res. Bull.* 260.

Scott, G. A. J. and Rotondo, G. M. 1983b: A model to explain the differences between Pacific Plate island-atoll types. *Coral Reefs* 1, pp. 139–50.

Semeniuk, V. 1980: Mangrove zonation along an eroding coastline in King Sound, north-western Australia. *J. Ecol.* 68, pp. 789–812.

Semeniuk, V. 1981: Sedimentology and the stratigraphic sequence of a tropical tidal flat, north-western Australia. *Sed. Geol.* 29, pp. 195–221.

Semeniuk, V. 1994: Predicting the effect of sea-level rise on mangroves in northwestern Australia. *J. Coastal Res.* 10, pp. 1050–76.

Seppala, M. and Linde, K. 1978: Wind tunnel studies of ripple formation. *Geografiska Annlr.* A 60, pp. 29–42.

Sexton, D. J., Dowdeswell, J. A., Solheim, A., and Elverhøi, A. 1992: Seismic architecture and sedimentation in northwest Spitsbergen fjords. *Marine Geol.* 103, pp. 53–68.

Seymour, R. J. and Aubrey, D. G. 1985: Rhythmic beach cusp formation: a conceptual synthesis. *Marine Geol.* 65, pp. 289–304.

Seymour, R. J. and Castel, D. 1989: Modelling cross-shore transport. In *Nearshore Sediment Transport*, R. J. Seymour (ed.), Plenum Press, New York, pp. 387–401.

Sha, L. P. and Van den Berg, J. H. 1993: Variation in ebb-tidal delta geometry along the coast of the Netherlands and the German Bight. *J. Coastal Res.* 9, pp. 730–46.

Shackleton, N. J. 1987: Oxygen isotopes, ice volume and sea level. *Quat. Sci. Rev.* 6, pp. 183–90.

Shackleton, N. J., Berger, A., and Peltier, W. R. 1990: An alternate astronomical calibration of the lower Pleistocene timescale based on ODP site 677. *Trans. Roy. Soc. Edinb., Earth Sci.* 81, pp. 251–61.

Shackleton, N. J. and Opdyke, N. D. 1973: Oxygen isotope and palaeomagnetic stratigraphy of equatorial Pacific core V28-238: oxygen isotope temperatures and ice volumes on a 10^5-year scale. *Quat. Res.* 3, pp. 39–55.

Shackleton, N. J. and Opdyke, N. D. 1976: Oxygen-isotope and paleomagnetic stratigraphy of Pacific core V28-239: late Pliocene to latest Pleistocene. In *Investigations of late Quaternary Paleoceanography and Paleoclimatology*, R. M. Cline and J. D. Hays (eds.), *Geol. Soc. Amer. Mem.* 145, pp. 449–64.

Sharma, P., Gardner, L. R., Moore, W. S., and Bollinger, M. S. 1987: Sedimentation and bioturbation in a salt marsh as revealed by ^{210}Pb, ^{137}Cs, and ^{7}Be studies. *Limnol. Oceanogr.* 32, pp. 313–26.

Sharp, R. P. 1963: Wind ripples. *J. Geol.* 71, pp. 617–636.

Shaw, J. 1985: Beach morphodynamics of an Atlantic coast embayment: Runkerry Strand, County Antrim. *Irish Geogr.* 18, pp. 51–8.

Shaw, J. and Forbes, D. L. 1992: Barriers, barrier platforms, and spillover deposits in St. George's Bay, Newfoundland: paraglacial sedimentation on the flanks of a deep coastal basin. *Marine Geol.* 105, pp. 119–40.

Shaw, J., Taylor, R. B., and Forbes, D. L. 1993: Impact of the Holocene transgression on the Atlantic coastline of Nova Scotia. *Géogr. Phys. Quat.* 47, pp. 221–38.

Sheldon, R. W. 1968: Sedimentation in the estuary of the

River Crouch, Essex, England. *Limnol. Oceanogr.* 13, pp. 72–83.

Shepard, F. P. and Kuhn, G. G. 1983: History of sea arches and remnant stacks of La Jolla, California, and their bearing on similar features elsewhere. *Marine Geol.* 51, pp. 139–61.

Shepherd, M. J. 1987: Sandy beach ridge system profiles as indicators of changing coastal processes. *Proc. 14th N.Z. Geogr. Conf.*, pp. 106–12.

Sherman, C. E., Glenn, C. R., Jones, A. T., Burnett, W. C., and Schwarcz, H. P. 1993: New evidence of two highstands of the sea during the last interglacial, oxygen isotope substage 5e. *Geology* 21, pp. 1079–82.

Sherman, D. J. and Bauer, B. O. 1993: Dynamics of beach-dune systems. *Progr. Phys. Geogr.* 17, pp. 413–47.

Sherman, D. J. and Greenwood, B. 1985: Wind shear and shore-parallel flows in the surf zone. *Proc. Can. Coastal Conf.*, D. L. Forbes (ed.), National Res. Council Can. (ACROSES), Ottawa, pp. 53–66.

Sherman, D. J. and Hotta, S. 1990: Aeolian sediment transport: theory and measurement. In *Coastal Dunes: Form and Process*, K. F. Nordstrom, N. P. Psuty, and R. W. G. Carter (eds.), Wiley, Chichester, pp. 17–37.

Sherman, D. J., Nordstrom, K. F., Jackson, N. L., and Allen, J. R. 1994: Sediment mixing-depths on a low-energy reflective beach. *J. Coastal Res.* 10, pp. 297–305.

Sherman, D. J., Orford, J. D., and Carter, R. W. G. 1993a: Development of cusp-related, gravel size and shape facies at Malin Head, Ireland. *Sedimentology* 40, pp. 1139–52.

Sherman, D. J., Short, A. D., and Takeda, I. 1993b: Sediment mixing depth and bedform migration in RIP channels. *J. Coastal Res.* Spec. Issue 15, pp. 39–48.

Shi, N. C. and Larsen, L. H. 1983/1984: Reverse sediment transport induced by amplitude-modulated waves. *Marine Geol.* 54, pp. 181–200.

Shibayama, T. and Horikawa, K. 1982: Sediment transport and beach transformation. *Proc. 18th Coastal Engnr. Conf.* pp. 1439–58.

Shibayama, T., Okayasu, A., and Kashiwagi, M. 1992: Long period wave and suspended sand transport in the surf zone. *Proc. 23rd Coastal Engnr. Conf.* pp. 2438–49.

Shigemura, T., Takasugi, J., and Komiya, Y. 1984: Formation of tombolo at the west coast of Iwo-Jima. *Proc. 19th Coastal Engnr. Conf.* pp. 1403–19.

Shih, S.-M. and Komar, P. D. 1994: Sediments, beach morphology and sea cliff erosion within an Oregon coast littoral cell. *J. Coastal Res.* 10, pp. 144–57.

Shinn, E. A. 1980: Geologic history of Grecian Rocks, Key Largo Coral Reef Marine sanctuary. *Bull. Mar. Sci.* 30, pp. 646–56.

Shinn, E. A. 1983: Tidal flat environment. In *Carbonate Depositional Environments*, P. A. Scholle, D. G.

Bebout, and C. H. Moore (eds.), *Amer. Assoc. Petrol. Geol.* pp. 172–210.

Shinn, E. A., Hudson, J. H., Robbin, D. M., and Lidz, B. 1981: Spurs and grooves revisted: construction versus erosion Looe Key Reef, Florida. *Proc. 4th Int. Coral Reef Symp., Manila.* 1, pp. 475–83.

Shlemon, R. J. 1975: Subaqueous delta formation—Atchafalaya Bay, Louisiana. In *Deltas—Models for Exploration*, M. L. Broussard (ed.), Houston Geol. Soc., Houston, Texas, pp. 209–21.

Short, A. D. 1975: Multiple offshore bars and standing waves. *J. Geophys. Res.* 80, pp. 3838–40.

Short, A. D. 1978: Wave power and beach-stages: a global model. *Proc. 16th Coastal Engnr. Conf.* pp. 1145–62.

Short, A. D. 1979: Three dimensional beach-stage model. *J. Geol.* 87, pp. 553–71.

Short, A. D. 1988a: The South Australian coast and Holocene sea-level transgression. *Geogr. Rev.* 78, pp. 119–36.

Short, A. D. 1988b: Holocene coastal dune formation in southern Australia. *Sed. Geol.* 55, pp. 121–42.

Short, A. D. 1991: Macro-meso tidal beach morphodynamics—an overview. *J. Coastal Res.* 7, pp. 417–36.

Short, A. D. and Aagaard, T. 1993: Single and multi-bar beach change models. *J. Coastal Res.* Spec. Issue 15, pp. 141–57.

Short, A. D. and Hesp, P. A. 1982: Wave, beach and dune inter-actions in southeastern Australia. *Marine Geol.* 48, pp. 259–84.

Shulmeister, J. and Head, J. 1993: Aspects of the emplacement, evolution and ^{14}C chronology of ridges on a coastal spit, Groote Eylandt, Northern Australia. *Marine Geol.* 111, pp. 159–69.

Shulmeister, J. and Kirk, R. M. 1993: Evolution of a mixed sand and gravel barrier system in North Canterbury, New Zealand, during Holocene sea-level rise and still-stand. *Sed. Geol.* 87, pp. 215–35.

Silva, E. F. 1972: Geomorphological observations and generalizations on the coasts of the South Shetland Islands and Antarctic Peninsula. In *Antarctic Geology and Geophysics*, R. J. Adie (ed.), Universitetsforlaget, Oslo, pp. 99–103.

Silvester, R. 1962: Sediment movement around the coastlines of the world. *Conf. Civ. Engnr. Problems Overseas*, Inst. Civ. Engnr. London, pp. 289–304.

Silvester, R. 1976: Headland defense of coasts. *Proc. 15th Coastal Engnr. Conf.* pp. 1394–406.

Simpson, J. H. and Turrell, W. R. 1986: Covergent fronts in the circulation of tidal estuaries. In *Estuarine Variability*, D. A. Wolfe (ed.), Academic Press, New York, pp. 139–52.

Skafel, M. G. 1995: Laboratory measurement of nearshore velocities and erosion of cohesive sediment (till) shorelines. *Coastal Engnr.* 24, pp. 343–9.

Skafel, M. G. and Bishop, C. T. 1994: Flume experiments on the erosion of till shores by waves. *Coastal Engnr.* 23, pp. 329–48.

Sleath, J. F. A. 1974a: Velocities above rough bed in oscillatory flow. *J. Waterw. Harbors Coastal Div. Amer. Soc. Civ. Engnr.* 100, pp. 287–304.

Sleath, J. F. A. 1974b: Stability of laminar flow at sea bed. *J. Waterw. Harbors Coastal Div. Amer. Soc. Civ. Engnr.* 100, pp. 105–22.

Sleath, J. F. A. 1975: Transition in oscillatory flow over rippled beds. *Proc. Inst. Civ. Engnr.* 59, pp. 309–22.

Sleath, J. F. A. 1978: Measurements of bed load in oscillatory flow. *J. Waterw. Port Coastal Ocean Div. Amer. Soc. Civ. Engnr.* 104, pp. 291–307.

Sleath, J. F. A. 1984: *Sea Bed Mechanics*, Wiley, Chichester.

Sleath, J. F. A. 1990: Seabed boundary layers. In *The Sea* Vol. 9 Part B (*Ocean Engnr. Sci.*), B. Le Méhauté and D. M. Hanes (eds.), Wiley, New York, pp. 693–727.

Slingerland, R. L. 1977: The effects of entrainment on the hydraulic equivalence relationships of light and heavy minerals in sands. *J. Sed. Petrol.* 47, pp. 753–70.

Smith, A. W. S. 1994: The coastal engineering literature and the field engineer. *J. Coastal Res.* 10, pp. iii–viii.

Smith, E. R. and Kraus, N. C. 1992: Laboratory study of wave transformation on barred beach profiles. *Proc. 23rd Coastal Engnr. Conf.* pp. 630–43.

Smith, J. M., Larson, M., and Kraus, N. C. 1993: Longshore current on a barred beach: field measurements and calculation. *J. Geophysical Res.* 98, pp. 22717–31.

Smith, S. V. 1978: Coral-reef area and the contributions of reefs to processes and resources of the world's oceans. *Nature* 273, pp. 225–6.

Smith, S. V. 1981: The Houtman Abrolhos Islands: carbon metabolism of coral reefs at high latitude. *Limnol. Oceanogr.* 26, pp. 612–21.

Smith, S. V. and Kinsey, D. W. 1976: Calcium carbonate production, coral reef growth, and sea level change. *Science* 194, pp. 937–9.

So, C. L. 1965: Coastal platforms of the Isle of Thanet, Kent. *Trans. Inst. Brit. Geogr.* 37, pp. 147–56.

Sondl, I., J. Mladen., and Pravdić, V. 1995: Sedimentation in a disequilibrium river-dominated estuary: the Raša River Estuary (Adriatic Sea, Croatia). *Sedimentology* 42, pp. 769–82.

Sonu, C. J. 1972: Field observation of nearshore circulation and meandering currents. *J. Geophys. Res.* 77, pp. 3232–47.

Sonu, C. J. 1973: Three-dimensional beach changes. *J. Geol.* 81, pp. 42–64.

Sonu, C. J. and van Beek, J. L. 1971: Systematic beach changes on the Outer Banks, North Carolina. *J. Geol.* 79, pp. 416–25.

Sørensen, M. 1991: An analytic model of wind-blown sand transport. In *Aeolian Grain Transport 1 Mechanics*, O. E. Barndorff-Nielsen and B. B. Willetts (eds.), Acta Mechanica Suppl. 1, Springer-Verlag, Wien, pp. 67–81.

Soulsby, R. L. 1983: The bottom boundary layer of shelf seas. In *Physical Oceanography of Coastal and Shelf Seas*, B. Johns (ed.), Elsevier, Amsterdam, pp. 189–266.

Soulsby, R. L. 1991: Sediment transport by strong wave-plus-current flows. *Coastal Sediments '91*, N. C. Kraus, K. J. Gingerich, and D. L. Kriebel (eds.), Amer. Soc. Civ. Engnr., pp. 405–17.

Soulsby, R. L., Hamm, L., Klopman, G., Myrhaug, D., Simons, R. R., and Thomas, G. P. 1993: Wave-current interaction within and outside the bottom boundary layer. *Coastal Engnr.* 21, pp. 41–69.

Speer, P. E. and Aubrey, D. G. 1985: A study of non-linear tidal propagation in shallow inlet/estuarine systems, part II theory. *Estuarine, Coastal and Shelf Sci.* 21, pp. 207–24.

Spenceley, A. P. 1976: Unvegetated saline tidal flats in north Queensland. *J. Tropical Geogr.* 42, pp. 78–85.

Spenceley, A. P. 1977: The role of pneumatophores in sedimentary processes. *Marine Geol.* 24, pp. M31–37.

Spencer, T. 1985: Rates of karst processes on raised reef limestones and their implications for coral reef histories. *Proc. 5th Int. Coral Reef Congr., Tahiti* 6, pp. 629–34.

Spencer, T. 1988a: Limestone coastal morphology. *Progr. Phys. Geogr.* 12, pp. 66–101.

Spencer, T. 1988b: Coastal biogeomorphology. In *Biogeomorphology*, H. A. Viles (ed.), Blackwell, Oxford, pp. 255–318.

Spencer, T. 1995: Potentialities, uncertainties and complexities in the response of coral reefs to future sea-level rise. *Earth Surface Processes and Landforms* 20, pp. 49–64.

Spencer, T., Stoddart, D. R., and Woodroffe, C. D. 1987: Island uplift and lithospheric flexure: observations and cautions from the South Pacific. *Z. Geomorph.* Suppl. Band 63, pp. 87–102.

Stearn, C. W. 1982: The shapes of Paleozoic and modern reef-builders: a critical review. *Paleobiol.* 8, pp. 228–41.

Steers, J. A. 1977: Physiography. In *Wet Coastal Ecosystems*, V. J. Chapman (ed.), Elsevier, Amsterdam, pp. 31–60.

Steers, J. A. and Stoddart, D. R. 1977: The origin of fringing reefs, barrier reefs, and atolls. In *Biology and Geology of Coral Reefs*, O. A. Jones and R. Endean (eds.), Academic Press, New York, 4, pp. 21–57.

Steidtmann, J. R. 1982: Size-density sorting of sand-size spheres during deposition from bedload transport and implications concerning hydraulic equivalence. *Sedimentology* 29, pp. 877–83.

Sternberg, R. W. 1971: Measurements of incipient motion of sediment particles in the marine environment. *Marine Geol.* 10, pp. 113–19.

Sternberg, R. W., Shi, N. C., and Downing, J. P. 1989: Suspended sediment measurements A. Continuous measurements of suspended sediment. In *Nearshore Sediment Transport*, R. J. Seymour (ed.), Plenum Press, New York, pp. 231–57.

Stevenson, J. C., Ward, L. G., and Kearney, M. S. 1986: Vertical accretion in marshes with varying rates of sea level rise. In *Estuarine Variability*, D. A. Wolfe (ed.), Academic Press, New York, pp. 241–59.

Stevenson, J. C., Ward, L. G., and Kearney, M. S. 1988: Sediment transport and trapping in marsh systems: implications of tidal flux studies. *Marine Geol.* 80, pp. 37–59.

Stoddart, D. R. 1965*a*: British Honduras cays and the Low Wooded Island problem. *Trans. Inst. Brit. Geogr.* 36, pp. 131–47.

Stoddart, D. R. 1965*b*: The shape of atolls. *Marine Geol.* 3, pp. 369–83.

Stoddart, D. R. 1969: Ecology and morphology of recent coral reefs. *Biol. Rev.* 44, pp. 433–98.

Stoddart, D. R. 1971*a*: Coral reefs and islands and catastrophic storms. In J. A. Steers (ed.), *Applied Coastal Geomorphology*, Macmillan, London, pp. 155–97.

Stoddart, D. R. 1971*b*: Environment and history in Indian Ocean reef morphology. In *Symposia of the Zoological Society of London*, D. R. Stoddart and M. Yonge (eds.), Academic Press, London, 28, pp. 3–38.

Stoddart, D. R. 1973: Coral reefs of the Indian Ocean. In *Biology and Geology of Coral Reefs*, O. A. Jones and R. Endean (eds.), Academic Press, New York, 1, Geology, pp. 51–92.

Stoddart, D. R. 1974: Post-hurricane changes on the British Honduras reefs: re-survey of 1972. *Proc. 2nd Int. Coral Reef Symp., Brisbane* 2, pp. 473–83.

Stoddart, D. R. 1976: Community and crisis in the reef community. *Micronesica* 12, pp. 1–9.

Stoddart, D. R. 1980: Mangroves as successional stages, inner reefs of the northern Great Barrier Reef. *J. Biogeogr.* 7, pp. 269–84.

Stoddart, D. R. 1985: Hurricane effects on coral reefs. *Proc. 5th Int. Coral Reef Congr., Tahiti*, 3, pp. 349–50.

Stoddart, D. R. 1990: Coral reefs and islands and predicted sea-level rise. *Progr. Phys. Geogr.* 14, pp. 521–36.

Stoddart, D. R., McLean, R. F., and Hopley, D. 1978: Geomorphology of reef islands, northern Great Barrier Reef. *Phil. Trans. Roy. Soc. London* B 284, pp. 39–61.

Stoddart, D. R., Reed, D. J., and French, J. R. 1989: Understanding salt-marsh accretion, Scolt Head Island, Norfolk, England. *Estuaries* 12, pp. 228–36.

Stoddart, D. R. and Scoffin, T. P. 1979: Micro-atolls: review of form, origin and terminology. *Atoll Res. Bull.* 224.

Stoddart, D. R., Spencer, T., and Scoffin, T. P. 1985: Reef growth and karst erosion on Mangaia, Cook Islands: a reinterpretation. *Z. Geomorph.* Suppl. Band 57, pp. 121–40.

Stoddart, D. R. and Steers, J. A. 1977: The nature and origin of coral reef islands. In *Biology and Geology of Coral Reefs*, O. A. Jones and R. Endean (eds.), Academic Press, New York, 4, Geology, pp. 59–105.

Stoddart, D. R. and Taylor, J. D. 1971: Geography and ecology of Diego Garcia Atoll, Chagos Archipelago. *Atoll Res. Bull.* 149.

Strecker, M. R., Bloom, A. L., Gilpin, L. M., and Taylor, F. W. 1986: Karst morphology of uplifted Quaternary coral limestone terraces: Santo Island, Vanuatu. *Z. Geomorph.* 30, pp. 387–405.

Streif, H. 1989: Barrier islands, tidal flats, and coastal marshes resulting from a relative rise of sea level in East Frisia on the German North Sea coast. In *Coastal Lowlands*, W. J. M. Van der Linden, S. A. P. L. Cloetingh, J. P. K. Kaasschieter, W. J. E. Van de Graaf, J. Vandenberghe, and J. A. M. Van der Gun (eds.), Kluwer, Dordrecht, Proc. Symp. Coastal Lowlands, Roy. Geol. Mining Soc. Netherlands, pp. 213–23.

Stride, A. H. (ed.), 1982: *Offshore Tidal Sands*, Chapman & Hall, London.

Stumpf, R. P. 1983: The process of sedimentation on the surface of a salt marsh. *Estuarine, Coastal and Shelf Sci.* 17, pp. 495–508.

Sudara, S. and Nateekarnchanalap, S. 1988: Impact of tourism development on the reef in Thailand. *Proc. 6th Int. Coral Reef Symp., Townsville* 2, pp. 273–8.

Sunamura, T. 1973: Coastal cliff erosion due to waves— field investigations and laboratory experiments. *J. Faculty Eng. Univ. Tokyo* 32, pp. 1–86.

Sunamura, T. 1977: A relationship between wave-induced cliff erosion and erosive force of wave. *J. Geol.* 85, pp. 613–18.

Sunamura, T. 1981: Bedforms generated in a laboratory wave tank. *Sci. Rept. Inst. Geosci., Univ. Tsukuba (Japan)* Section A, 2, pp. 31–43.

Sunamura, T. 1988: Beach morphologies and their change. In *Nearshore Dynamics and Coastal Processes*, K. Horikawa (ed.), Univ. Tokyo Press, Tokyo, pp. 136–52.

Sunamura, T. 1989: Sandy beach geomorphology elucidated by laboratory modeling. In *Applications in Coastal Modeling*, V. C. Lakhan and A. S. Trenhaile (eds.), Elsevier, Amsterdam, pp. 159–213.

Sunamura, T. 1992: *The Geomorphology of Rocky Coasts*, Wiley, Chichester.

Sunamura, T. and Horikawa, K. 1974: Two-dimensional beach transformation due to waves. *Proc. 14th Coastal Engnr. Conf.* pp. 920–38.

Sunamura, T. and Kraus, N. C. 1985: Prediction of average mixing depth of sediment in the surf zone. *Marine Geol.* 62, pp. 1–12.

Sunamura, T. and Maruyama, K. 1987: Wave-induced geomorphic response of eroding beaches—with special reference to seaward migrating bars. *Coastal Sediments '87*, N. C. Kraus (ed.), Amer. Soc. Civ. Engnr., pp. 788–801.

Sunamura, T. and Mizuno, O. 1987: A study on depositional shoreline forms behind an island. *Annual Rept. Inst. Geosci. Univ. Tsukuba* 13, pp. 71–3.

Sunamura, T. and Takeda, I. 1993: Bar movement and shoreline change: predictive relations. *J. Coastal Res.* Spec. Issue 15, pp. 125–40.

Sutherland, R. A. and Lee, C.-T. 1994: Application of

the log-hyperbolic distribution to Hawai'ian beach sands. *J. Coastal Res.* 10, pp. 251–62.

Suzuki, T., Yakahashi, K., Sunamura, Y., and Terada, M. 1970: Rock mechanisms on the formation of washboard-like relief on wave-cut benches at Arasaki, Miura Peninsula, Japan. *Geogr. Rev. Japan.* 43, pp. 211–22.

Svendsen, I. A. and Hansen, J. B. 1988: Cross-shore currents in surf-zone modelling. *Coastal Engnr.* 12, pp. 23–42.

Svendsen, I. A. and Madsen, P. A. 1984: A turbulent bore on a beach. *J. Fluid Mech.* 148, pp. 73–96.

SWAMP Group 1985: *Ocean Wave Modeling*, Plenum Press, New York.

Sweet, M. L. and Kocurek, G. 1990: An empirical model of aeolian dune lee-face airflow. *Sedimentology* 37, pp. 1023–38.

Swift, D. J. P., Niedoroda, A. W., Vincent, C. E., and Hopkins, T. S. 1985: Barrier island evolution, middle Atlantic shelf, USA. Part 1: shoreface dynamics. *Marine Geol.* 63, pp. 331–61.

SWIM Group 1985: A shallow water intercomparison of three numerical wave prediction models (SWIM). *Quart. J. Roy. Meteor. Soc.* 111, pp. 1087–112.

Symonds, G., Black, K. O., and Young, I. R. 1995: Wave-driven flow over shallow reefs. *J. Geophys. Res.* 100, pp. 2639–48.

Symonds, G. and Bowen, A. J. 1984: Interactions of nearshore bars with incoming wave groups. *J. Geophys. Res.* 89, pp. 1953–9.

Symonds, G. and Huntley, D. A. 1980: Waves and currents over nearshore bar systems. *Proc. Can. Coastal Conf.*, National Res. Council Can. (ACROSES), Ottawa, pp. 64–78.

Symonds, G., Huntley, D. A., and Bowen, A. J. 1982: Two-dimensional surf beat: long wave generation by a time-varying breakpoint. *J. Geophys. Res.* 87, pp. 492–8.

Syvitski, J. P. M., Burrell, D. C., and Skei, J. M. 1987: *Fjords, Processes and Products*, Springer-Verlag, New York, 379 pp.

Takahashi, T. 1977: *Shore Platforms in Southwestern Japan—Geomorphological Study* (Coastal Landform Study Soc. of Southwestern Japan, Osaka).

Takahashi, T., Koba, M., and Kan, H. 1988: Relationship between reef growth and sea level on the northwest coast of Kume Island, The Ryukyus: data from drill holes on the Holocene coral reef. *Proc. 6th Int. Coral Reef Symp., Townsville* 3, pp. 491–6.

Takeda, I. 1984: Beach changes by waves. *Sci. Rept. Inst. Geosci., Univ. Tsukuba (Japan)* Sect. A, 5, pp. 29–63.

Takeda, I., Terasaki, T., and Sunamura, T. 1986: Formation and spacing of beach cusps on a laboratory beach. *Annual Rept. Inst. Geosci., Univ. Tsukuba (Japan)* 12, pp. 55–8.

Tallent, J. R., Yamashita, T., and Tsuchiya, Y. 1990: Transformation characteristics of breaking water waves. In *Water Wave Kinematics*, Tørum, A. and Gudmestad, O. T. (eds.), Kluwer, Dordrecht, pp. 509–23.

Tanaka, H. and Shuto, N. 1992: Field investigation at a mouth of small river. *Proc. 23rd Coastal Engnr. Conf.* pp. 2486–99.

Tanner, K. and Walsh, P. 1984: *Hallsands: a Pictorial History*, Tanner and Walsh, Woodleigh, Kingsbridge.

Tanner, W. F. 1960: Bases for coastal classification. *Southeastern Geol.* 2, pp. 13–22.

Tanner, W. F. 1977: Discontinuities in the 'a-b-c...' model. In *Coastal Sedimentology*, W. F. Tanner (ed.), Dept. Geol., Florida State Univ., Tallahassee, Florida, pp. 69–80.

Tanner, W. F. 1988: Beach ridge data and sea level history from the Americas. *J. Coastal Res.* 4, pp. 81–91.

Tanner, W. F. and Stapor, F. W. 1972: Precise control of wave run-up in beach ridge construction. *Z. Geomorph.* 16, pp. 393–9.

Tatavarti, R. V. S. N., Huntley, D. A., and Bowen, A. J. 1988: Incoming and outgoing wave interactions on beaches. *Proc. 21st Coastal Engnr. Conf.* pp. 136–50.

Tayama, R. 1952: Coral reefs in the South Seas. *Bull. Hydrograph. Off.*, Maritime Safety Agency, Tokyo, Japan 11, Publ. 941 (In English, pp. 182–292).

Taylor, R. B. and McCann, S. B. 1976: The effect of sea and nearshore ice on coastal processes in Canadian Arctic Archipelago. *Rev. Géogr. Montréal* 30, pp. 123–32.

Taylor, R. B. and McCann, S. B. 1983: Coastal depositional landforms in northern Canada. In *Shorelines and Isostasy*, D. E. Smith and A. Dawson (eds.), *Inst. Brit. Geogr. (Academic Press, London)* Special Publ. 16, pp. 53–75.

Taylor, R. B., Wittmann, S. L., Milne, M. J., and Kober, S. M. 1985: Beach morphology and coastal changes at selected sites, mainland Nova Scotia. *Geol. Surv. Can. Paper* 85–12.

Teas, H. J. (ed.) 1983: *Biology and Ecology of Mangroves*, Junk, The Hague, The Netherlands.

Terchunian, A. V. 1990: Performance of beachface dewatering; the Stabeach system at Sailfish Point (Stuart), Florida. *Proc. National Conf. Beach Preservation Technology.* pp. 185–201.

Terpstra, P. D. and Chrzastowski, M. J. 1992: Geometric trends in the evolution of a small log-spiral embayment on the Illinois shore of Lake Michigan. *J. Coastal Res.* 8, pp. 603–17.

Terzaghi, K. 1962: Stability of steep slopes on hard unweathered rock. *Géotechnique* 12, pp. 251–70.

Thom, B. G. 1964: Origin of sand beach ridges. *Austr. J. Sci.* 26, pp. 351–2.

Thom, B. G. 1967: Mangrove ecology and deltaic geomorphology: Tabasco, Mexico. *J. Ecol.* 55, pp. 301–43.

Thom, B. G. 1983: Transgressive and regressive strati-graphies of coastal sand barriers in southeast Australia. *Marine Geol.* 56, pp. 137–58.

Thom, B. G., Bowman, G. M., and Roy, P. S. 1981: Late Quaternary evolution of coastal sand barriers, Port Stephens–Myall Lakes area, central New South Wales, Australia. *Quat. Res.* 15, pp. 345–64.

Thom, B. G. and Hall, W. 1991: Behaviour of beach profiles during accretion and erosion dominated periods. *Earth Surface Processes and Landforms* 16, pp. 113–27.

Thom, B. G., Orme, G. R., and Polach, H. A. 1978: Drilling investigation of Bewick and Stapleton Islands. *Phil. Trans. Roy. Soc. London* A 291, pp. 37–54.

Thom, B. G., Wright, L. D., and Coleman, J. M. 1975: Mangrove ecology and deltaic-estuarine geomorphology: Cambridge Gulf–Ord River, Western Australia. *J. Ecol.* 63, pp. 203–32.

Thomas, M. C., Wiltshire, R. J., and Williams, A. T. 1995: The use of Fourier descriptors in the classification of particle shape. *Sedimentology* 42, pp. 635–45.

Thompson, R. W. 1968: Tidal flat sedimentation on the Colorado River Delta, northwestern Gulf of California, *Geol. Soc. Amer. Mem.* 107.

Thorne, P. D., Williams, J. J., and Heathershaw, A. D. 1989: *In situ* acoustic measurements of marine gravel threshold and transport. *Sedimentology* 36, pp. 61–74.

Thornton, E. B. and Kim, C. S. 1993: Longshore current and wave height modulation at tidal frequency inside the surf zone. *J. Geophysical Res.* 98, pp. 16509–19.

Thurber, D. L., Broecker, W. S., Blanchard, R. L., and Potranz, H. A. 1965: Uranium series ages of Pacific atoll corals. *Science* 149, pp. 55–8.

Tippie, V. K. 1984: An environmental characterization of Chesapeake Bay and a framework for action. In *The Estuary as a Filter*, V. S. Kennedy (ed.), Academic Press, New York, pp. 467–87.

Tjia, H. D. 1985: Notching by abrasion on a limestone coast. *Z. Geomorph.* 29, pp. 367–72.

Tomlinson, P. B. 1986: *The Botany of Mangroves*, Cambridge Univ. Press, Cambridge.

Tooley, M. J. and Jelgersma, S. (eds.) 1992: Impacts of Sea-Level Rise on European Coastal Lowlands. *Inst. Brit. Geogr. (Blackwell, Oxford)* Spec. Publ. 27.

Tracey, J. I., Ladd, H. S., and Hoffmeister, J. E. 1948: Reefs of Bikini, Marshall Islands. *Geol. Soc. Amer. Bull.* 59, pp. 861–78.

Trask, C. B. and Hand, B. M. 1985: Differential transport of fall-equivalent sand grains, Lake Ontario, New York. *J. Sed. Petrol.* 55, pp. 226–34.

Trenhaile, A. S. 1972: The shore platforms of the Vale of Glamorgan, Wales. *Trans. Inst. Brit. Geogr.* 56, pp. 127–44.

Trenhaile, A. S. 1974: The geometry of shore platforms in England and Wales. *Trans. Inst. Brit. Geogr.* 62, pp. 129–42.

Trenhaile, A. S. 1978: The shore platforms of Gaspé, Québec. *Ann. Assoc. Amer. Geogr.* 68, pp. 95–114.

Trenhaile, A. S. 1980: Shore platforms: a neglected coastal feature. *Progr. Phys. Geogr.* 4, pp. 1–23.

Trenhaile, A. S. 1983*a*: The development of shore platforms in high latitudes. In *Shorelines and Isostasy*, D. E. Smith and A. G. Dawson (eds.), *Inst. Brit. Geogr.* (Academic Press, London) Spec. Publ. 16, pp. 77–93.

Trenhaile, A. S. 1983*b*: The width of shore platforms; a theoretical approach. *Geografiska Annlr.* 65A, pp. 147–58.

Trenhaile, A. S. 1987: *The Geomorphology of Rock Coasts*, Cavendish—Oxford Univ. Press, Oxford.

Trenhaile, A. S. 1989: Sea level oscillations and the development of rock coasts. In *Applications in Coastal Modeling*, V. C. Lakhan and A. S. Trenhaile (eds.), Elsevier, Amsterdam, pp. 271–95.

Trenhaile, A. S. 1990: *The Geomorphology of Canada*, Oxford Univ. Press, Toronto.

Trenhaile, A. S. and Byrne, M. L. 1986: A theoretical investigation of rock coasts with particular reference to shore platforms. *Geografiska Annlr.* A 68, pp. 1–14.

Trenhaile, A. S. and Dumala, R. 1978: The geomorphology and origin of Point Pelee, southwestern Ontario. *Can. J. Earth Sci.* 15, pp. 963–70.

Trenhaile, A. S. and Layzell, M. G. J. 1981: Shore platform morphology and the tidal duration factor. *Trans. Inst. Brit. Geogr.* 6, pp. 82–102.

Trenhaile, A. S. and Mercan, D. W. 1984: Frost weathering and the saturation of coastal rocks. *Earth Surface Processes and Landforms* 9, pp. 321–31.

Trenhaile, A. S., Van Der Nol, L. V., and La Valle, P. D. in press: Sand grain roundness and transport in the swash zone. *J. Coastal Res.*

Trowbridge, J. H. and Kineke, G. C. 1994: Structure and dynamics of fluid muds on the Amazon continental shelf. *J. Geophys. Res.* 99, pp. 865–74.

Trudgill, S. T. 1976: The marine erosion of limestone on Aldabra Atoll, Indian Ocean. *Z. Geomorph.* Suppl. Band 26, pp. 164–200.

Trudgill, S. T. 1987: Bioerosion of intertidal limestone, Co. Clare, Eire—3: zonation, process and form. *Marine Geol.* 74, pp. 111–21.

Tsoar, H. 1990: Trends in the development of sand dunes along the southeastern Mediterranean coast. In *Dunes of the European Coasts*, W. Bakker, P. D. Jungerius, and J. A. Klijn (eds.), *Catena* Suppl. 18, pp. 51–60.

Tsoar, H. and Blumberg, D. 1991: The effect of sea cliffs on inland encroachment of aeolian sand. In *Aeolian Grain Transport 2 The Erosional Environment*, O. E. Barndorff-Nielsen and B. B. Willetts (eds.), Acta Mechanica Suppl. 2, Springer-Verlag, Wien, pp. 131–46.

Tsujimoto, H. 1987: Dynamic conditions for shore platform initiation. *Sci. Rept. Inst. Geosci. Univ. Tsukuba* (Japan) A 8, pp. 45–93.

Tubbs, C. R. 1977: Muddy foreshores. In *The Coastline*, R. S. K. Barnes (ed.), Wiley, London, pp. 83–92.

Tucker, M. E. 1973: The sedimentary environments of tropical African estuaries: Freetown Peninsula, Sierra Leone. *Geol. Mijnbouw*, 52, pp. 203–15.

Turmel, R. J. and Swanson, R. G. 1976: The development of Rodriguez Bank, a Holocene mudbank in the Florida reef tract. *J. Sed. Petrol.* 46, pp. 497–518.

Turner, I. L. 1993: The total water content of sandy beaches. *J. Coastal Res.* Spec. Issue 15, pp. 11–26.

Turner, I. L. 1995: Simulating the influence of groundwater seepage on sediment transported by the sweep of the swash zone across macro-tidal beaches. *Marine Geol.* 125, pp. 153–74.

Tye, R. S. and Coleman, J. M. 1989*a*: Evolution of Atchafalaya lacustrine deltas, south-central Louisiana. *Sed. Geol.* 65, pp. 95–112.

Tye, R. S. and Coleman, J. M. 1989*b*: Depositional processes and stratigraphy of fluvially dominated lacustrine deltas: Mississippi delta plain. *J. Sed. Petrol.* 59, pp. 973–96.

Uncles, R. J., Barton, M. L., and Stephens, J. A. 1994: Seasonal variability of fine-sediment concentrations in the turbidity maximum region of the Tamar Estuary. *Estuarine, Coastal and Shelf Sci.* 38, pp. 19–39.

Urish, D. W. 1989: The effect of groundwater on beach erosion. *Coastal Zone '89*, Amer. Soc. Civ. Engnr., pp. 4613.

Ustach, J. F., Kirby-Smith, W. W., and Barber, R. T. 1986: Effect of watershed modification on a small coastal plain estuary. In *Estuarine Variability*, D. A. Wolfe (ed.), Academic Press, New York, pp. 177–92.

Uusinoka, R. and Matti, E. 1979: On weathering depressions and their occurrence in Finland. *Terra* 91, pp. 81–6.

Vagin, N. F. 1970: The laws of ice processes in deltas. In *Hydrology of Deltas*, Gentbrugge, Paris, Proc. Bucharest Symp. May 1969, IASH/AIHS—UNESCO, ii. 296–304.

Vale, C. and Sundby, B. 1987: Suspended sediment fluctuations in the Tagus Estuary on semi-diurnal and fortnightly time scales. *Estuarine, Coastal and Shelf Sci.* 25, pp. 495–508.

Valencia, M. J. 1977: Christmas Island, Pacific Ocean: reconnaissance geologic observations. *Atoll Res. Bull.* 197.

Valentine, J. W. 1971: Plate tectonics and shallow marine diversity and endemism, an actualistic model. *Systematic Zoology.* 20, pp. 253–64.

Van Bohemen, H. D. and Meesters, H. J. N. 1992: Ecological engineering and coastal defence. In *Coastal Dunes*, R. W. G. Carter, T. G. F. Curtis, and M. J. Sheehy-Skeffington (eds.), Proc. 3rd European Dune Congr., Balkema, Rotterdam, pp. 369–78.

van den Berg, J. H. 1977: Morphodynamic development and preservation of physical sedimentary structures in two prograding recent ridge and runnel beaches along the Dutch coast. *Geol. Mijnbouw* 56, pp. 185–202.

Van Der Ancker, J. A. M., Jungerius, P. D., and Mur, L. R. 1985: The role of algae in the stabilization of coastal dune blowouts. *Earth Surface Processes and Landforms* 10, pp. 189–92.

Van Der Meulen, F. and Jungerius, P. D., 1989*a*: Landscape development in Dutch coastal dunes: the breakdown and restoration of geomorphological and geohydrological processes. In *Coastal Sand Dunes*, C. H. Gimingham, W. Ritchie, B. B. Willetts, and A. J. Willis (eds.), *Proc. Roy. Soc. Edinb.* B 96, pp. 219–29.

Van Der Meulen, F. and Jungerius, P. D., 1989*b*: The decision environment of dynamic dune management. In *Perspectives in Coastal Dune Management*, F. Van Der Meulen, P. D. Jungerius, and J. Visser (eds.), SPB Academic Publishing, The Hague, The Netherlands, pp. 133–40.

Van Der Meulen, F., Jungerius, P. D., and Visser, J. (eds.) 1989: *Perspectives in Coastal Dune Management*, SPB Academic Publishing, The Hague, The Netherlands.

Van Der Meulen, F. and Van Der Maarel, E. 1989: Coastal defense alternatives and nature development perspectives. In *Perspectives in Coastal Dune Management*, F. Van Der Meulen, P. D. Jungerius, and J. Visser (eds.), SPB Academic Publishing, The Hague, The Netherlands, pp. 183–95.

Van Dijk, H. W. J. 1989: Ecological impact of drinking-water production in Dutch coastal dunes. In *Perspectives in Coastal Dune Management*, F. Van Der Meulen, P. D. Jungerius, and J. Visser (eds.), SPB Academic Publishing, The Hague, The Netherlands, pp. 163–82.

Van Eerdt, M. M. 1987: The influence of basic soil and vegetation parameters on salt marsh cliff strength. In *International Geomorphology*, V. Gardiner (ed.), Wiley, Chichester, 1, pp. 1073–86.

van Gelder, A., van den Berg, J. H., Cheng, G., and Xue, C. 1994: Overbank and channelfill deposits of the modern Yellow River delta. *Sed. Geol.* 90, pp. 293–305.

Van Heerden, I. Ll. and Roberts, H. H. 1988: Facies development of Atchafalaya Delta, Louisiana: a modern bayhead delta. *Amer. Assoc. Petrol. Geol. Bull.* 72, pp. 439–53.

Van Leussen, W. 1988: Aggregation of particles, settling velocity of mud flocs: a review. In *Physical Processes in Estuaries*, J. Dronkers and W. Van Leussen (eds.), Springer-Verlag, Berlin, pp. 347–403.

Van Leussen, W. and van Velzen, E. 1989: High concentration suspensions: their origin and importance in Dutch estuaries and coastal waters. *J. Coastal Res.* Spec. Issue 5, pp. 1–22.

Vann, J. H. 1980: Shoreline changes in mangrove areas. *Z. Geomorph.* Suppl. Band 34, pp. 255–61.

van Rijn, L. C. 1989: *Handbook Sediment Transport by Currents and Waves*, Delft Hydraul. Rept. H 461, Delft, The Netherlands.

Van Straaten, L. M. J. U. and Kuenen, Ph. H. 1957: Accumulation of fine grained sediments in the Dutch Wadden Sea. *Geol. Mijnbouw* 19, pp. 329–54.

Van 'T Hof, T. 1985: The economic benefits of marine parks and protected areas in the Caribbean region. *Proc. 5th Int. Coral Reef Congr., Tahiti* 6, pp. 551–6.

Veron, J. E. N. 1985: Aspects of the biogeography of hermatypic corals. *Proc. 5th Int. Coral Reef Congr., Tahiti* 4, pp. 83–8.

Vincent, C. E. and Green, M. O. 1990: Field measurements of the suspended sand concentration profiles and fluxes and of the resuspension coefficient y_o '*gamma*' over a rippled bed. *J. Geophys. Res.* 95, pp. 11591–601.

Vincent, C. E., Hanes, D. M., and Bowen, A. J. 1991: Acoustic measurements of suspended sand on the shoreface and the control of concentration by bed roughness. *Marine Geol.* 96, pp. 1–18.

Vincent, C. E., Young, R. A., and Swift, D. J. P. 1981: Bed-load transport under waves and currents. *Marine Geol.* 39, pp. M71–80.

Vita-Finzi, C. and Cornelius, P. F. S. 1973: Cliff sapping by molluscs. *J. Sed. Petrol.* 43, pp. 31–2.

Vitale, P. 1979: Sand bed friction factors for oscillatory flows. *J. Waterw. Port, Coastal Ocean Div. Amer. Soc. Civ. Engnr.* 105, pp. 229–45.

Volker, A. 1966: Tentative classification and comparison with deltas of other climatic regions. In *Scientific Problems of the Humid Tropical Zone Deltas and their Implications*, UNESCO, Paris, Proc. Dacca Symp. Feb./Mar. 1964, pp. 399–408.

Vongvisessomjai, S. 1984: Oscillatory ripple geometry. *J. Hydraul. Engnr. Amer. Soc. Civ. Engnr.* 110, pp. 247–66.

Waddell, E. 1976: Swash-groundwater-beach profile interactions. *Beach and Nearshore Sedimentation*, R. A. Davis and R. L. Ethington (eds.), *Soc. Econ. Paleont. Mineral.* Spec. Publ. 24, pp. 115–25.

Waddell, E. 1980: Wave forcing of beach groundwater. *Proc. 17th Coastal Engnr. Conf.* pp. 1436–52.

Walker, H. J. 1991: Bluff erosion at Barrow and Wainwright, Arctic Alaska. *Z. Geomorph.* Suppl. band 81, pp. 53–61.

Walker, J. R., Everts, C. H., Schmelig, S., and Demirel, V. 1991: Observations of a tidal inlet on a shingle beach. *Coastal Sediments '91*, N. C. Kraus, K. J. Gingerich, and D. L. Kriebel (eds.), Amer. Soc. Civ. Engnr., pp. 975–96.

Walsh, G. E. 1974: Mangroves: a review. In *Ecology of Halophytes*, R. J. Reimold and W. H. Queen (eds.), Academic Press, London, pp. 51–174.

Walton, T. L. and Chiu, T. Y. 1979: A review of analytical techniques to solve the sand transport equation and some simplified solutions. *Coastal Structures 79*, Amer. Soc. Civ. Engnr. pp. 809–37.

WAMDI Group 1988: The WAM model—a third generation ocean wave prediction model. *J. Phys. Oceanogr.* 18, pp. 1775–810.

Wang, F. C. 1984: The dynamics of a river-bay-delta system. *J. Geophys. Res.* 89, pp. 8054–60.

Wang, Y. 1983: The mudflat system of China. *Can. J. Fish. Aquat. Sci.* 40 (Suppl. 1), pp. 160–71.

Wang, Y. and Ke, X. 1989: Cheniers on the east coastal plain of China. *Marine Geol.* 90, pp. 321–35.

Ward, G. H. 1980: Hydrography and circulation processes of Gulf estuaries. In *Estuarine and Wetland Processes*, P. Hamilton and K. B. Macdonald (eds.), Plenum Press, New York, pp. 183–215.

Warme, J. E. 1975: Borings as trace fossils and the processes of marine bioerosion. In *The Study of Trace Fossils*, R. W. Frey (ed.), Springer-Verlag, Berlin, pp. 181–227.

Warne, A. G. and Stanley, D. J. 1993: Late Quaternary evolution of the northwest Nile Delta and adjacent coast in the Alexandria Region, Egypt. *J. Coastal Res.* 9, pp. 26–64.

Warrick, R. A. 1993: Climate and sea level change: a synthesis. In *Climate and Sea Level Change: Observations, Projections and Implications*, R. A. Warrick, E. M. Barrow, and T. M. L. Wigley (eds.), Cambridge Univ. Press, Cambridge, pp. 3–21.

Warrick, R. A., Barrow, E. M., and Wigley, T. M. L. (eds.) 1993: *Climate and Sea Level Change: Observations, Projections and Implications*, Cambridge Univ. Press, Cambridge.

Warrick, R. A. and Farmer, G. 1990: The greenhouse effect, climatic change and rising sea level: implications for development. *Trans. Inst. Brit. Geogr.* 15, pp. 5–20.

Watanabe, A. 1992: Total rate and distribution of longshore sand transport. *Proc. 23nd Coastal Engnr. Conf.* pp. 2528–41.

Watanabe, A. and Isobe, M. 1990: Sand transport rate under wave-current action. *Proc. 22nd Coastal Engnr. Conf.* pp. 2495–507.

Watanabe, A., Riho, Y., and Horikawa, K. 1980: Beach profiles and on–offshore sediment transport. *Proc. 17th Coastal Engnr. Conf.* pp. 1106–21.

Watson, G. and Peregrine, D. H. 1992: Low frequency waves in the surf zone. *Proc. 23rd Coastal Engnr. Conf.* pp. 818–31.

Watts, S. H. 1985: A scanning electron microscope study of bedrock micro-fractures in granites under high Arctic conditions. *Earth Surface Processes and Landforms* 10, pp. 161–72.

Wayne, C. J. 1976: The effect of sea and marsh grass on wave energy. *Coastal Res. Notes* 4, pp. 6–8.

Weber, P. 1994: Coral reefs in decline. In *Vital Signs*, L. R. Brown, H. Kane, and D. M. Roodman (eds.), W. W. Norton, New York, pp. 122–3.

Weedman, S. D. and Slingerland, R. 1985: Experimental study of sand streaks formed in turbulent boundary layers. *Sedimentology* 32, pp. 133–45.

Weirich, F. H. 1989: The generation of turbidity currents by subaerial debris flows, California. *Geol. Soc. Amer. Bull.* 101, pp. 278–91.

Weiss, M. P. and Goddard, D. A. 1977: Man's impact on coastal reefs—an example from Venezuela. In *Reefs and Related Carbonates—Ecology and Sedimentology*, S. H. Frost, M. P. Weiss, and J. B. Saunders (eds.), *Amer. Assoc. Petrol. Geol. Studies in Geol.* 4, pp. 111–24.

Wells, J. T., Adams, C. E., Park, Y.-A., and Frankenberg, E. W. 1990: Morphology, sedimentology and tidal channel processes on a high-tide-range mudflat, west coast of South Korea. *Marine Geol.* 95, pp. 111–30.

Wells, J. T. and Coleman, J. M. 1981: Periodic mudflat progradation, northeastern coast of South America: a hypothesis. *J. Sed. Petrol.* 51, pp. 1069–75.

Wells, J. W. 1956: Scleractinia. In *Treatise on Invertebrate Paleontology*, R. C. Moore (ed.), Part F. Coelenterata, F328–F444.

Wells, J. W. 1957: Corals. In *Treatise on Marine Ecology and Paleoecology*, J. W. Hedgpeth (ed.), Vol. 1 (Ecology), *Geol. Soc. Amer. Mem.* 67, pp. 1087–104.

Weng, W. S., Hunt, J. C. R., Carruthers, D. J., Warren, A., Wiggs, G. F. S., Livingstone, I., and Castro, I. 1991: Air flow and sand transport over sand-dunes. In *Aeolian Grain Transport 2 The Erosional Environment*, O. E. Barndorff-Nielsen and B. B. Willetts (eds.), Acta Mechanica Suppl. 2, Springer-Verlag, Wien, pp. 1–22.

Werner, B. T. 1990: A steady-state model of wind-blown sand transport. *J. Geol.* 98, pp. 1–17.

Werner, B. T. and Fink, T. M. 1993: Beach cusps as self-organized patterns. *Science* 260, pp. 968–71.

West, R. C. 1956: Mangrove swamps of the Pacific coast of Columbia. *Ann. Assoc. Amer. Geogr.* 46, pp. 98–121.

West, R. C. 1977: Tidal salt-marsh and mangal formations of Middle and South America. In *Wet Coastal Ecosystems*, V. J. Chapman (ed.), Elsevier, Amsterdam, pp. 193–213.

Westhoff, V. 1989: Dunes and dune management along the North Sea coasts. In *Perspectives in Coastal Dune Management*, F. Van Der Meulen, P. D. Jungerius, and J. Visser (eds.), SPB Academic Publishing, The Hague, The Netherlands, pp. 41–51.

Whalley, W. B. 1981: Physical properties. In *Geomorphological Techniques*, A. Goudie (ed.), Allen and Unwin, London, pp. 80–103.

White, T. E. and Inman, D. L. 1989a: Measuring longshore transport with tracers. In *Nearshore Sediment Transport*, R. J. Seymour (ed.), Plenum Press, New York, pp. 287–312.

White, T. E. and Inman, D. L. 1989b: Transport determination by tracers B. Application of tracer theory to NSTS experiments. In *Nearshore Sediment Transport*, R. J. Seymour (ed.), Plenum Press, New York, pp. 115–28.

Whitehouse, R. J. S. 1992: Combined flow sand transport: field measurements. *Proc. 23rd Coastal Engnr. Conf.* pp. 2542–55.

Whitehouse, R. J. S. and Hardisty, J. 1988: Experimental assessment of two theories for the effect of bedslope on the threshold of bedload transport. *Marine Geol.* 79, pp. 135–9.

Wiberg, P. L. and Harris, C. K. 1994: Ripple geometry in wave-dominated environments. *J. Geophys. Res.* 99, pp. 775–89.

Wiegel, R. L. 1964: *Oceanographical Engineering*, Prentice-Hall, New Jersey.

Wiens, H. J. 1962: *Atoll Environment and Ecology*, Yale Univ. Press, New Haven.

Wigley, T. M. L. and Raper, S. C. B. 1992: Implications for climate and sea level of revised IPCC emissions scenarios. *Nature* 357, pp. 293–300.

Wilkinson, B. H. 1975: Matagorda Island, Texas, the evolution of a Gulf coast barrier complex. *Geol. Soc. Amer. Bull.* 86, pp. 959–67.

Willetts, B. B. 1989: Physics of sand movement in vegetated dune systems. In *Coastal Sand Dunes*, C. H. Gimingham, W. Ritchie, B. B. Willetts, and A. J. Willis (eds.), *Proc. Roy. Soc. Edinburgh* B 96, pp. 37–49.

Willetts, B. B., McEwan, I. K., and Rice, M. A. 1991: Initiation of motion of quartz sand grains. In *Aeolian Grain Transport 1 Mechanics*, O. E. Barndorff-Nielsen and B. B. Willetts (eds.), Acta Mechanica Suppl. 1, Springer-Verlag, Wien, pp. 123–34.

Willetts, B. B. and Rice, M. A. 1983: Practical representation of characteristic grain shape of sands: a comparison of methods. *Sedimentology* 30, pp. 557–65.

Willetts, B. B. and Rice, M. A. 1989: Collisions of quartz grains with a sand bed: the influence of incident angle. *Earth Surface Processes and Landforms* 14, pp. 719–30.

Williams, A. T. 1973: The problem of beach cusp development. *J. Sed. Petrol.* 43, pp. 857–66.

Williams, A. T. and Caldwell, N. E. 1988: Particle size and shape in pebble-beach sedimentation. *Marine Geol.* 82, pp. 199–215.

Williams, A. T. and Davies, P. 1980: Man as a geological agent: the sea cliffs of Llantwit Major, Wales, UK. *Z. Geomorph.* Suppl. Band 34, pp. 129–41.

Williams, A. T. and Davies, P. 1987: Rates and mechanisms of coastal cliff erosion in Lower Lias rocks. *Coastal Sediments '87*, N. C. Kraus (ed.), Amer. Soc. Civ. Engnr., pp. 1855–70.

Williams, A. T. and Jones, D. G. 1991: Mechanisms of coastal cliff erosion in Ceredigion, west Wales, UK. *Coastal Sediments '91*, N. C. Kraus, K. J. Gingerich, and D. L. Kriebel (eds.), Amer. Soc. Civ. Engnr., pp. 1571–85.

Williams, A. T. and Morgan, P. 1988: Quartz grain S.E.M. textural variations of the beach/dune interface, Long Island, U.S.A. *J. Coastal Res.* Spec. Issue 3, pp. 37–45.

Williams, E. H. and Bunkley-Williams, L. 1988: Bleaching of Caribbean coral reef symbionts in 1987–88. *Proc. 6th Int. Coral Reef Symp., Townsville* 3, pp. 313–18.

Williams, J. J., Butterfield, G. R., and Clark, D. G. 1990: Rates of aerodynamic entrainment in a developing boundary layer. *Sedimentolgy* 37, pp. 1039–48.

Williams, J. J., Butterfield, G. R., and Clark, D. G. 1994: Aerodynamic entrainment threshold: effects of boundary layer flow conditions. *Sedimentology* 41, pp. 309–28.

Williams, J. J., Thorne, P. D., and Heathershaw, A. D. 1989*a*: Measurements of turbulence in the benthic boundary layer over a gravel bed. *Sedimentology* 36, pp. 959–71.

Williams, J. J., Thorne, P. D., and Heathershaw, A. D. 1989*b*: Comparisons between acoustic measurements and predictions of the bedload transport of marine gravels. *Sedimentology* 36, pp. 973–9.

Williams, R. B. G. and Robinson, D. A. 1981: Weathering of sandstone by the combined action of frost and salt. *Earth Surface Processes and Landforms* 6, pp. 1–9.

Williams, S. J., Penland, S., Sallenger, A. H., McBridge, R. A., and Kindinger, J. L. 1991: Geologic controls on the formation and evolution of Quaternary coastal deposits of the northern Gulf of Mexico. *Coastal Sediments '91*, N. C. Kraus, K. J. Gingerich, and D. L. Kriebel (eds.), Amer. Soc. Civ. Engnr., pp. 1082–93.

Willis, A. J. 1989: Coastal sand dunes as biological systems. In *Coastal Sand Dunes*, C. H. Gimingham, W. Ritchie, B. B. Willetts, and A. J. Willis (eds.), *Proc. Roy. Soc. Edinb.* B 96, pp. 17–36.

Wilson, I. G. 1972: Aeolian bedforms—their development and origins. *Sedimentology* 19, pp. 173–210.

Wilson, K. C. 1989: Friction of wave-induced sheet flow. *Coastal Engnr.* 12, pp. 371–79.

Wilson, P. 1990: Coastal dune chronology in the north of Ireland. In *Dunes of the European Coasts*, W. Bakker, P. D. Jungerius, and J. A. Klijn (eds.), *Catena* Suppl. 18, pp. 71–9.

Wind, H. G. 1994: An analytical model of crenulate shaped beaches. *Coastal Engnr.* 23, pp. 243–53.

Winkelmolen, A. M. 1971: Rollability, a functional shape property of sand grains. *J. Sed. Petrol.* 41, pp. 703–14.

Wolanski, E. 1986: An evaporation-driven salinity maximum zone in Australian tropical estuaries. *Estuarine, Coastal and Shelf Sci.* 22, pp. 415–24.

Wolanski, E., Jones, M., and Bunt, J. S. 1980: Hydrodynamics of a tidal creek-mangrove swamp system. *Austr. J. Marine and Freshwater Res.* 31, pp. 431–50.

Wolanski, E., Mazda, Y., and Ridd, P. 1992: Mangrove hydrodynamics. In *Tropical Mangrove Ecosystems*, A. I. Robertson and D. M. Alongi (eds.), Amer. Geophys. Union, Washington, DC, pp. 43–62.

Wolfe, S. A. and Nickling, W. G. 1993: The protective role of sparse vegetation in wind erosion. *Progr. Phys. Geogr.* 17, pp. 50–68.

Wood, A. 1959: The erosional history of the cliffs around Aberystwyth. *Liverpool and Manchester Geol. J.* 2, pp. 271–87.

Woodley, J. D. and nineteen others 1981: Hurricane Allen's impact on Jamaican coral reefs. *Science* 214, pp. 749–55.

Woodroffe, C. D. 1983: Development of mangrove forests from a geological perspective. In *Biology and Ecology of Mangroves*, H. J. Teas (ed.), Junk, The Hague, The Netherlands, pp. 1–17.

Woodroffe, C. D. 1987: Pacific island mangroves: distribution and environmental settings. *Pacific Sci.* 41, pp. 166–85.

Woodroffe, C. D. 1990: The Impact of sea-level rise on mangrove shorelines. *Progr. Phys. Geogr.* 14, pp. 483–520.

Woodroffe, C. D. 1993: Late Quaternary evolution of coastal and lowland riverine plains of southeast Asia and northern Australia: an overview. *Sed. Geol.* 83, pp. 163–75.

Woodroffe, C. D. 1995: Response of tide-dominated mangrove shorelines in northern Australia to anticipated sea-level rise. *Earth Surface Processes and Landforms* 20, pp. 65–85.

Woodroffe, C. D. and Chappell, J. 1993: Holocene emergence and evolution of the McArthur River Delta, southwestern Gulf of Carpentaria, Australia. *Sed. Geol.* 83, pp. 303–17.

Woodroffe, C. D., Chappell, J., Thom, B. G., and Wallensky, E. 1989: Depositional model of a macrotidal estuary and floodplain, South Alligator River, Northern Australia. *Sedimentology* 36, pp. 737–56.

Woodroffe, C. D., Curtis, R. J., and McLean, R. F. 1983: Development of a chenier plain, Firth of Thames, New Zealand. *Marine Geol.* 53, pp. 1–22.

Woodroffe, C. D. and McLean, R. 1990: Microatolls and recent sea level change on coral atolls. *Nature* 344, pp. 531–4.

Woodroffe, C. D., McLean, R., Polach, H., and Wallensky, E. 1990: Sea level and coral atolls: late Holocene emergence in the Indian Ocean. *Geology* 18, pp. 62–6.

Woodroffe, C. D., Mulrennan, M. E., and Chappell, J. 1993: Estuarine infill and coastal progradation, southern van Diemen Gulf, northern Australia. *Sed. Geol.* 83, pp. 257–75.

Woodroffe, C. D., Stoddart, D. R., Harmon, R. S., and Spencer, T. 1983*b*: Coastal morphology and Late Quaternary history, Cayman Islands, West Indies. *Quat. Res.* 19, pp. 64–84.

Worsley, T. R., Nance, D., and Moody, J. B. 1984: Global tectonics and eustasy for the past 2 billion years. *Marine Geol.* 58, pp. 373–400.

Wright, L. D. 1977: Sediment transport and deposition

at river mouths: a synthesis. *Geol. Soc. Amer. Bull.* 88, pp. 857–68.

Wright, L. D. 1980: Beach cut in relation to surf zone morphodynamics. *Proc. 17th Coastal Engnr. Conf.* pp. 978–96.

Wright, L. D. 1985: River deltas. In *Coastal Sedimentary Environments*, R. A. Davis (ed.), Springer-Verlag, New York, pp. 1–76.

Wright, L. D. 1987: Shelf-surfzone coupling: diabathic shoreface transport. *Coastal Sediments '87*, N. C. Kraus (ed.), Amer. Soc. Civ. Engnr., pp. 25–40.

Wright, L. D. 1989: Benthic boundary layers of estuarine and coastal environments. *Rev. Aquatic Sci.* 1, pp. 75–95.

Wright, L. D. 1993: Micromorphodynamics of the inner continental shelf: a middle Atlantic Bight case study. *J. Coastal Res.* Spec. Issue 15, pp. 93–124.

Wright, L. D., Boon, J. D., Kim, S. C., and List, J. H. 1991: Modes of cross-shore sediment transport on the shoreface of the middle Atlantic Bight. *Marine Geol.* 96, pp. 19–51.

Wright, L. D., Chappell, J., Thom, B. G., Bradshaw, M. P., and Cowell, P. 1979: Morphodynamics of reflective and dissipative beach and inshore systems: southeastern Australia. *Marine Geol.* 32, pp. 105–40.

Wright, L. D. and Coleman, J. M. 1973: Variations in morphology of major river deltas as functions of ocean wave and river discharge regimes. *Amer. Assoc. Petrol. Geol. Bull.* 57, pp. 370–98.

Wright, L. D., Coleman, J. M., and Thom, B. G. 1973: Processes of channel development in a high-tide-range environment: Cambridge Gulf–Ord River Delta, Western Australia. *J. Geol.* 81, pp. 15–41.

Wright, L. D., Nielsen, P., Shi, N. C., and List, J. H. 1986: Morphodynamics of a bar-trough surf zone. *Marine Geol.* 70, pp. 251–85.

Wright, L. D., Nielsen, P., Short, A. D., and Green, M. O. 1982: Morphodynamics of a macrotidal beach. *Marine Geol.* 50, pp. 97–128.

Wright, L. D. and Short, A. D. 1984: Morphodynamic variability of surf zones and beaches: a synthesis. *Marine Geol.* 56, pp. 93–118.

Wright, L. D., Short, A. D., Boon, J. D. III, Hayden, B., Kimball, S., and List, J. H. 1987: The morphodynamic effects of incident wave groupiness and tide range on an energetic beach. *Marine Geol.* 74, pp. 1–20.

Wright, L. D., Short, A. D., and Green, M. O. 1985: Short-term changes in the morphodynamic states of beaches and surf zones: an empirical predictive model. *Marine Geol.* 62, pp. 339–64.

Wright, L. D. and Sonu, C. J. 1975: Processes of sediment transport and tidal delta development in a stratified tidal inlet. In *Estuarine Research*, L. E. Cronin (ed.), Academic Press, New York, ii. 63–76.

Wright, L. D., Thom, B. G., and Higgins, R. J. 1980: Wave influences on river-mouth depositional processes:

examples from Australia and Papua New Guinea. *Estuarine and Coastal Marine Sci.* 11, pp. 263–77.

Wright, P. 1984: Facies development on a barred (ridge and runnel) coastline, the case of southwest Lancashire (Merseyside), U.K. In *Coastal Research: UK Perspectives*, M. W. Clark (ed.), Geo Books, Norwich, pp. 105–18.

Wu, B. L. 1985: Coral reefs of Hainan Island in the South China Sea. *Proc. 5th Int. Coral Reef Congr., Tahiti*, 2, p. 412.

Xitao, Z. 1989: Cheniers in China: an overview. *Marine Geol.* 90, pp. 311–20.

Xu, J. P., Wright, L. D., and Boon, J. D. 1994: Estimation of bottom stress and roughness in lower Chesapeake Bay by the inertial dissipation method. *J. Coastal Res.* 10, pp. 329–38.

Xue, C. 1993: Historical changes in the Yellow River delta, China. *Marine Geol.* 113, pp. 321–9.

Yalin, M. S. 1963: An expression for bed-load transportation. *J. Hydraul. Div. Amer. Soc. Civ. Engnr.* 89, pp. 221–50.

Yasso, W. E. 1982: Headland bay beach. In *The Encyclopedia of Beaches and Coastal Environments*, M. L. Schwartz (ed.), Hutchinson Ross, Stroudsburg, Pennsylvania, pp. 460–1.

Yeh, H. H., Ghazali, A., and Marton, I. 1989: Experimental study of bore run-up. *J. Fluid Mech.* 206, pp. 563–78.

Young, I. R. 1989: Wave transformation over coral reefs. *J. Geophys. Res.* 94, pp. 9779–89.

Young, R. W. and Bryant, E. A. 1993: Coastal rock platforms and ramps of Pleistocene and Tertiary age in southern New South Wales, Australia. *Z. Geomorph.* 37, pp. 257–72.

Yu, Z., Niemeyer, H. D., and Bakker, W. T. 1990: Site investigation on sand concentration in the sheet flow layer. *Proc. 22nd Coastal Engnr. Conf.* pp. 2360–71.

Zaitlin, B. A. and Shultz, B. C. 1990: Wave-influenced estuarine sand body, Senlac heavy oil pool, Saskatchewan, Canada. In *Sandstone Petroleum Reservoirs*, J. H. Barwis, J. G. McPherson, and R. J. Studlick (eds.), Springer-Verlag, New York, pp. 363–87.

Zampol, J. A. and Inman, D. L. 1989: Suspended sediment measurements B. Discrete measurements of suspended sediment. In *Nearshore Sediment Transport*, R. J. Seymour (ed.), Plenum Press, New York, pp. 259–72.

Zarillo, G. A. 1985: Tidal dynamics and substrate response in a salt-marsh estuary. *Marine Geol.* 67, pp. 13–35.

Zeman, A. J. 1986: Erodibility of Lake Erie undisturbed tills. *Proc. Symp. Cohesive Shores*, M. G. Skafel (ed.), National Res. Council Can. (ACROSES), Ottawa, pp. 150–69.

Zenkovitch, V. P. 1959: On the genesis of cuspate spits along lagoon shores. *J. Geol.* 67, pp. 269–77.

Zenkovitch, V. P. 1967: *Processes of Coastal Development*, Oliver & Boyd, Edinburgh.

Zezza, F. 1981: Morfogenesi litorale e fenomeni d'instabilità della costa a falesia del Gargano tra Vieste e Manfredonia. *Geol. Applic. Idrogedogia* (Bari Univ. Inst. Geol. Applicata) 16, pp. 193–226.

Zhang, D. P. 1994: Wave flume experiments on the formation of longshore bars produced by breaking waves. *Sci. Rept. Inst. Geosci. Univ. Tsukuba* (Japan) A 15, pp. 47–105.

Zhuang, W.-Y. and Chappell, J. 1991: Effects of seagrass beds on tidal flat sedimentation, Corner Inlet, southeast Australia. In *Clastic Tidal Sedimentology*, D. G. Smith, G. E. Reinson, B. A. Zaitlin, and R. A. Rahmani (eds.), *Can. Soc. Petrol. Geol. Mem.* 16, pp. 291–300.

Zingg, A. W. 1953: Wind-tunnel studies of the movement of sedimentary material. *Proc. 5th Hydraul. Conf. Univ. Iowa Studies in Engnr. Bull.* 24, pp. 111–35.

Location Index

Subject Index

3 1862 015 518 091
University of Windsor Libraries

University